Telecommunication Circuits and Technology

Telecommunication Circuits and Technology

Andrew Leven
BSc (Hons), MSc, CEng, MIEE, MIP

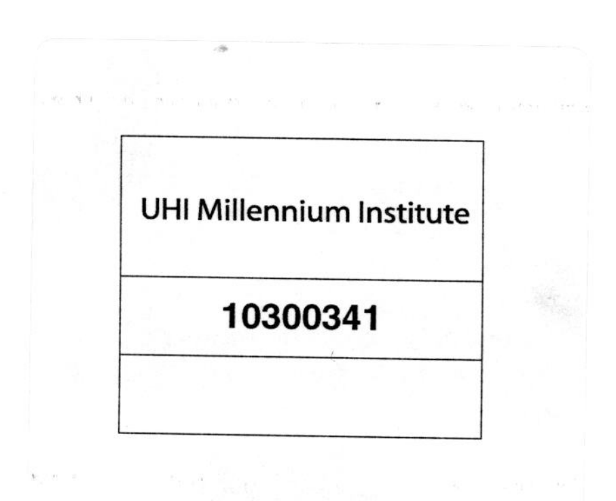

OXFORD AUCKLAND BOSTON JOHANNESBURG MELBOURNE NEW DELHI

Butterworth-Heinemann
Linacre House, Jordan Hill, Oxford OX2 8DP
225 Wildwood Avenue, Woburn, MA 01801-2041
A division of Reed Educational and Professional Publishing Ltd

A member of the Reed Elsevier plc group

First published 2000

While the author has attempted to mention all parties, if we have failed to acknowledge use of information or product in the text, our apologies and acknowledgement.

British Library Cataloguing in Publication Data
A catalogue record for this book is available from the British Library

ISBN 0 7506 5045 1

Typeset in 10/12pt Times by Replika Press Pvt Ltd, Delhi 110 040, India
Printed and bound by MPG Books, Bodmin Cornwall

To my wife Lorna and the siblings, Roddy, Bruce, Stella and Russell.
They have all inspired me

Contents

1

Oscillators

1.1 Introduction

Communication systems consist of an input device, transmitter, transmission medium, receiver and output device, as shown in Fig. 1.1. The input device may be a computer, sensor or oscillator, depending on the application of the system, while the output device could be a speaker or computer. Irrespective of whether a data communications or telecommunications system is used, these elements are necessary.

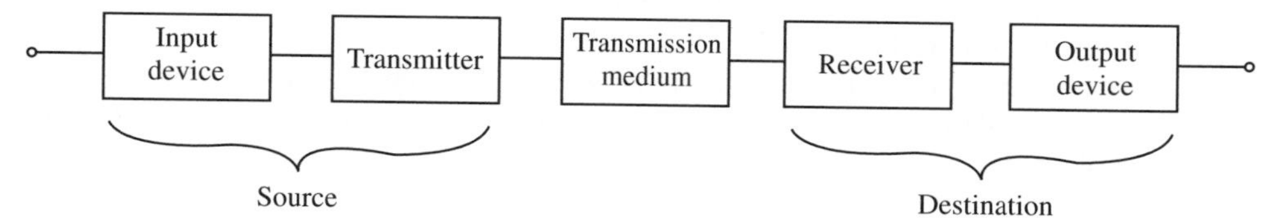

Fig. 1.1

The source section produces two types of signal, namely the information signal, which may be speech, video or data, and a signal of constant frequency and constant amplitude called the carrier. The information signal mixes with the carrier to produce a complex signal which is transmitted. This is discussed further in Chapter 2.

The destination section must be able to reproduce the original information, and the receiver block does this by separating the information from the carrier. The information is then fed to the output device.

The transmission medium may be a copper cable, such as a co-axial cable, a fibre-optic cable or a waveguide. These are all guided systems in which the signal from the transmitter is directed along a solid medium. However, it is often the case with telecommunication systems that the signal is unguided. This occurs if an antenna system is used at the output of the transmitter block and the input of the receiver block.

Both the transmitter block and the receiver block incorporate many amplifier and processing stages, and one of the most important is the oscillator stage. The oscillator in the transmitter is generally referred to as the master oscillator as it determines the channel at which the transmitter functions. The receiver oscillator is called the local oscillator as it produces a local carrier within the receiver which allows the incoming carrier from the transmitter to be modified for easier processing within the receiver.

Figure 1.2 shows a radio communication system and the role played by the oscillator. The master oscillator generates a constant-amplitude, constant-frequency signal which is used to carry the audio or intelligence signal. These two signals are combined in the modulator, and this stage produces an output carrier which varies in sympathy with the audio signal or signals. This signal is low-level and must be amplified before transmission.

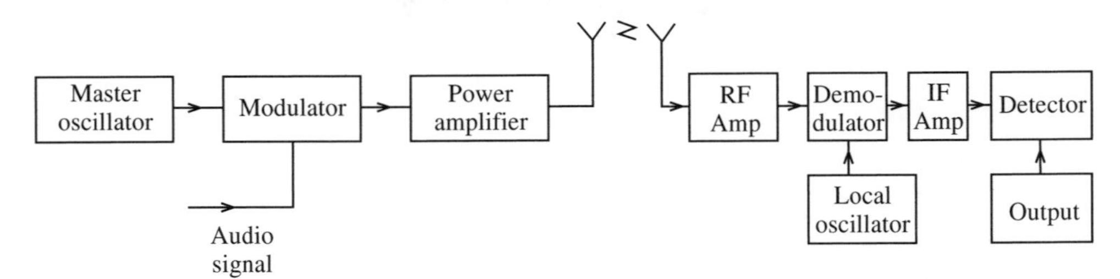

Fig. 1.2

The receiver amplifies the incoming signal, extracts the intelligence and passes it on to an output transducer such as a speaker. The local oscillator in this case causes the incoming radio frequency (RF) signals to be translated to a fixed lower frequency, called the intermediate frequency (IF), which is then passed on to the following stages. This common IF means that all the subsequent stages can be set up for optimum conditions and do not need to be readjusted for different incoming RF channels. Without the local oscillator this would not be possible.

It has been stated that an oscillator is a form of frequency generator which must produce a constant frequency and amplitude. How these oscillations are produced will now be explained.

1.2 The principles of oscillation

A small signal voltage amplifier is shown in Fig. 1.3.

In Fig. 1.3(a) the operational amplifier has no external components connected to it and

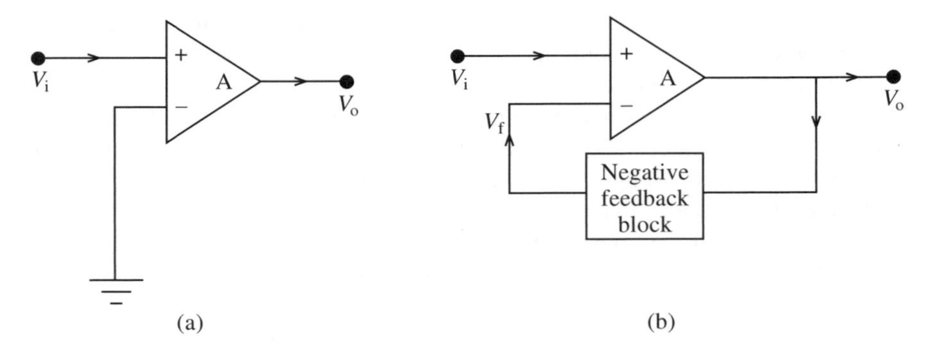

Fig. 1.3

the signal is fed in as shown. The operational amplifier has an extremely high gain under these circumstances and this leads to saturation within the amplifier. As saturation implies working in the non-linear section of the characteristics, harmonics are produced and a ringing pattern may appear inside the chip. As a result of this, a square wave output is produced for a sinusoidal input. The amplifier has ceased to amplify and we say it has become unstable. There are many reasons why an amplifier may become unstable, such as temperature changes or power supply variations, but in this case the problem is the very high gain of the operational amplifier.

Figure 1.3(b) shows how this may be overcome by introducing a feedback network between the output and the input. When feedback is applied to an amplifier the overall gain can be reduced and controlled so that the operational amplifier can function as a linear amplifier. Note also that the signal fedback has a phase angle, due to the inverting input, which is in opposition to the input signal (V_i).

Negative feedback can therefore be defined as the process whereby a part of the output voltage of an amplifier is fed to the input with a phase angle that opposes the input signal. Negative feedback is used in amplifier circuits in order to give stability and reduced gain. Bandwidth is generally increased, noise reduced and input and output resistances altered. These are all desirable parameters for an amplifier, but if the feedback is overdone then the amplifier becomes unstable and will produce a ringing effect.

In order to understand stability, instability and its causes must be considered. From the above discussion, as long as the feedback is negative the amplifier is stable, but when the signal feedback is **in phase** with the input signal then positive feedback exists. Hence positive feedback occurs when the total phase shift through the operational amplifier (op-amp) and the feedback network is 360° (0°). The feedback signal is now in phase with the input signal (V_i) and oscillations take place.

1.3 The basic structure and requirements of an oscillator

Any oscillator consists of three sections, as shown in Fig. 1.4.

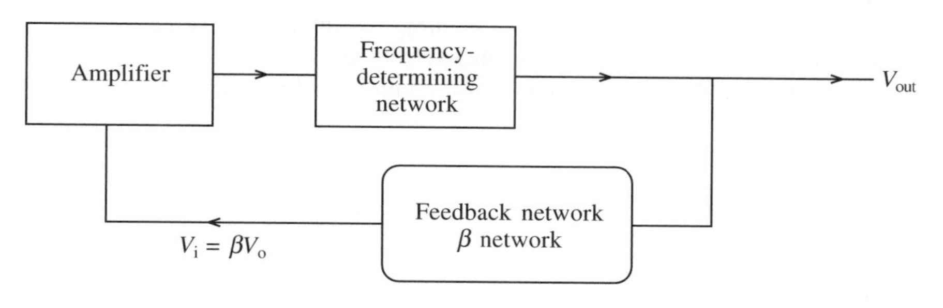

Fig. 1.4

The frequency-determining network is the core of the oscillator and deals with the generation of the specified frequency. The desired frequency may be generated by using an inductance–capacitance (*LC*) circuit, a resistance–capacitance (*RC*) circuit or a piezo-electric crystal. Each of these networks produces a particular frequency depending on the values of the components and the cut of the crystal. This frequency is known as the

resonant or **natural frequency** of the network and can be calculated if the values of components are known.

Each of these three different networks will produce resonance, but in quite different ways. In the case of the *LC* network, a parallel arrangement is generally used which is periodically fed a pulse of energy to keep the current circulating in the parallel circuit. The current circulates in one direction and then in the other as the magnetic and electric fields of the coil and capacitor interchange their energies. A constant frequency is therefore generated.

The *RC* network is a **time-constant** network and as such responds to the charge and discharge times of a capacitor. The frequency of this network is determined by the values of *R* and *C*. The capacitor and resistor cause phase shift and produce positive feedback at a particular frequency. Its advantage is the absence of inductances which can be difficult to tune.

For maximum stability a crystal is generally used. It resonates when a pressure is applied across its ends so that mechanical energy is changed to electrical energy. The crystal has a large *Q* factor and this means that it is highly selective and stable.

The amplifying device may be a bipolar transistor, a field-effect transistor (FET) or operational amplifier. This block is responsible for maintaining amplitude and frequency stability and the correct d.c. bias conditions must apply, as in any simple discrete amplifier, if the output frequency has to be undistorted. The amplifier stage is generally class C biased, which means that the collector current only flows for part of the feedback cycle (less than 180° of the input cycle).

The feedback network can consist of pure resistance, reactance or a combination of both. The feedback factor (β) is derived from the output voltage. It is as well to note at this point that the product of the feedback factor (β) and the open loop gain (A) is known as the loop gain. The term **loop gain** refers to the fact that the product of all the gains is taken as one travels around the loop from the amplifier input, through the amplifier and through the feedback path. It is useful in predicting the behaviour of a feedback system. Note that this is different from the **closed-loop gain** which is the ratio of the output voltage to the input voltage of an amplifier.

When considering oscillator design, the important characteristics which must be considered are the range of frequencies, frequency stability and the percentage distortion of the output waveform. In order to achieve these characteristics two necessary requirements for oscillation are that the loop gain (βA) must be unity and the loop phase shift must be zero.

Consider Fig. 1.5. We have

$$V_f = \beta V_o = -\beta A_V \cdot V_i$$

but

$$V_f = V_i$$

therefore

$$V_i = -\beta A_V \cdot V_i$$

or

$$V_i(1 + \beta A_V) = 0$$

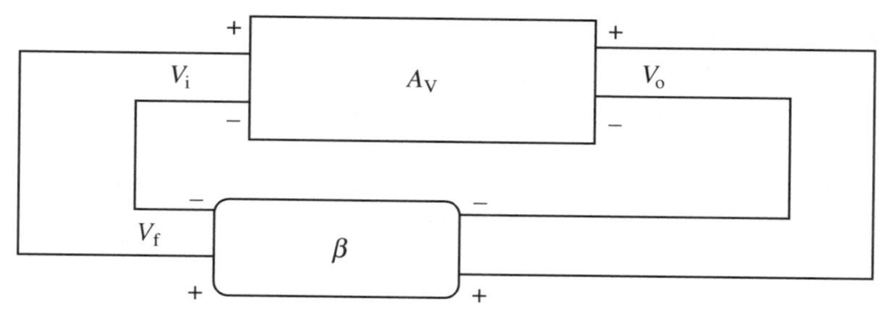

Fig. 1.5

since

$$V_i = 0$$

or

$$1 + \beta A_V = 0$$

then we have

$$\beta A_V = -1 + j0 \tag{1.1}$$

Thus the requirements for oscillation to occur are:

(i) $A_V = 1$.

(ii) The phase shift around the closed loop must be an integral multiple of 2π, i.e. 2π, 4π, 6π, etc.

These requirements constitute the Barkhausen criterion and an oscillating amplifier self-adjusts to meet them.

The gain must initially provide $\beta A_V > 1$ with a switching surge at the input to start operation. An output voltage resulting from this input pulse propagates back to the input and appears as an amplified output. The process repeats at greater amplitude and as the signal reaches saturation and cut-off the average gain is reduced to the level required by equation (1.1).

If $\beta A_V > 1$ the output increases until non-linearity limits the amplitude. If $\beta A_V < 1$ the oscillation will be unable to sustain itself and will stop. Thus $\beta A_V > 1$ is a necessary condition for oscillation to start. $\beta A_V = 1$ is a necessary condition for oscillation to be maintained.

There are many types of oscillator but they can be classified into four main groups: resistance–capacitance oscillators; inductance–capacitance oscillators; crystal oscillators; and integrated circuit oscillators. In the following sections we look at each of these types in turn.

1.4 *RC* oscillators

There are three functional types of *RC* oscillator used in telecommunications applications: the phase-shift oscillator; the Wien bridge oscillator; and the twin-T oscillator.

Phase-shift oscillators

Figure 1.6 shows the phase-shift oscillator using a bipolar junction transistor (BJT). Each of the *RC* networks in the feedback path can provide a maximum phase shift of almost 60°. Oscillation occurs at the output when the *RC* ladder network produces a 180° phase shift. Hence three *RC* networks are required, each providing 60° of phase shift. The transistor produces the other 180°. Generally $R_5 = R_6 = R_7$ and $C_1 = C_2 = C_3$.

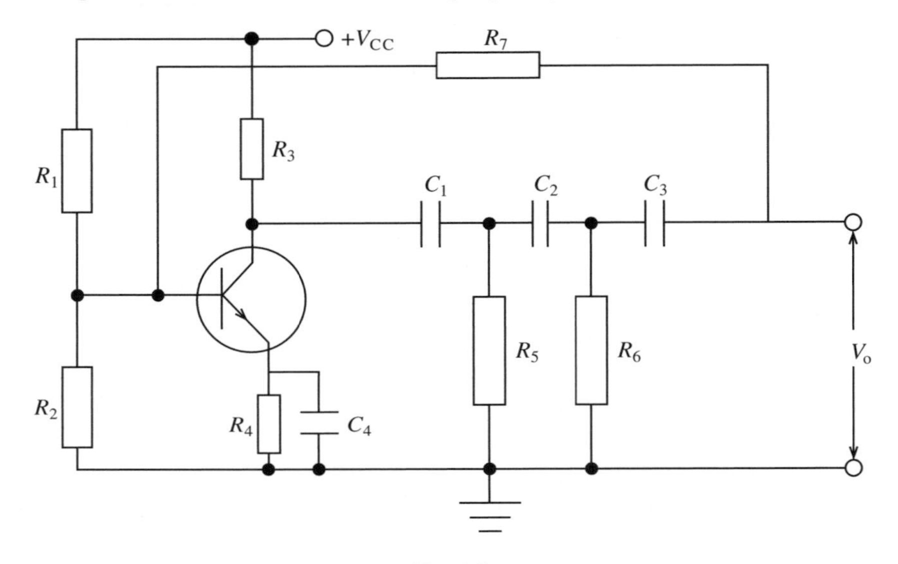

Fig. 1.6

The output of the feedback network is shunted by the low input resistance of the transistor to provide voltage–voltage feedback.

It can be shown that the closed-loop voltage gain should be $A_V = 29$. Hence

$$\beta = \frac{1}{29} \tag{1.2}$$

Also the oscillatory frequency is given as

$$f = \frac{1}{2\pi RC\sqrt{6 + \dfrac{4R_3}{R}}} \tag{1.3}$$

The derivation of this formula, as with other formulae in this section, is beyond the requirement of this book and may be found in any standard text. The application of the formula is important in simple design.

Exactly the same circuit as Fig. 1.6 may be used when the active device is an FET. As before the loop gain $A_V = 29$ but the frequency, because of the high input resistance of the FET, is now given by

$$f = \frac{1}{2\pi CR\sqrt{6}} \tag{1.4}$$

Figure 1.7 shows the use of an op-amp version of this type of oscillator. Formulae (1.2) and (1.4) apply in this design.

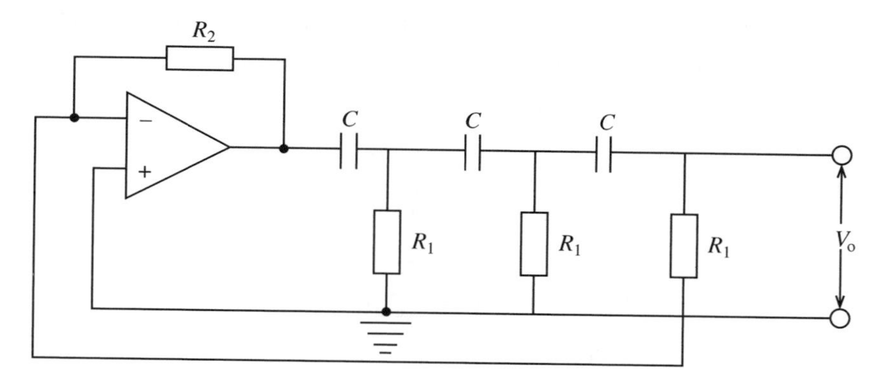

Fig. 1.7

One final point should be mentioned when designing a phase-shift oscillator using a transistor. It is essential that the h_{fe} of the transistor should have a certain value in order to ensure oscillation. This may be determined by using an equivalent circuit and performing a matrix analysis on it. However, for the purposes of this book the final expression is

$$h_{fe} > 4\left(\frac{R_3}{R}\right) + 23 + 29\left(\frac{R}{R_3}\right) \tag{1.5}$$

Example 1.1

A phase-shift oscillator is required to produce a fixed frequency of 10 kHz. Design a suitable circuit using an op-amp.

Solution

$$f = \frac{1}{2\pi CR_1\sqrt{6}}$$

Select $C = 22$ nF. Rearranging as expression for f, we obtain

$$R_1 = \frac{1}{2\pi Cf\sqrt{6}} = \frac{1}{2\pi \times 22 \times 10^{-9} \times 10^4 \times \sqrt{6}} = 295.3\ \Omega$$

As this value is critical in this type of oscillator, a potentiometer should be used and set to the required value. Since

$$A = \frac{R_2}{R_1} = 29$$

$$R_2 = AR_1 = 29 \times 295.3 = 8.56\ \text{k}\Omega$$

A value slightly greater than this should be chosen to ensure oscillation.

Wien bridge oscillator

This circuit (Fig. 1.8) uses a balanced bridge network as the frequency-determining network. R_2 and R_3 provide the gain which is

$$A_V = 3 \tag{1.6}$$

The frequency is given by

$$f = \frac{1}{2\pi RC} \tag{1.7}$$

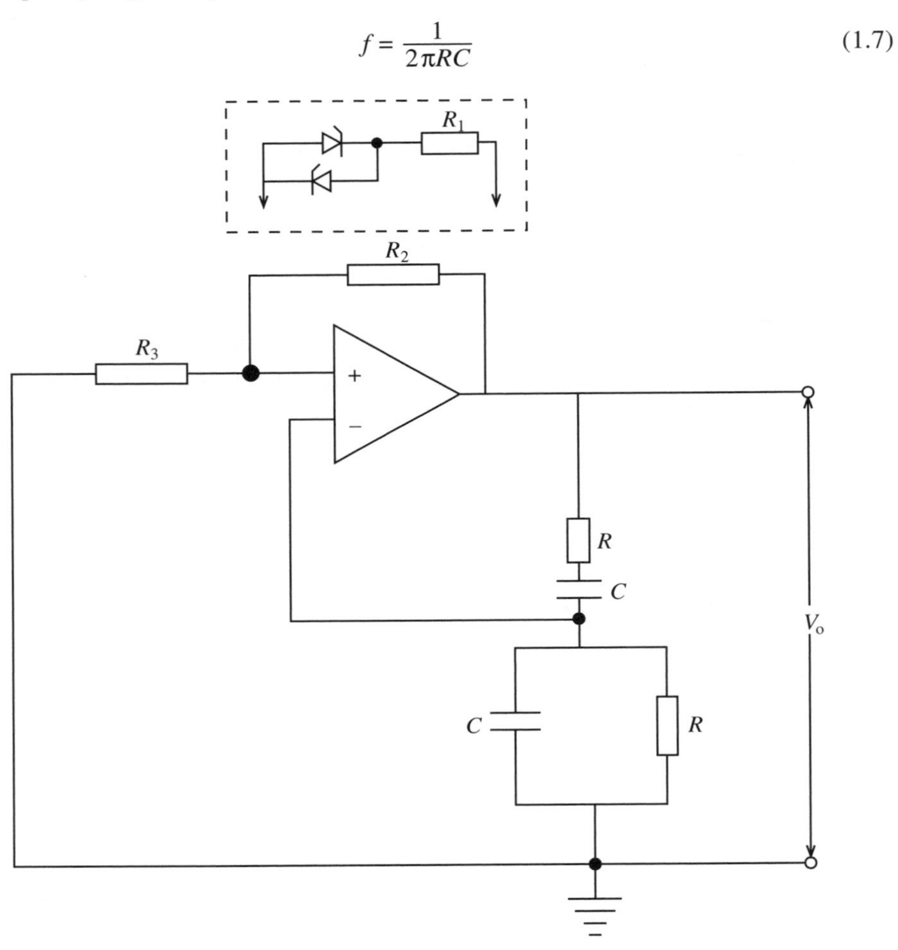

Fig. 1.8

The following points should be noted about this oscillator:

(i) R and C may have different values in the bridge circuit, but it is customary to make them equal.

(ii) This oscillator may be made variable by using variable resistors or capacitors.

(iii) If a BJT or FET is used then two stages must be used in cascade to provide the 360° phase shift between input and output.

(iv) The amplitude of the output waveform is dependent on how much the loop gain $A\beta$ is greater than unity. If the loop gain is excessive, saturation occurs. In order to prevent this, the zener diode network shown in Fig. 1.8 should be connected across R_2.

(v) The closed loop gain must be 3.

Example 1.2

A Wien bridge oscillator has to operate at 10 kHz. The diagram is shown in Fig. 1.9. A diode circuit is used to keep the gain between 2.5 and 3.5. Calculate all the components if a 311 op-amp is used.

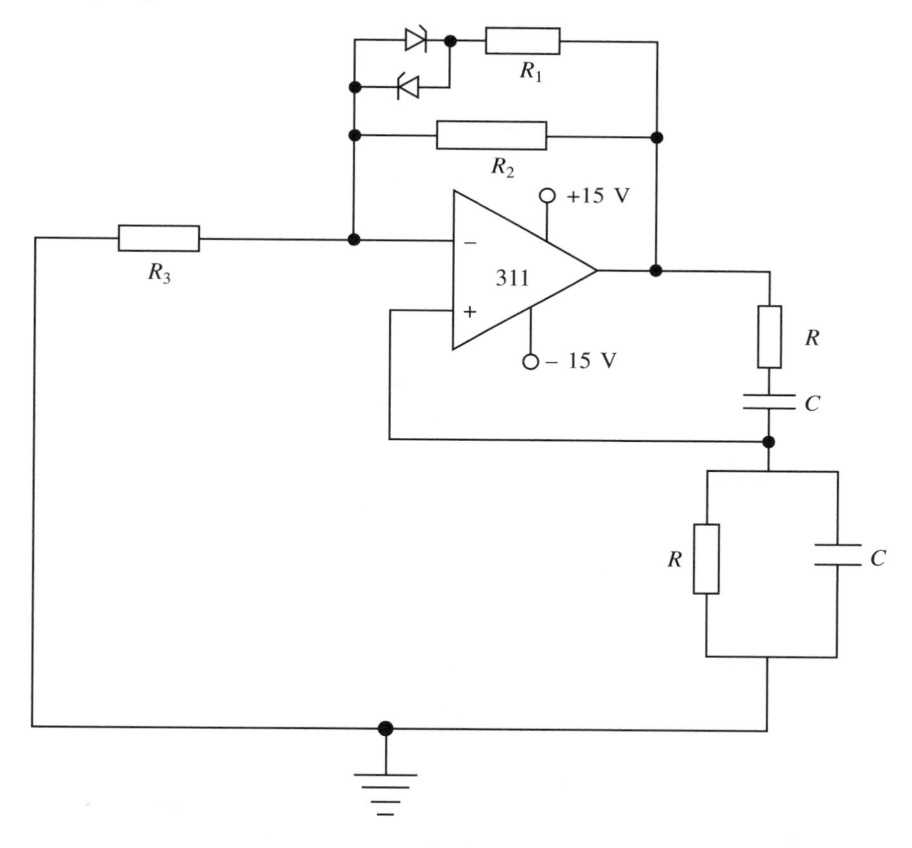

Fig. 1.9

Solution

When the op-amp is operating with a gain of 3, R_2 and R_3 may be calculated by using

$$A_V = 1 + \frac{R_2}{R_3}$$

However, for practical purposes this gain is dependent on the current flowing through R_2 and this should be very much larger than the maximum bias current, say 2000 times. The maximum bias current for the 311 is 250 nA. Also the voltage swing of the op-amp must be known and this is generally one or two volts below the supply voltage.

Hence, by Ohm's law,

$$R_2 + R_3 = \frac{14 \times 10^9}{5 \times 10^5} = 28\ \text{k}\Omega$$

$$R_3 = 9.3\ \text{k}\Omega \text{ and } R_2 = 18.6\ \text{k}\Omega$$

The nearest available value for $R_2 = 18.6$ kΩ. However, as the oscillator is subject to gain variation, the zener diode circuit will alter the value of R_2 if the amplitude of the oscillations increases.

The zeners are virtually open-circuited when the amplitude is stable and under this condition

$$3.5 = 1 + \frac{R_2}{9.3}$$

Hence $R_2 = 23.25$ kΩ, for which the nearest available value is 27 kΩ. Also,

$$2.5 = 1 + \frac{R_T}{9.3}$$

where

$$R_T = \frac{R_1 R_2}{R_1 + R_2} = \frac{23.25 \times 13.95}{23.25 - 13.95}$$

$$= 34.8\ \text{k}\Omega$$

The nearest available value is $R_1 = 33$ kΩ.

When the diodes are open

$$A_V = 1 + \frac{R_2}{R_3} = 1 + \frac{27}{8.6} = 3.23$$

If the amplitude of the oscillations increases the zener diodes will conduct and this places R_1 in parallel with R_2, thus reducing the gain:

$$R_T = \frac{34.8 \times 23.25}{34.8 + 23.25} = 13.93\ \text{k}\Omega$$

The nearest available value is 13.6 kΩ.

$$A_V = 1 + \frac{R_T}{R_3} = 1 + \frac{13.6}{8.6} = 2.6$$

Finally, the frequency is given by

$$f = \frac{1}{2\pi RC}$$

Select $C = 100$ nF.

$$R = \frac{1}{2\pi f C} = \frac{10^9}{10^4 \times 2\pi \times 100} = 159.2\ \Omega$$

Two 1 kΩ potentiometers could be set to this value using a Wayne–Kerr bridge. Note that this is a frequency-determining bridge which uses the principle of the Wheatstone bridge configuration. Alternating current bridges are a natural extension of this principle, with one of the impedance arms being the unknown component value. The Wayne–Kerr bridge is available commercially and is a highly accurate instrument containing a powerful processor capable of determining resistance, capacitance, self-inductance and mutual inductance values. It can also select batches of components having exactly the same value, which is useful in such circuits as the Wien bridge oscillator where similar component values are used.

The twin-T oscillator

This oscillator is shown in Fig. 1.10(a) and is, strictly speaking, a notch filter. It is used in problems where a narrow band of noise frequencies of a single-frequency component has to be attenuated. It consists of a low-pass and high-pass filter, both of which have a sharp cut-off at the rejected frequency or narrow band of frequencies. This response is shown in Fig. 1.10(b). The notch frequency (f_o) is attenuated sharply as shown. Frequencies immediately on either side of the notch are also attenuated, while the characteristic responses of the low and high-pass filters will pass all other frequencies in their flat passbands.

This type of oscillator provides good frequency stability due to the notch filter effect. There are two feedback paths, the negative feedback path of the twin-T network and the positive feedback path caused by the voltage divider R_5 and R_4. One of the T-networks is low-pass (R, $2C$) and the other is high-pass (C, $R/2$).

The function of these two filters is to produce a notch response with a centre frequency which is the desired frequency. Oscillation will not occur at frequencies above or below this frequency. At the oscillatory frequency the negative feedback is virtually zero and the positive feedback produced by the voltage divider permits oscillation.

The frequency of operation is given by

$$f = \frac{1}{2\pi RC} \tag{1.8}$$

and the gain is set by R_1 and R_2.

The main problem with this oscillator is that the components must be closely matched to about 1% or less. They should also have a low temperature coefficient to give a deep notch.

The twin-T filter is generally used for a fixed frequency as it is difficult to tune because of the number of components involved.

A more practical circuit is shown in Fig. 1.11, as fine-tuning of the oscillator can be achieved due to the potentiometer which is part of the low-pass network, Also Fig. 1.10(a) functions more like a filter, while Fig. 1.11 ensures suitable loop gain and phase shift, due to the output being strapped to the input, to ensure a stable notch frequency.

Once again matching of components is required but tuning over a range of frequencies can be achieved by a single potentiometer R_2/R_3. Note that

$$R_1 = 6(R_2 + R_3) \tag{1.9}$$

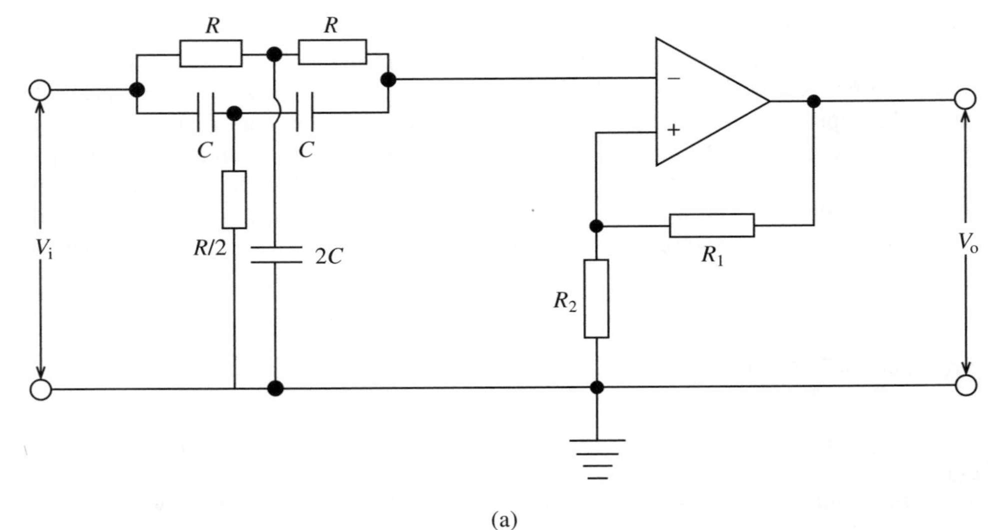

(a)

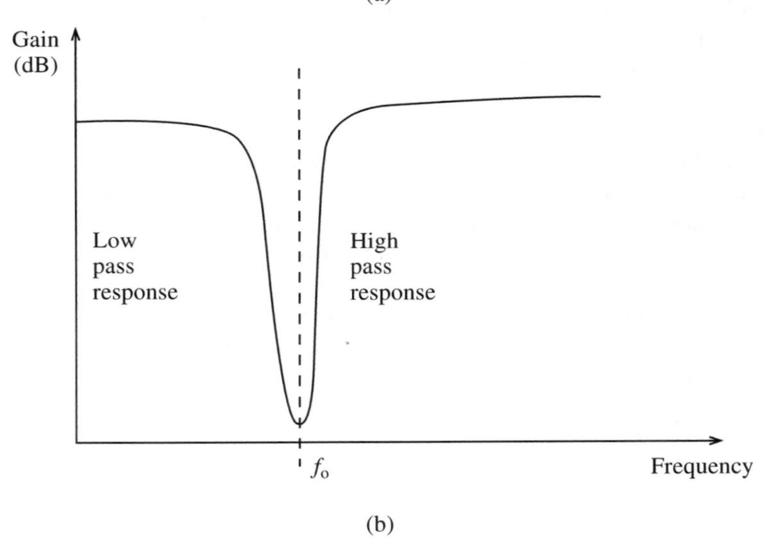

(b)

Fig. 1.10

and

$$f = \frac{1}{2\pi C\sqrt{3R_2R_3}} \tag{1.10}$$

Example 1.3

A notch oscillator has to be designed using an op-amp to eliminate 50 Hz in a radio receiver. Design such a filter using a twin-T network and a modified network.

Solution

If a 741 op-amp is used, its maximum input bias current is 500 nA and its voltage swing

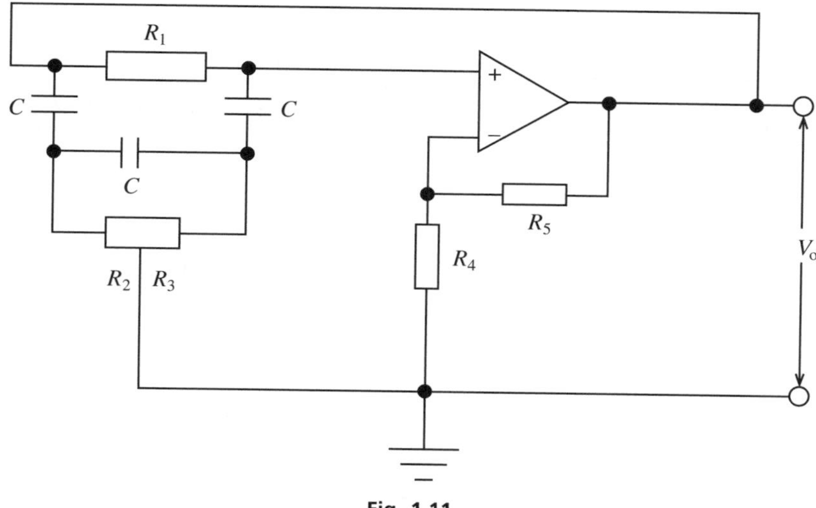

Fig. 1.11

is ±14 V for a ±15 V supply. As the gain is dependent on the current passing through R_5, this current must be large, say $2000 \times 500 \times 10^{-9}$ nA = 1 mA. Hence

$$R_1 + R_2 = \frac{14 \times 10^{-9}}{10^{-6}} = 14 \text{ k}\Omega$$

Select R_1 = 8.2 kΩ1% so R_2 = 5.6 kΩ1%; select C = 1 μF. Hence

$$R = \frac{1}{2\pi fC} = \frac{10^6}{2\pi \times 50 \times 1} = 3.18 \text{ k}\Omega$$

Use a 5 kΩ potentiometer. If the modified circuit is used then, with reference to Fig. 1.9, R_5 = 8.2 kΩ 1 and R_4 = 5.6 kΩ. Select a potentiometer of $R_2 + R_3$ = 10 kΩ, so $R_1 = 6(R_2 + R_3)$ = 60 kΩ. Select a 100 kΩ potentiometer. Hence, if R_2 = 40 kΩ and R_3 = 20 kΩ, then

$$C = \frac{1}{2\pi f\sqrt{3R_2R_3}} = \frac{10^3}{6.28 \times 50\sqrt{3 \times 20 \times 40}} = 6.5 \text{ μF}$$

1.5 *LC* oscillators

These oscillators have a greater operational range than *RC* oscillators which are generally stable up to 1 MHz. Also the very small values of *R* and *C* in *RC* oscillators become impractical. In this section we discuss Colpitts, Hartley, Clapp and Armstrong oscillators in turn.

The Colpitts oscillator

This oscillator consists of a basic amplifier with an *LC* feedback circuit as shown in Fig. 1.12. The oscillator uses a split capacitance configuration. The approximate frequency is given by

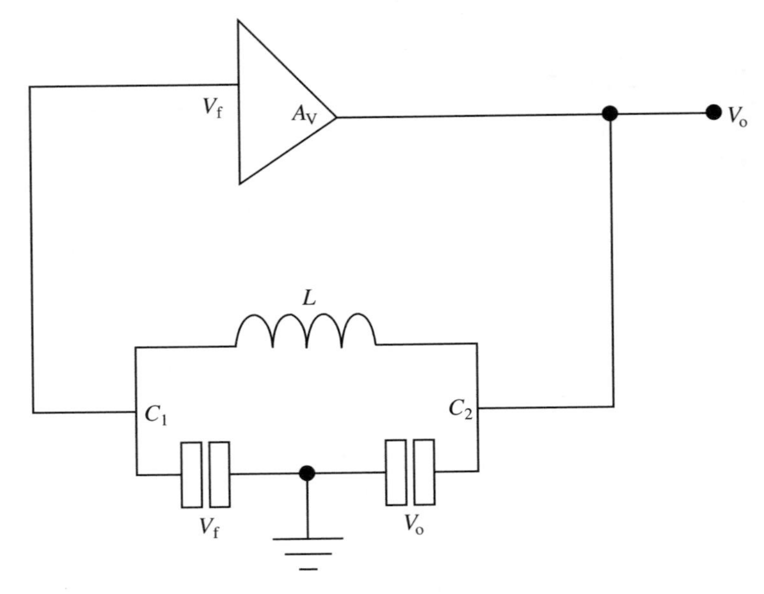

Fig. 1.12

$$f = \frac{1}{2\pi\sqrt{LC_T}} \tag{1.11}$$

where C_T is the total capacitance. This can be calculated by appreciating that the two capacitors are effectively in series.

The β factor can be derived by using Fig. 1.5:

$$\beta = \frac{V_f}{V_o} = \frac{IX_{C1}}{IX_{C2}} = \frac{X_{C1}}{X_{C2}} = \frac{1}{2\pi fC_1} \bigg/ \frac{1}{2\pi fC_2} = \frac{C_2}{C_1} \tag{1.12}$$

As $A\beta = 1$ for oscillation

$$A = \frac{C_1}{C_2} \tag{1.13}$$

In practice, $A > C_1/C_2$ for start up conditions.

Two practical circuits are shown in Fig. 1.13. Input and output resistances have an effect on the Q factor and hence the stability of these circuits. Figure 1.13(a) has the input resistance (h_{ie}) of the transistor in parallel with the tuned load and this will reduce the Q factor substantially.

Some further points should be noted concerning the design of this oscillator as well as the other oscillators discussed later.

(a) The input resistance to the transistor configuration shown in Fig. 1.13(a) is normally between 1 kΩ and 1.5 kΩ. Hence this will load the tuned circuit.

(b) If a load is connected to the output of the oscillator in Fig. 1.13(a) the Q factor may fall if the load resistance is small. One way of overcoming this is to include a buffer stage, such as an emitter follower, or else use transformer coupling.

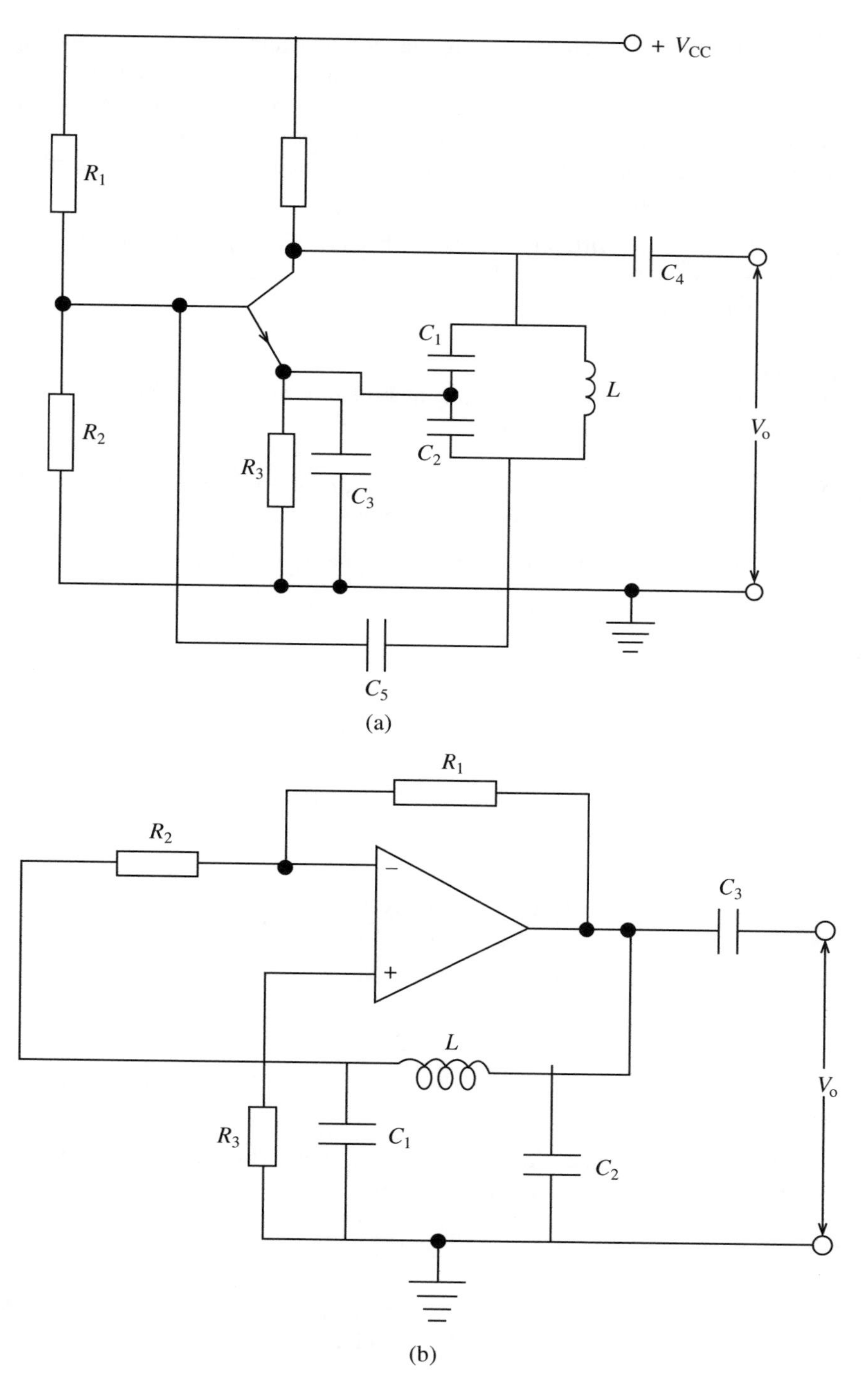

Fig. 1.13

(c) The effects of input loading can be minimized by using an FET or an op-amp, but if either is used C_2 in Fig. 1.13(b) will be in parallel with the output resistance, which is characteristically about 10–100 Ω. Consequently, the reactance of C_2 should be larger than this so that more of the signal voltage may be developed across it. The reactance should have a minimum value of at least ten times the value of the output resistance.

(d) In Fig. 1.13(b) R_2 is virtually across C_1, because the high input resistance at the oscillator frequency is very small compared to R_2. The theoretical gain of $A = C_1/C_2$ is more realistic.

Example 1.4

A transistor Colpitts oscillator has to operate at a fixed frequency of 1 MHz. A 25 μH coil is available which has a d.c. resistance of 2 Ω.

(a) Determine the values of C_1 and C_2 if the h_{ie} of the transistor is ignored. Hence determine the gain and show how frequency stable this circuit should be.

(b) Determine the frequency of the oscillator if the h_{ie} is 1 kΩ.

Solution

(a) We rearrange equation (1.11) to obtain

$$C_T = \frac{1}{(2\pi f)^2 L} = \frac{10^6}{(6.28 \times 10^6)^2 \times 25} = \frac{10^6}{985.96 \times 10^{12}} = 1013 \text{ pF}$$

Select $C_2 = 250$ pF. Then

$$C_T = \frac{C_1 C_2}{C_1 + C_2}$$

$$\therefore \quad C_1 = \frac{C_T C_2}{C_2 - C_T} = \frac{1013 \times 250}{250 - 1013} = 331.9 \text{ pF}$$

As h_{ie} is ignored, the coil is unloaded.

$$Q = \frac{\omega_o L}{r} = \frac{6.28 \times 10^6 \times 25}{10^6 \times 2}$$

$$= 78.5$$

Thus $Q > 10$, hence the assumption is that the frequency will vary very little.

(b) In this case the coil is loaded by 1 kΩ. So

$$Z_o = \frac{L}{Cr} = \frac{25 \times 10^{12}}{10^6 \times 1014 \times 2} = 12.33 \text{ k}\Omega$$

$$Z_o \parallel h_{ie} = R = \frac{12.33 \times 1}{12.33 + 1} = \frac{12.33}{13.33} = 925$$

So

$$Q = \frac{R}{\omega_o L} = \frac{925 \times 10^6}{2\pi \times 10^6 \times 25} = 5.9$$

The frequency would be variable. The value of frequency can be determined by using the relationship for a resonant circuit:

$$f = \frac{1}{\pi\sqrt{LC_T}}\sqrt{\frac{Q^2}{Q^2+1}} = \frac{1}{2\pi\sqrt{\frac{25}{10^6}\times\frac{1013}{10^{12}}}} \times 0.986 = 986\text{ kHz}$$

This example shows how an op-amp or FET would be more suitable.

Example 1.5

A Colpitts oscillator is designed to operate at 800 kHz using an op-amp with an output resistance (R_o) of 100 Ω and an inductance of 100 μH. Determine all the component values.

Solution

$$f = \frac{1}{2\pi\sqrt{LC_T}}$$

∴

$$C_T = \frac{1}{(2\pi f)^2 L} = \frac{10^6}{(2\pi \times 8 \times 10^5)^2 \times 100} = 395.8\text{ pF}$$

Since $R_o = 100\ \Omega$, then $X_{C2} = 10 \times 100 = 1000\ \Omega$. Hence

$$C_2 = \frac{1}{2\pi f X_{C2}} = \frac{1}{2\pi \times 8 \times 10^5 \times 1000} = 199\text{ pF}$$

This gives $C_1 = 399$ pF.

$$R_2 = 10X_{C1}$$

$$= \frac{10 \times 10^{12}}{399 \times 2\pi \times 8 \times 10^5} \simeq 5\text{ k}\Omega$$

As

$$A_V > \frac{C_1}{C_2} = \frac{399}{199} > 2$$

We select $A_V = 3$. Since

$$A_V = \frac{R_1}{R_2}$$

we have

$$R_1 = 3 \times 5 = 15\text{ k}\Omega$$

Finally,

$$R_3 = \frac{R_2 R_1}{R_2 + R_1} = \frac{15 \times 5}{15 + 5} = 3.75\text{ k}\Omega$$

The Hartley oscillator

This oscillator is very similar to the Colpitts except that it has a split inductance. It is represented in a similar way to the Colpitts, as seen in Fig. 1.14. It may be designed using a similar approach to the Colpitts but it has the disadvantages of mutual inductance between the coils, which causes unpredictable frequencies, and also the inductance is more difficult to vary.

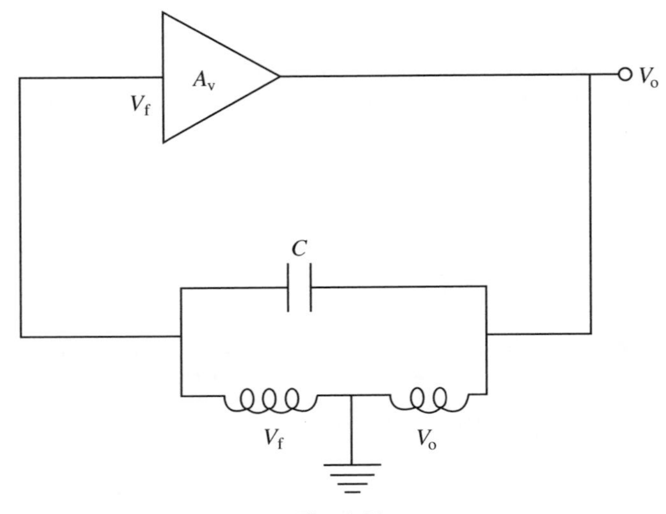

Fig. 1.14

When two coils are placed in close proximity to one another the flux due to the magnetic field of one interacts with the other. Hence an induced voltage is applied to the second coil due to the rate of change of flux. Similarly, flux due to the magnetic field of the second coil may cut the first coil, also inducing a voltage in it. This is referred to as mutual induction, in contrast to self-inductance which is caused by lines of magnetic force cutting a single coil. Hence the rate of change of flux in one coil affects the other. Splitting a single coil causes similar effects and mutual inductance exists between the two parts. As can be seen from equations (1.14), (1.15) and (1.16), the gain and frequency are dependent on the mutual inductance, and these parameters may be difficult to achieve as the tapping point has to be precise.

Two practical circuits are shown in Fig. 1.15. In both circuits the frequency is given by

$$f = \frac{1}{2\pi\sqrt{L_T C}} \tag{1.14}$$

where $L_T = L_1 + L_2 + 2M$ as both coils are virtually in series; note that M is the mutual inductance. The β factor and gain are

$$\beta = \frac{L_1 + M}{L_2 + M} \tag{1.15}$$

$$A > \frac{L_2 + M}{L_1 + M} \tag{1.16}$$

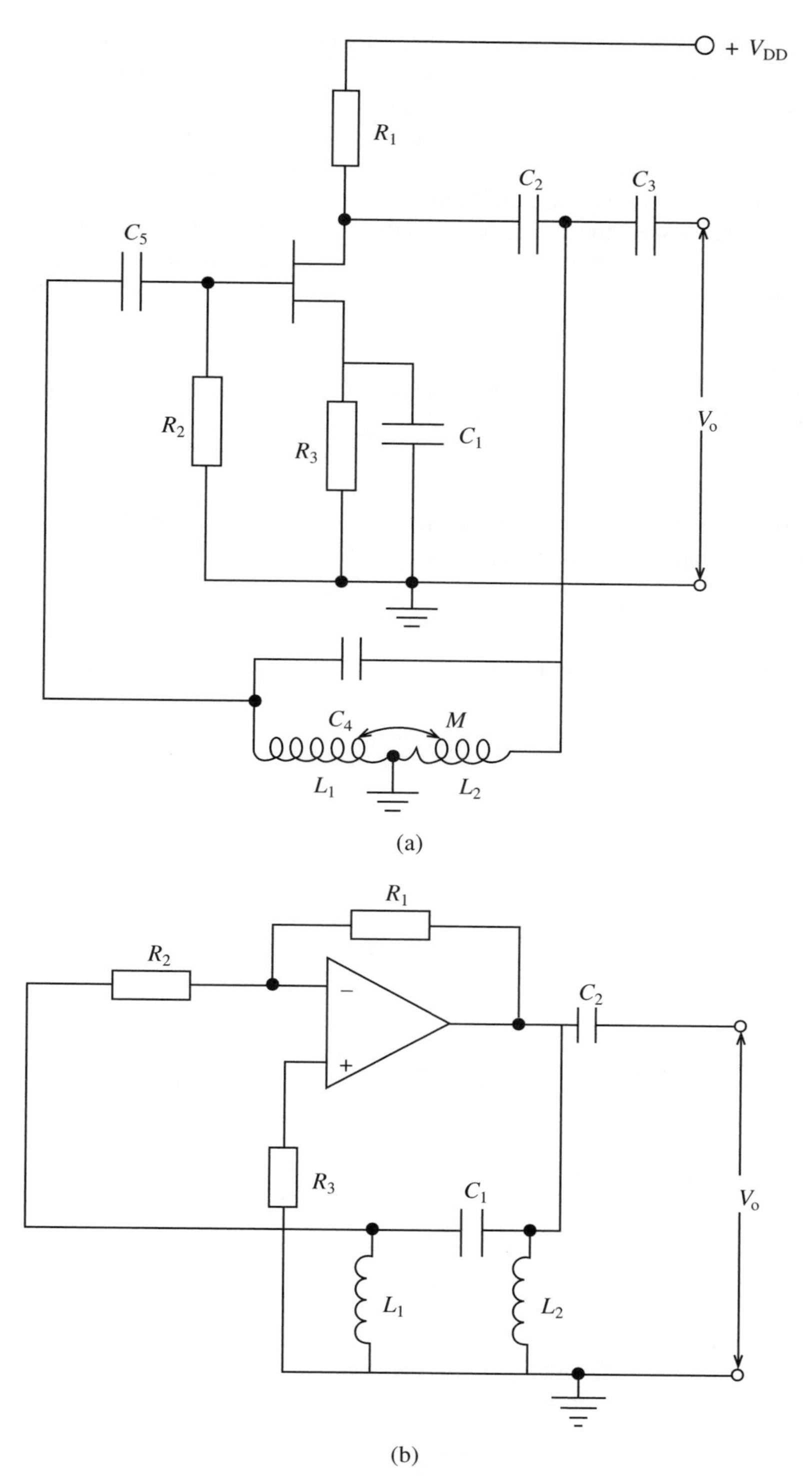

Fig. 1.15

The remarks made earlier concerning loading and Q factors also apply here.

While the Hartley and Colpitts oscillators have a similar design, the Hartley is easier to tune while the Colpitts requires two ganged capacitors. An advantage of using a Colpitts oscillator is the reduction in low-capacitance paths which can cause spurious oscillations at high frequencies. This is mainly due to the inter-electrode capacitance of the semiconductors. The Hartley oscillator, on the other hand, can produce several LC combinations due to the capacitance between the turns of the coil and thus cause spurious oscillations. It is for this reason that the Colpitts oscillator is often used as the local oscillator in receivers.

Example 1.6

Design a Hartley oscillator having a frequency of 25 kHz and $Q > 10$. Assume that the coupling coefficient is unity.

Solution

For $Q > 10$ a 741 op-amp is chosen. The mutual inductance is given by $M = k\sqrt{L_1 L_2}$, but since the coupling coefficient k is unity we have

$$M = \sqrt{L_1 L_2}$$

Substituting for L_T in equation (1.14), we have

$$f = \frac{1}{2\pi\sqrt{C(L_1 + L_2 + 2M)}}$$

Select $L_2 = 6$ mH and $L_1 = 3$ mH. Then

$$M = \sqrt{\frac{6 \times 3}{10^6}} = 4.24 \text{ mH}$$

$$f = \frac{1}{2\pi\sqrt{L_T C}}$$

$$C = \frac{1}{(2\pi F)^2 (L_1 + L_2 + 2M)} = \frac{10^3 \times 10^9}{24649 \times 10^6 \times 13.24} = 3.06 \text{ nF}$$

$$A = \frac{R_1}{R_2}$$

and note that

$$A > \frac{L_2}{L_1} = \frac{6}{3} = 2$$

Hence selecting $R_2 = 1$ kΩ and a gain of 3 will give $R_1 = 3$ kΩ. Either select the nearest value or use a potentiometer. Finally,

$$R_3 = \frac{3 \times 1}{3 + 1} \text{ k}\Omega = 750\ \Omega$$

The Clapp oscillator

This oscillator is a modified Colpitts, as can be seen from Fig. 1.16. If C_4 is substantially smaller than C_1 and C_2, the frequency can be controlled virtually by C_4. Once again,

$$f = \frac{1}{2\pi\sqrt{LC_T}} \tag{1.17}$$

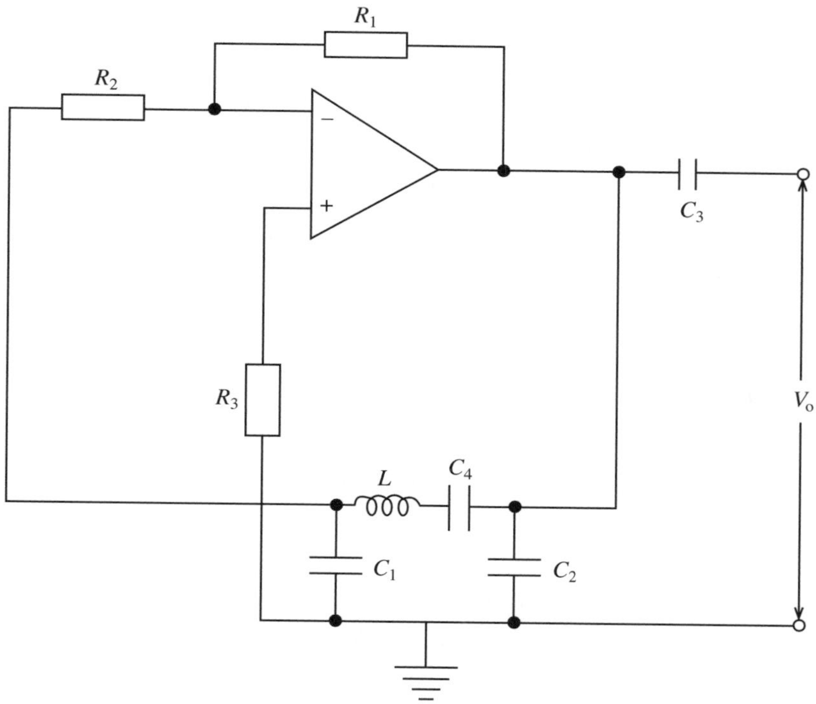

Fig. 1.16

If C_4 is much smaller than C_1 or C_2 then

$$f = \frac{1}{2\pi\sqrt{LC_4}} \tag{1.18}$$

Also

$$A = \frac{C_1}{C_2} \tag{1.19}$$

The inclusion of C_4 has the advantage that it is not affected by stray or junction capacitance which may appear across C_1 and C_2 thus altering the tuning.

Example 1.7

A Clapp oscillator has to be used as a test oscillator in a telephone system using frequency division multiplexing. Four carrier frequencies are required (1.8 MHz, 1.92 MHz,

2.09 MHz and 2.21 MHz). Determine the range of C_4 if it is made variable and also suitable values for all components if the gain has to be 2.5 and $L = 100\ \mu H$.

Solution
If C_4 is much smaller than C_1 or C_2,

$$f = \frac{1}{2\pi\sqrt{LC_4}}$$

so

$$C_4 = \frac{1}{(2\pi f)^2 L}$$

For a 1.8 MHz carrier frequency,

$$C_4 = \frac{10^6}{127.78 \times 10^{12} \times 100} = 78.26 \text{ pF}$$

For 1.92 MHz,

$$C_4 = \frac{10^6}{145.4 \times 10^{12} \times 100} = 68.78 \text{ pF}$$

For 2.09 MHz,

$$C_4 = \frac{10^6}{172.3 \times 10^{12} \times 100} = 58 \text{ pF}$$

Finally, for 2.21 MHz,

$$C_4 = \frac{10^6}{192.2 \times 10^{12} \times 100} = 51.92 \text{ pF}$$

Hence C_4 should be variable between 40 and 100 pF to ensure correct tuning.
Since the gain has to be 2.5,

$$A = \frac{C_1}{C_2} = 2.5$$

$\therefore$

$$C_1 = 2.5C_2$$

Selecting the 1.8 MHz frequency

$$f = \frac{1}{2\pi\sqrt{LC_T}}$$

and

$$\frac{1}{C_T} = \frac{1}{C_1} + \frac{1}{C_2} + \frac{1}{C_4}$$

$$\frac{1}{C_T} = \frac{1}{2.5C_2} + \frac{1}{C_2} + \frac{1}{C_4}$$

$$\frac{1}{C_T} = \frac{3.5}{2.5C_2} + \frac{1}{C_4}$$

Also

$$C_T = \frac{1}{(2\pi f)^2 L} = \frac{10^6}{(6.28 \times 18)\, 2 \times 10^{12} \times 100} = 78.3 \text{ pF}$$

Which is, as expected, close to C_4. Hence

$$\frac{1}{78.3} = \frac{3.5}{2.5C_2} + \frac{1}{78.25}$$

This gives

$$C_2 = 1389.7 \text{ pF}$$

$$C_1 = 1389.7 \times 2.5 = 3474.2 \text{ pF}$$

Finally, we rearrange $A = R_1/R_2$ as

$$R_1 = 2.5\, R_2$$

Select $R_2 = 10$ kΩ; then $R_1 = 25$ kΩ. These should be close-tolerance resistors.

The Armstrong oscillator

This oscillator uses transformer coupling to feed back a portion of the output voltage. A simple design is shown in Fig. 1.17. The frequency can be found from the expression

$$f = \frac{1}{2\pi\sqrt{L_1 C_3}} \tag{1.20}$$

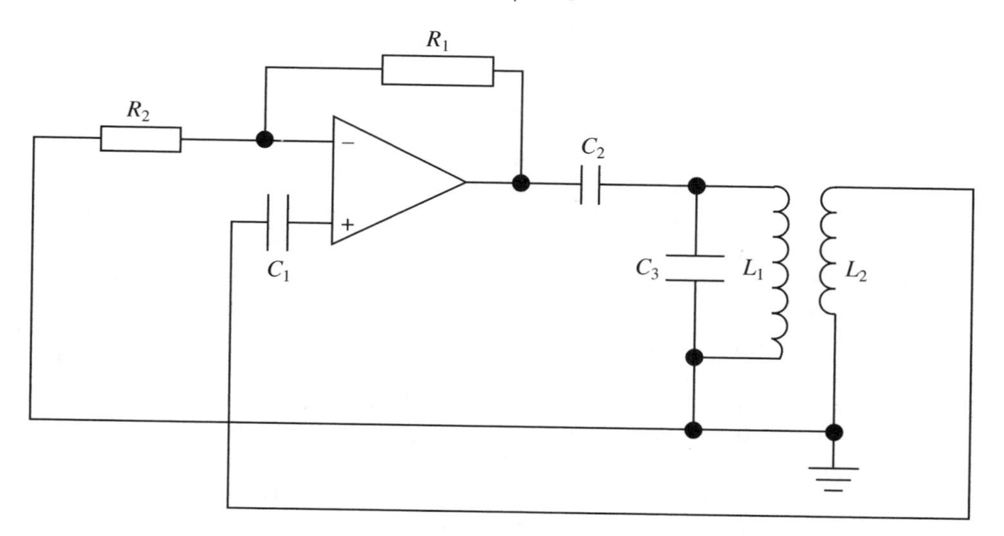

Fig. 1.17

This oscillator is used in high-frequency, long-distance communications because of its power-handling capabilities. The high-frequency part of the spectrum generally uses the curvature of the earth for transmission and this requires high-power circuits at the transmission end. However, because of the transformer size and cost, it is not as common as the other oscillators discussed in this chapter.

1.6 Crystal oscillators

These are amongst the most stable of all oscillators and are generally used in broadcasting and telecommunication systems where high stability is required.

A crystal used in telecommunications work is generally made from quartz. It uses the piezo-electric principle, whereby the application of a voltage across its axis causes the crystal to change shape. The converse is also true. This property is useful because the properties of quartz are very stable with temperature. In certain applications, such as the fabrication of silicon wafers where radio frequency methods are used, and also in military applications, crystals are kept in a temperature-regulated oven which is microprocessor-controlled.

The equivalent circuit of a crystal is shown in Fig. 1.18. C_1 represents the package capacitance (usually 5–30 pF), L the mechanical inertia of the crystal which has its electrical analogue in inductance (usually 10–100 H), and C_2 the mechanical compliance of the crystal (usually 0.05 pF). R represents the losses, which are normally very small (of the order of 50 Ω).

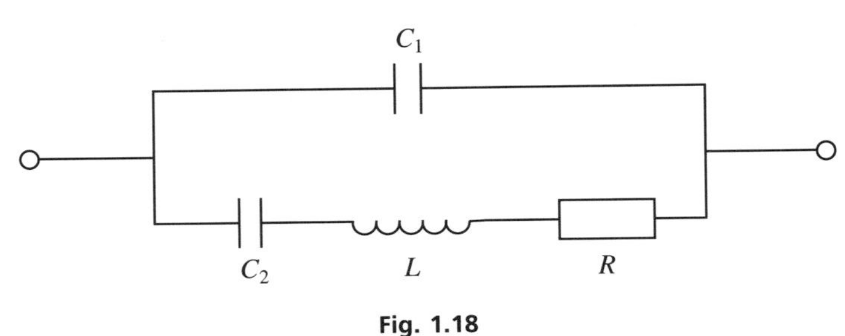

Fig. 1.18

In any LC resonant circuit, if the L/C ratio is large then the Q factor is high. L is extremely high for quartz crystals, while C_2 is very small. A second contributing factor is the low damping resistance, which gives Q factors of 10^6 for crystals. If the resistance of any LC circuit is small the circuit has a series resonant frequency called the undamped or natural frequency. This frequency is related to the Q factor of the LC circuit, but the Q factor in turn is inversely proportional to the damping resistance. Hence the smaller the resistance of a crystal (the damping resistance) the higher the Q factor.

From Fig. 1.18 it should be appreciated that there are two possible frequencies for the crystal; one for the series mode and one for the parallel mode. They are generally separated by about 1 kHz, and the crystal is usually operated between the two frequencies. Adding capacitance in parallel with the crystal decreases its parallel resonant frequency, while adding capacitance in series increases the parallel resonant frequency. Series-mode crystals normally operate with zero load capacitance, while parallel-mode crystals operate with a specific capacitance load.

Every crystal has a maximum rating, which might lie between 20 and 250 mA.

Overloading of the crystal may cause temperature increase and change in frequency. The most common cause of overloading is excessive feedback. Finally, a d.c. voltage applied to a crystal can also cause crystal damage due to the crystal being twisted out of shape.

Operation at higher frequencies is limited by how thin the crystal may be cut, but because of the mechanical resonances involved specially fabricated crystals may be obtained commercially which work at different overtones.

1.7 Crystal cuts

The crystal slices used in oscillator circuits are cut from whole or 'mother' crystals which have the general appearance of hexagonal prisms with each end capped by a hexagonal pyramid. The actual crystal used is commonly in the form of a slice cut at some specific angle to the whole crystal.

The crystal has three major axes, labelled *X*, *Y* and *Z*, the *X* and *Y* axes being at right angles to the *Z* axis. The crystal sections used in oscillators are cut on either the *X* or *Y* axis or at some angle to one of them. A slice cut with its larger surfaces perpendicular to an *X* axis is known as an *X*-cut slice, and a *Y*-cut slice is cut so that its major surfaces are perpendicular to the *Y* axis. Crystals are also cut at various angles with respect to the *Z* axis, and this gives a range of different frequency values.

The quartz crystal, when caused to vibrate, has a tendency to do so in parts so that harmonics of the fundamental vibration frequency are also produced. A crystal also tends to vibrate along its other axes as well as the *Y* axis, but the two principal vibrations occur in the *X* direction and in the *Y* direction. The vibration frequency in each direction is determined by the dimensions of the crystal in that direction and is dependent on the width and thickness of the slice in that direction. Hence the terms width vibration and thickness vibration are used.

The frequency temperature coefficient is the same for both of these vibrations and the crystal can be made to vibrate at either of these frequencies merely by tuning the load to a frequency slightly above the frequency desired.

The width vibration of *X*- and *Y*-cut crystals is commonly employed for low-frequency oscillators and the thickness vibration for high-frequency oscillators.

1.8 Types of crystal oscillator

Most of the oscillators already discussed may be adapted for crystal oscillations.

The Colpitts oscillator shown in Fig. 1.13 may have the inductor *L* replaced with a crystal, or a crystal may be incorporated in the feedback path as shown in Fig. 1.19. In this circuit the tuned network provides the narrow band output while the crystal provides positive feedback. The crystal in this case will work at its series resonant mode, which is the same frequency as the tuned circuit.

One point should be noted here. As has already been mentioned, the crystal has an equivalent circuit which includes the package capacitance (C_1 in Fig. 1.18). At higher frequencies, this capacitance can detune the oscillator and for this reason a compensatory

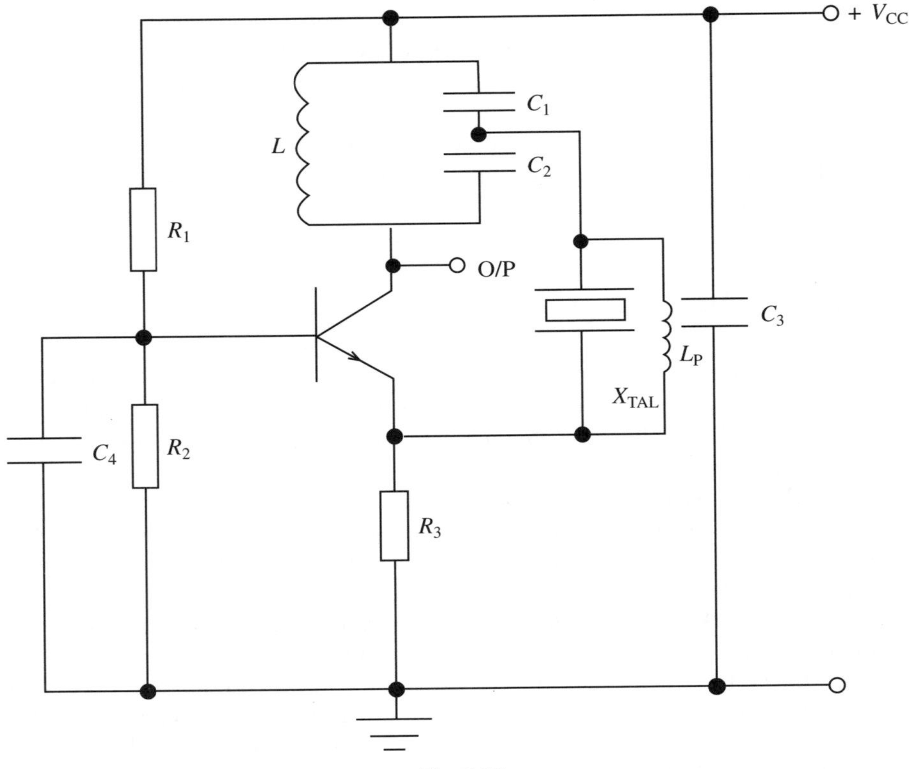

Fig. 1.19

inductance is sometimes placed in parallel with the crystal. This cancels out the effect of C_1. It can easily be calculated by using the expression

$$L_p = \frac{1}{(2\pi f)^2 C_1} \tag{1.20}$$

where L_p is the neutralizing inductor.

The oscillator shown in Fig. 1.20 is called a Pierce oscillator, and it uses a single crystal in conjunction with C_1 and C_2. Because a parallel LC tuned circuit is not used, crystals can be switched in without altering the other circuit components. This oscillator uses the characteristic inductance of the crystal to provide feedback at the correct phase. C_1 and C_2 also form part of the LC network, while R_1 is generally chosen large enough to give sufficient gain. All other components perform the usual functions.

A Wien bridge oscillator is shown in Fig. 1.21. This oscillator functions in the usual way, but the crystal adds stability to the bridge network. This network is tuned to the resonant frequency of the crystal.

1.9 Oscillator frequency stability

All oscillators suffer from frequency drift, noise and harmonic content. When the frequency

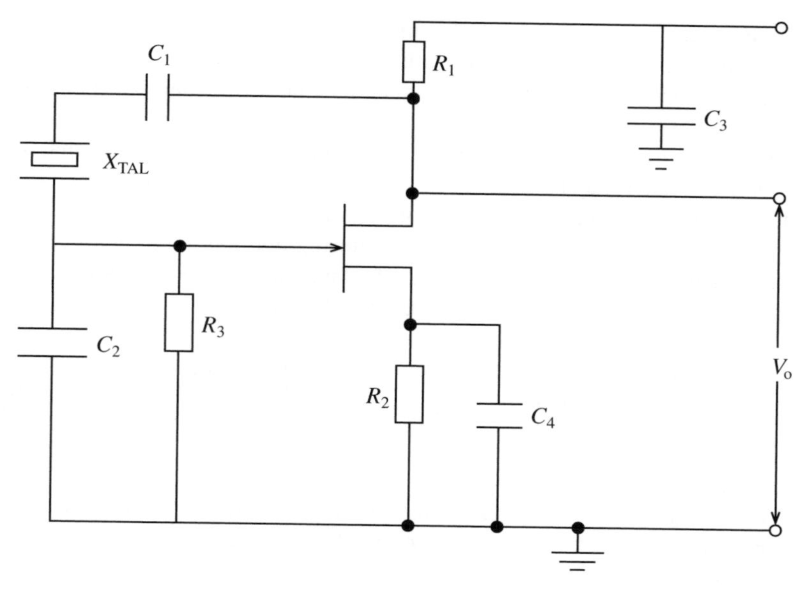

Fig. 1.20

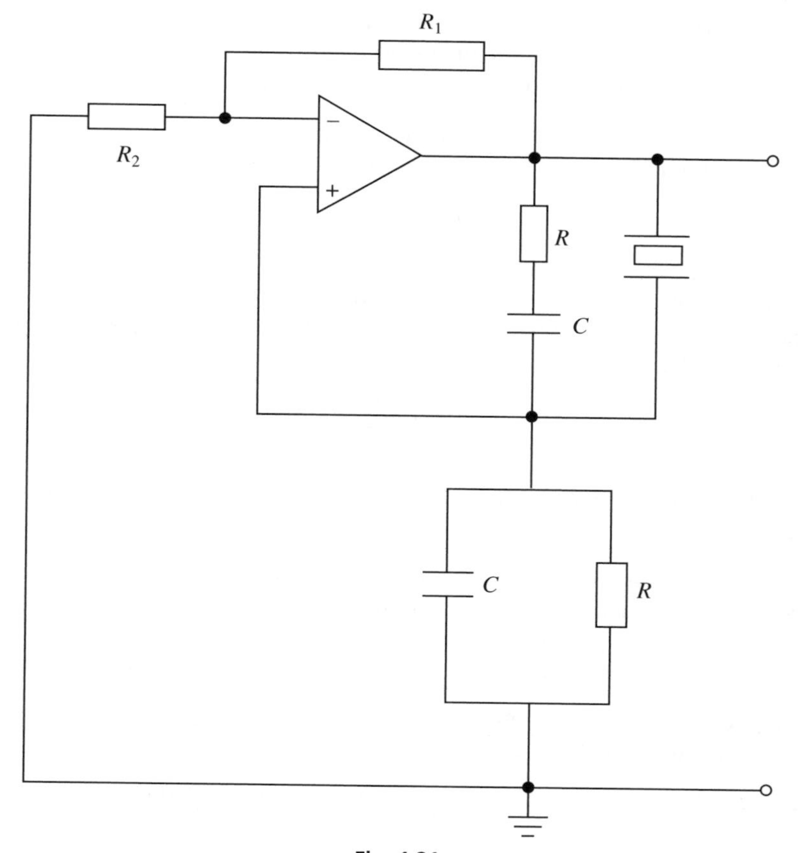

Fig. 1.21

of an oscillator varies from its specified value, it is said to have drifted. This is generally expressed as a percentage or, as temperature may be involved, as so many hertz per degree Celsius. Noise may be introduced into an oscillator externally or internally. As the oscillator is a radiator it can also pick up unwanted signals, some of which may be noise. Harmonics are multiples of a fundamental frequency and it is possible that second or third harmonic or higher may be generated by an oscillator which is not properly calibrated or designed. In most telecommunication transmitters and receivers harmonic content and other unwanted signals can be eliminated by filtering and automatic gain control.

Frequency drift or stability is the most important parameter when designing an oscillator, and the factors which generally affect it are as follows:

(i) Loading effects. Often an oscillator will function without a load, but load changes may cause frequency drift due to lack of matching. This can be remedied by means of a buffer stage between load and oscillator. An op-amp in buffer mode may be used.

(ii) Power supply coupling. The oscillator should be operated at low power in order to prevent ripple content coupling to the oscillator input. Decoupling capacitors may also be used to overcome this problem.

(iii) Temperature variations. These may be counteracted by using components which have known temperature coefficients. This is particularly applicable to capacitors and for this reason negative temperature coefficient capacitors should be used to compensate for positive temperature coefficient tuned circuits.

Associated with temperature stability is the temperature coefficient parameter. This is the small change in the parameter for each degree change in temperature.

$$TC = \Delta f_o / f_o \tag{1.21}$$

Generally the change is small and is expressed as parts per million (ppm). This is shown in Table 1.1, where a short list of crystals is given with some of their characteristics. If the 6 MHz crystal is selected, it has a temperature coefficient of ±100 ppm. This means that:

$$TC = \frac{100\text{ Hz}}{1\text{ MHz}} = \frac{600\text{ Hz}}{6\text{ MHz}}$$

So the output can vary as much as ± 600 Hz/°C. At a temperature change of 20°C this would be ± 12 kHz.

Note that crystals are generally cut in the *X* or *Y* axis. *X*-cut crystals cause a decrease in frequency with temperature increase and vice versa. *Y*-cut crystals cause an increase in frequency with temperature. However, crystals are generally cut at angles between the *X* and *Y* axis to give lower temperature coefficients.

(iv) Component selection. Components with close tolerances should be used where possible, and if suitable a crystal should always be used.

The crystal-controlled Colpitts oscillator shown in Fig. 1.22 illustrates the application of these points.

Table 1.1

Operating frequency (MHz)	Manufacturer	Load capacitance (pF)	Temperature stability (ppm)	Tolerance at 25°C (ppm)
4.9152	IQD	16	± 100	± 50
5	EUR	7	± 50	± 30
5	IQD	30	± 50	± 20
5.0688	EUR	series	± 50	20
5.0688	IQD	series	± 50	± 20
6.0	AEL	30	± 50	20
6.000	AEL	30	± 30	± 20
6.00	AEL	30	± 100	± 50
6.000	SNY	30	± 50	± 50
6.000	IQD	30	± 50	± 30
6.000	IQD	30	± 50	± 30
6	IQD	30	± 100	± 50
6.144	IQD	30	± 50	± 20
6.4	IQD	30	± 30	± 20
6.5536	IQD	12	± 30	± 20
7.3728	AEL	30	± 50	20
7.3728	AEL	30	± 100	± 50
7.3728	EUR	7	± 50	± 50
7.3728	SNY	30	± 50	± 50
7.3728	IQD	30	± 50	± 20
7.3728	IQD	18	± 30	± 15
7.3728	IQD	16	± 100	± 50

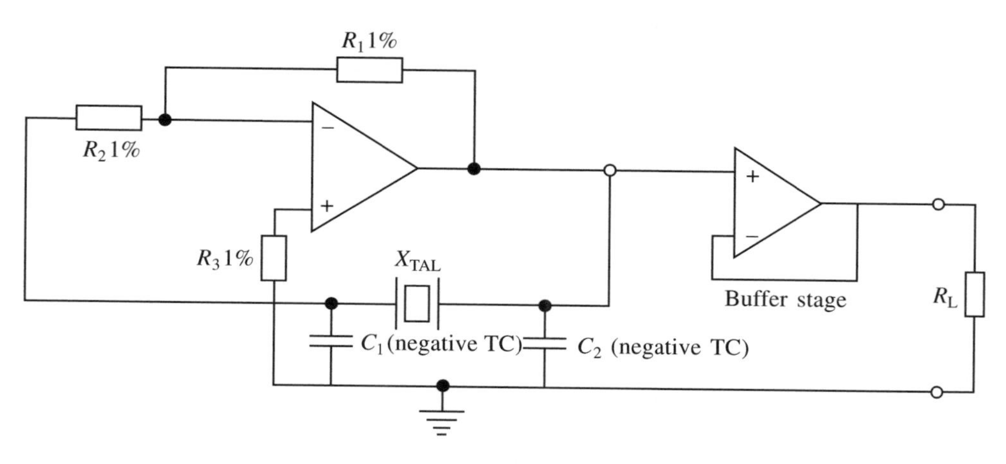

Fig. 1.22

Example 1.8

Design a Wien bridge oscillator working at a frequency of 5 MHz which has to be crystal-controlled with a temperature stability of ± 50 ppm. Use a crystal operating in the parallel mode.

Solution
Recall that

$$f = \frac{1}{2\pi RC}$$

Select $C = 100$ pF. Hence

$$R = \frac{1}{2\pi fC} = \frac{10^{12}}{6.28 \times 5 \times 10^{6} \times 10^{2}} = 318.3\ \Omega$$

From Table 1.1 there is a choice of crystals. The one manufactured by IQD has been selected as it has a slightly better tolerance. However, the load capacitance is 30 pF and a trimmer capacitor may have to be connected in series with the crystal.

The gain setting resistors are selected in the usual way subject to op-amp bias current:

$$R_1 = 2R_2$$

Select $R_2 = 10$ kΩ, so that $R_1 = 20$ kΩ. The final circuit is shown in Fig. 1.23.

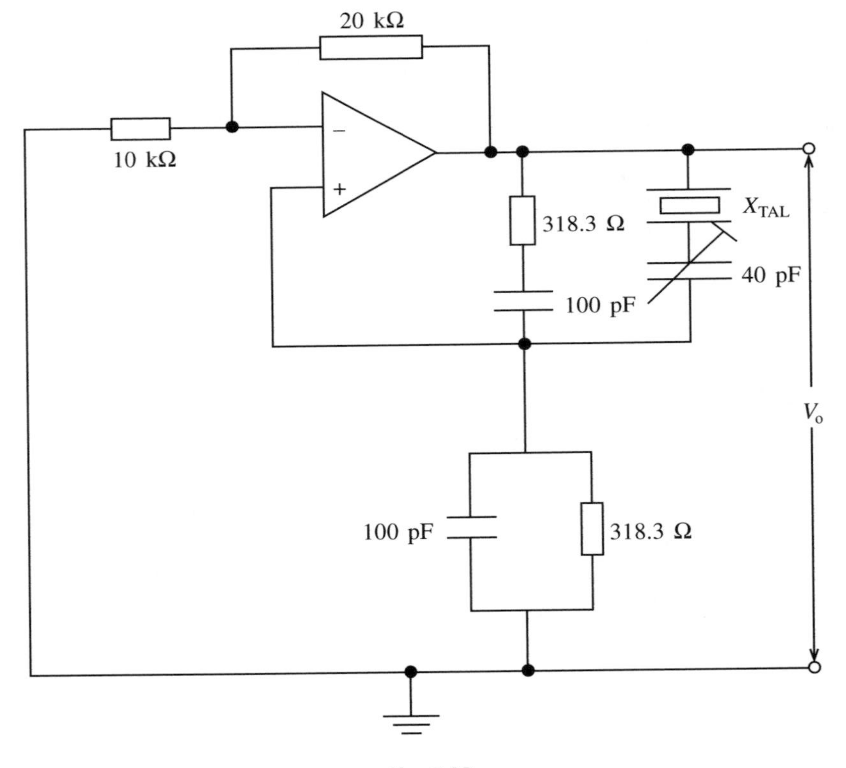

Fig. 1.23

Example 1.9
Design a crystal-controlled Colpitts oscillator operating at 30 MHz. Assume the crystal selected operates in its series mode and has a package capacitance of 12 pF.

Solution
A transistorized circuit similar to Fig. 1.16 will be designed. Thus

$$f = \frac{1}{2\pi\sqrt{LC_T}}$$

Select $L = 0.5$ μH (a few windings on a former). Rearranging and substituting,

$$C_T = \frac{1}{(2\pi f)^2 L} = \frac{10^6}{(6.28 \times 30 \times 10^6)^2 \times 0.5} = 56.4 \text{ pF}$$

from which $C_2 = 191$ pF if C_1 is selected as 80 pF. As the crystal has a package capacitance (C) of 12 pF, this has to be neutralized by shunting the crystal with a small inductance:

$$L_P = \frac{1}{(2\pi f_o)^2 C} = \frac{10^{12}}{(2\pi \times 3 \times 10^7)^2 \times 12} = 2.3 \text{ μH}$$

All other components may be evaluated in the same way using low-tolerance temperature coefficient components.

1.10 Integrated circuit oscillators

Most integrated circuits used in telecommunications systems currently incorporate such stages as the modulator, RF amplifier, IF amplifier and local oscillator. However, it is informative to look at the Harris HA7210 which is a low-power complementary metal-oxide semiconductor (CMOS) crystal oscillator capable of giving an output range from 10 kHz to 10 MHz.

The data sheets at the end of this chapter give all the relevant information required to produce a range of crystal-controlled frequencies using a Pierce oscillator. The output from this oscillator is non-sinusoidal and, unlike such integrated circuits as the 555 timer, it produces a highly stable output with very few components. It is also ideal for certain data communications circuits. The use of this chip is better explained by an example.

Example 1.10
An HA7210 chip has to be used to produce a specific frequency of 4.218 MHz. If a series-mode crystal having a resonance frequency of 4 MHz is used, design an oscillator suitable for this specification, taking into account all design considerations. Assume $C_M = 1.2$ pF and $C_0 = 4$ pF.

Solution
As the crystal used is a series-mode type, pullability will be involved to achieve the specific frequency of 4.218 MHz. From the data sheets,

$$F_P = F_S\left[1 + \frac{C_M}{2(C_0 + C_{CL})}\right]$$

Transposing gives

$$C_{CL} = \frac{C_M + 2C_0\left(1 - \frac{F_P}{F_S}\right)}{2\left(\frac{F_P}{F_S} - 1\right)} = \frac{1.2 + 2 \times 4\left(1 - \frac{4.218}{4}\right)}{2\left(\frac{4.218}{4} - 1\right)} = 7 \text{ pF}$$

Hence $C_1 = C_2 = 14$ pF

The timer capacitor (L) may be selected for a range of 5–200 pF but this will be trial and error for the final frequency. A 0.1 μF decoupling capacitor is connected across the supply. The circuit is shown in Fig. 1.24.

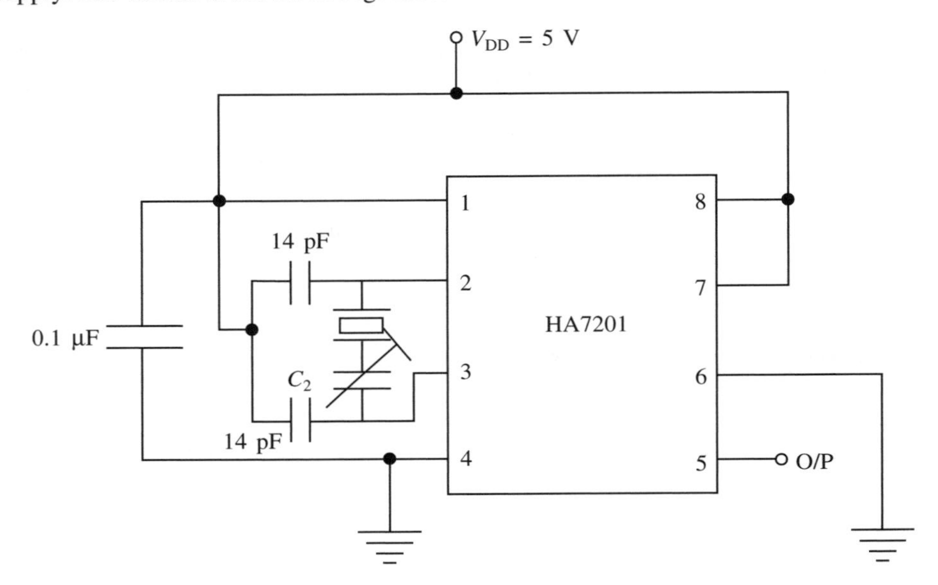

Fig. 1.24

Example 1.11

Using an HA7210 chip, design a crystal-controlled oscillator operating at 50 kHZ.

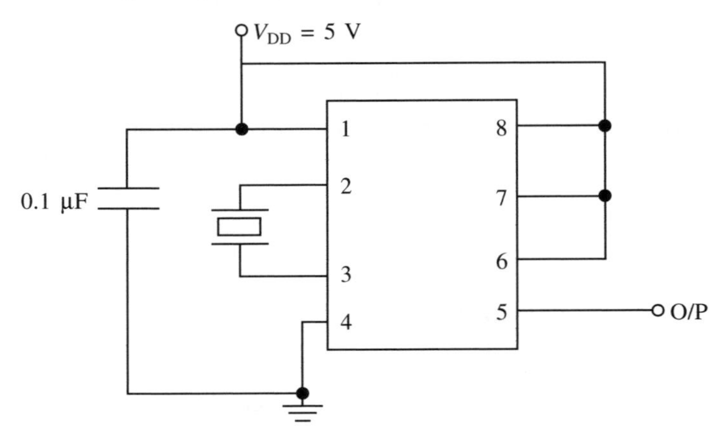

Fig. 1.25

Solution
A parallel-mode crystal with a loading capacitance of 7.5 pF is ideal as the specified frequency falls within the first frequency range. Hence the circuit would be as in Fig. 1.25.

1.11 Further problems

1. A phase-shift oscillator has to provide a frequency of 15 kHz. Design a suitable circuit similar to Fig. 1.5. Select C = 12 nF.
   Answer: R_1 = 361, a range of gain setting resistors
2. Design a Wien bridge oscillator capable of giving a constant frequency of 2.5 kHz. Select C = 0.01 μF.
   Answer: R = 6.369 kΩ; select R_1 and R_2 for A_v = 3
3. A Wien bridge oscillator has to operate at 75 kHz using Fig. 1.9. To ensure stability a zener diode network is used to keep the gain between 2.7 and 3.2. Calculate all the components if a 741 op-amp is used. Select C = 1200 pF and a supply voltage of ±10 V. Assume the maximum current through R_2 is 500 times the bias current.
   Answer: R = 1.77 kΩ, R_3 = 33 kΩ, R_2 = 72.6 kΩ, R_1 = 87 kΩ, R_T = 39.6 kΩ
4. Design a twin-T oscillator similar to Fig. 1.11 which uses a 741 op-amp and notches at 100 Hz. Take maximum current as 1000 times the bias current. Select a value for C and resistors with a 0.1% tolerance.
5. Design a Colpitts oscillator operating at 750 kHz. A 12 μH coil is available having a d.c. resistance of 1 Ω. Determine: (a) all the component values if the h_{ie} of the transistor can be ignored, and also the Q factor; (b) the frequency of the oscillator if h_{ie} = 1.2 kΩ.
   Answer: C_T = 3,756 pF, Q_a = 56.5, Q_b = 15.4, f = 747.7% kHz
6. A Colpitts oscillator operates at 2.5 MHz using an op-amp with an output resistance (R_o) of 50 Ω. If a suitable inductance of 40 μH is available, determine all the component values to the nearest available values using series parallel networks.
   Answer: C_1 = 127.3 pF, C_2 = 488 pF, R_2 = 5 kΩ, R_1 = 5 kΩ
7. A Hartley oscillator uses an op-amp to give an output frequency of 100 kHz. Design this oscillator assuming k = 0.4, L_1 = 10 mH and L_2 = 25 mH.
   Answer: C = 53.2 pF, A = 3
8. A Clapp oscillator is designed with an op-amp as shown in Fig. 1.16. Determine the frequency if C_1 = 1000 pF, C_2 = 4000 pF, C_4 = 100 pF and L = 80 μH.
   Answer: 1.8 MHz
9. An Armstrong oscillator is used in a high-frequency transmitter to produce a frequency of 800 kHz. If it is used in a transistorized configuration, determine the approximate value of L_1, given C_3 = 2.2 nF. What effect would transistor loading have on the frequency?
   Answer: 17.9 μH
10. The crystal oscillator shown in Fig. 1.19 has C_1 = 150 pF, C_2 = 100 pF and L = 60 μH. Ignore L_p. Determine:

(a) the crystal required for this circuit;
(b) the oscillator frequency at 40°C if the temperature stability is ±30 ppm;
(c) the neutralizing inductance required across the crystal.
Answer: 2.6525 MHz ± 3.2 kHz

11. A Pierce oscillator is shown in Fig. 1.26. C_2, C_3 and the crystal form the tuned circuit. From Table 1.1 the 6.4 MHz crystal is used. Determine suitable values for C_2 and C_3 and suggest values for R_1 and C_4.

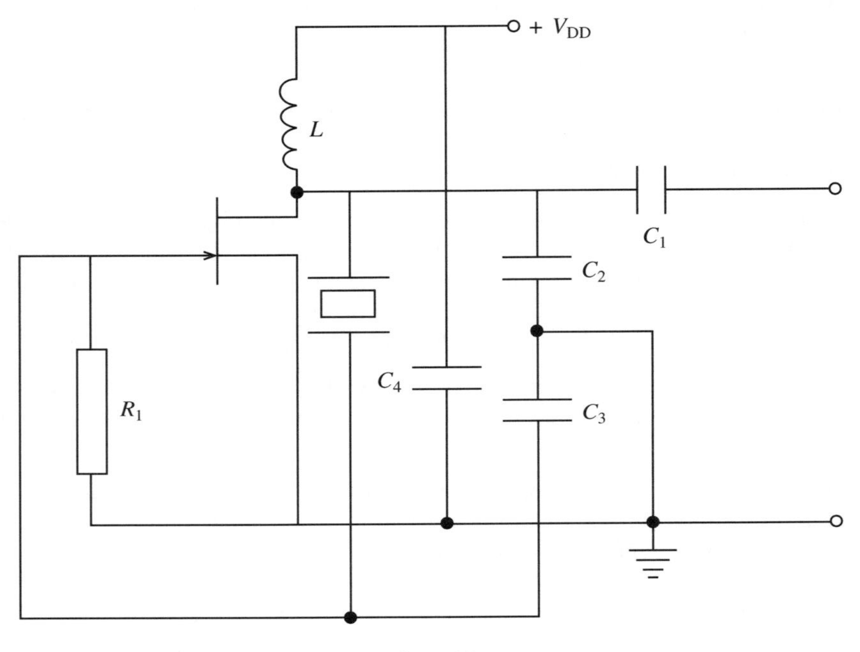

Fig. 1.26

12. An HA7210 integrated circuit is used to produce a frequency of 2 MHz. If a parallel mode crystal with a load has to be used, design a suitable oscillator for this specification.

13. A clock frequency of 7.3728 MHz is required for a data communications network. If the crystal chosen has a load capacitance of 16 pF, design the oscillator if an HA7210 chip is used.

14. Design a Wien bridge oscillator similar to Fig. 1.18 for a frequency of 5 MHz. Suggest any modifications that may be required to give fine-tuning.
Answer: R_1 = 36 kΩ, R_2 = 18 kΩ, R = 10 kΩ, C = 31.8 pF

Data sheets

intersil

HA7210

Data Sheet | February 1999 | File Number 3389.8

10kHz to 10MHz, Low Power Crystal Oscillator

The HA7210 is a very low power crystal-controlled oscillators that can be externally programmed to operate between 10kHz and 10MHz. For normal operation it requires only the addition of a crystal. The part exhibits very high stability over a wide operating voltage and temperature range.

The HA7210 also features a disable mode that switches the output to a high impedance state. This feature is useful for minimizing power dissipation during standby and when multiple oscillator circuits are employed.

Ordering Information

PART NUMBER (BRAND)	TEMP. RANGE (^{o}C)	PACKAGE	PKG. NO.
HA7210IP	-40 to 85	8 Ld PDIP	E8.3
HA7210IB (H7210I)	-40 to 85	8 Ld SOIC	M8.15
HA7210Y	-40 to 85	DIE	

Pinout

HA7210
(PDIP, SOIC)
TOP VIEW

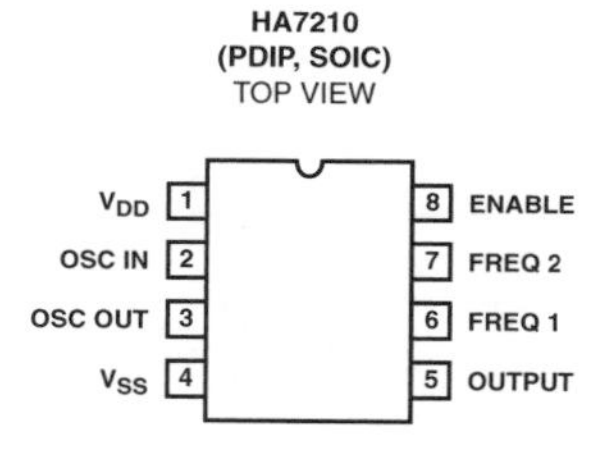

Features

- Single Supply Operation at 32kHz . . . . 2V to 7V
- Operating Frequency Range . . . . 10kHz to 10MHz
- Supply Current at 32kHz 5μA
- Supply Current at 1MHz. 130μA
- Drives 2 CMOS Loads
- Only Requires an External Crystal for Operation

Applications

- Battery Powered Circuits
- Remote Metering
- Embedded Microprocessors
- Palm Top/Notebook PC
- Related Literature
  - AN9334, Improving HA7210 Start-Up Time

Typical Application Circuit

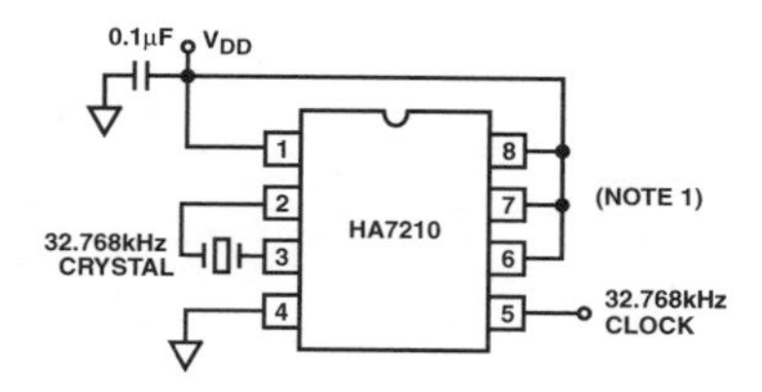

32.768kHz MICROPOWER CLOCK OSCILLATOR

NOTE:

1. Internal pull-up resistors provided on EN, FREQ1, and FREQ2 inputs.

Simplified Block Diagram

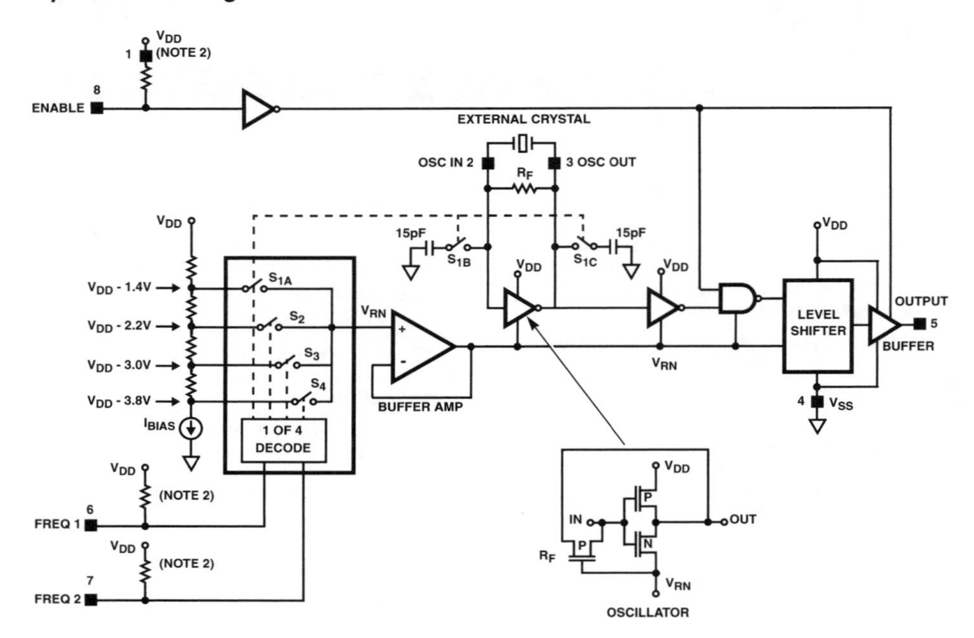

FREQUENCY SELECTION TRUTH TABLE

ENABLE	FREQ 1	FREQ 2	SWITCH	OUTPUT RANGE
1	1	1	S_{1A}, S_{1B}, S_{1C}	10kHz - 100kHz
1	1	0	S_2	100kHz - 1MHz
1	0	1	S_3	1MHz - 5MHz
1	0	0	S_4	5MHz - 10MHz+
0	X	X	X	High Impedance

NOTE:

2. Logic input pull-up resistors are constant current source of 0.4μA.

HA7210

Absolute Maximum Ratings

Supply Voltage 10V
Voltage (Any Pin) . . . . V_{SS} -0.3V to V_{DD} +0.3V
ESD Rating
Human Body Model (Per MIL-STD-883 Method 3015.7) . . . 4000V

Operating Conditions

Temperature Range (Note 3) . . . . -40°C to 85°C

Thermal Information

Thermal Resistance (Typical, Note 4) θ_{JA} (°C/W)
PDIP Package 125
SOIC Package 170
Maximum Junction Temperature (Plastic Package) . . . . 150°C
Maximum Storage Temperature Range . . . . -65°C to 150°C
Maximum Lead Temperature (Soldering 10s) . . . . 300°C
(SOIC - Lead Tips Only)

CAUTION: Stresses above those listed in "Absolute Maximum Ratings" may cause permanent damage to the device. This is a stress only rating and operation of the device at these or any other conditions above those indicated in the operational sections of this specification is not implied.

NOTES:

3. This product is production tested at 25°C only.
4. θ_{JA} is measured with the component mounted on an evaluation PC board in free air.

Electrical Specifications V_{SS} = GND, T_A = 25°C, Unless Otherwise Specified

PARAMETER	TEST CONDITIONS	V_{DD} = 5V			V_{DD} = 3V			UNITS
		MIN	TYP	MAX	MIN	TYP	MAX	
V_{DD} Supply Range	f_{OSC} = 32kHz	2	5	7	-	-	-	V
I_{DD} Supply Current	f_{OSC} = 32kHz, EN = 0 (Standby)	-	5.0	9.0	-	-	-	µA
	f_{OSC} = 32kHz, C_L = 10pF (Note 5), EN = 1, Freq1 = 1, Freq2 = 1	-	5.2	10.2	-	3.6	6.1	µA
	f_{OSC} = 32kHz, C_L = 40pF, EN = 1, Freq1 = 1, Freq2 = 1	-	10	15	-	6.5	9	µA
	f_{OSC} = 1MHz, C_L = 10pF (Note 5), EN = 1, Freq1 = 0, Freq2 = 1	-	130	200	-	90	180	µA
	f_{OSC} = 1MHz, C_L = 40pF, EN = 1, Freq1 = 0, Freq2 = 1	-	270	350	-	180	270	µA
V_{OH} Output High Voltage	I_{OUT} = -1mA	4.0	4.9	-	-	2.8	-	V
V_{OL} Output Low Voltage	I_{OUT} = 1mA	-	0.07	0.4	-	0.1	-	V
I_{OH} Output High Current	$V_{OUT} \geq 4V$	-	-10	-5	-	-	-	mA
I_{OL} Output Low Current	$V_{OUT} \leq 0.4V$	5.0	10.0	-	-	-	-	mA
Three-State Leakage Current	V_{OUT} = 0V, 5V, T_A = 25°C, -40°C	-	0.1	-	-	-	-	nA
	V_{OUT} = 0V, 5V, T_A = 85°C	-	10	-	-	-	-	nA
I_{IN} Enable, Freq1, Freq2 Input Current	V_{IN} = V_{SS} to V_{DD}	-	0.4	1.0	-	-	-	µA
V_{IH} Input High Voltage Enable, Freq1, Freq2		2.0	-	-	-	-	-	V
V_{IL} Input Low Voltage Enable, Freq1, Freq2		-	-	0.8	-	-	-	V
Enable Time	C_L = 18pF, R_L = 1kΩ	-	800	-	-	-	-	ns
Disable Time	C_L = 18pF, R_L = 1kΩ	-	90	-	-	-	-	ns
t_r Output Rise Time	10% - 90%, f_{OSC} = 32kHz, C_L = 40pF	-	12	25	-	12	-	ns
t_f Output Fall Time	10% - 90%, f_{OSC} = 32kHz, C_L = 40pF	-	12	25	-	14	-	ns
Duty Cycle, Packaged Part Only (Note 6)	C_L = 40pF, f_{OSC} = 1MHz	40	54	60	-	-	-	%
Duty Cycle, (See Typical Curves)	C_L = 40pF, f_{OSC} = 32kHz	-	41	-	-	44	-	%
Frequency Stability vs Supply Voltage	f_{OSC} = 32kHz, V_{DD} = 5V, C_L = 10pF	-	1	-	-	-	-	ppm/V
Frequency Stability vs Temperature	f_{OSC} = 32kHz, V_{DD} = 5V, C_L = 10pF	-	0.1	-	-	-	-	ppm/°C
Frequency Stability vs Load	f_{OSC} = 32kHz, V_{DD} = 5V, C_L = 10pF	-	0.01	-	-	-	-	ppm/pF

NOTES:

5. Calculated using the equation $I_{DD} = I_{DD}$ (No Load) + (V_{DD}) (f_{OSC})(C_L)
6. Duty cycle will vary with supply voltage, oscillation frequency, and parasitic capacitance on the crystal pins.

Test Circuit

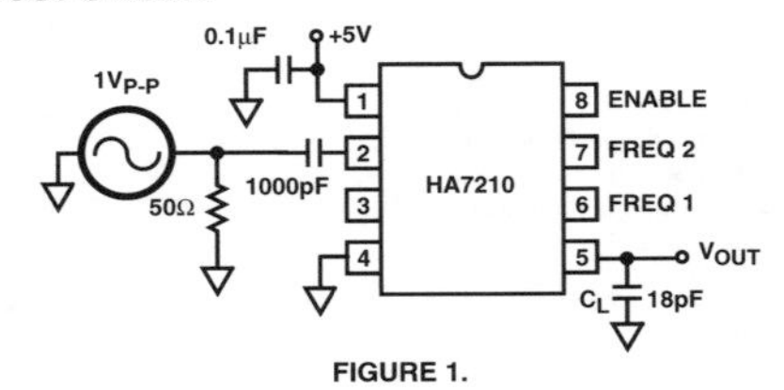

FIGURE 1.

In production the HA7210 is tested with a 32kHz and a 1MHz crystal. However for characterization purposes data was taken using a sinewave generator as the frequency determining element, as shown in Figure 1. The $1V_{P-P}$ input is a smaller amplitude than what a typical crystal would generate so the transitions are slower. In general the Generator data will show a "worst case" number for I_{DD}, duty cycle, and rise/fall time. The Generator test method is useful for testing a variety of frequencies quickly and provides curves which can be used for understanding performance trends. Data for the HA7210 using crystals has also been taken. This data has been overlaid onto the generator data to provide a reference for comparison.

Application Information

Theory Of Operation

The HA7210 is a Pierce Oscillator optimized for low power consumption, requiring no external components except for a bypass capacitor and a Parallel Mode Crystal. The Simplified Block Diagram shows the Crystal attached to pins 2 and 3, the Oscillator input and output. The crystal drive circuitry is detailed showing the simple CMOS inverter stage and the P-channel device being used as biasing resistor R_F. The inverter will operate mostly in its linear region increasing the amplitude of the oscillation until limited by its transconductance and voltage rails, V_{DD} and V_{RN}. The inverter is self biasing using R_F to center the oscillating waveform at the input threshold. Do not interfere with this bias function with external loads or excessive leakage on pin 2. Nominal value for R_F is 17MΩ in the lowest frequency range to 7MΩ in the highest frequency range.

The HA7210 optimizes its power for 4 frequency ranges selected by digital inputs Freq1 and Freq2 as shown in the Block Diagram. Internal pull up resistors (constant current 0.4µA) on Enable, Freq1 and Freq2 allow the user simply to leave one or all digital inputs not connected for a corresponding "1" state. All digital inputs may be left open for 10kHz to 100kHz operation.

A current source develops 4 selectable reference voltages through series resistors. The selected voltage, V_{RN}, is buffered and used as the negative supply rail for the oscillator section of the circuit. The use of a current source in the reference string allows for wide supply variation with minimal effect on performance. The reduced operating voltage of the oscillator section reduces power consumption and limits transconductance and bandwidth to the frequency range selected. For frequencies at the edge of a range, the higher range may provide better performance.

The OSC OUT waveform on pin 3 is squared up through a series of inverters to the output drive stage. The Enable function is implemented with a NAND gate in the inverter string, gating the signal to the level shifter and output stage. Also during Disable the output is set to a high impedance state useful for minimizing power during standby and when multiple oscillators are OR'ed to a single node.

Design Considerations

The low power CMOS transistors are designed to consume power mostly during transitions. Keeping these transitions short requires a good decoupling capacitor as close as possible to the supply pins 1 and 4. A ceramic 0.1µF is recommended. Additional supply decoupling on the circuit board with 1µF to 10µF will further reduce overshoot, ringing and power consumption. The HA7210, when compared to a crystal and inverter alone, will speed clock transition times, reducing power consumption of all CMOS circuitry run from that clock.

Power consumption may be further reduced by minimizing the capacitance on moving nodes. The majority of the power will be used in the output stage driving the load. Minimizing the load and parasitic capacitance on the output, pin 5, will play the major role in minimizing supply current. A secondary source of wasted supply current is parasitic or crystal load capacitance on pins 2 and 3. The HA7210 is designed to work with most available crystals in its frequency range with no external components required. Two 15pF capacitors are internally switched onto crystal pins 2 and 3 on the HA7210 to compensate the oscillator in the 10kHz to 100kHz frequency range.

The supply current of the HA7210 may be approximately calculated from the equation:

$I_{DD} = I_{DD}(\text{Disabled}) + V_{DD} \times f_{OSC} \times C_L$ where:

- I_{DD} = Total supply current
- V_{DD} = Total voltage from V_{DD} (pin 1) to V_{SS} (pin 4)
- f_{OSC} = Frequency of Oscillation
- C_L = Output (pin 5) load capacitance

EXAMPLE #1:

V_{DD} = 5V, f_{OSC} = 100kHz, C_L = 30pF
I_{DD}(Disabled) = 4.5µA (Figure 10)
I_{DD} = 4.5µA + (5V)(100kHz)(30pF) = 19.5µA
Measured I_{DD} = 20.3µA

EXAMPLE #2:

V_{DD} = 5V, f_{OSC} = 5MHz, C_L = 30pF
I_{DD} (Disabled) = 75µA (Figure 9)
I_{DD} = 75µA + (5V)(5MHz)(30pF) = 825µA
Measured I_{DD} = 809µA

Crystal Selection

For general purpose applications, a Parallel Mode Crystal is a good choice for use with the HA7210. However for applications where a precision frequency is required, the designer needs to consider other factors.

Crystals are available in two types or modes of oscillation, Series and Parallel. Series Mode crystals are manufactured to operate at a specified frequency with zero load capacitance and appear as a near resistive impedance when oscillating. Parallel Mode crystals are manufactured to operate with a specific capacitive load in series, causing the crystal to operate at a more inductive impedance to cancel the load capacitor. Loading a crystal with a different capacitance will "pull" the frequency off its value.

The HA7210 has 4 operating frequency ranges. The higher three ranges do not add any loading capacitance to the oscillator circuit. The lowest range, 10kHz to 100kHz, automatically switches in two 15pF capacitors onto OSC IN and OSC OUT to eliminate potential start-up problems. These capacitors create an effective crystal loading capacitor equal to the series combination of these two capacitors. For the HA7210 in the lowest range, the effective loading capacitance is 7.5pF. Therefore the choice for a crystal, in this range, should be a Parallel Mode crystal that requires a 7.5pF load.

In the higher 3 frequency ranges, the capacitance on OSC IN and OSC OUT will be determined by package and layout parasitics, typically 4 to 5pF. Ideally the choice for crystal should be a Parallel Mode set for 2.5pF load. A crystal manufactured for a different load will be "pulled" from its nominal frequency (see Crystal Pullability).

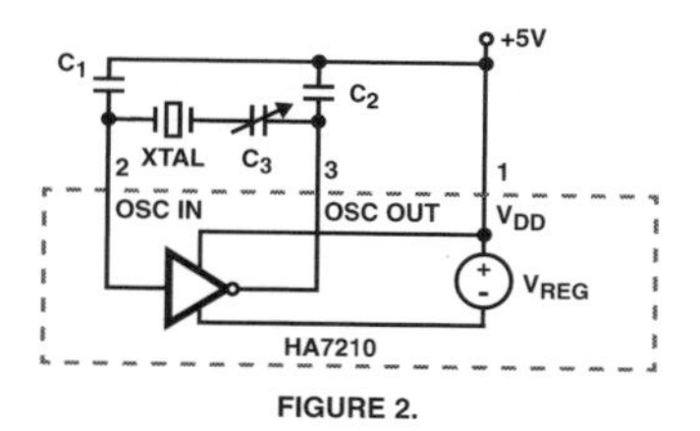

FIGURE 2.

Frequency Fine Tuning

Two Methods will be discussed for fine adjustment of the crystal frequency. The first and preferred method (Figure 2), provides better frequency accuracy and oscillator stability than method two (Figure 3). Method one also eliminates start-up problems sometimes encountered with 32kHz tuning fork crystals.

For best oscillator performance, two conditions must be met: the capacitive load must be matched to both the inverter and crystal to provide ideal conditions for oscillation, and the frequency of the oscillator must be adjustable to the desired frequency. In Method two these two goals can be at odds with each other; either the oscillator is trimmed to frequency by de-tuning the load circuit, or stability is increased at the expense of absolute frequency accuracy.

Method one allows these two conditions to be met independently. The two fixed capacitors, C_1 and C_2, provide the optimum load to the oscillator and crystal. C_3 adjusts the frequency at which the circuit oscillates without appreciably changing the load (and thus the stability) of the system. Once a value for C_3 has been determined for the particular type of crystal being used, it could be replaced with a fixed capacitor. For the most precise control over oscillator frequency, C_3 should remain adjustable.

This three capacitor tuning method will be more accurate and stable than method two and is recommended for 32kHz tuning fork crystals; without it they may leap into an overtone mode when power is initially applied.

Method two has been used for many years and may be preferred in applications where cost or space is critical. Note that in both cases the crystal loading capacitors are connected between the oscillator and V_{DD}; do not use V_{SS} as an AC ground. The Simplified Block Diagram shows that the oscillating inverter does not directly connect to V_{SS} but is referenced to V_{DD} and V_{RN}. Therefore V_{DD} is the best AC ground available.

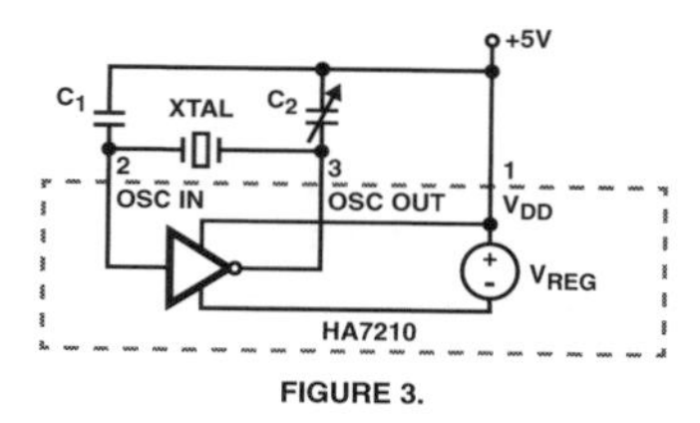

FIGURE 3.

Typical values of the capacitors in Figure 2 are shown below. Some trial and error may be required before the best combination is determined. The values listed are total capacitance including parasitic or other sources. Remember that in the 10kHz to 100kHz frequency range setting the HA7210 switches in two internal 15pF capacitors.

CRYSTAL FREQUENCY	LOAD CAPS C_1, C_2	TRIMMER CAP C_3
32kHz	33pF	5pF to 50pF
1MHz	33pF	5pF to 50pF
2MHz	25pF	5pF to 50pF
4MHz	22pF	5pF to 100pF

Crystal Pullability

Figure 4 shows the basic equivalent circuit for a crystal and its loading circuit.

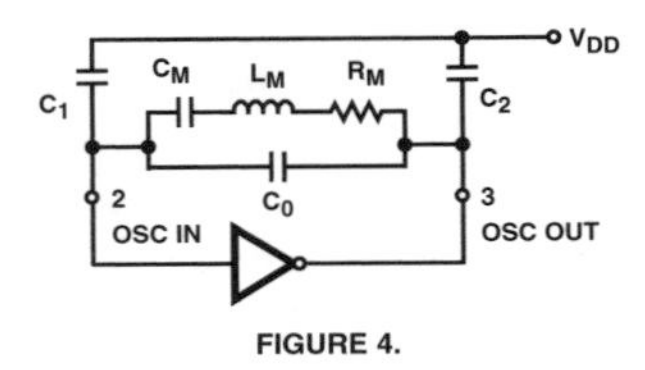

FIGURE 4.

Where:
C_M = Motional Capacitance
L_M = Motional Inductance
R_M = Motional Resistance
C_O = Shunt Capacitance

$$C_{CL} = \frac{1}{\left(\frac{1}{C_1} + \frac{1}{C_2}\right)} = \text{Equivalent Crystal Load}$$

If loading capacitance is connected to a Series Mode Crystal, the new Parallel Mode frequency of resonance may be calculated with the following equation:

$$f_P = f_S\left[1 + \frac{C_M}{2(C_O + C_{CL})}\right]$$

Where:
f_P = Parallel Mode Resonant Frequency
f_S = Series Mode Resonant Frequency

In a similar way, the Series Mode resonant frequency may be calculated from a Parallel Mode crystal and then you may calculate how much the frequency will "pull" with a new load.

Layout Considerations

Due to the extremely low current (and therefore high impedance) the circuit board layout of the HA7210 must be given special attention. Stray capacitance should be minimized. Keep the oscillator traces on a single layer of the PCB. Avoid putting a ground plane above or below this layer. The traces between the crystal, the capacitors, and the OSC pins should be as short as possible. Completely surround the oscillator components with a thick trace of V_{DD} to minimize coupling with any digital signals. The final assembly must be free from contaminants such as solder flux, moisture, or any other potential source of leakage. A good solder mask will help keep the traces free of moisture and contamination over time.

Further Reading

Al Little "HA7210 Low Power Oscillator: Micropower Clock Oscillator and Op Amps Provide System Shutdown for Battery Circuits". Harris Semiconductor Application Note AN9317.

Robert Rood "Improving Start-Up Time at 32kHz for the HA7210 Low Power Crystal Oscillator". Harris Semiconductor Application Note AN9334.

S. S. Eaton "Timekeeping Advances Through COS/MOS Technology". Harris Semiconductor Application Note ICAN-6086.

E. A. Vittoz, et. al. "High-Performance Crystal Oscillator Circuits: Theory and Application". IEEE Journal of Solid-State Circuits, Vol. 23, No. 3, June 1988, pp774-783.

M. A. Unkrich, et. al. "Conditions for Start-Up in Crystal Oscillators". IEEE Journal of Solid-State Circuits, Vol. 17, No. 1, Feb. 1982, pp87-90.

Marvin E. Frerking "Crystal Oscillator Design and Temperature Compensation". New York: Van Nostrand-Reinhold, 1978. Pierce Oscillators Discussed pp56-75.

Typical Performance Curves

C_L = 40pF, f_{OSC} = 5MHz, V_{DD} = 5V, V_{SS} = GND
1.0V/DIV. 20.0ns/DIV.

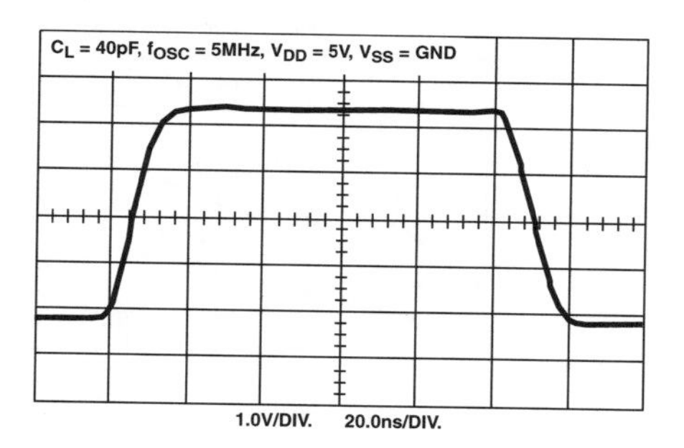

FIGURE 5. OUTPUT WAVEFORM (C_L = 40pF)

C_L = 18pF, f_{OSC} = 5MHz, V_{DD} = 5V, V_{SS} = GND
1.0V/DIV. 20.0ns/DIV.

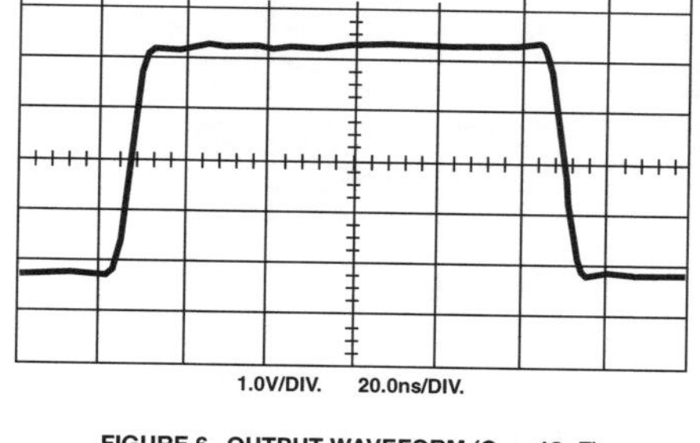

FIGURE 6. OUTPUT WAVEFORM (C_L = 18pF)

f_{IN} = 5MHz, EN = 1, F1 = 0, F2 = 0, C_L = 30pF, V_{DD} = 5V
GENERATOR (1V_{P-P}) (NOTE)
X_{TAL} AT 25°C
SUPPLY CURRENT (μA)
1050
1000
950
900
850
800
750
-100
-50
0
50
100
150
TEMPERATURE (°C)

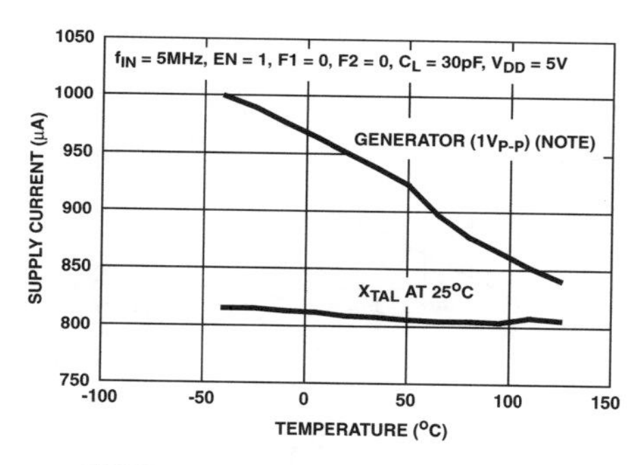

FIGURE 7. SUPPLY CURRENT vs TEMPERATURE

EN = 1, F1 = 1, F2 = 1, f_{IN} = 100kHz, C_L = 30pF, V_{DD} = 5V
GENERATOR (1V_{P-P}) (NOTE)
X_{TAL} AT 25°C
SUPPLY CURRENT (μA)
26
25
24
23
22
21
20
19
18
-100
-50
0
50
100
150
TEMPERATURE (°C)

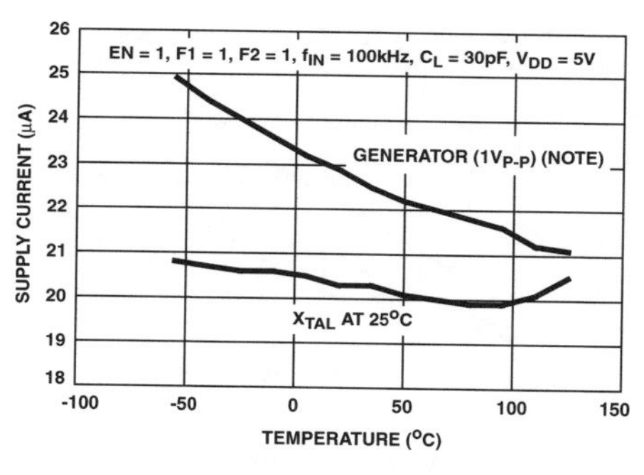

FIGURE 8. SUPPLY CURRENT vs TEMPERATURE

f_{IN} = 5MHz, EN = 0, F1 = 0, F2 = 0, V_{DD} = 5V
GENERATOR (1V_{P-P}) (NOTE)
X_{TAL} AT 25°C
SUPPLY CURRENT (μA)
350
300
250
200
150
100
50
0
-100
-50
0
50
100
150
TEMPERATURE (°C)

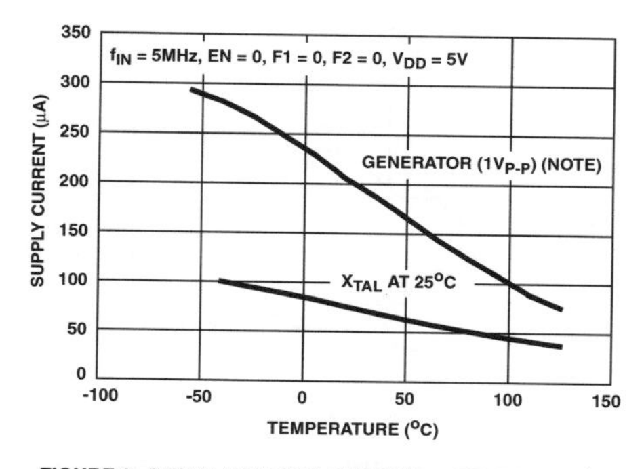

FIGURE 9. DISABLE SUPPLY CURRENT vs TEMPERATURE

EN = 0, F1 = 1, F2 = 1, f_{IN} = 100kHz, V_{DD} = 5V
GENERATOR (1V_{P-P}) (NOTE)
X_{TAL} AT 25°C
SUPPLY CURRENT (μA)
7.5
7
6.5
6
5.5
5
4.5
4
-100
-50
0
50
100
150
TEMPERATURE (°C)

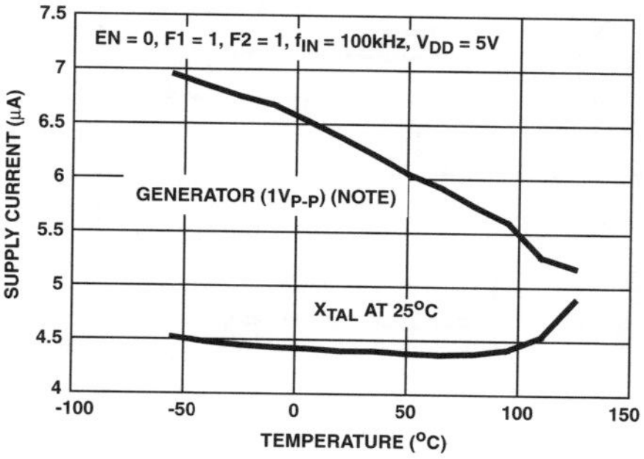

FIGURE 10. DISABLE SUPPLY CURRENT vs TEMPERATURE

NOTE: Refer to Test Circuit (Figure 1).

Typical Performance Curves (Continued)

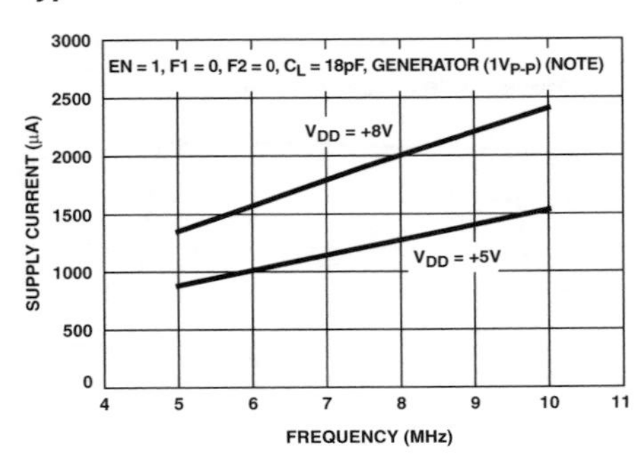

FIGURE 11. SUPPLY CURRENT vs FREQUENCY

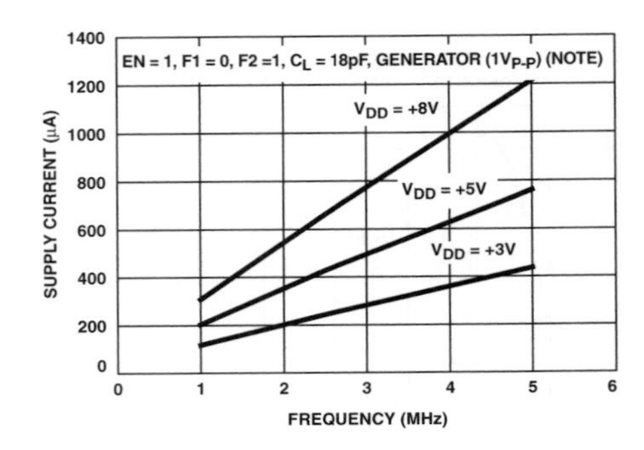

FIGURE 12. SUPPLY CURRENT vs FREQUENCY

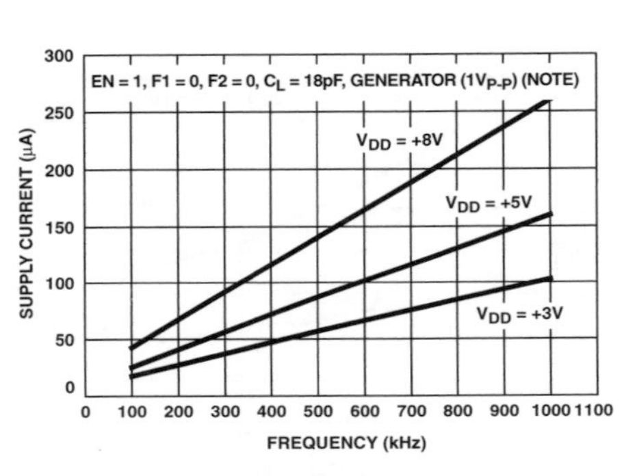

FIGURE 13. SUPPLY CURRENT vs FREQUENCY

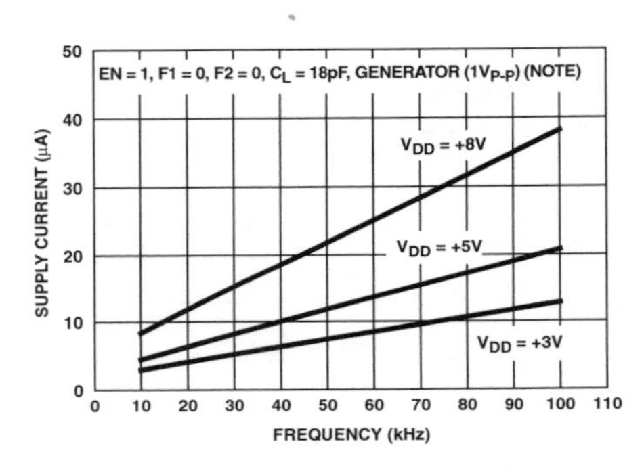

FIGURE 14. SUPPLY CURRENT vs FREQUENCY

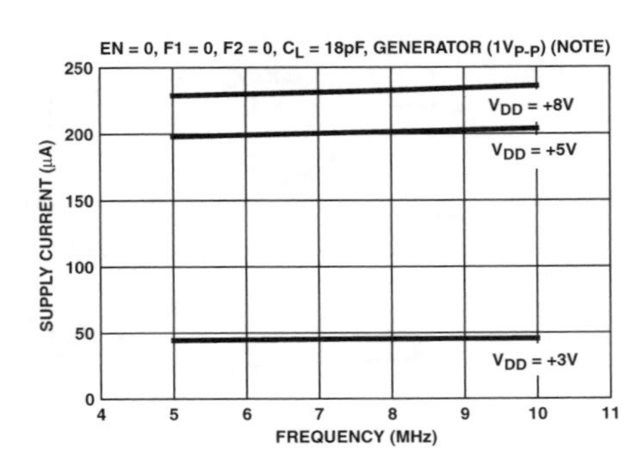

FIGURE 15. DISABLED SUPPLY CURRENT vs FREQUENCY

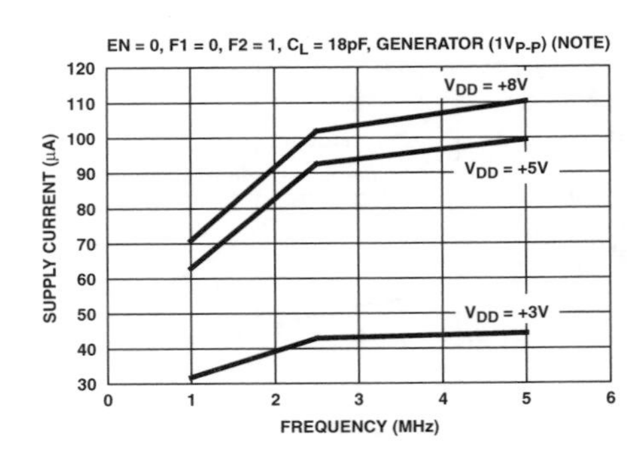

FIGURE 16. DISABLE SUPPLY CURRENT vs FREQUENCY

NOTE: Refer to Test Circuit (Figure 1).

Typical Performance Curves (Continued)

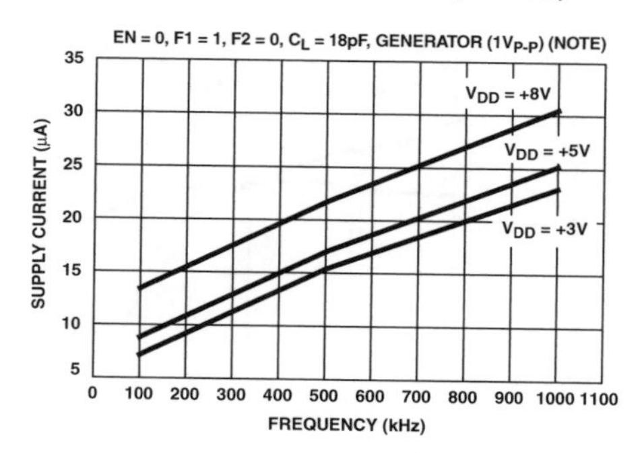

FIGURE 17. DISABLE SUPPLY CURRENT vs FREQUENCY

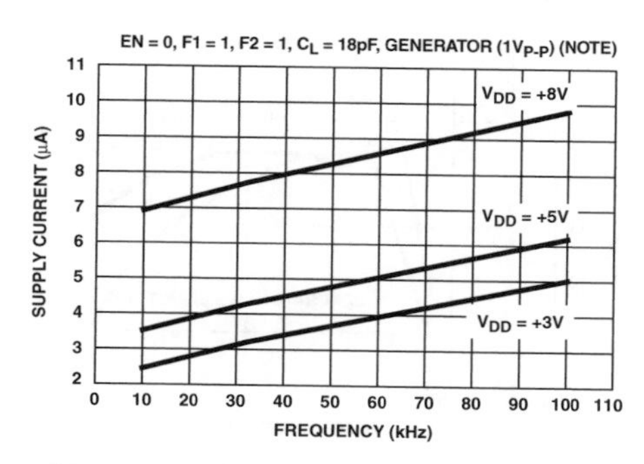

FIGURE 18. DISABLE SUPPLY CURRENT vs FREQUENCY

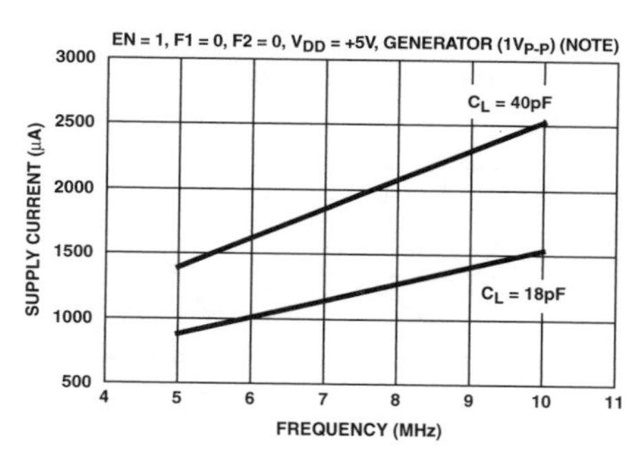

FIGURE 19. SUPPLY CURRENT vs FREQUENCY

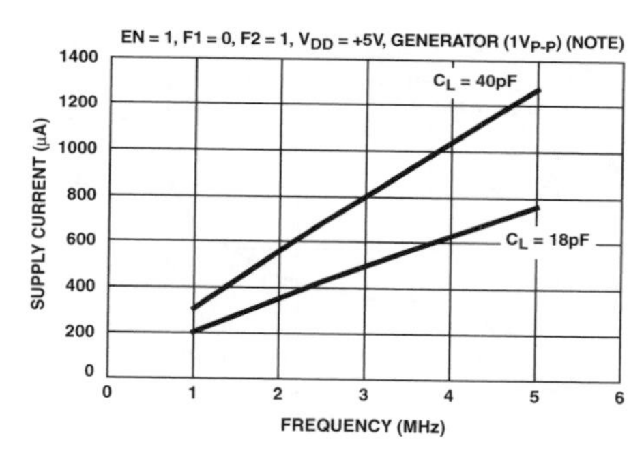

FIGURE 20. SUPPLY CURRENT vs FREQUENCY

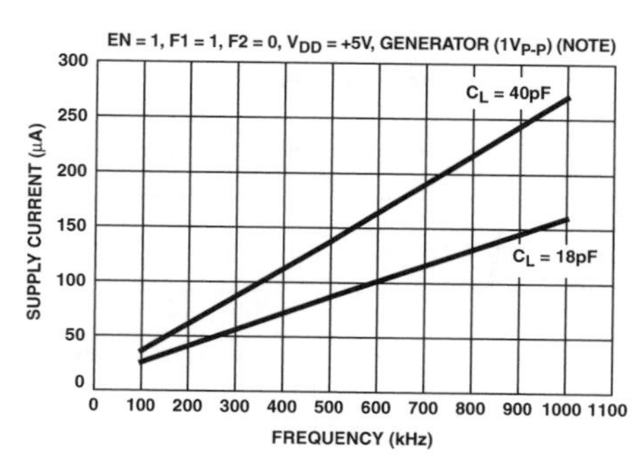

FIGURE 21. SUPPLY CURRENT vs FREQUENCY

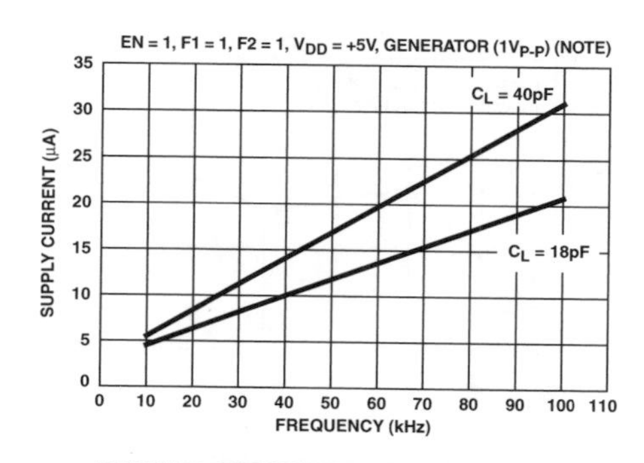

FIGURE 22. SUPPLY CURRENT vs FREQUENCY

NOTE: Refer to Test Circuit (Figure 1).

Typical Performance Curves (Continued)

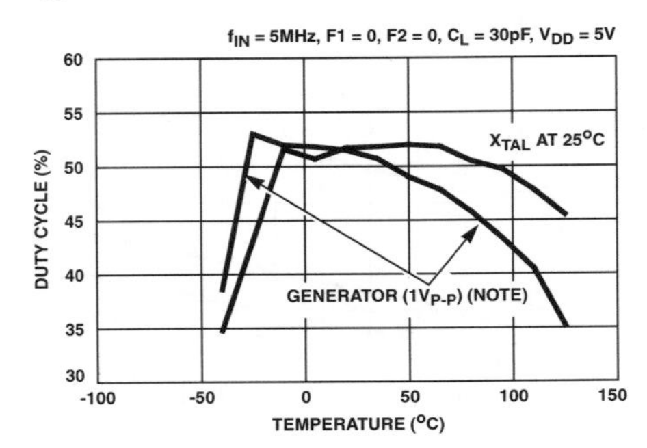

FIGURE 23. DUTY CYCLE vs TEMPERATURE

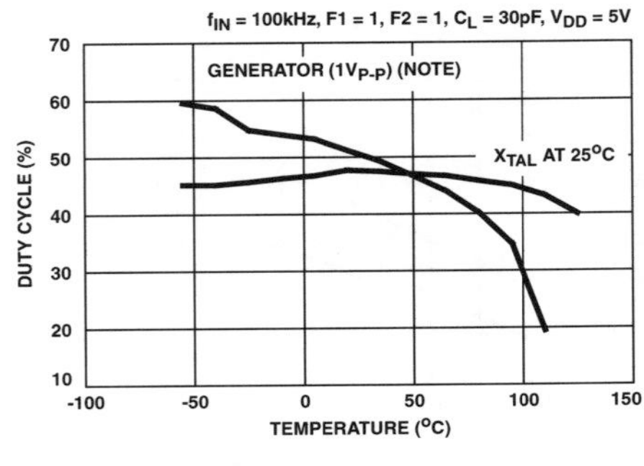

FIGURE 24. DUTY CYCLE vs TEMPERATURE

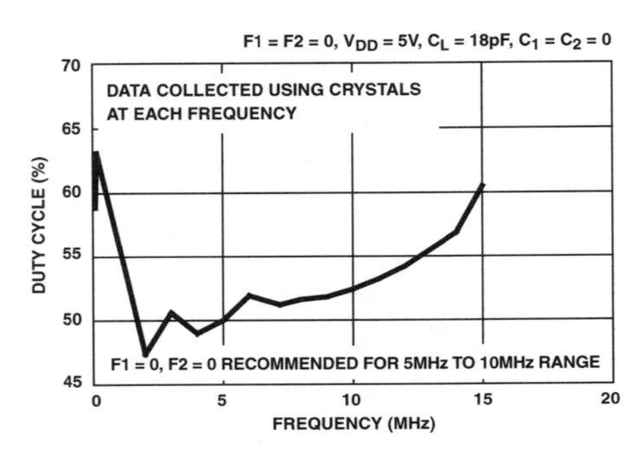

FIGURE 25. DUTY CYCLE vs FREQUENCY

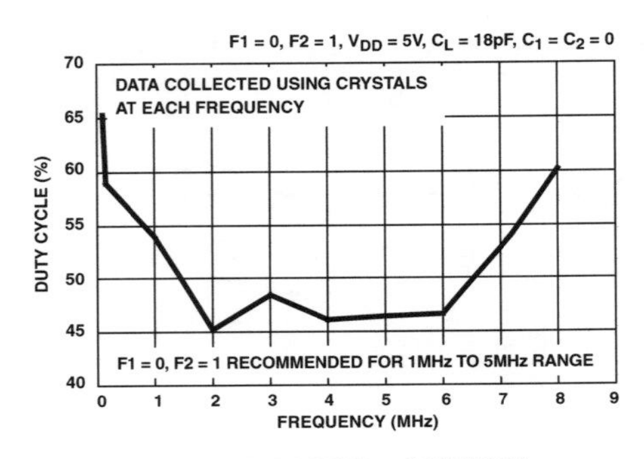

FIGURE 26. DUTY CYCLE vs FREQUENCY

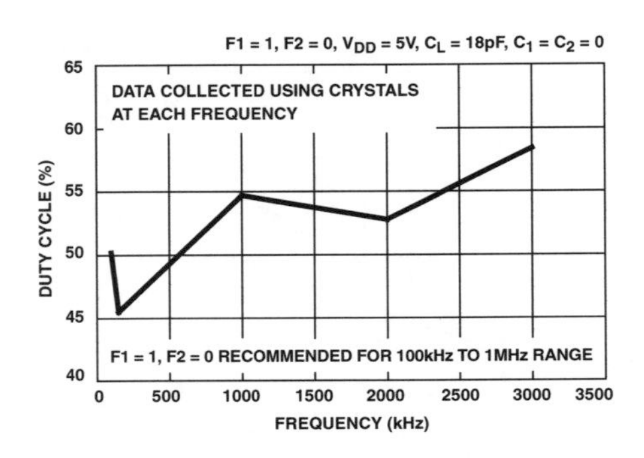

FIGURE 27. DUTY CYCLE vs FREQUENCY

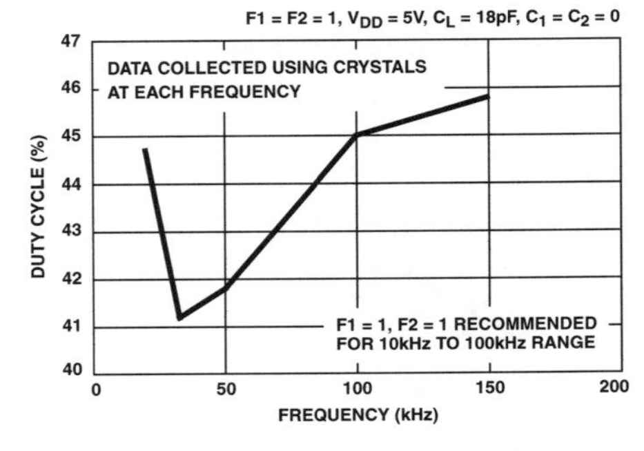

FIGURE 28. DUTY CYCLE vs FREQUENCY

NOTE: Refer to Test Circuit (Figure 1).

Typical Performance Curves (Continued)

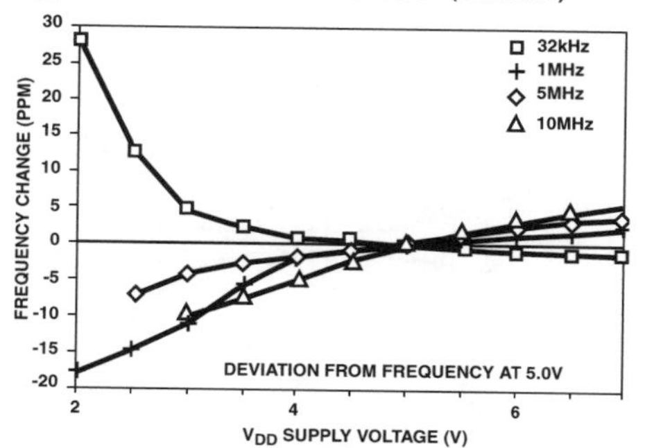

FIGURE 29. FREQUENCY CHANGE vs V_{DD}

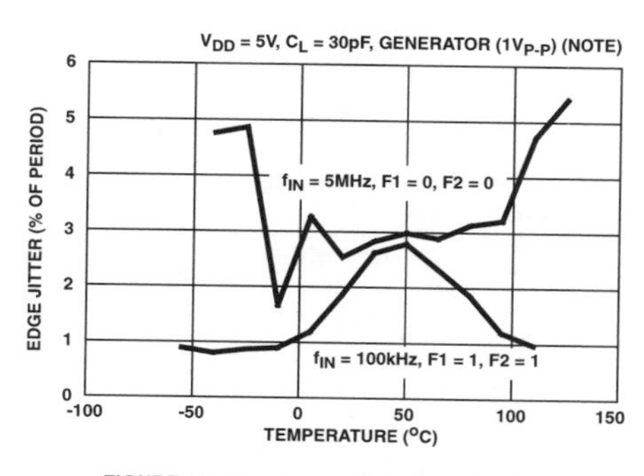

FIGURE 30. EDGE JITTER vs TEMPERATURE

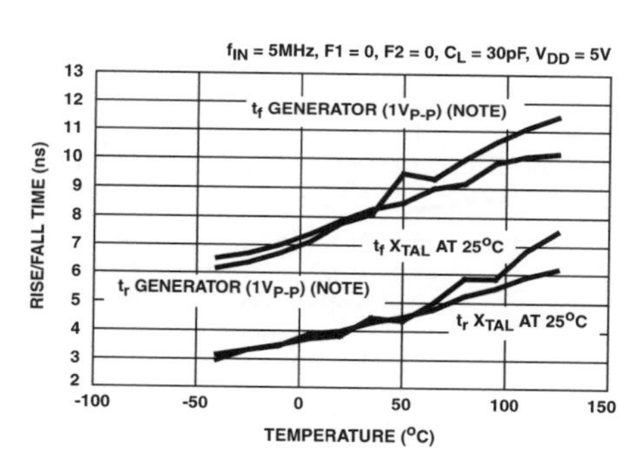

FIGURE 31. RISE/FALL TIME vs TEMPERATURE

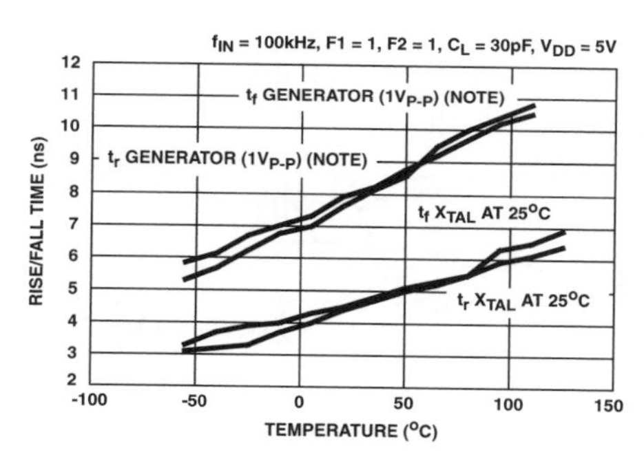

FIGURE 32. RISE/FALL TIME vs TEMPERATURE

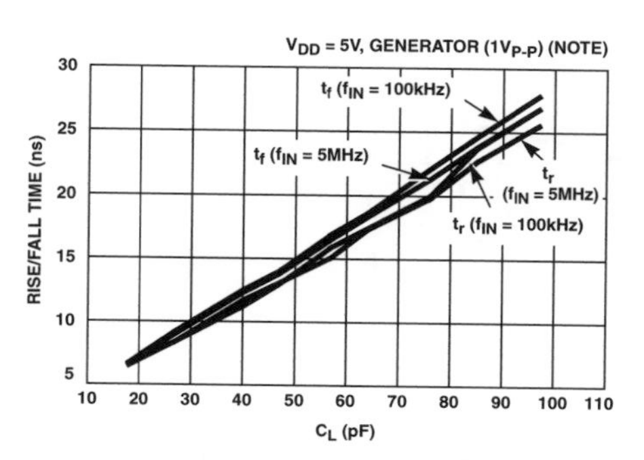

FIGURE 33. RISE/FALL TIME vs C_L

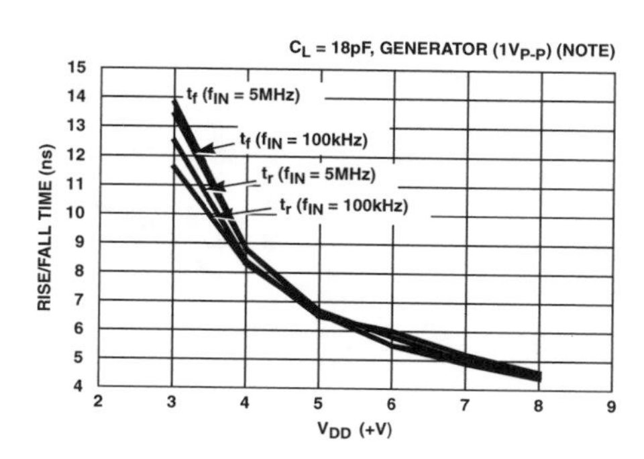

FIGURE 34. RISE/FALL TIME vs V_{DD}

NOTE: Refer to Test Circuit (Figure 1).

Typical Performance Curves (Continued)

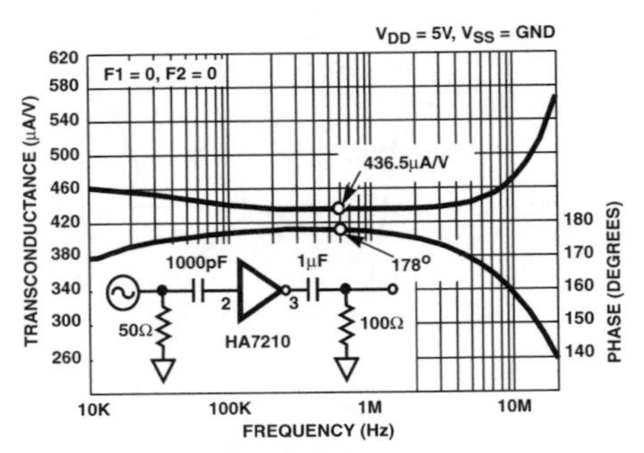

FIGURE 35. TRANSCONDUCTANCE vs FREQUENCY

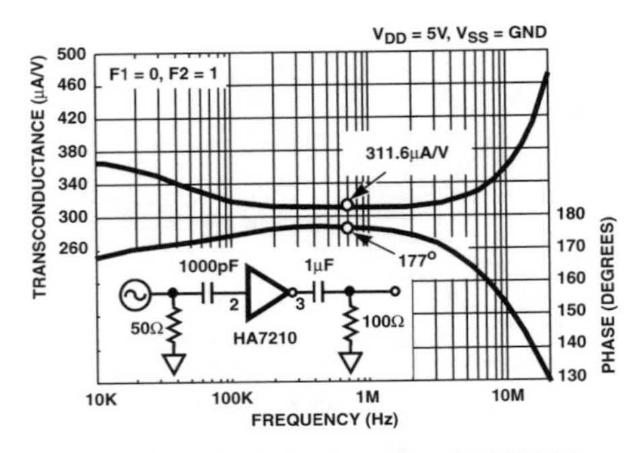

FIGURE 36. TRANSCONDUCTANCE vs FREQUENCY

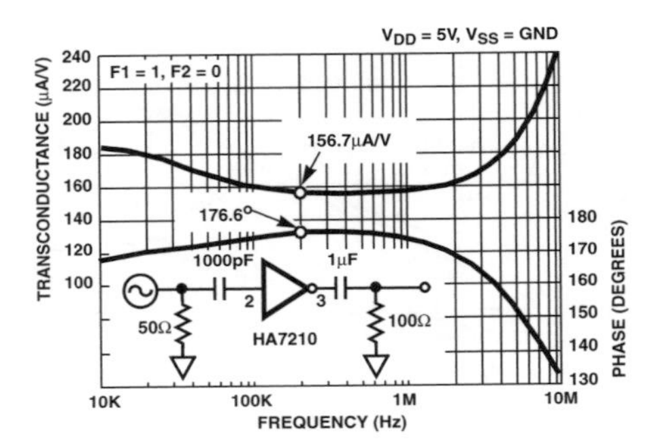

FIGURE 37. TRANSCONDUCTANCE vs FREQUENCY

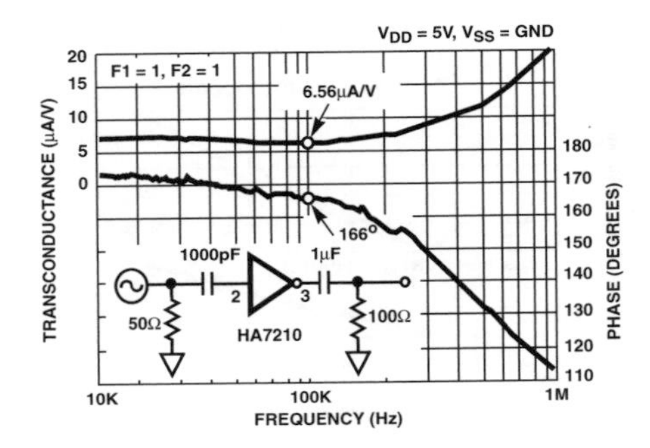

FIGURE 38. TRANSCONDUCTANCE vs FREQUENCY

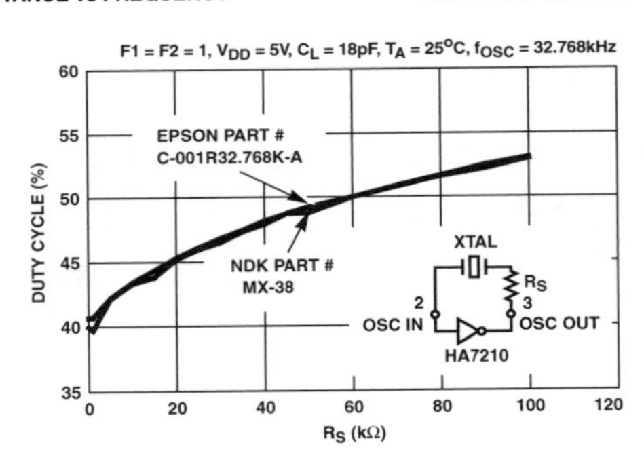

NOTE: Figure 39 (Duty Cycle vs R_S at 32kHz) should only be used for 32kHz crystals. R_S may be used at other frequencies to adjust Duty Cycle but experimentation will be required to find an appropriate value. The R_S value will be proportional to the effective series resistance of the crystal being used.

FIGURE 39. DUTY CYCLE vs R_S at 32kHz

NOTE: Refer to Test Circuit (Figure 1).

HA7210

Die Characteristics

DIE DIMENSIONS:

68 mils x 64 mils x 14 mils

METALLIZATION:

Type: SiAl
Thickness: 10kÅ ±1kÅ

SUBSTRATE POTENTIAL:

V_{SS}

PASSIVATION:

Type: Nitride (Si_3N_4) Over Silox (SiO_2, 3% Phos)
Silox Thickness: 7kÅ ±1kÅ
Nitride Thickness: 8kÅ ±1kÅ

Metallization Mask Layout

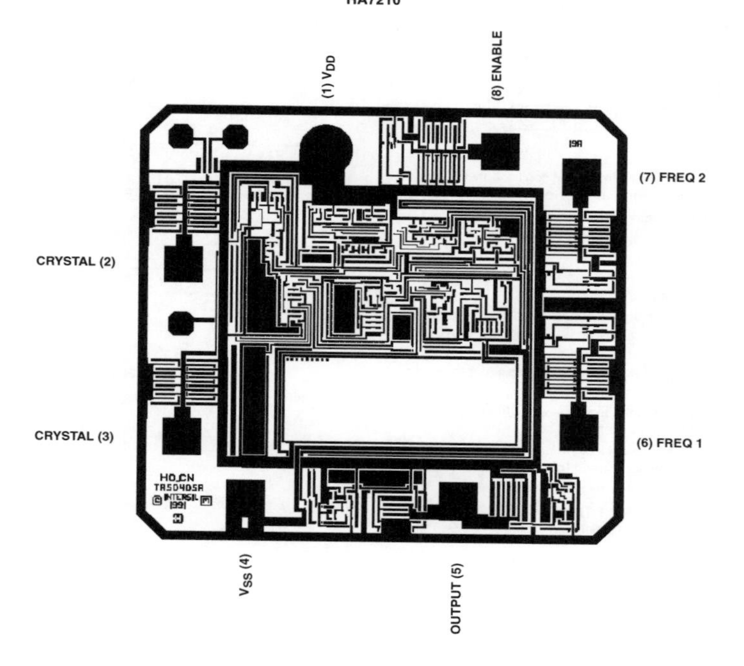

Dual-In-Line Plastic Packages (PDIP)

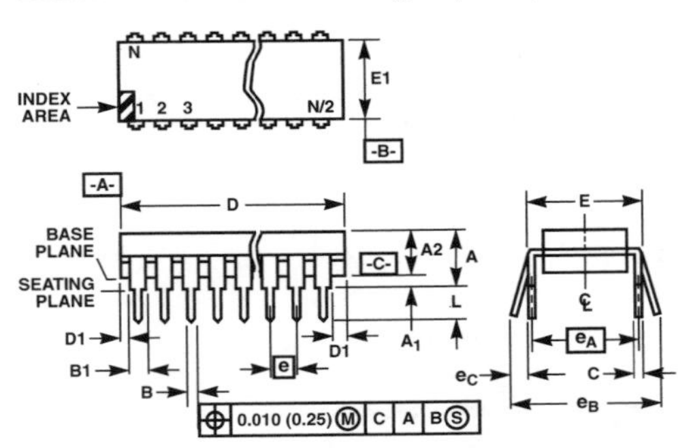

NOTES:

1. Controlling Dimensions: INCH. In case of conflict between English and Metric dimensions, the inch dimensions control.
2. Dimensioning and tolerancing per ANSI Y14.5M-1982.
3. Symbols are defined in the "MO Series Symbol List" in Section 2.2 of Publication No. 95.
4. Dimensions A, A1 and L are measured with the package seated in JEDEC seating plane gauge GS-3.
5. D, D1, and E1 dimensions do not include mold flash or protrusions. Mold flash or protrusions shall not exceed 0.010 inch (0.25mm).
6. E and e_A are measured with the leads constrained to be perpendicular to datum -C-.
7. e_B and e_C are measured at the lead tips with the leads unconstrained. e_C must be zero or greater.
8. B1 maximum dimensions do not include dambar protrusions. Dambar protrusions shall not exceed 0.010 inch (0.25mm).
9. N is the maximum number of terminal positions.
10. Corner leads (1, N, N/2 and N/2 + 1) for E8.3, E16.3, E18.3, E28.3, E42.6 will have a B1 dimension of 0.030 - 0.045 inch (0.76 - 1.14mm).

E8.3 (JEDEC MS-001-BA ISSUE D)
8 LEAD DUAL-IN-LINE PLASTIC PACKAGE

SYMBOL	INCHES MIN	INCHES MAX	MILLIMETERS MIN	MILLIMETERS MAX	NOTES
A	-	0.210	-	5.33	4
A1	0.015	-	0.39	-	4
A2	0.115	0.195	2.93	4.95	-
B	0.014	0.022	0.356	0.558	-
B1	0.045	0.070	1.15	1.77	8, 10
C	0.008	0.014	0.204	0.355	-
D	0.355	0.400	9.01	10.16	5
D1	0.005	-	0.13	-	5
E	0.300	0.325	7.62	8.25	6
E1	0.240	0.280	6.10	7.11	5
e	0.100 BSC		2.54 BSC		-
e_A	0.300 BSC		7.62 BSC		6
e_B	-	0.430	-	10.92	7
L	0.115	0.150	2.93	3.81	4
N	8		8		9

Rev. 0 12/93

Small Outline Plastic Packages (SOIC)

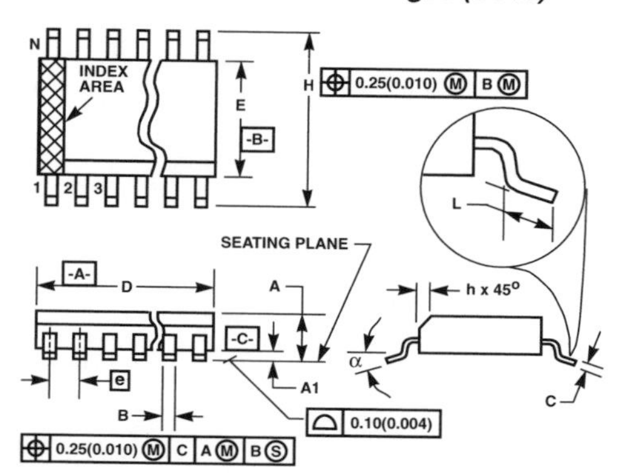

NOTES:

1. Symbols are defined in the "MO Series Symbol List" in Section 2.2 of Publication Number 95.
2. Dimensioning and tolerancing per ANSI Y14.5M-1982.
3. Dimension "D" does not include mold flash, protrusions or gate burrs. Mold flash, protrusion and gate burrs shall not exceed 0.15mm (0.006 inch) per side.
4. Dimension "E" does not include interlead flash or protrusions. Interlead flash and protrusions shall not exceed 0.25mm (0.010 inch) per side.
5. The chamfer on the body is optional. If it is not present, a visual index feature must be located within the crosshatched area.
6. "L" is the length of terminal for soldering to a substrate.
7. "N" is the number of terminal positions.
8. Terminal numbers are shown for reference only.
9. The lead width "B", as measured 0.36mm (0.014 inch) or greater above the seating plane, shall not exceed a maximum value of 0.61mm (0.024 inch).
10. Controlling dimension: MILLIMETER. Converted inch dimensions are not necessarily exact.

M8.15 **(JEDEC MS-012-AA ISSUE C)**
8 LEAD NARROW BODY SMALL OUTLINE PLASTIC PACKAGE

SYMBOL	INCHES		MILLIMETERS		NOTES
	MIN	MAX	MIN	MAX	
A	0.0532	0.0688	1.35	1.75	-
A1	0.0040	0.0098	0.10	0.25	-
B	0.013	0.020	0.33	0.51	9
C	0.0075	0.0098	0.19	0.25	-
D	0.1890	0.1968	4.80	5.00	3
E	0.1497	0.1574	3.80	4.00	4
e	0.050 BSC		1.27 BSC		-
H	0.2284	0.2440	5.80	6.20	-
h	0.0099	0.0196	0.25	0.50	5
L	0.016	0.050	0.40	1.27	6
N	8		8		7
α	0°	8°	0°	8°	-

Rev. 0 12/93

All Intersil semiconductor products are manufactured, assembled and tested under **ISO9000** quality systems certification.

For information regarding Intersil Corporation and its products, see web site **http://www.intersil.com**

Sales Office Headquarters

NORTH AMERICA
Intersil Corporation
P. O. Box 883, Mail Stop 53-204
Melbourne, FL 32902
TEL: (407) 724-7000
FAX: (407) 724-7240

EUROPE
Intersil SA
Mercure Center
100, Rue de la Fusee
1130 Brussels, Belgium
TEL: (32) 2.724.2111
FAX: (32) 2.724.22.05

ASIA
Intersil (Taiwan) Ltd.
7F-6, No. 101 Fu Hsing North Road
Taipei, Taiwan
Republic of China
TEL: (886) 2 2716 9310
FAX: (886) 2 2715 3029

2

Modulation systems

2.1 Introduction

Figure 2.1 shows a simplified block diagram of a coloured television receiver. We will refer to it throughout this book. It shows that any receiver must be capable of extracting information from the incoming channels to which it is tuned. The shaded blocks show examples of where demodulation or detection occur for the video and audio signals. Generally the sound uses frequency modulation (FM), while the video signal uses amplitude modulation (AM). The age of digital television and modern data communications will use other techniques. All these methods will be discussed in this chapter.

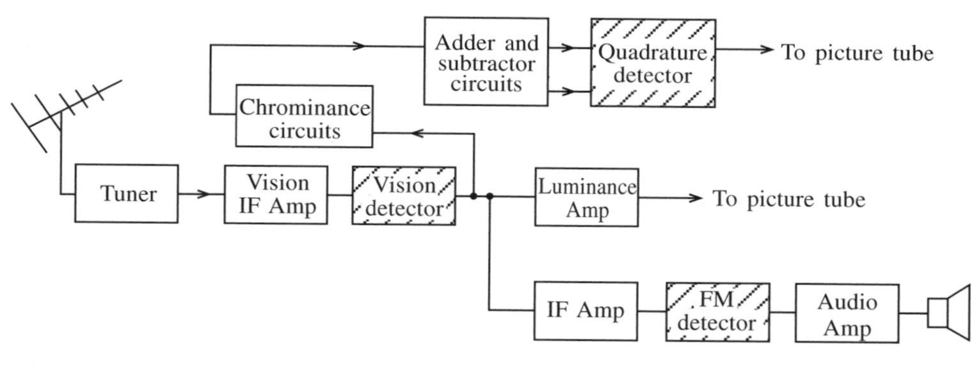

Fig. 2.1

The requirements for modulation are threefold. First, all channels must be separated from one another to avoid interference in the form of intermodulation distortion and crosstalk. Crosstalk occurs when one channel spills over into an adjacent channel, causing interference. Intermodulation distortion occurs when two signals at frequencies f_1 and f_2 are amplified by a non-linear device. Second-order products ($2f_1, f_1 + f_2$ and $f_1 - f_2$) are produced. This might only be troublesome in a broadband system where these products fall within the band. However, third-order components ($2f_1 + f_2$ and $2f_2 - f_1$) usually fall within a system bandwidth, i.e. a particular range of frequencies over which the system operates with good linearity, flat response and minimum distortion, and again cause intermodulation distortion.

In order to achieve good channel separation and avoid interference data, audio and video are generally superimposed on a carrier signal. Each station may have a different carrier or use sophisticated techniques like polarization or frequency sharing, but the point is that frequency translation takes place, with the information signals being shifted to a new frequency.

Second, the physical size of half-wavelength antenna systems would be prohibitive if higher frequencies were not used. In order to understand this, it is convenient to consider the properties of a quarter-wavelength ($\lambda/4$) transmission line. Figure 2.2(a) shows an open-ended $\lambda/4$ transmission line and its voltage and current distributions. If this is opened out as in Fig. 2.2(b), then a $\lambda/2$ radiator is produced with the voltage and current distributions as shown. It can be seen that the current is at its maximum at the centre while the voltage is at its minimum. This is equivalent to a low-resistance series resonant circuit which can be tuned to the required transmitted or received channels. However, the point here is that the antenna has an electrical length of half the operating wavelength and is referred to as a $\lambda/2$ dipole. (In practice, it is actually 5% shorter than this theoretical value.)

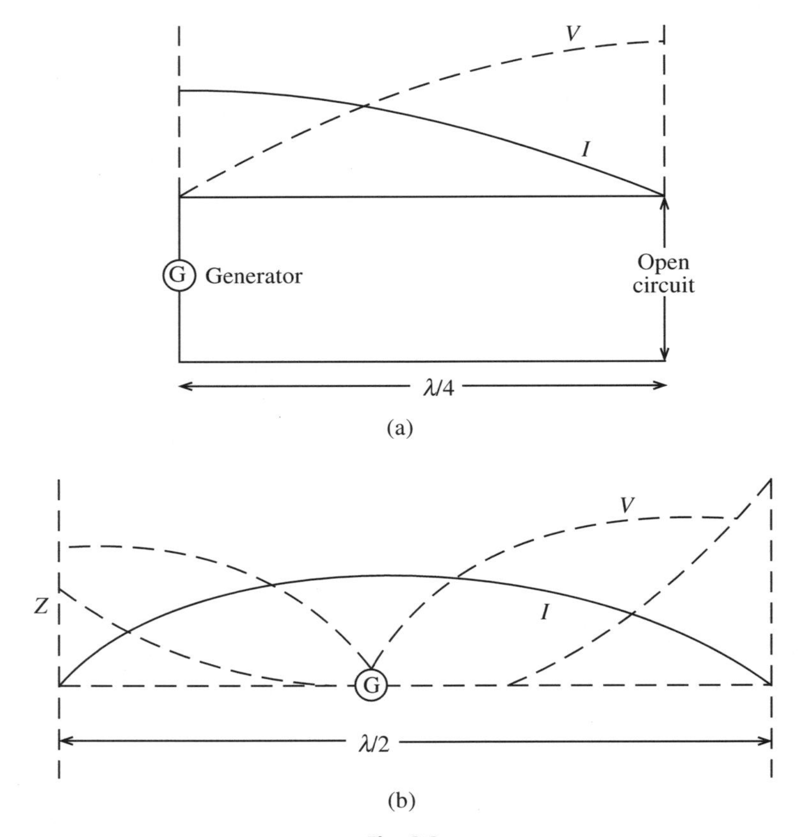

Fig. 2.2

Consider the case of the speech band being transmitted. This is generally from 300 Hz to 3.4 kHz. The two-dimensional wave equation is used to determine the wavelength:

$$\lambda_1 = \frac{v}{f_1} = \frac{3 \times 10^8}{3 \times 10} = 10^6 \text{ m}$$

$$\lambda_2 = \frac{v}{f_2} = \frac{3 \times 10^8}{3.4 \times 10^3} = 1.13 \times 10^5 \text{ m}$$

Here v is the velocity of light (3×10^8 m/s). Obviously using a much higher frequency would solve this problem by making the wavelength shorter.

Third, transmitting information in raw form, normally known as the baseband, would be impractical due to the low energy content. Losses between transmission and reception would soon attenuate the signals, with a resultant loss in reception. Modulating the signal by analogue or digital methods increases the power to the information and gives a higher signal-to-noise ratio.

In this chapter the following modulation techniques will be discussed together with suitable circuits: amplitude modulation (AM); frequency modulation (FM); frequency shift keying (FSK); phase shift keying (PSK); and quadrature phase shift keying (QPSK).

2.2 Analogue modulation techniques

Amplitude modulation

When the amplitude of a carrier signal is varied in accordance with the information signal, amplitude modulation is produced. This method is mainly used where large power outputs are required for long-distance communications.

Figure 2.3 shows a constant-amplitude, constant-frequency carrier being modulated by a single tone. In practice, many modulating signals may be used.

The general expression for the waveform in Fig. 2.3(a) is

$$v_c = V_c \sin(\omega_c + \theta) \tag{2.1}$$

where v_c is the instantaneous carrier voltage and V_c is the peak amplitude; and ω_c is the frequency of the carrier in radians. θ is the phase of the carrier but this will be ignored in the following analysis.

The modulating signal in Fig. 2.3(b) is given by

$$v_m = V_m \sin \omega_m t \tag{2.2}$$

where v_m is the instantaneous amplitude of the modulating signal and V_m is the peak amplitude.

The amplitude-modulated wave as shown in Fig. 2.3(c) is given by

$$v = (V_c + V_m \sin \omega_m t) \sin \omega_c t \tag{2.3}$$

$$= V_c \sin \omega_c t + V_m \sin \omega_m t \sin \omega_c t \tag{2.4}$$

Using the trigonometric identity

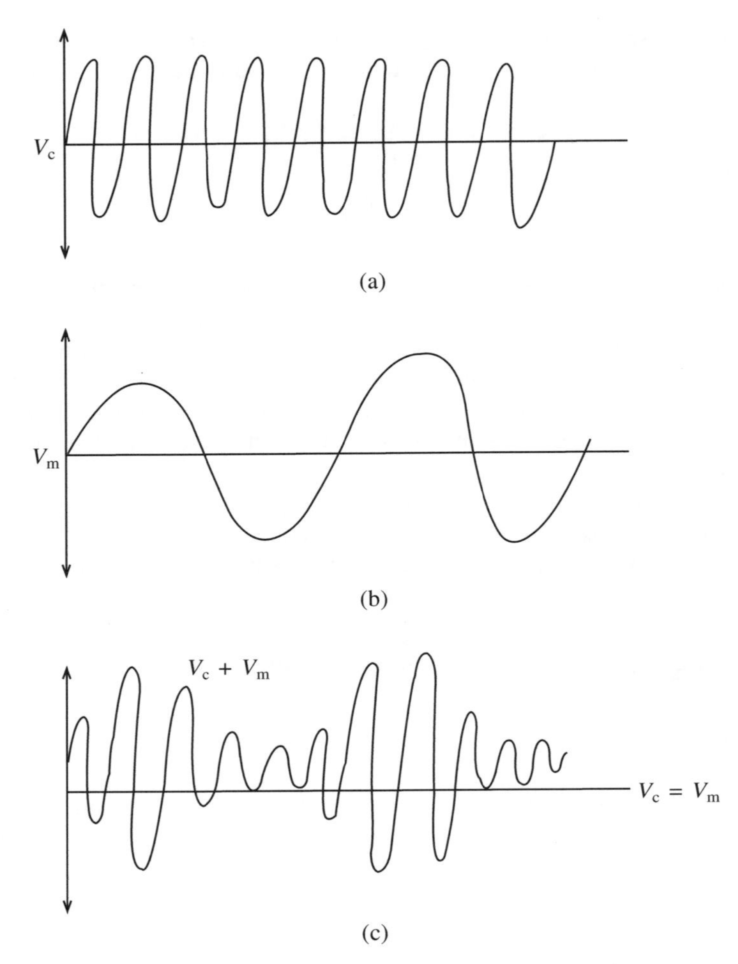

Fig. 2.3

$$\sin A \sin B = \frac{1}{2}\cos(A - B) - \frac{1}{2}\cos(A + B)$$

equation (2.4) becomes

$$v = V_c \sin \omega_c t + \frac{V_m}{2}\cos(\omega_c - \omega_m)t - \frac{V_m}{2}\cos(\omega_c + \omega_m)t \tag{2.5}$$

The modulated wave has three frequency components, namely the carrier frequency (f_c), the lower sideband ($f_c - f_m$) and the upper sideband ($f_c + f_m$). These components are represented in the form of a line or spectrum diagram as shown in Fig. 2.4. If several modulating tones were present as in the speech band they would be as shown in Fig. 2.5.

Figure 2.3(c) shows two important factors used in practice: the modulating factor and the depth of modulation. The modulating factor (m) is given by

$$m = \frac{(V_c + V_m) - (V_c - V_m)}{(V_c + V_m) + (V_c - V_m)} = \frac{V_m}{V_c} \tag{2.6}$$

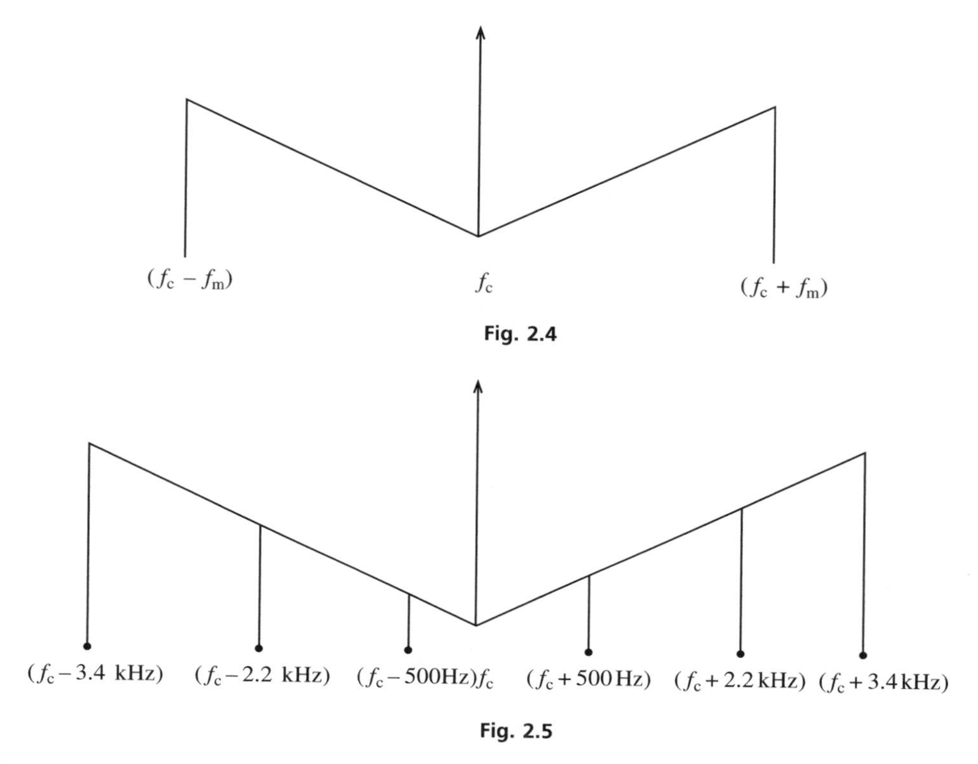

Fig. 2.4

Fig. 2.5

Expressed as a percentage, this is known as the *depth of modulation*. Hence the depth to which the carrier is modulated depends on the amplitude of the carrier and the modulating voltage. The maximum modulation factor used is unity. Exceeding this causes overmodulation and break-up of the signal, and hence some figure less than unity is used in practice.

Power distribution in an AM wave

The power which is coupled to an antenna by an AM wave is developed across its resistance. An antenna must be coupled to a transmitter by means of a transmission line or waveguide in order to be excited and hence produce radiation. The antenna input impedance which the feeder 'sees' must be known in order to achieve efficient coupling, and this requires a knowledge of transmission line theory (see Chapter 7).

Note, however, that the antenna input impedance generally has a resistive and reactive part. The reactive element originates from the inherent inductance and capacitance in the antenna. The resistive element of the input impedance originates from the numerous losses in the antenna. The radiated loss (radiation resistance) is the actual power transmitted and is a necessary loss caused by the modulated wave generating power in the antenna. However, other losses are present such as ohmic losses and those due to currents lost in the ground. Because of this it is important that the radiation resistance be much greater than all the other losses. The radiation resistance is generally defined as the equivalent resistance that would dissipate an amount of power equal to the total radiated power when the current through the resistance is equal to the current at the antenna input terminals.

Rearranging equation (2.6) as

$$V_m = mV_c$$

we can rewrite equation (2.5) as

$$v = V_c \sin \omega_c t + \frac{1}{2} mV_c \,[\cos (\omega_c - \omega_m)t - \cos (\omega_c + \omega_m)t] \tag{2.7}$$

The r.m.s. power developed across the antenna resistance (R_a) by the carrier and two sidebands is therefore

$$P_c \left(\frac{V_c}{\sqrt{2}}\right)^2 \frac{1}{R_a} = \frac{V_c^2}{2R_a}$$

$$P_m = \left(\frac{mV_c}{\sqrt{2}}\right)^2 \frac{2}{R_a} = \frac{m^2 V_c^2}{4R_a} \qquad \text{(sidebands)}$$

Total power is

$$P_T = \frac{V_c^2}{2R_a} + \frac{m^2 V_c^2}{4R_a}$$

$$P_T = \frac{V_c^2}{2R_a}\left[1 + \frac{m^2}{2}\right] \tag{2.8}$$

If several pairs of sidebands are involved, equation (2.8) becomes

$$P_T = P_c\left[1 + \frac{m_1^2}{2} + \frac{m_2^2}{2} + \ldots\right] \tag{2.9}$$

Example 2.1

The carrier of an AM transmitter is 50 W and, when modulated by a sinusoidal tone, the power increases to 59 W. Calculate:

(a) the depth of modulation;

(b) the ratio of maximum to minimum values of the wave envelope.

Solution

(a) From equation (2.9)

$$P_T = P_c\left(1 + \frac{m^2}{2}\right)$$

so

$$m = \sqrt{2\left(\frac{P_T}{P_c} - 1\right)} = \sqrt{2\left(\frac{59}{50} - 1\right)} = 0.6 \text{ or } 60\%$$

(b) From the wave envelope

$$P_c(1 + m) = 50(1 + 0.6) = 80 \text{ W}$$

$$P_c(1 - m) = 50(1 - 0.6) = 20 \text{ W}$$

Hence

$$\frac{P_{max}}{P_{min}} = 4$$

Example 2.2

An AM transmitter radiates 2 kW when the carrier is unmodulated and 2.25 kW when the carrier is modulated. When a second modulating signal is applied giving a modulation factor of 0.4, calculate the total radiated power with both signals applied.

Solution

$$m = \sqrt{2\left[\frac{P_T}{P_c} - 1\right]} = \sqrt{2\left[\frac{2.25}{2} - 1\right]} = 0.5$$

As the carrier power for the unmodulated wave is unchanged,

$$P_T = 2\left(1 + \frac{0.5^2}{2} + \frac{0.4^2}{2}\right) = 2.4 \text{ kW}$$

Example 2.3

An AM signal has a 25 V/100 kHz carrier and is modulated by a 5 kHz tone to a modulation depth of 95%.

(a) Sketch the spectrum diagram of this modulated wave, showing all values.

(b) Determine the bandwidth required.

(c) Calculate the power delivered to a 75 Ω load.

Solution

(a) Using our familiar rearrangement of equation (2.6),

$$V_m = 0.95 \times 25 = 23.75 \text{ V}$$

The amplitude of the sidebands, from (2.7), is

$$\frac{V_m}{2} = \frac{23.75}{2} = 11.875 \text{ V}$$

(See Fig. 2.6).

(b) As there is a double sideband, the bandwidth is 10 kHz.

(c) We have

$$P_c = \left(\frac{V_c}{\sqrt{2}}\right)^2 \frac{1}{R_a} = \left(\frac{25}{\sqrt{2}}\right)^2 \frac{1}{75} = 4.17 \text{ W}$$

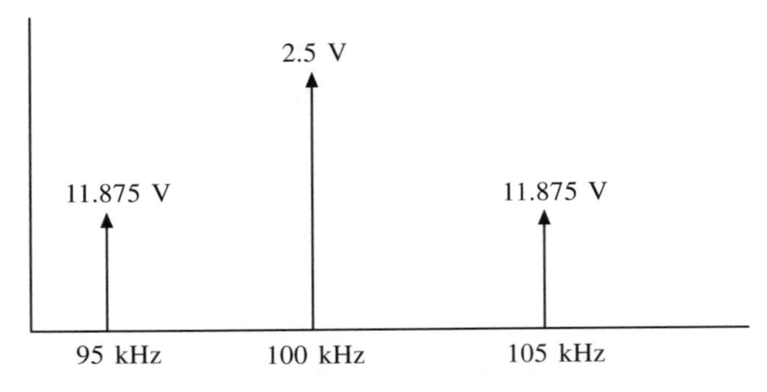

Fig. 2.6

$$P_{\mathrm{m}} = \frac{m^2 V_{\mathrm{c}}^2}{4R_{\mathrm{a}}} = \frac{(0.95 \times 25)^2}{4 \times 75} = 1.88 \text{ W}$$

Hence total power is 6.05 W.

Amplitude modulation techniques

The method of amplitude modulation previously discussed is known as double sideband modulation (d.s.b.). However, this method has a number of disadvantages which can be overcome by filtering out the carrier, one of the sidebands or both. Such a system would have the following advantages:

(a) reduced bandwidth, hence less noise;

(b) more channels available;

(c) increase in efficiency, as power is only transmitted when information is sent;

(d) selective fading is reduced as there is no carrier component to fade below the sideband level and cause sideband beating which would produce unwanted components;

(e) non-linearity is reduced as the carrier amplitude is the largest of all the components and this can cause saturation.

Double sideband suppressed carrier modulation (d.s.b.s.c.) requires the carrier to be reinserted at the receiver with the correct phase and frequency. Single sideband suppressed carrier modulation (s.s.b.s.c.) only requires the frequency of the reinserted carrier to be correct.

The basic principle of s.s.b.s.c. is shown in Fig. 2.7. The carrier and modulating signal are applied to a balanced modulator (which will be discussed later). The output of the modulator consists of the upper and lower sidebands, but the carrier is suppressed. The bandpass filter then removes one of the sidebands.

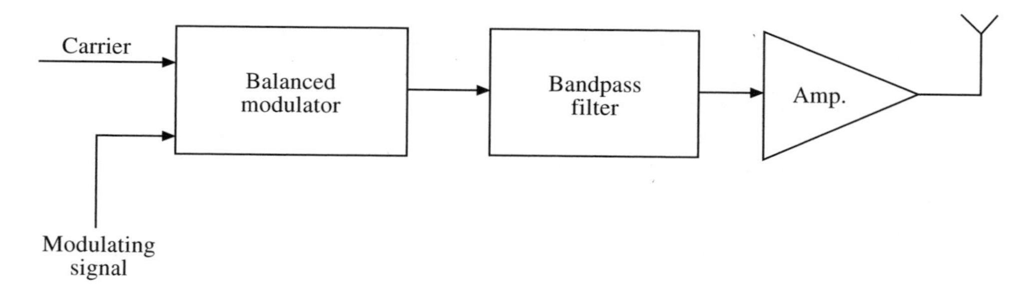

Fig. 2.7

Example 2.4

An AM transmitter is modulated by the audio range 20–15 000 Hz. If the carrier frequency is 820 kHz with a voltage level of 150 V, determine

(a) the modulating factor,

(b) the d.s.b. power and

(c) the s.s.b.s.c. power

for the frequency components 400 Hz/80 V, 1 kHz/50 V and 10 kHz/20 V if the antenna load is 120 Ω.

Solution

(a)
$$m_1 = \frac{V_{m1}}{V_c} = \frac{80}{150} = 0.53$$

$$m_2 = \frac{V_{m2}}{V_c} = \frac{50}{150} = 0.33$$

$$m_3 = \frac{V_{m3}}{V_c} = \frac{20}{150} = 0.13$$

(b)
$$P_T = \frac{V_c^2}{2R_a} + \frac{m_1^2 V_c^2}{4R_a} + \frac{m_2 V_c^2}{4R_a} + \frac{m_3^2 V_c^2}{4R_a}$$

$$= \frac{150^2}{2 \times 120} + \frac{(0.53 \times 150)^2}{4 \times 120} + \frac{(0.33 \times 150)^2}{4 \times 120} + \frac{(0.13 \times 150)^2}{4 \times 120}$$

$$= 93.75 + 13.33 + 5.21 + 0.83 = 113.1 \text{ W}$$

(c)
$$P_T = \left(\frac{m_1 V_c}{2\sqrt{2}}\right)^2 \frac{1}{R_a} + \left(\frac{m_2 V_c}{2\sqrt{2}}\right)^2 \frac{1}{R_a} + \left(\frac{m_3 V_c}{2\sqrt{2}}\right)^2 \frac{1}{R_a}$$

$$= \left(\frac{0.53 \times 150}{2\sqrt{2}}\right)^2 \frac{1}{120} + \left(\frac{0.33 \times 150}{2\sqrt{2}}\right)^2 \frac{1}{120} + \left(\frac{0.13 \times 150}{2\sqrt{2}}\right)^2 \frac{1}{120}$$

$$= 6.67 + 2.60 + 0.42 = 9.69 \text{ W}$$

2.3 The balanced modulator/demodulator

The function of a modulator, as has been shown, is to superimpose the baseband signals on to a carrier while the demodulator provides the reverse role by extracting a carrier, known as the intermediate frequency, and leaving the baseband.

There are many modulators and demodulators commercially sold on the market as integrated chips or as part of a front end receiver chip containing other stages such as the tuner and detector. However, it is informative to look at the balanced modulator which is used in the majority of AM and other circuits.

A common method of obtaining a single sideband wave is illustrated in Fig. 2.7. The output of this circuit differs from a conventional AM output in that it does not include the original radio frequency signal, but only the two sidebands. The single sideband is obtained by a highly selective filter.

An integrated circuit which is commonly used is the Philips MC 1496. This is a modulator/demodulator chip which uses a monolithic transistor array. It has many applications such as AM and suppressed carrier modulators, AM and FM demodulators and phase detectors. The basic theory of operation is shown in the data sheets at the end of this chapter. Figures 2.8 and 2.9 show the application of this chip as an AM modulator and demodulator, respectively.

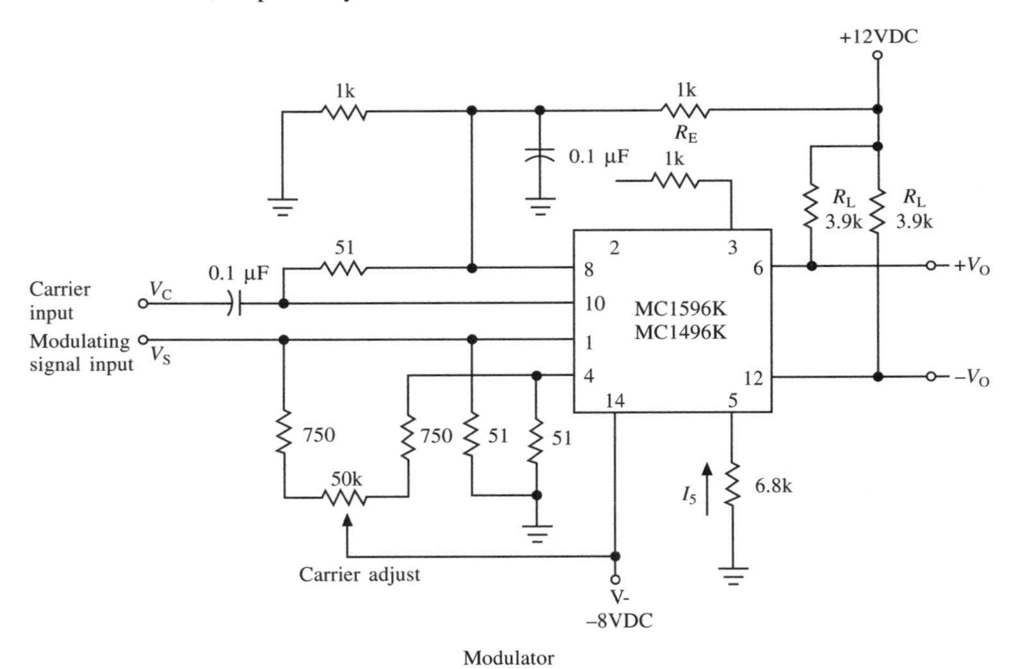

Modulator

Fig. 2.8

The AM modulator shown in Fig. 2.8 allows no carrier at the output; by adding a variable offset voltage to the differential pairs at the carrier input the carrier level changes and its amplitude is determined by the AM modulation.

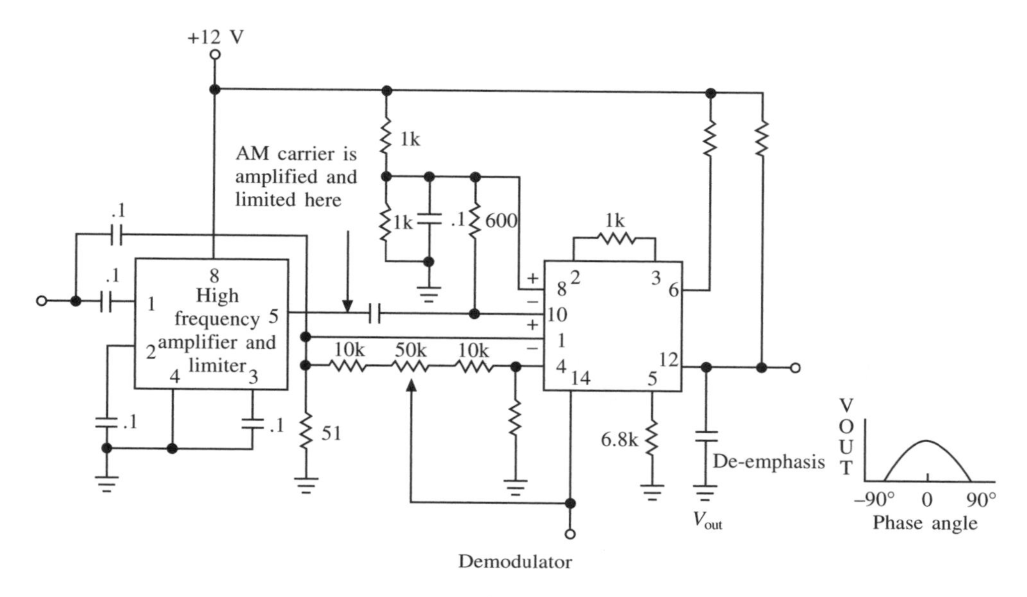

Fig. 2.9

The frequency spectrum is shown in the data sheets; it can be seen that undesired sidebands appear if the modulation or carrier levels are high. These need to be filtered and a fourth-order Butterworth is ideal. Note also that the modulation levels may be varied by means of R_E connected between pins 2 and 3 in Fig. 2.8.

As can be seen from equation (1) in the data sheets, the output of the balanced modulator is a cosine function of the phase between the signal and carrier inputs. If the carrier input is driven hard enough, a switching action occurs and the output becomes a function of the input voltage. The output amplitude is maximized when the phase difference is zero.

A typical demodulator is shown in Fig. 2.9. In this case the carrier is amplified by an intermediate frequency chip which provides a limited gain of 55 dB or higher at 400 μV. The carrier is then applied to the demodulator where the carrier frequency is attenuated. Output filtering is required to remove the high-frequency unwanted components.

2.4 Frequency modulation and demodulation

With frequency modulation the frequency (rather than the amplitude) of a constant-amplitude, constant-frequency sinusoidal carrier is made to vary in proportion to the amplitude of the applied modulating signal. This is shown in Fig. 2.10, where a constant-amplitude carrier is frequency-modulated by a single tone. Note how the frequency of the carrier changes.

Frequency modulation can be understood by considering Fig. 2.11. This shows that a modulating square or sine wave may be used for this type of modulation. The frequency of the frequency-modulated carrier remains constant, and this indicates that the modulating process does not increase the power of the carrier wave. For FM the instantaneous frequency ω is made to vary as

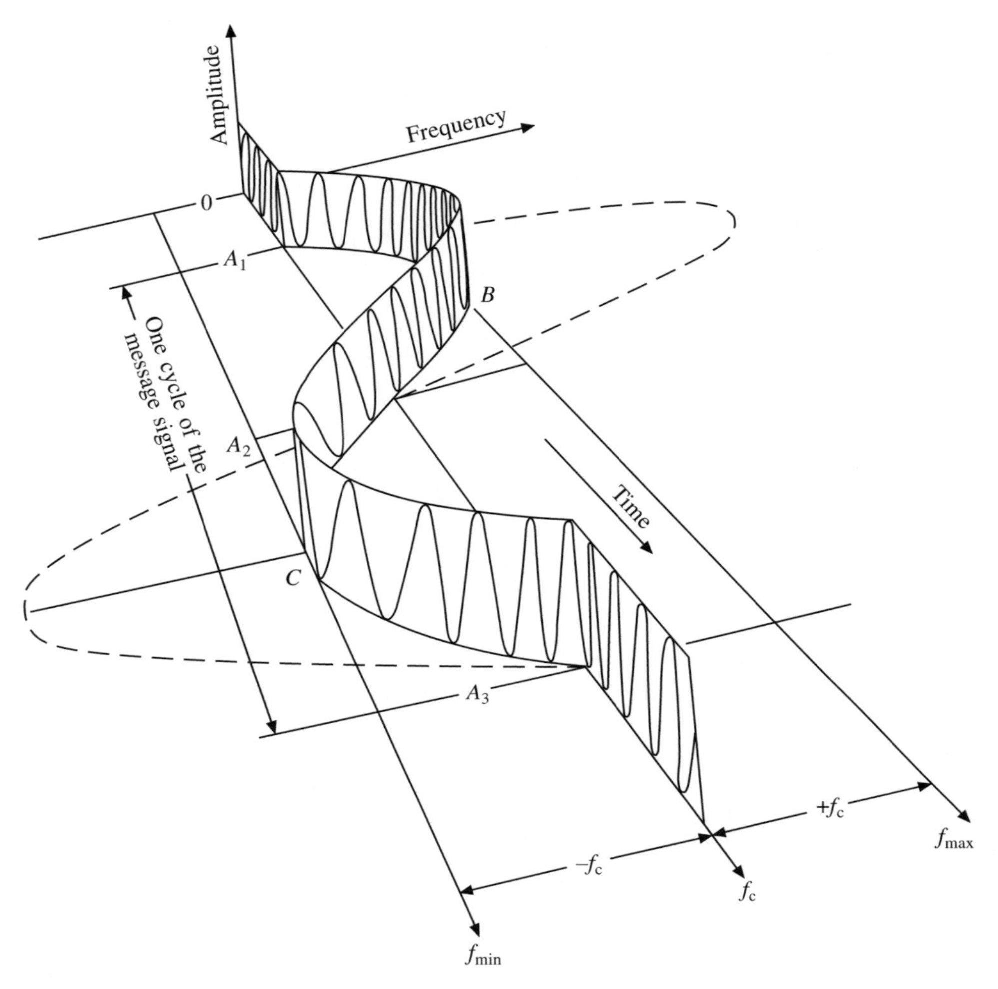

Fig. 2.10

$$\omega = \omega_c + kV_m \sin \omega_m t \tag{2.10}$$

In this equation

$$kV_m = \Delta\omega \tag{2.11}$$

and we have

$$\theta = \left(\omega_c t - \frac{\Delta f}{f_m} \cos \omega_m t \right) \tag{2.12}$$

Substituting (2.12) into (2.11) gives

$$v = V_c \sin\left(\omega_c t - \frac{\Delta f}{f_m} \cos \omega_m t \right) \tag{2.13}$$

for the peak angular frequency shift for the modulating signal.

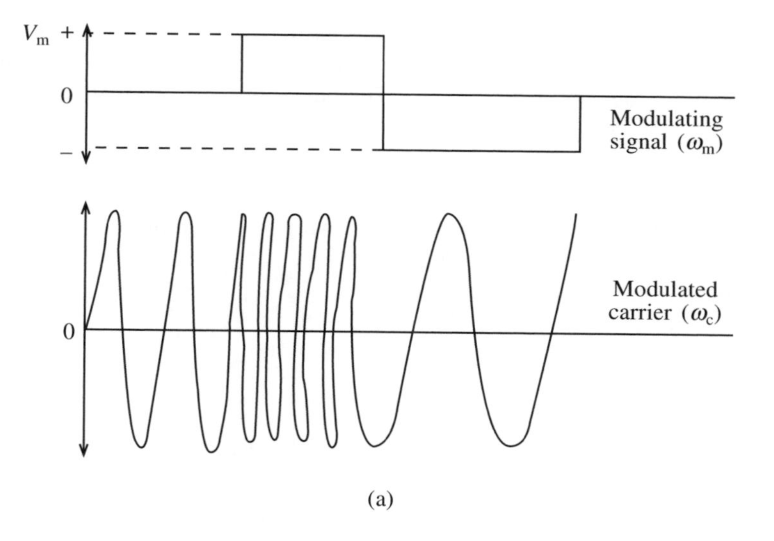

(a)

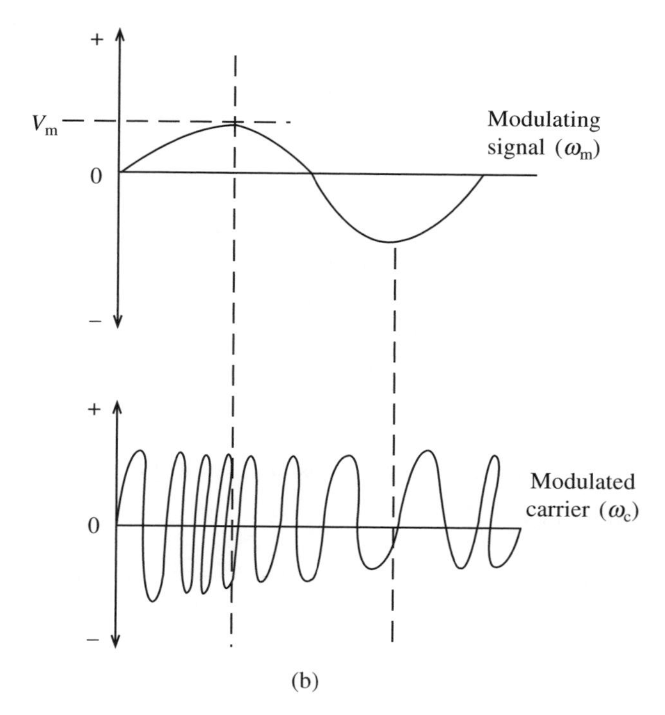

(b)

Fig. 2.11

The modulating index is given by

$$\beta = \frac{\Delta f}{f_m} = \frac{\text{peak frequency shift}}{\text{modulating frequency}} \tag{2.14}$$

for a constant frequency, constant amplitude modulating signal. In practice the modulating signal varies in amplitude and frequency. This leads to two further parameters: a maximum value of the modulating signal ($f_{m(max)}$); and a maximum allowable frequency shift, which is defined as the frequency deviation (f_d). The deviation ratio (δ) is then defined as:

$$\delta = \frac{f_d}{f_{m(max)}} \qquad (2.15)$$

For any given FM system the frequency will swing to a maximum value of frequency deviation known as the rated system derivation. This parameter determines the maximum allowable modulating signal voltage. Equation (2.15) applies for this condition of rated system deviation.

Finally, the frequency-modulated wave can be written as

$$v = V_c \sin(\omega_c t - \beta \cos \omega_m t) \qquad (2.16)$$

for a constant-amplitude, constant-frequency modulating signal such as a square wave, and

$$v = V_c \sin(\omega_c t - \delta \cos \omega_m t) \qquad (2.17)$$

for a variable-amplitude, variable-frequency modulating signal such as a sinusoidal wave.

Expanding (2.17) using the identity

$$\sin(A + B) = \sin A \cos B - \cos A \sin B$$

gives

$$\sin \omega_c t \cos(\delta \cos \omega_m t) - \cos \omega_c t \sin(\delta \cos \omega_m t)$$

The second factor of each term expands into an infinite series whose coefficients are a function of δ. These coefficients are called **Bessel functions**, denoted by $J_n(\delta)$, which vary as δ varies. More specifically, they are Bessel functions of the first kind and of order n.

Expanding the second factor gives

$$\cos(\delta \cos \omega_m t) = J_0(\delta) - 2J_2(\delta) \cos 2\omega_m t + 2J_4(\delta) \cos 4\omega_m t - \ldots$$

and

$$\sin(\delta \cos \omega_m t) = 2J_1(\delta) \cos \omega_m t - 2J_3(\delta) \cos 3\omega_m t + \ldots$$

Using the relationships

$$\sin A \cos B = \frac{1}{2}[\sin(A + B) + \sin(A - B)]$$

$$\cos A \sin B = \frac{1}{2}[\sin(A + B) - \sin(A - B)]$$

we obtain

$$\begin{aligned} v = V_c\{ & J_0(\delta) \sin \omega_c t + J_1(\delta)[\sin(\omega_c + \omega_m)t - \sin(\omega_c - \omega_m)t] \\ & - J_2(\delta)[\sin(\omega_c + 2\omega_m)t + \sin(\omega_c - \omega_m)t] \\ & + J_3(\delta)[\sin(\omega_c + 3\omega_m)t - \sin(\omega_c - 3\omega_m)t] - \ldots\} \end{aligned}$$

Thus the modulated wave consists of a carrier and an infinite number of upper and lower side-frequencies spaced at intervals equal to the modulation frequency. Also, since the amplitude of the unmodulated and modulated waves are the same, the powers in the unmodulated and modulated waves are equal.

The Bessel coefficients can be determined either from graphs or tables. Both are given in Appendix A.

Example 2.5

Determine the values of the amplitudes of the carrier and side frequencies if f_d is 5 kHz, $f_{m(max)}$ is 5 kHz and the carrier amplitude is 10 V.

Solution

The deviation ratio is unity and the sideband amplitudes and carrier amplitude may be obtained from Table A.1. These are:

$$J_0 = 0.77 \qquad J_1 = 0.44 \qquad J_2 = 0.11 \qquad J_3 = 0.02 \qquad J_4 = 0$$

As the carrier has an amplitude of 10 V, each component in the spectrum diagram will have the values shown.

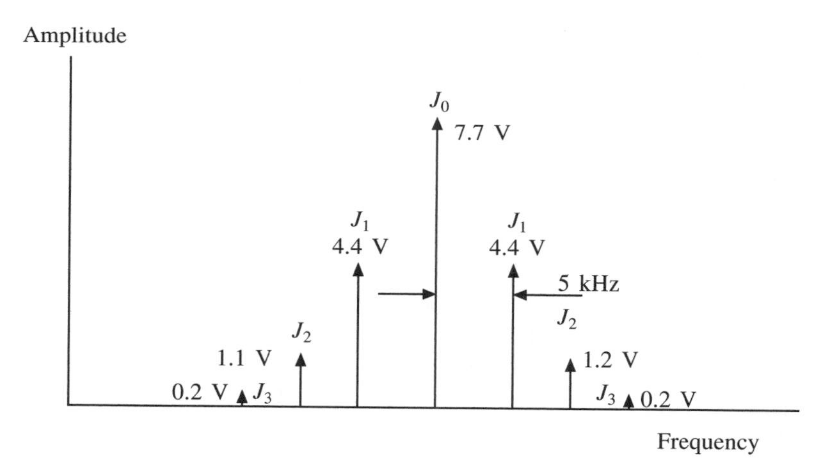

Fig. 2.12

Example 2.6

Determine the amplitudes of the side frequencies generated by an FM transmitter having a deviation ratio of 10 kHz, a modulating frequency of 5 kHz and a carrier level of 10 V.

Solution

The deviation ratio is 2 for this system, so once again from Table A.1 the following amplitudes are obtained.

$$J_0 = 0.22 \qquad J_1 = 0.58 \qquad J_2 = 0.35 \qquad J_3 = 0.13 \qquad J_4 = 0.03 \qquad J_5 = 0.01$$

The spectrum diagram is shown in Fig. 2.13. There are more side frequencies in this case and hence the quality of the transmission would be improved. However, not all the side frequencies are relevant, as can be seen from their amplitudes.

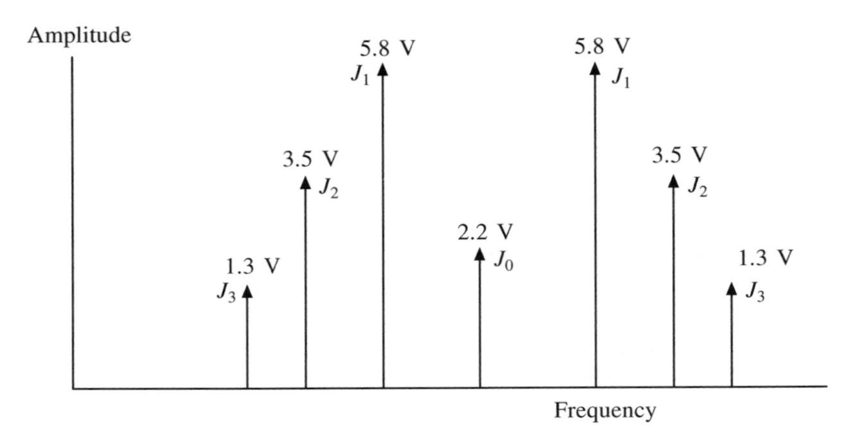

Fig. 2.13

Bandwidth and Carson's rule

It has already been mentioned that not all the side frequencies are necessary for satisfactory performance. Generally an acceptable performance can be obtained with a finite number of side frequencies, and this may be considered satisfactory when not less than 98% of the power is contained in the carrier and its adjacent frequencies. Since the amplitude of the nth side frequency is given as $J_n V_c$, the power dissipated in a load (R) by the modulated wave is

$$P = \frac{(J_0 V_c)^2}{R} + 2\left[\frac{(J_1 V_c)^2}{R} + \frac{(J_2 V_c)^2}{R} + \frac{(J_3 V_c)^2}{R}\right]$$

$$\therefore \qquad P = \frac{V_c^2}{R} = [J_0^2 + 2(J_1^2 + J_2^2 + J_3^2 + \ldots)] \qquad (2.18)$$

$$J_0^2 + (J_1^2 + J_2^2 + J_3^2 + \ldots) > 0.98$$

Thus for 98% of the power to be contained in the carrier plus the side frequencies the following applies.

Example 2.7

An FM transmitter transmits with a rated system deviation of 60 kHz and a maximum modulating frequency of 15 kHz. If the carrier amplitude is 25 V, determine the number of side frequencies required to ensure that 98% of the power is contained in the carrier and side frequencies. Sketch the spectrum diagram.

Solution

The frequency deviation is given by

$$\delta = \frac{60}{15} = 4$$

From the Bessel tables,

$$J_0 = -0.3971 \qquad J_1 = -0.0660 \qquad J_2 = 0.3641 \qquad J_3 = 0.4302$$

$$J_4 = 0.2811 \qquad J_5 = 0.1321 \qquad J_6 = 0.0491 \qquad J_7 = 0.0152$$

It will be seen that only J_2, J_3, J_4, J_5 and J_0 are required:

$$J_0^2 + 2(J_2^2 + J_3^2 + J_4^2 + J_5^2 + J_7^2) > 0.98$$

The spectrum diagram is sketched in Figure 2.14.

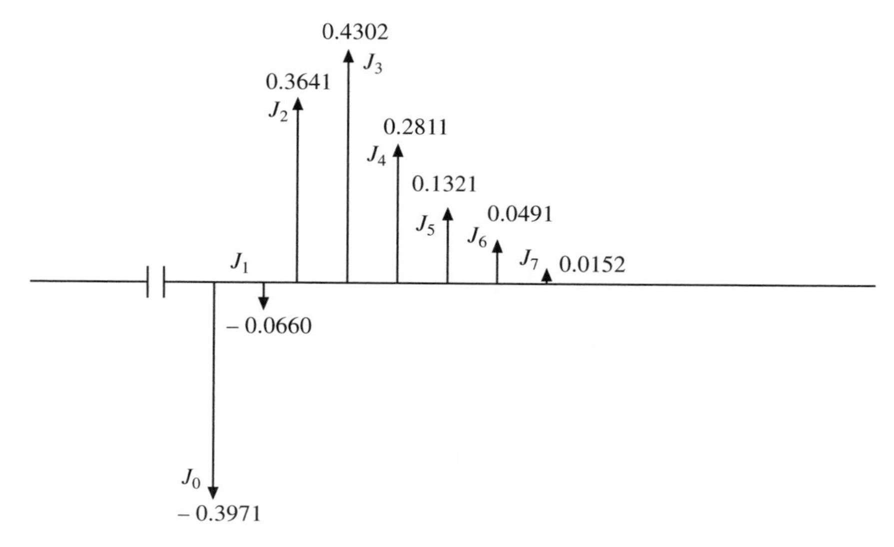

Fig. 2.14

The Bessel tables show negative and positive values for the Bessel functions J_0, J_1, J_2 etc. This can be explained with reference to Figure A.1 in Appendix A. This figure can be used to determine the amplitudes of the sidebands relative to the carrier wave. It can be seen that the carrier and sidebands reach a positive maximum value, pass through zero and then reach a maximum negative (i.e. minimum) value.

When the deviation ratio is zero the carrier is unmodulated and has its maximum value. For any other modulation index the energy levels are distributed between the sidebands and carrier. As the deviation ratio increases the number of sidebands increases, as does the number of negative values. The negative values indicate a 180° phase shift, and this can be seen from the Bessel graphs where the carrier and each sideband behave as sinusoids as frequency deviation takes place. Negative signs are usually ignored in practice, since only the magnitude of the carrier and each sideband is required. Squaring the negative values produces positive or magnitude quantities.

Example 2.7 shows that not all of the side frequencies are necessary for a high-fidelity output in an FM system. The bandwidth can therefore be determined by considering only the useful side frequencies with the higher amplitudes. Note that the required bandwidth, when all the side frequencies are considered, is given as

$$BW = (6 \times 15)2 = 180 \text{ kHz}$$

This is a considerable saving in bandwidth and hence a reduction in noise in the modulator circuits.

For a rated system deviation the required bandwidth for Example 2.7 is

$$BW = (5 \times 15)2 = 150 \text{ kHz}$$

Also since $\delta = 4$,

$$\delta + 1 = 5$$

This indicates the required number of pairs of side frequencies for the 98% criterion. Since $BW = (\delta + 1)$ pairs of side frequencies, i.e.

$$BW = 2f_{m(max)} \times \text{pairs of side frequencies} \qquad \text{(Carson's rule)} \qquad (2.19)$$

Since

$$\delta = \frac{f_d}{f_{m(max)}}$$

substituting in (2.19) gives

$$BW = 2f_{m(max)}\left(\frac{f_d}{f_{m(max)}} + 1\right)$$

$$\therefore \qquad BW = 2(f_d + f_{m(max)}) \qquad (2.20)$$

Equations (2.19) and (2.20) express a relationship known as Carson's rule for determining the bandwidth of an FM system requiring the requisite number of side frequencies to satisfy the 98% criterion. Also δ indicates the rated system deviation, while β is used for values less than this.

Example 2.8

(a) An FM system uses a carrier frequency of 100 MHz with an amplitude of 100 V. It is modulated by a 10 kHz signal and the rated system deviation is 80 kHz. Determine the amplitude of the centre frequency.

(b) An FM station with a maximum modulating frequency of 15 kHz and a deviation ratio of 6 operates at a centre frequency (f_c) of 10 MHz. Determine the 3 dB bandwidth of the stage following the modulator which would pass 98% of the power in the modulated wave. Also determine the Q factor of this circuit.

Solution

(a)

$$\delta = \frac{f_d}{f_{m(max)}} = \frac{80}{10} = 8$$

For $\delta = 8$, the Bessel tables give $J_0 = 0.1717$. Therefore the amplitude of the centre frequency is

$$100 \times 0.1717 = 17.17 \text{ V}$$

(b) Assume rated system deviation. Thus

$$BW = 2(f_d + f_{m(max)})$$

Also

$$\delta = \frac{f_d}{f_{m(max)}} = 6$$

Hence

$$f_d = 6 \times f_{m(max)} = 6 \times 15 = 90 \text{ kHz}$$

$$BW = 2(90 + 15) = 210 \text{ kHz}$$

so

$$Q = \frac{f}{BW} = \frac{10 \times 10^3}{210} = 47.6$$

Example 2.9
An FM system has a rated system deviation of 65 kHz. Determine the maximum permitted value of the modulating signal voltage if the modulator has a sensitivity of 5 kHz/V.

Solution
Since the maximum swing is 65 kHz, then

$$65 = 5 \times V_m$$

$$V_m = \frac{65}{5} = 13 \text{ V}$$

Example 2.10
An FM broadcast station is assigned a channel between 92.1 and 92.34 MHz. If the maximum modulating frequency is 15 kHz determine:

(a) the maximum permissible value of the deviation ratio;

(b) the number of side frequencies.

Solution
(a)

$$BW = 92.34 - 92.1 \text{ MHz} = 0.24 \text{ MHz} = 240 \text{ kHz}$$

$$BW = 2(f_d + f_{m(max)})$$

$$240 = 2(f_d + 15)$$

$$\therefore \quad f_d = 105 \text{ kHz}$$

Also

$$\delta = \frac{f_d}{f_{m(max)}} = \frac{105}{15} = 7$$

(b) From the Bessel tables, this will give 10 pairs of sidebands.

2.5 FM modulators

The most frequently used modulator in FM systems is the reactance modulator, which incorporates some method of varying the reactance across the oscillator circuit. This can be done by incorporating a device which changes either its inductive or capacitive reactance, depending on the oscillator involved. With a Colpitts oscillator some type of capacitive modulator would be used; a Hartley oscillator would use an inductive modulator.

The capacitance of a simple signal diode depends on the width of its depletion layer when forward or reverse biased. A particular diode, called a variable reactance diode (varactor for short), is fabricated in such a way that its capacitance is a function of the voltage applied across it.

If a varactor diode is connected across the tuned circuit of an oscillator, and the voltage across the diode is varied, the variation of the diode capacitance will cause a variation in the oscillator's frequency. The circuit virtually functions as a voltage-to-frequency convertor.

A typical diode FM modulator is shown in Fig. 2.15. In this diagram the modulating signal is fed to a transformer which is coupled to a Colpitts oscillator in this case. The varactor diode C_4 is reverse biased by V_g, as shown, to a practical point on its characteristics. This reverse bias voltage varies the modulating signal voltage, causing C_4 to vary, thus varying the oscillator frequency.

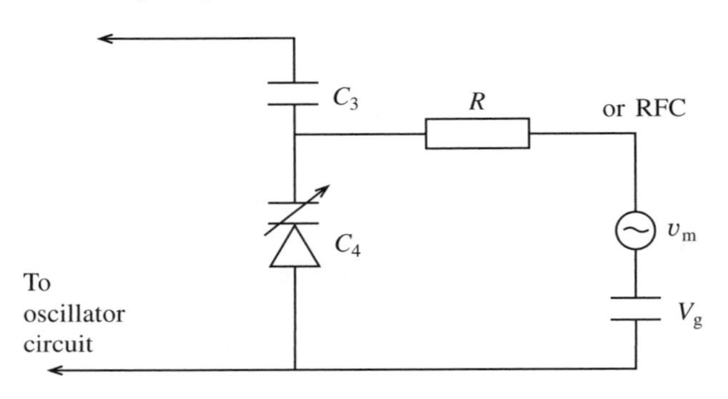

Fig. 2.15

The radio frequency choke is necessary so that the RF voltage across C_4 is not shorted out by the sources v_m and V_g. C_3 is used to block any d.c. voltage from the oscillator.

Example 2.11

A reactance modulator is used in an FM transmitter. It consists of a Colpitts oscillator and audio injection circuit as shown in Fig. 2.16. The two varactor diodes have a tuneable range from 2.4 to 3 pF. Determine the tuning range of the modulator if the inductance value is $L = 2$ μH.

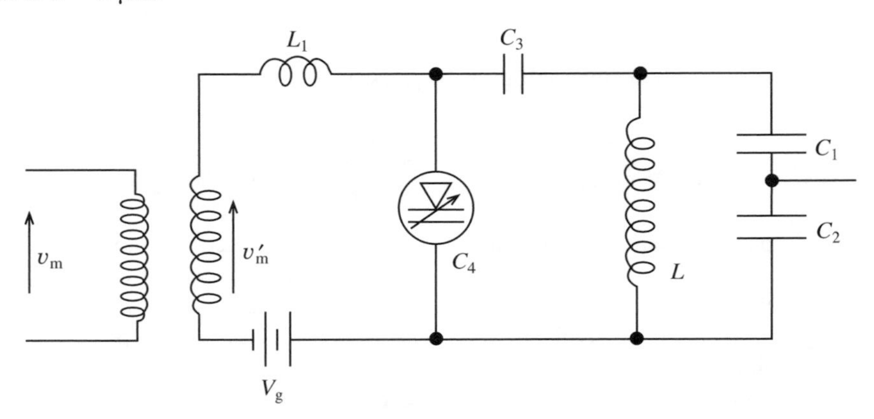

Fig. 2.16 Diode FM modulator

Solution

$$C_T = \frac{2.4 \times 2.4}{2.4 + 2.4} = 1.2 \text{ pF} \qquad f = \frac{1}{2\pi\sqrt{LC_T}} = \frac{10^9}{6.28\sqrt{2 \times 1.2}} = 103 \text{ MHz}$$

$$C_T = \frac{3 \times 3}{3 + 3} = 1.5 \text{ pF} \qquad f = \frac{10^9}{6.28\sqrt{2 \times 1.5}} = 92 \text{ MHz}$$

2.6 FM demodulators

The FM demodulator performs the reverse operation to modulation in that it converts variations in frequency into variations in amplitude. The frequency-to-voltage transfer may be non-linear over the operating range, and several methods are used in practice. However, this text will discuss only two common types of demodulator, namely the phase-looked loop (PLL) demodulator and the ratio detector.

The phase-locked loop demodulator

The PLL will be discussed in Chapter 6, but for the purpose of this particular application the block diagram shown in Fig. 2.17 will be used. This is the simplest type of demodulator and is frequently used in data communications systems. It consists of a phase comparator which has two input signals, one from the voltage controlled oscillator (f_2) and the other being the FM signal (f_1). The phase comparator compares the phase of the VCO with the incoming FM signal, giving an output proportional to the difference in phase. This is then filtered to remove unwanted high-frequency components and the output from the filter is used to control the frequency of the voltage-controlled oscillator (VCO), locking it to the incoming signal. Hence the VCO should be capable of tracking the incoming FM signal within the frequency deviation of the system. The output from the low pass filter, i.e. the error voltage, is used to obtain the demodulated output, and the linearity of the output depends only on the linearity of the voltage-to-frequency characteristics of the VCO.

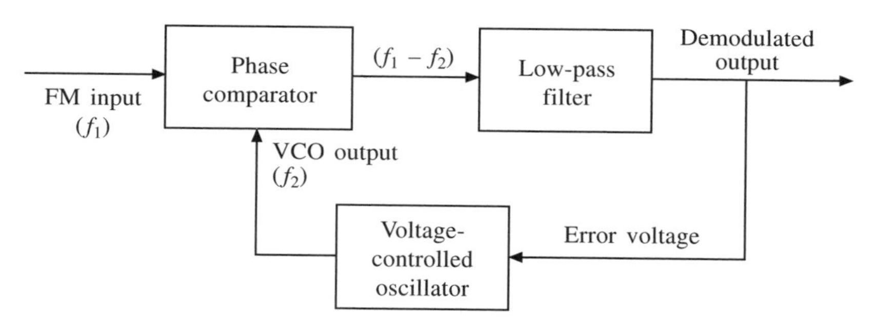

Fig. 2.17

The ratio detector

Figure 2.18 shows a circuit which is commonly used in telecommunications applications for FM demodulation. Receivers which use this circuit generally have a bandpass limiter as the previous stage. This improves the filtering before demodulation takes place as ratio detectors have a low-input Q factor which causes the input voltage V_1 to change.

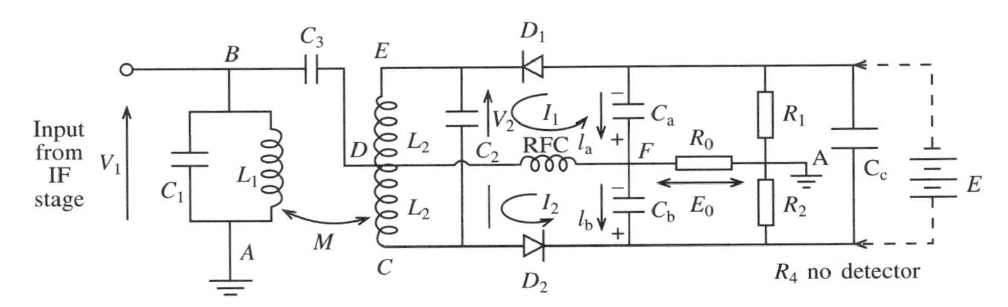

Fig. 2.18

The circuit C_1L_1 and C_2L_2 are tuned to the carrier frequency. The rectifying diodes D_1 and D_2 are connected such that the d.c. voltages across C_a and C_b are of the same polarity. C_3 electrically connects B to D and must have negligible reactance. The operation of the circuit hinges on the fact that the voltage V_2 is 90° out of phase with V_1 at resonance.

Consider initially the double tuned circuit shown in Fig. 2.19. Since the circuit is in resonance, V_1 is in phase with I and I_L will lag V_1 by 90°. From Fig. 2.16, V_{EC} will lead $V_{BA} = V_{DA}$ by 90°. This is shown in the phasor diagram shown in Fig. 2.20(a). As the frequency rises above the centre frequency (f_c), the secondary becomes more inductive and V_{EC} shifts clockwise, as shown in Fig. 2.20(b). If the frequency falls below f_c the secondary becomes more capacitive and V_{EC} will shift anticlockwise from the 90° position. This is shown in Fig. 2.20(c). Note that in Fig. 2.20 $V_{BA} = V_{EC}$ remains constant as the frequency varies.

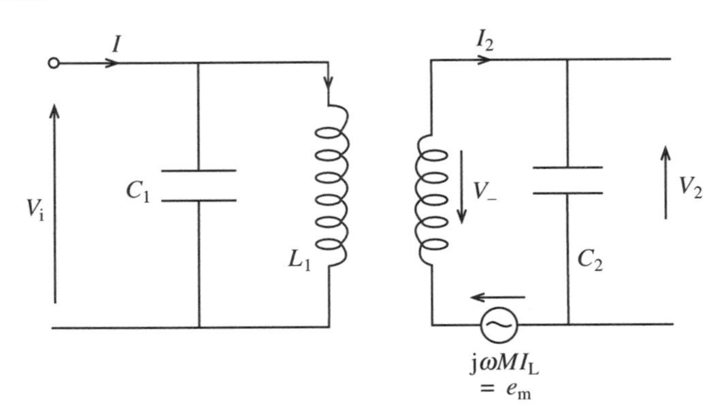

Fig. 2.19

Consider a d.c. voltage source E_X replacing the capacitor C_C in Fig. 2.18. When the peak-to-peak value of the incoming signal is less than E_B, D_1 and D_2 will not conduct and

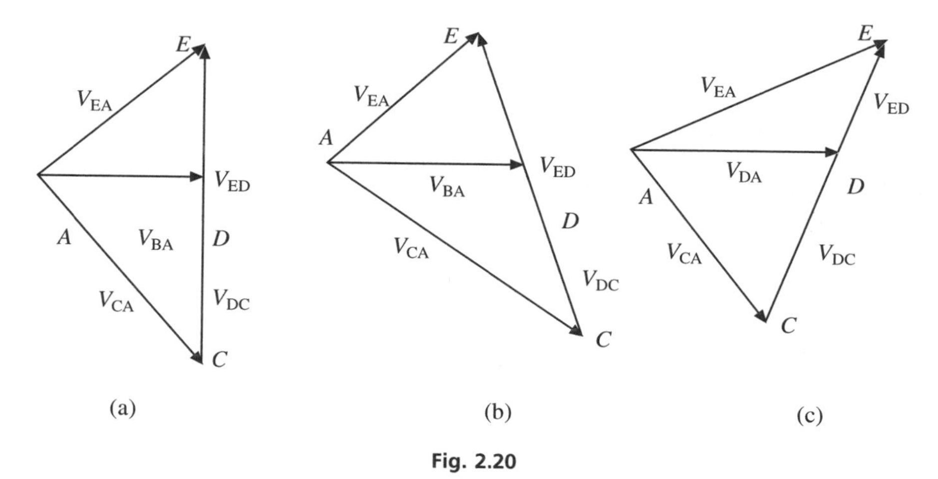

Fig. 2.20

the output voltage is zero. An output voltage will only appear if the peak-to-peak value of the input signal is greater than E_X. Also the potential across R_1 and R_2 is clamped to E_X so that E_X acts as an amplitude limiter removing variations in amplitude of the modulated input signal. When E_X is replaced by C_C the large time constant $C_C(R_1 + R_2)$ serves to maintain a constant voltage across R_1 and R_2 and hence C_a and C_b.

If e_a is the voltage across C_a which is proportional to V_{EA}, e_b is the voltage across C_b which is proportional to V_{CA} and $e_a + e_b = E_X$ then the voltage at A, the junction of R_1 and R_2, is a constant, i.e. $E_X/2$. There are three cases to consider. If $f = f_c$ then $V_{EA} = V_{CA}$, hence $e_a = e_b$ and the output voltage is zero. If $f > f_c$ then $V_{EA} > V_{CA}$, hence $e_a > e_b$, point F rises in potential above point A and the d.c. output voltage goes positive. Finally, if $f < f_c$ then $V_{EA} < V_{CA}$, hence $e_b > e_c$ and the potential at F falls below the potential of A so that the d.c. output voltage goes negative. Thus the value of the output voltage depends on the frequency shift from f_c, and the polarity of the output voltage will be determined by whether $f > f_c$ or $f < f_c$.

Note, finally, that only the ratio $e_a : e_b$ changes, which is why the circuit is known as a ratio detector.

2.7 Digital modulation techniques

In the last few sections methods of transmitting analogue information using analogue signals were explained. This section will consider methods of transmitting digital data using analogue signals.

The most familiar use of these methods is in data communications, where modems and telephone networks are used; because integrated circuits are generally used a block diagram approach will be considered.

Frequency shift keying

In some situations data can be transmitted directly without any modulation technique being necessary. This is applicable over short distances where the baseband signal may be sent in a raw form. However, where distance is involved more sophisticated methods are required.

One of the modulation methods most frequently used is frequency shift keying (FSK). In FSK the transmitted signal is switched between two frequencies every time there is a change in the level of the modulating data stream. The higher frequency may be used to represent a high level (1) and the lower frequency used for the low level (0). This results in a waveform similar to the one shown in Fig. 2.21.

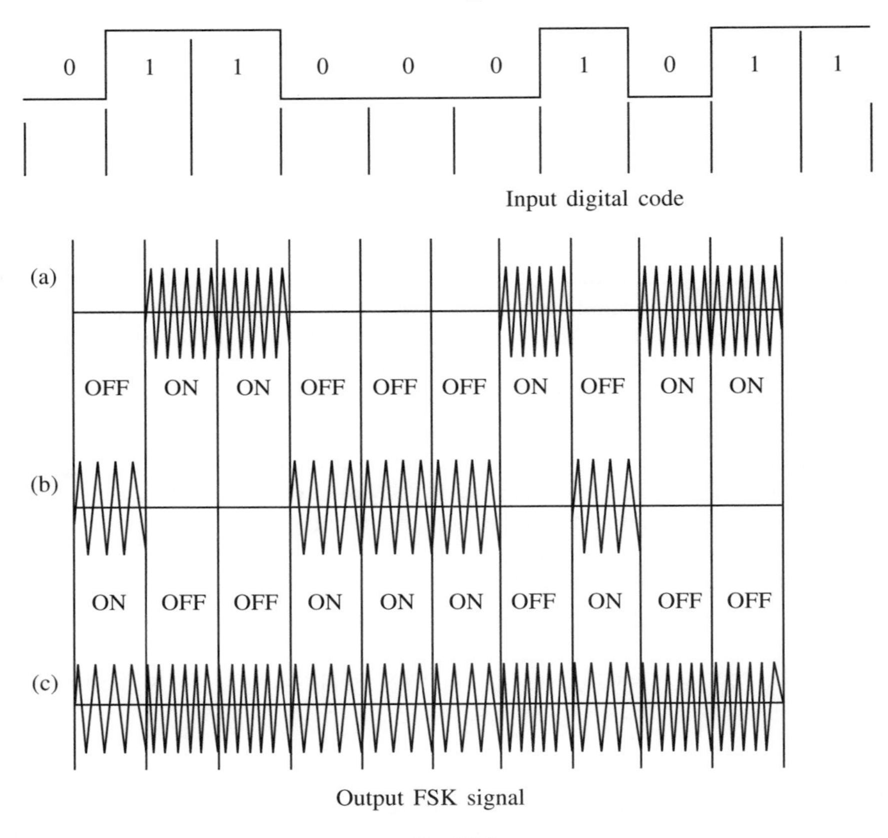

Fig. 2.21

Generally the frequencies used in FSK depend on the system application. Most modems traditionally use frequencies within the voice range (300–3400 Hz), while much higher frequencies would be used for satellite or radio relay systems. No matter what system is used there are fundamental blocks which are necessary for successful operation.

A balanced modulator is necessary to generate the required waveforms. This device has been mentioned earlier; it simply multiplies two signals together at its two inputs, the output voltage being the product of these two voltages. One of the inputs (the carrier input) is generally a.c. coupled, while the other (the digital data input) is d.c. coupled. The block diagram is shown in Fig. 2.22. Note that the data stream is inverted in modulator 2 in order to switch to the second carrier frequency.

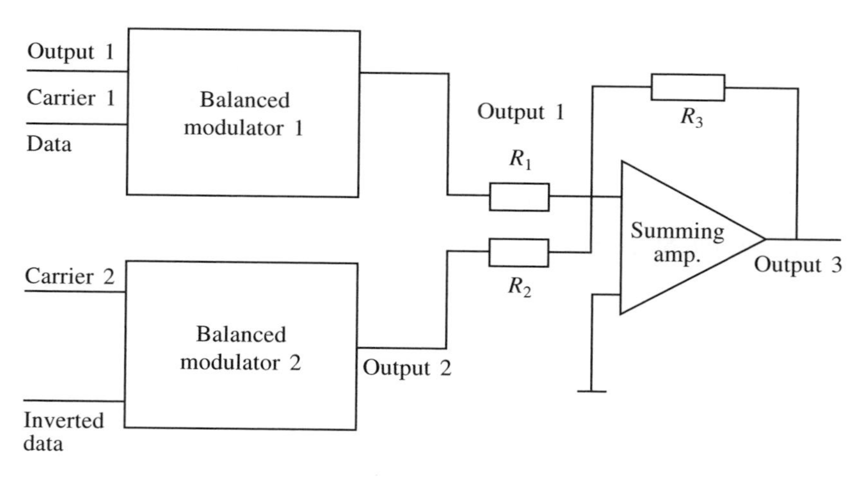

Fig. 2.22

At the receiver end, the FSK waveform has to be demodulated or, more specifically, decoded. One approach used for this is shown in Fig. 2.23. The demodulation of FSK signals can be accomplished by means of a ratio detector or a PLL, but for modem applications the PLL is generally preferred as it can be used for both modulation and demodulation. The data stream consists of marks (1) and spaces (0) which are each allocated a switched frequency. The space is normally allocated the higher frequency. The rate at which the carrier frequency is switched is known as the **baud rate**, and this is the same as the digital data rate for FSK. This is not always the case for other demodulation methods. As will be discussed in Chapter 6, the PLL has a free running frequency of its own and this is normally set between the mark and space carriers when designing such a system.

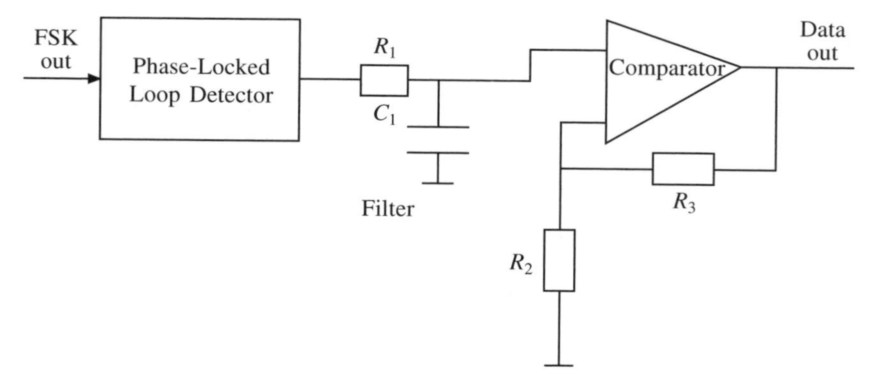

Fig. 2.23

The output of the PLL in Fig. 2.23 contains numerous components due to the interaction of the two frequencies, and hence a low-pass filter is used. However, the output of the filter produces rounded waveforms instead of oblong-shaped pulses, and this is modified by including a comparator.

Example 2.12
An FSK receiver uses a PLL as part of its demodulation circuitry, to receive digital data at the rate of 1200 bps. If the mark frequency is 1 kHz and the space frequency is 1.72 kHz, determine:

(a) the free running frequency of the PLL

(b) the bandwidth of the receiver.

Solution

(a) Since the mark and space carriers are separated by 720 Hz, the centre frequency is 1.36 kHz and this will be the free running frequency of the VCO in the PLL.

(b) Since the data rate is 1200 bps, the frequency is 600 Hz. Also the input to the PLL has to swing between 1 kHz and 1.72 kHz, i.e. ± 360 Hz. The deviation ratio is thus

$$\delta = \frac{360}{600} = 0.6$$

From the Bessel tables,

$$J_0 = 0.9120 \quad J_1 = 0.2867 \quad J_2 = 0.0437$$

Hence two side frequencies are available and the bandwidth is

$$2(2 \times f_{m(max)}) = 4 \times 600 = 2.4 \text{ kHz}$$

This falls within the baseband range of 300–3400 Hz, and the output will use a filter and comparator as shown in Fig. 2.23.

Phase shift keying (BPSK)

High-speed modems operating at bit rates of up to 56 kbps require phase shift keying or quadrature phase shift keying. It is also the preferred modulation method for satellite and space technology. Unlike FSK, phase shift keying uses one carrier frequency which is modulated by the data stream. It is a modulation system in which only discrete phase states are allowed. Usually 2^n phase states are used, and when $n = 1$ this gives two-phase changes. This is sometimes called binary phase shift keying (BPSK). When $n = 2$, four phase changes are produced, and this is called quadrature phase shift keying (QPSK).

BPSK (Fig. 2.24), which will be considered here, is a two-phase modulation method in which a carrier is transmitted to indicate a mark (1) or the phase is reversed (shifted through 180°) to indicate a space (0). Note that the phase shift does not have to be 180°, but this allows for the maximum separation of the digital states between 1 and 0, which is important when noise is prevalent.

The block diagram shown in Fig. 2.25 indicates the stages necessary to produce the modulator section. A balanced modulator is used with the carrier applied as shown. The digital input passes through a unipolar–bipolar convertor to ensure that the digital signal passed to the balanced modulator is unipolar.

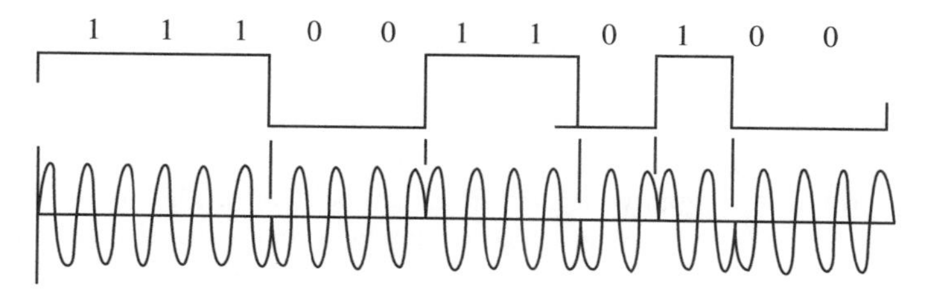

Fig. 2.24

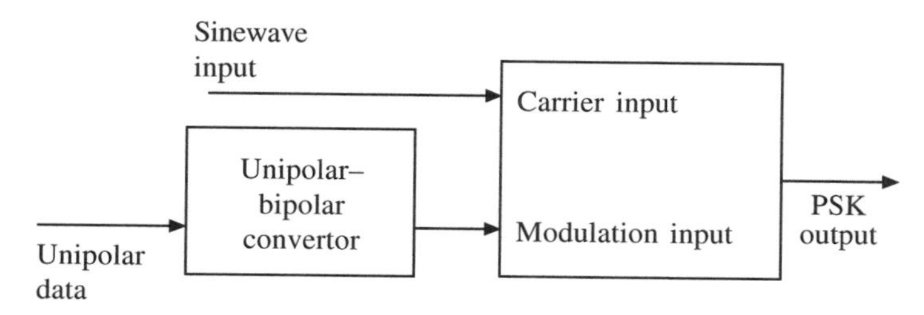

Fig. 2.25

It can be seen from Fig. 2.24 that when the modulation input is positive, the modulator multiplies the carrier input by this constant positive level so that the output signal is simply the carrier sine wave. Note this is in phase with the carrier input. When the digital input data is negative, the modulator multiplies the carrier input by this constant negative level. This causes an output sine wave which is 180° out of phase with the carrier input. The result is that the sine wave at the output is inverted in phase every time the data input changes and produces a transition from 1 to 0 or 0 to 1. The consequence of this action is that the sine wave is inverted each time the modulation input undergoes a transition.

In order to demodulate a BPSK waveform, the demodulator must have an internal signal whose frequency is exactly equal to the incoming carrier. The PLL on its own is unsuitable in this case because of the sudden phase reversals which cannot produce a discrete carrier component to lock on to. One circuit which overcomes this is shown in Fig. 2.26.

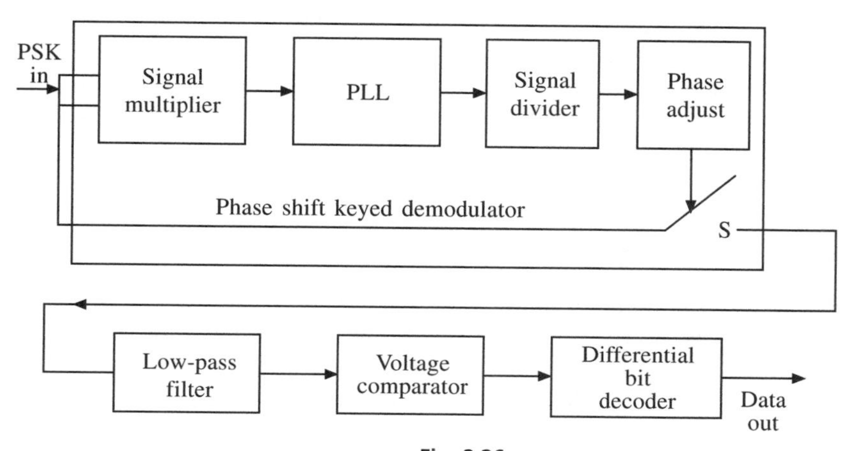

Fig. 2.26

The input signal is applied to a signal multiplier which is a square-law device. In this case a balanced demodulator is used, with its inputs tied together. The output from this stage is a signal at twice the original frequency and having phase changes of 0° and 360°. Hence the signal multiplier has removed any phase changes from the original BPSK signal. It now provides a signal which the PLL can lock on to. The output from the PLL is then passed to a divide-by-two network which produces the original BPSK signal. The phase of this signal is then adjusted to the phase of the original BPSK signal. Finally, this output is used to activate an FET switch. When the phase adjust output is high (1) the switch is closed, and the initial BPSK signal is switched through to the demodulator's output. If the phase adjust output is low (0) then the switch is open and the demodulator's output drops to ground potential.

The output is then passed to a low-pass filter to remove unwanted signal components. This is followed by a comparator which squares the output and produces clean positive and negative half-cycles.

One final stage is necessary in order to produce the original data. At the output of the comparator the receiver has to look for level changes, and this has to be done by a differential decoder block which gives an output (1) when a level change is sensed and no output (0) when no level change takes place. Hence the original data stream is reproduced.

Quadrature phase shift keying

This type of modulation method has wide application in high-speed data transmission systems. It has two distinct advantages: it produces twice as much data with the same number of phase changes as BPSK, and this also means that the bandwidth is virtually decreased for the same amount of data being transmitted. In order to understand this, it is informative to look at single sideband modulation, which was mentioned in Section 2.2.

In quadrature phase shift keying each pair of consecutive data bits in a data stream is considered a two-bit code called a dibit. This is used to switch the carrier at the transmitter between one of four phases, instead of two as was the case with BPSK. The phases selected are 45°, 135°, 225° and 315°, lagging relative to the phase of the original unmodulated carrier. The system is shown in Fig. 2.27. This diagram is clarified by looking at the typical QPSK transmitter block diagram shown in Fig. 2.28.

The two carrier signals shown in Fig. 2.28 have the same frequency but differ in phase by 90°. The 0° phase carrier is called the in-phase (I) carrier, while the 90° phase carrier is called the quadrature (Q) carrier. The output from the first modulator is a BPSK signal which has phases of 0° and 90° relative to the I carrier while the output of the second modulator is a BPSK signal with phases of 90° and 270° relative to the I carrier. These two signals are then applied to the summing amplifier, but note that there is always a ±90° phase difference between the two modulator outputs.

The phase of the summing amplifier's output, relative to the I carrier, can take one of four phase values as shown in Fig. 2.26, but this will depend on the dibit code applied to the balanced modulator inputs. When the dibit changes, the phase of the QPSK output changes by 0°, 90°, 180° or 270° from its previous phase position.

It is necessary to include a differentially encoded dibit (DED) sequence at the transmitter in order to avoid phase ambiguity at the receiver. In order to achieve this, two blocks are

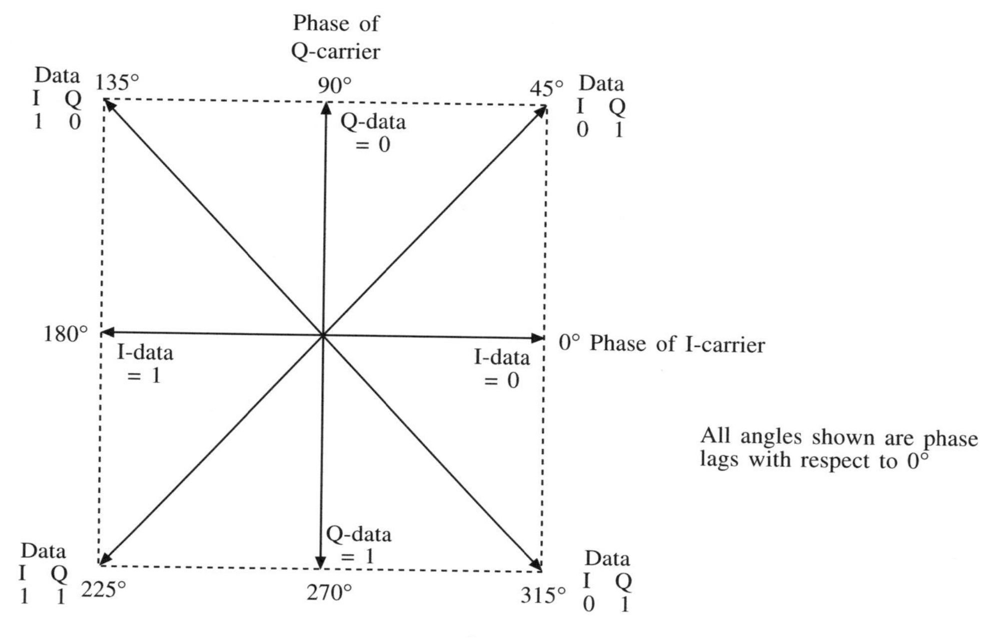

Fig. 2.27

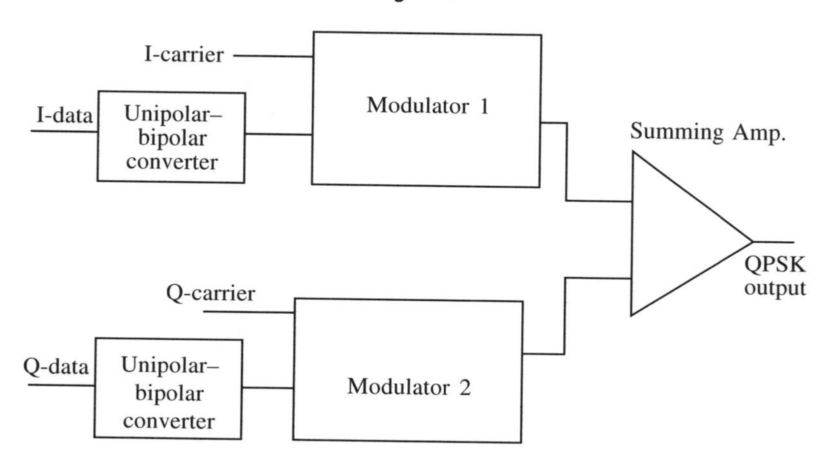

Fig. 2.28

connected at the input of the unipolar–bipolar converters of Fig. 2.28. These blocks cause each pair of consecutive dibits from the data stream to be encoded as a change in the code at the two outputs of the DED. These outputs are then used to drive the modulator inputs and the original dibits cause the appropriate phase changes.

As with the BPSK receiver, the circuitry is fairly complicated but integrated circuitry enables the block diagram of Fig. 2.29 to be drawn.

The QPSK signal goes to the first stage of the QPSK demodulator, which is a signal squarer. This multiplies the incoming signal by itself causing phase changes of 0° and 180°. (The four original phase changes have been doubled.) The output from this block is then passed on to a second signal squarer and the output from this stage only incorporates

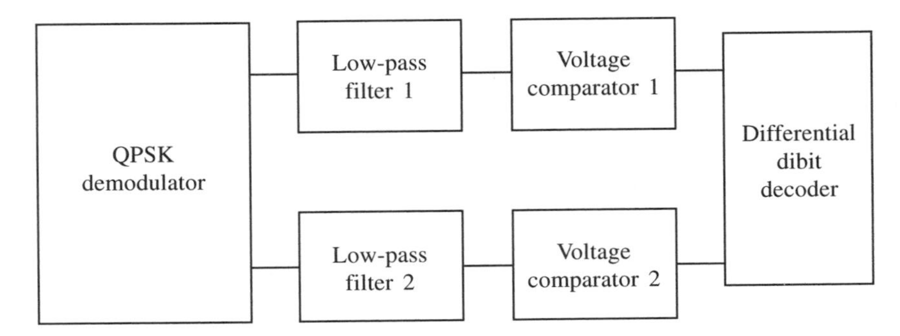

Fig. 2.29

a single phase change of 0°, but the frequency is now four times the original. The next stage is a PLL, which locks on to the incoming signal and outputs a clean square wave. This output is then passed to a divide-by-four circuit which outputs the original frequency and passes it on to a phase-changing circuit which generates two square waves at the same frequency but separated by a 90° phase difference. Finally, these outputs are used to operate two FET switches so that when the output is high one switch is closed and the original QPSK signal passes through to the demodulator. If the output is low, one of the FET switches will open and the demodulator input drops to zero.

Figure 2.27 shows how the outputs from the demodulator are arranged with reference to the I and Q signals. Note that the information concerning the original dibit code is incorporated in the average levels of the demodulator's I and Q outputs, hence low-pass filtering is used next to extract the average levels. This is then passed on to two voltage comparators to produce clean square waves. Finally, the change in the dibit code is determined by the differential dibit decoder, which encodes this change as the original dibit pair.

2.8 Further problems

1. An AM transmitter has a 75 V carrier when operating at 1.245 MHz. This is modulated to a depth of 40% by a 5 kHz tone and applied to a 57 Ω antenna system. Determine:
   (a) the sidebands;
   (b) the voltage in each sideband.
   Answer: 1.25 MHz, 1.24 MHz, 15 V

2. A carrier wave of 1 MHz and amplitude 10 V is amplitude-modulated by a sinusoidal modulating signal. If the lower sideband is 999 kHz and its voltage is 20 dB below the carrier amplitude, calculate the amplitude and frequency of the modulating signal.
   Answer: 2 V, 1 kHz

3. An AM waveform has a carrier frequency of 810 kHz and an upper sideband that extends from 812.5 kHz to 829 kHz. Determine the frequency range of the lower sideband.
   Answer: 791–807.5 kHz

4. The power dissipated by an AM wave is 70 W when its depth of modulation is 70%. Determine the modulation depth if the power has to be increased to 80 W.
   Answer: 0.9

5. A receiver is receiving a signal amplitude modulated to a depth of 80%. If the depth of modulation is reduced to 30%, what will be the change in the receiver output power in decibels?
   Answer: 1.0145 dB

6. (a) An AM transmitter generates 350 W into a 70 Ω antenna. Determine:
   (i) the carrier power component if the carrier voltage is 200 V;
   (ii) the depth of modulation;
   (iii) the sideband power.
   (b) If a filter is incorporated after the balanced modulator so that the carrier frequency component is reduced by 20 dB, determine the total power presented to the antenna.
   Answer: 285 W, 67.5%, 95.77 W, 326.7 W

7. The rated system deviation of an FM system is 50 kHz for a modulation amplitude of 10 V. Use the information given in Fig. A.1 to determine the frequency spectrum when the carrier is modulated in turn by audio signals of
   (i) 3 V at 15 kHz
   (ii) 4 V at 8 kHz.
   Assume the carrier frequency is 80 MHz and determine which signal requires the larger bandwidth.

8. Determine the frequency deviation produced by a modulation signal of 3 V peak and frequency 2.5 kHz given that a signal of 1.5 V peak and frequency 1 kHz produces a frequency deviation of 10 kHz.

9. A 50 MHz carrier wave has its frequency modulated by a 10 kHz sine wave which produces a frequency deviation of 4 kHz. Calculate the deviation ratio and show to a first approximation that the frequency spectrum of the FM wave consists of a carrier and two pairs of side frequencies.
   Answer: 0.04

10. Given that an unmodulated wave has an amplitude of 100 mV, calculate the amplitude of the components of the resultant FM wave in Question 9. Estimate the bandwidth needed to transmit this wave.
    Answer: 96 mV, 20 mV, 2 mV, 28 kHz

11. A 1 MHz carrier is amplitude modulated by a 10 kHz sine wave. What frequencies are contained in the modulated waveforms if the system is
    (i) double sideband modulated;
    (ii) double sideband suppressed carrier modulated;
    (iii) single sideband suppressed carrier modulated?

12. An FM transmitter, operating at rated system deviation, is shown in Fig. 2.30. If a test tone of 3 kHz is applied to the reactance modulator and the crystal oscillator has a deviation of 1 kHz at the test tone frequency, determine:
    (i) the number of side frequencies generated;
    (ii) the bandwidth.

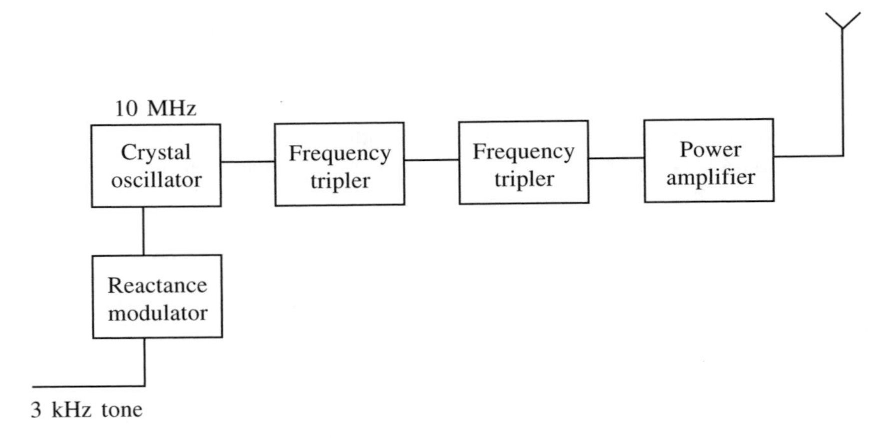

Fig. 2.30

Draw the spectrum diagram if the carrier level is 15 V.
Answer: 6, 24 kHz

13. An FSK receiver uses a phase-locked loop as part of its demodulation circuitry, to receive digital data at the rate of 1200 bps. If the mark frequency is 900 Hz and the space frequency is 2.1 kHz, determine:
    (i) the free running frequency of the PLL;
    (ii) the bandwidth of the receiver.
    Answer: 1.5 kHz, 3.6 kHz

14. An FSK receiver uses a phase-locked loop as part of the demodulator stage. It receives data at 2400 bps and the two carriers are 600 Hz and 3 kHz. Determine:
    (a) the free running frequencies of the PLL;
    (b) the bandwidth of the receiver.
    Answer: 1.2 kHz, 6 kHz

Data sheets

MOTOROLA

Order this document by MC1496/D

MC1496, B

Balanced Modulators/ Demodulators

These devices were designed for use where the output voltage is a product of an input voltage (signal) and a switching function (carrier). Typical applications include suppressed carrier and amplitude modulation, synchronous detection, FM detection, phase detection, and chopper applications. See Motorola Application Note AN531 for additional design information.

- Excellent Carrier Suppression –65 dB typ @ 0.5 MHz
  –50 dB typ @ 10 MHz
- Adjustable Gain and Signal Handling
- Balanced Inputs and Outputs
- High Common Mode Rejection –85 dB typical

This device contains 8 active transistors.

BALANCED MODULATORS/DEMODULATORS

SEMICONDUCTOR TECHNICAL DATA

D SUFFIX
PLASTIC PACKAGE
CASE 751A
(SO–14)

P SUFFIX
PLASTIC PACKAGE
CASE 646

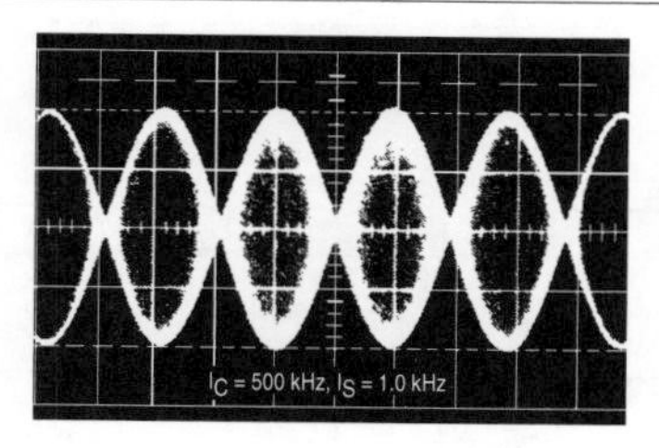

Figure 1. Suppressed Carrier Output Waveform

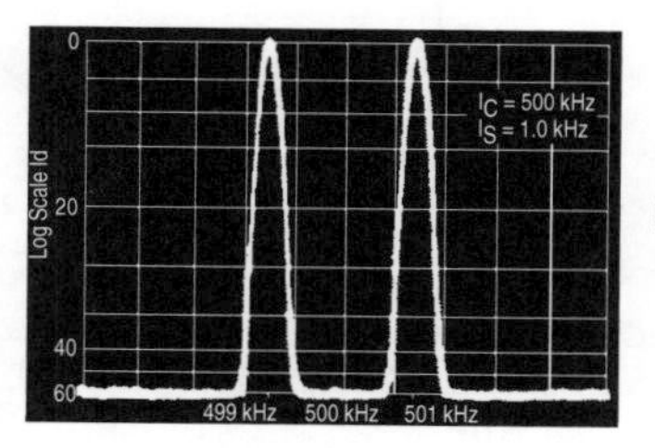

Figure 2. Suppressed Carrier Spectrum

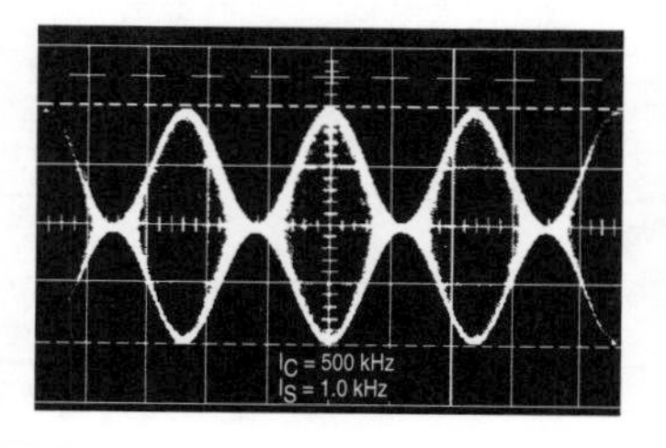

Figure 3. Amplitude Modulation Output Waveform

PIN CONNECTIONS

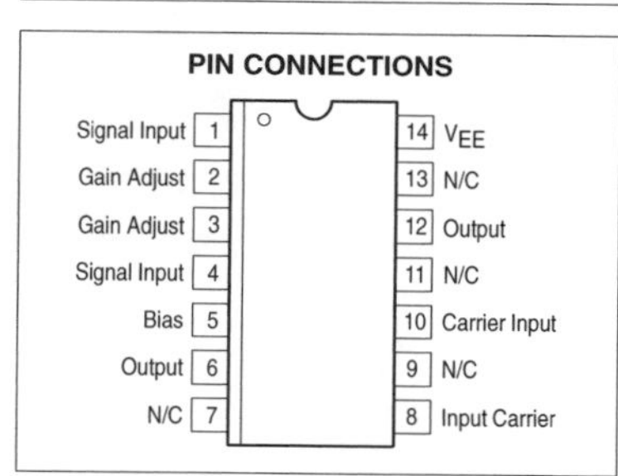

ORDERING INFORMATION

Device	Operating Temperature Range	Package
MC1496D	T_A = 0°C to +70°C	SO–14
MC1496P		Plastic DIP
MC1496BP	T_A = –40°C to +125°C	Plastic DIP

Figure 4. Amplitude–Modulation Spectrum

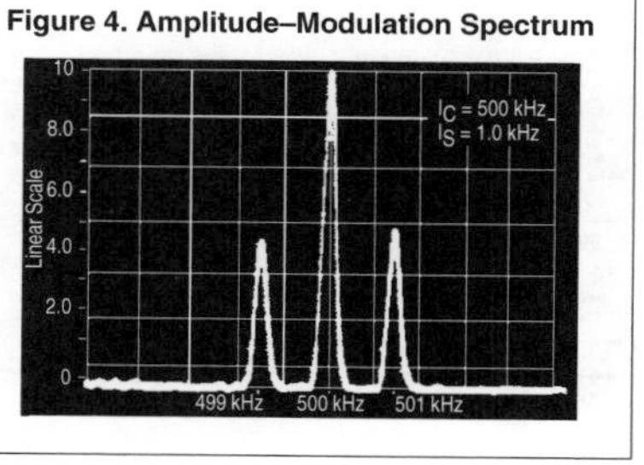

MC1496, B

MAXIMUM RATINGS (T_A = 25°C, unless otherwise noted.)

Rating	Symbol	Value	Unit
Applied Voltage (V6 – V8, V10 – V1, V12 – V8, V12 – V10, V8 – V4, V8 – V1, V10 – V4, V6 – V10, V2 – V5, V3 – V5)	ΔV	30	Vdc
Differential Input Signal	V8 – V10 V4 – V1	+5.0 $\pm(5 + I5R_e)$	Vdc
Maximum Bias Current	I_5	10	mA
Thermal Resistance, Junction–to–Air Plastic Dual In–Line Package	$R_{\theta JA}$	100	°C/W
Operating Temperature Range	T_A	0 to +70	°C
Storage Temperature Range	T_{stg}	–65 to +150	°C

NOTE: ESD data available upon request.

ELECTRICAL CHARACTERISTICS (V_{CC} = 12 Vdc, V_{EE} = –8.0 Vdc, I5 = 1.0 mAdc, R_L = 3.9 kΩ, R_e = 1.0 kΩ, T_A = T_{low} to T_{high}, all input and output characteristics are single–ended, unless otherwise noted.)

Characteristic	Fig.	Note	Symbol	Min	Typ	Max	Unit
Carrier Feedthrough	5	1	V_{CFT}				μVrms
V_C = 60 mVrms sine wave and offset adjusted to zero, f_C = 1.0 kHz				–	40	–	
f_C = 10 MHz				–	140	–	
V_C = 300 mVpp square wave: offset adjusted to zero, f_C = 1.0 kHz				–	0.04	0.4	mVrms
offset not adjusted, f_C = 1.0 kHz				–	20	200	
Carrier Suppression, f_S = 10 kHz, 300 mVrms	5	2	V_{CS}				dB
f_C = 500 kHz, 60 mVrms sine wave				40	65	–	
f_C = 10 MHz, 60 mVrms sine wave				–	50	–	k
Transadmittance Bandwidth (Magnitude) (R_L = 50 Ω)	8	8	BW_{3dB}				MHz
Carrier Input Port, V_C = 60 mVrms sine wave, f_S = 1.0 kHz, 300 mVrms sine wave				–	300	–	
Signal Input Port, V_S = 300 mVrms sine wave, $\|V_C\|$ = 0.5 Vdc				–	80	–	
Signal Gain (V_S = 100 mVrms, f = 1.0 kHz; $\|V_C\|$= 0.5 Vdc)	10	3	A_{VS}	2.5	3.5	–	V/V
Single–Ended Input Impedance, Signal Port, f = 5.0 MHz	6	–					
Parallel Input Resistance			r_{ip}	–	200	–	kΩ
Parallel Input Capacitance			c_{ip}	–	2.0	–	pF
Single–Ended Output Impedance, f = 10 MHz	6	–					
Parallel Output Resistance			r_{op}	–	40	–	kΩ
Parallel Output Capacitance			c_{oo}	–	5.0	–	pF
Input Bias Current $I_{bS} = \frac{I1 + I4}{2}$; $I_{bC} = \frac{I8 + I10}{2}$	7	–					μA
			I_{bS}	–	12	30	
			I_{bC}	–	12	30	
Input Offset Current I_{ioS} = I1–I4; I_{ioC} = I8–I10	7	–					μA
			$\|I_{ioS}\|$	–	0.7	7.0	
			$\|I_{ioC}\|$	–	0.7	7.0	
Average Temperature Coefficient of Input Offset Current (T_A = –55°C to +125°C)	7	–	$\|TC_{Iio}\|$	–	2.0	–	nA/°C
Output Offset Current (I6–I9)	7	–	$\|I_{oo}\|$	–	14	80	μA
Average Temperature Coefficient of Output Offset Current (T_A = –55°C to +125°C)	7	–	$\|TC_{Ioo}\|$	–	90	–	nA/°C
Common–Mode Input Swing, Signal Port, f_S = 1.0 kHz	9	4	CMV	–	5.0	–	Vpp
Common–Mode Gain, Signal Port, f_S = 1.0 kHz, $\|V_C\|$= 0.5 Vdc	9	–	ACM	–	–85	–	dB
Common–Mode Quiescent Output Voltage (Pin 6 or Pin 9)	10	–	V_{out}	–	8.0	–	Vpp
Differential Output Voltage Swing Capability	10	–	V_{out}	–	8.0	–	Vpp
Power Supply Current I6 +I12	7	6	I_{CC}	–	2.0	4.0	mAdc
I14			I_{EE}	–	3.0	5.0	
DC Power Dissipation	7	5	P_D	–	33	–	mW

MC1496, B

GENERAL OPERATING INFORMATION

Carrier Feedthrough

Carrier feedthrough is defined as the output voltage at carrier frequency with only the carrier applied (signal voltage = 0).

Carrier null is achieved by balancing the currents in the differential amplifier by means of a bias trim potentiometer (R1 of Figure 5).

Carrier Suppression

Carrier suppression is defined as the ratio of each sideband output to carrier output for the carrier and signal voltage levels specified.

Carrier suppression is very dependent on carrier input level, as shown in Figure 22. A low value of the carrier does not fully switch the upper switching devices, and results in lower signal gain, hence lower carrier suppression. A higher than optimum carrier level results in unnecessary device and circuit carrier feedthrough, which again degenerates the suppression figure. The MC1496 has been characterized with a 60 mVrms sinewave carrier input signal. This level provides optimum carrier suppression at carrier frequencies in the vicinity of 500 kHz, and is generally recommended for balanced modulator applications.

Carrier feedthrough is independent of signal level, V_S. Thus carrier suppression can be maximized by operating with large signal levels. However, a linear operating mode must be maintained in the signal–input transistor pair – or harmonics of the modulating signal will be generated and appear in the device output as spurious sidebands of the suppressed carrier. This requirement places an upper limit on input–signal amplitude (see Figure 20). Note also that an optimum carrier level is recommended in Figure 22 for good carrier suppression and minimum spurious sideband generation.

At higher frequencies circuit layout is very important in order to minimize carrier feedthrough. Shielding may be necessary in order to prevent capacitive coupling between the carrier input leads and the output leads.

Signal Gain and Maximum Input Level

Signal gain (single–ended) at low frequencies is defined as the voltage gain,

$$A_{VS} = \frac{V_o}{V_S} = \frac{R_L}{R_e + 2r_e} \quad \text{where } r_e = \frac{26\ \text{mV}}{I5(\text{mA})}$$

A constant dc potential is applied to the carrier input terminals to fully switch two of the upper transistors "on" and two transistors "off" ($V_C = 0.5$ Vdc). This in effect forms a cascode differential amplifier.

Linear operation requires that the signal input be below a critical value determined by R_E and the bias current I5.

$$V_S \leq I5\ R_E \text{ (Volts peak)}$$

Note that in the test circuit of Figure 10, V_S corresponds to a maximum value of 1.0 V peak.

Common Mode Swing

The common–mode swing is the voltage which may be applied to both bases of the signal differential amplifier, without saturating the current sources or without saturating the differential amplifier itself by swinging it into the upper switching devices. This swing is variable depending on the particular circuit and biasing conditions chosen.

Power Dissipation

Power dissipation, P_D, within the integrated circuit package should be calculated as the summation of the voltage–current products at each port, i.e. assuming V12 = V6, I5 = I6 = I12 and ignoring base current, P_D = 2 I5 (V6 – V14) + I5) V5 – V14 where subscripts refer to pin numbers.

Design Equations

The following is a partial list of design equations needed to operate the circuit with other supply voltages and input conditions.

A. Operating Current

The internal bias currents are set by the conditions at Pin 5. Assume:

$$I5 = I6 = I12,$$
$$I_B << I_C \text{ for all transistors}$$

then :

$$R5 = \frac{V - -\phi}{I5} - 500\ \Omega$$

where: R5 is the resistor between Pin 5 and ground
$\phi = 0.75$ at $T_A = +25°C$

The MC1496 has been characterized for the condition $I_5 = 1.0$ mA and is the generally recommended value.

B. Common–Mode Quiescent Output Voltage

$$V6 = V12 = V+ - I5\ R_L$$

Biasing

The MC1496 requires three dc bias voltage levels which must be set externally. Guidelines for setting up these three levels include maintaining at least 2.0 V collector–base bias on all transistors while not exceeding the voltages given in the absolute maximum rating table;

$$30\ \text{Vdc} \geq [(V6, V12) - (V8, V10)] \geq 2\ \text{Vdc}$$
$$30\ \text{Vdc} \geq [(V8, V10) - (V1, V4)] \geq 2.7\ \text{Vdc}$$
$$30\ \text{Vdc} \geq [(V1, V4) - (V5)] \geq 2.7\ \text{Vdc}$$

The foregoing conditions are based on the following approximations:

$$V6 = V12,\ V8 = V10,\ V1 = V4$$

Bias currents flowing into Pins 1, 4, 8 and 10 are transistor base currents and can normally be neglected if external bias dividers are designed to carry 1.0 mA or more.

Transadmittance Bandwidth

Carrier transadmittance bandwidth is the 3.0 dB bandwidth of the device forward transadmittance as defined by:

$$\gamma 21C = \frac{i_o \text{ (each sideband)}}{v_s \text{ (signal)}} \Bigg|_{V_o = 0}$$

Signal transadmittance bandwidth is the 3.0 dB bandwidth of the device forward transadmittance as defined by:

$$\gamma 21S = \frac{i_o \text{ (signal)}}{v_s \text{ (signal)}} \Bigg|_{V_c = 0.5\ \text{Vdc},\ V_o = 0}$$

MC1496, B

Coupling and Bypass Capacitors

Capacitors C1 and C2 (Figure 5) should be selected for a reactance of less than 5.0 Ω at the carrier frequency.

Output Signal

The output signal is taken from Pins 6 and 12 either balanced or single–ended. Figure 11 shows the output levels of each of the two output sidebands resulting from variations in both the carrier and modulating signal inputs with a single–ended output connection.

Negative Supply

V_{EE} should be dc only. The insertion of an RF choke in series with V_{EE} can enhance the stability of the internal current sources.

Signal Port Stability

Under certain values of driving source impedance, oscillation may occur. In this event, an RC suppression network should be connected directly to each input using short leads. This will reduce the Q of the source–tuned circuits that cause the oscillation.

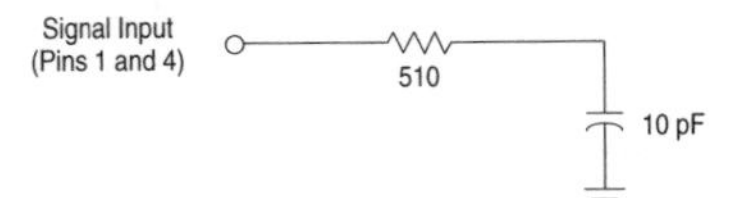

An alternate method for low–frequency applications is to insert a 1.0 kΩ resistor in series with the input (Pins 1, 4). In this case input current drift may cause serious degradation of carrier suppression.

TEST CIRCUITS

Figure 5. Carrier Rejection and Suppression

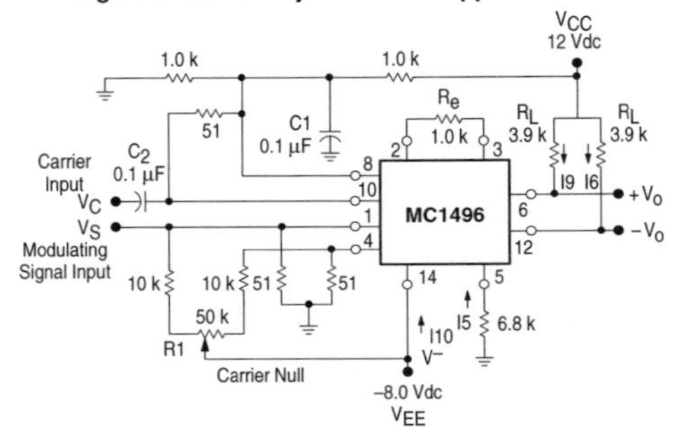

Figure 6. Input–Output Impedance

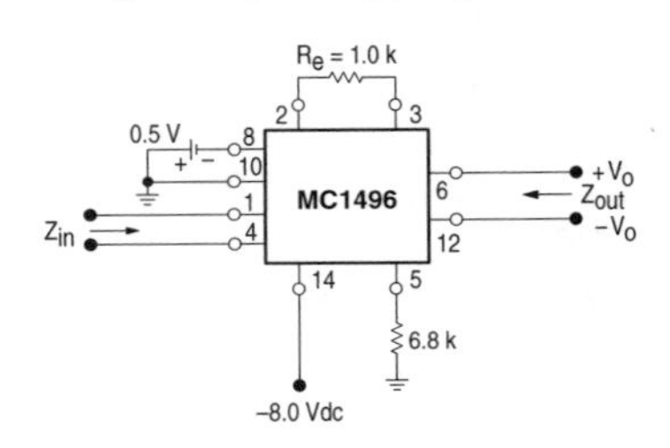

NOTE: Shielding of input and output leads may be needed to properly perform these tests.

Figure 7. Bias and Offset Currents

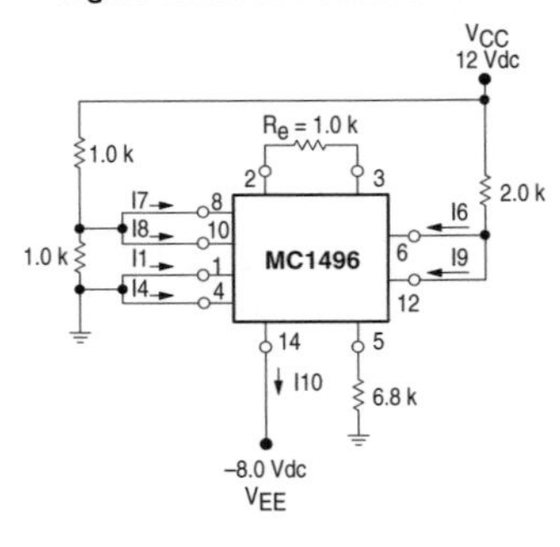

Figure 8. Transconductance Bandwidth

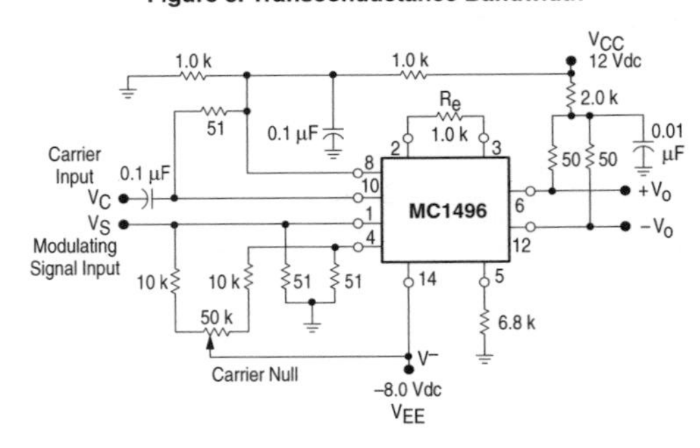

MC1496, B

Figure 9. Common Mode Gain

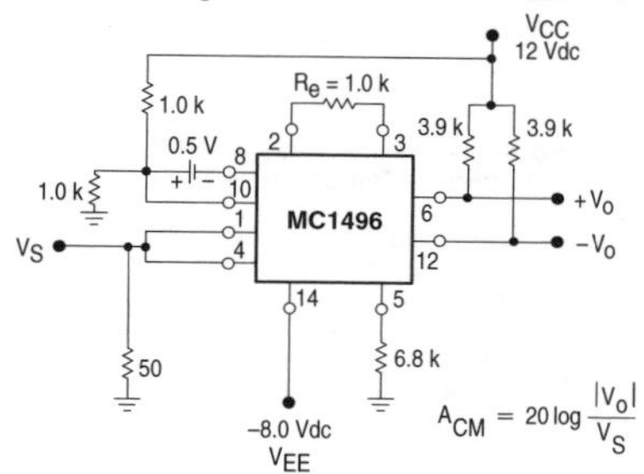

Figure 10. Signal Gain and Output Swing

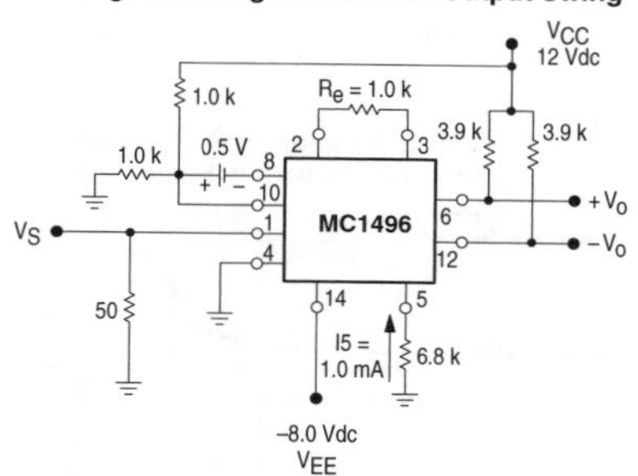

TYPICAL CHARACTERISTICS

Typical characteristics were obtained with circuit shown in Figure 5, f_C = 500 kHz (sine wave), V_C = 60 mVrms, f_S = 1.0 kHz, V_S = 300 mVrms, T_A = 25°C, unless otherwise noted.

Figure 11. Sideband Output versus Carrier Levels

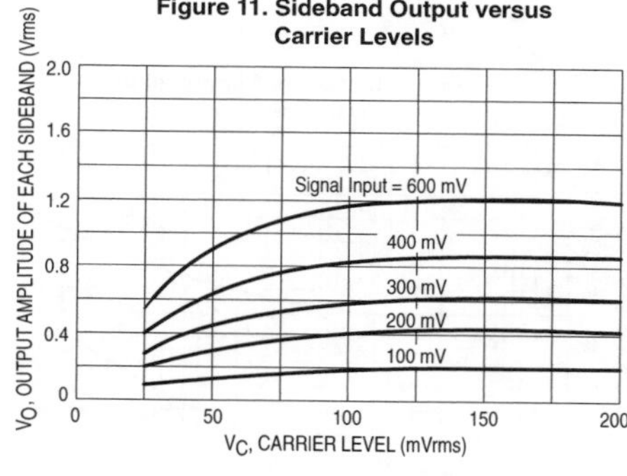

Figure 12. Signal–Port Parallel–Equivalent Input Resistance versus Frequency

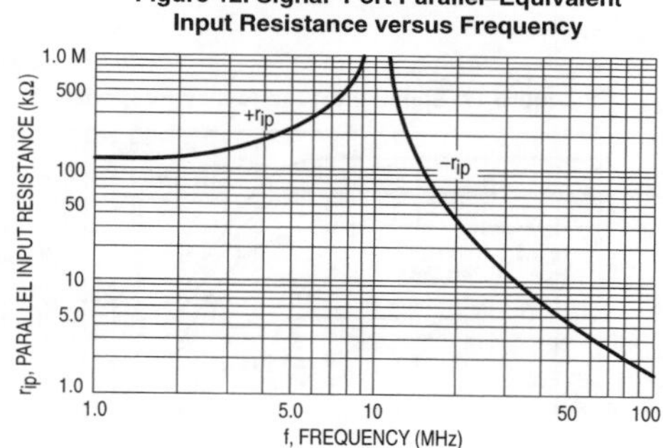

Figure 13. Signal–Port Parallel–Equivalent Input Capacitance versus Frequency

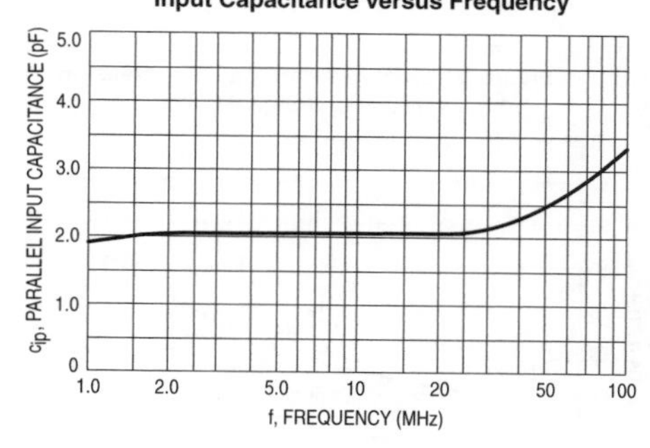

Figure 14. Single–Ended Output Impedance versus Frequency

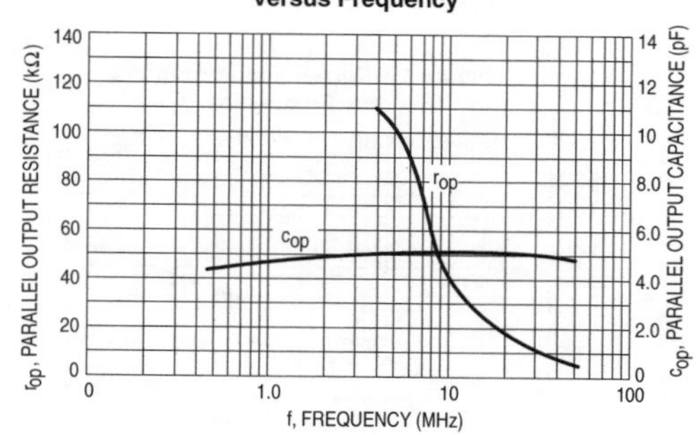

MC1496, B

TYPICAL CHARACTERISTICS (continued)

Typical characteristics were obtained with circuit shown in Figure 5, f_C = 500 kHz (sine wave), V_C = 60 mVrms, f_S = 1.0 kHz, V_S = 300 mVrms, T_A = 25°C, unless otherwise noted.

Figure 15. Sideband and Signal Port Transadmittances versus Frequency

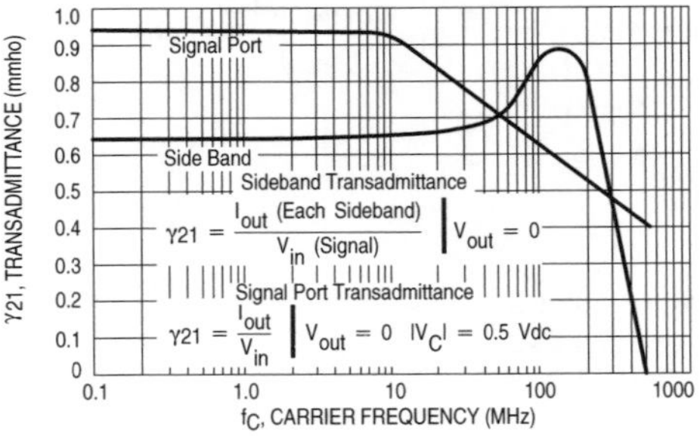

Figure 16. Carrier Suppression versus Temperature

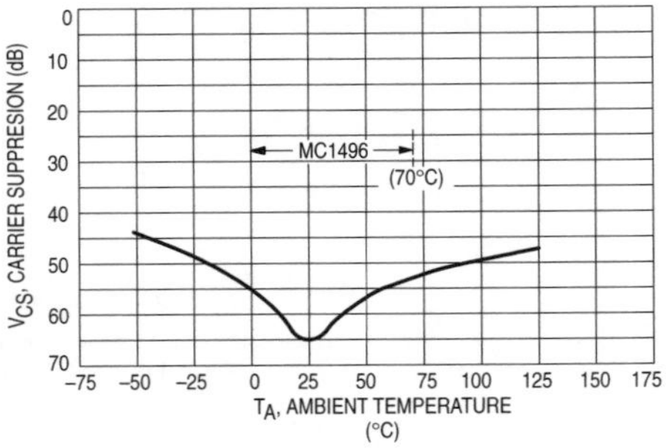

Figure 17. Signal–Port Frequency Response

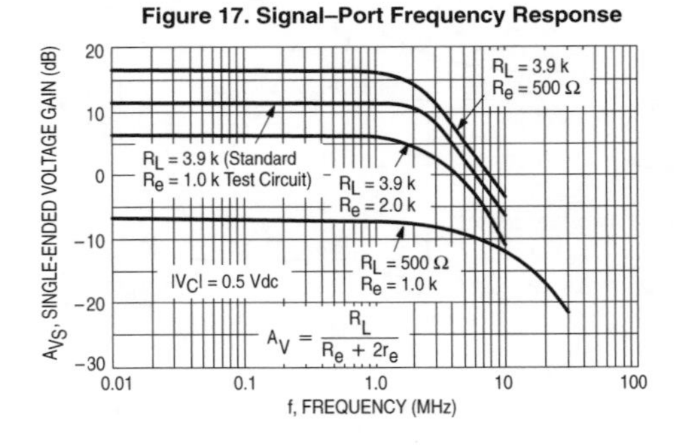

Figure 18. Carrier Suppression versus Frequency

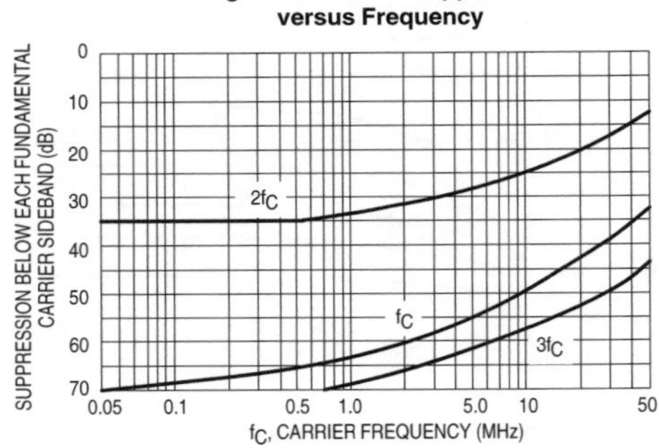

Figure 19. Carrier Feedthrough versus Frequency

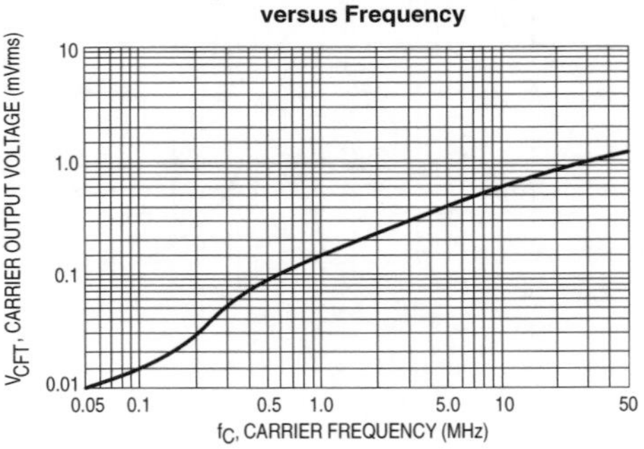

Figure 20. Sideband Harmonic Suppression versus Input Signal Level

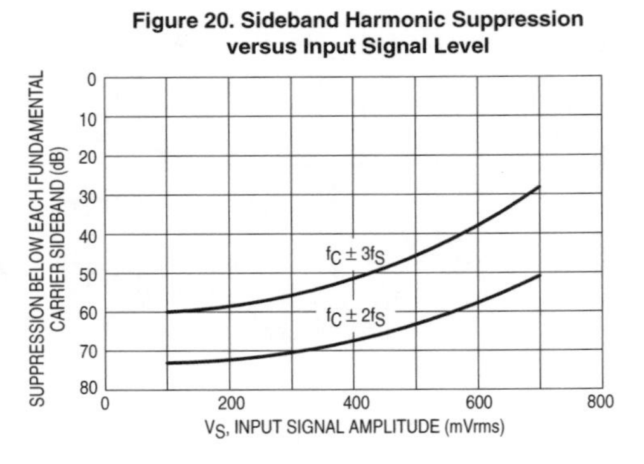

MC1496, B

Figure 21. Suppression of Carrier Harmonic Sidebands versus Carrier Frequency

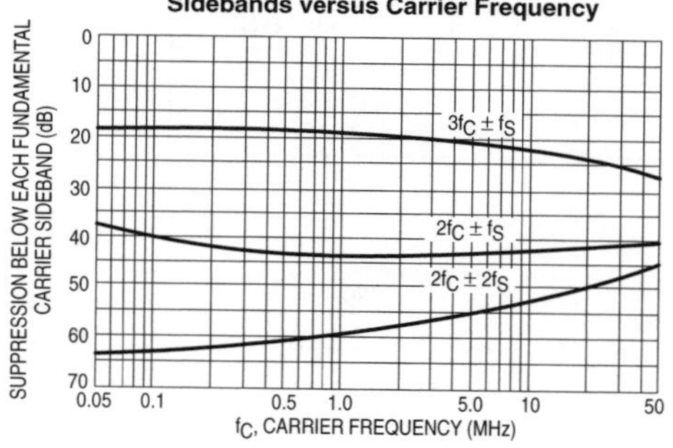

Figure 22. Carrier Suppression versus Carrier Input Level

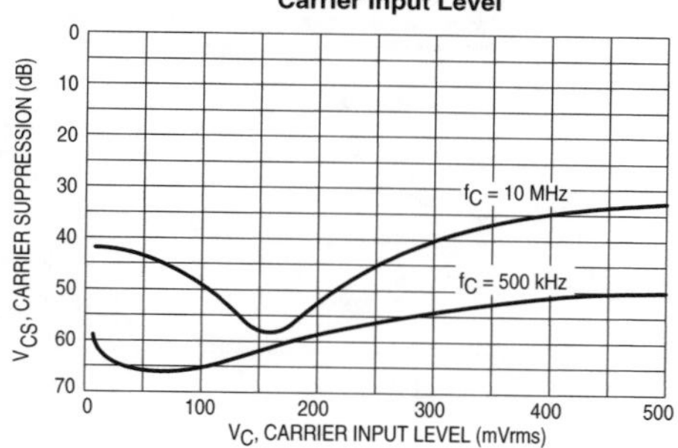

OPERATIONS INFORMATION

The MC1496, a monolithic balanced modulator circuit, is shown in Figure 23.

This circuit consists of an upper quad differential amplifier driven by a standard differential amplifier with dual current sources. The output collectors are cross–coupled so that full–wave balanced multiplication of the two input voltages occurs. That is, the output signal is a constant times the product of the two input signals.

Mathematical analysis of linear ac signal multiplication indicates that the output spectrum will consist of only the sum and difference of the two input frequencies. Thus, the device may be used as a balanced modulator, doubly balanced mixer, product detector, frequency doubler, and other applications requiring these particular output signal characteristics.

The lower differential amplifier has its emitters connected to the package pins so that an external emitter resistance may be used. Also, external load resistors are employed at the device output.

Signal Levels

The upper quad differential amplifier may be operated either in a linear or a saturated mode. The lower differential amplifier is operated in a linear mode for most applications.

For low–level operation at both input ports, the output signal will contain sum and difference frequency components and have an amplitude which is a function of the product of the input signal amplitudes.

For high–level operation at the carrier input port and linear operation at the modulating signal port, the output signal will contain sum and difference frequency components of the modulating signal frequency and the fundamental and odd harmonics of the carrier frequency. The output amplitude will be a constant times the modulating signal amplitude. Any amplitude variations in the carrier signal will not appear in the output.

The linear signal handling capabilities of a differential amplifier are well defined. With no emitter degeneration, the maximum input voltage for linear operation is approximately 25 mV peak. Since the upper differential amplifier has its emitters internally connected, this voltage applies to the carrier input port for all conditions.

Since the lower differential amplifier has provisions for an external emitter resistance, its linear signal handling range may be adjusted by the user. The maximum input voltage for linear operation may be approximated from the following expression:

$$V = (I5)(R_E) \text{ volts peak.}$$

This expression may be used to compute the minimum value of R_E for a given input voltage amplitude.

Figure 23. Circuit Schematic

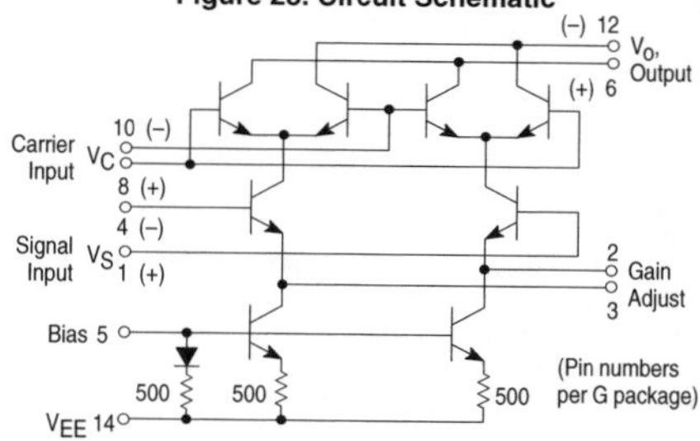

Figure 24. Typical Modulator Circuit

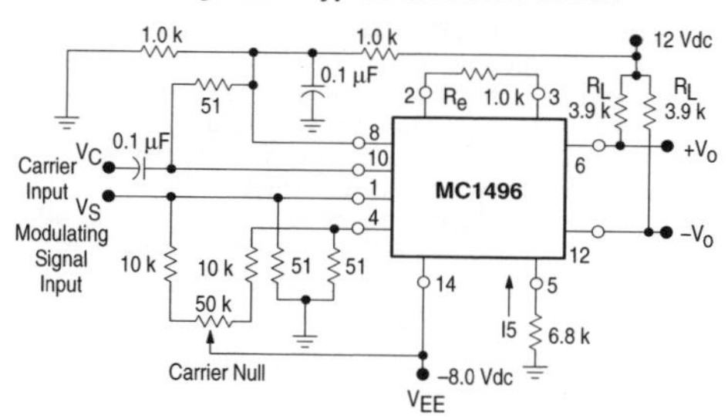

MC1496, B

Figure 25. Voltage Gain and Output Frequencies

Carrier Input Signal (V_C)	Approximate Voltage Gain	Output Signal Frequency(s)
Low–level dc	$\frac{R_L V_C}{2(R_E + 2r_e)\left(\frac{KT}{q}\right)}$	f_M
High–level dc	$\frac{R_L}{R_E + 2r_e}$	f_M
Low–level ac	$\frac{R_L V_C(rms)}{2\sqrt{2}\left(\frac{KT}{q}\right)(R_E + 2r_e)}$	$f_C \pm f_M$
High–level ac	$\frac{0.637 R_L}{R_E + 2r_e}$	$f_C \pm f_M, 3f_C \pm f_M, 5f_C \pm f_M, \ldots$

NOTES: 1. Low–level Modulating Signal, V_M, assumed in all cases. V_C is Carrier Input Voltage.
2. When the output signal contains multiple frequencies, the gain expression given is for the output amplitude of each of the two desired outputs, $f_C + f_M$ and $f_C - f_M$.
3. All gain expressions are for a single–ended output. For a differential output connection, multiply each expression by two.
4. R_L = Load resistance.
5. R_E = Emitter resistance between Pins 2 and 3.
6. r_e = Transistor dynamic emitter resistance, at 25°C;

$$re \approx \frac{26\text{ mV}}{I_5\text{ (mA)}}$$

7. K = Boltzmann's Constant, T = temperature in degrees Kelvin, q = the charge on an electron.

$$\frac{KT}{q} \approx 26\text{ mV at room temperature}$$

The gain from the modulating signal input port to the output is the MC1496 gain parameter which is most often of interest to the designer. This gain has significance only when the lower differential amplifier is operated in a linear mode, but this includes most applications of the device.

As previously mentioned, the upper quad differential amplifier may be operated either in a linear or a saturated mode. Approximate gain expressions have been developed for the MC1496 for a low–level modulating signal input and the following carrier input conditions:

1) Low–level dc
2) High–level dc
3) Low–level ac
4) High–level ac

These gains are summarized in Figure 25, along with the frequency components contained in the output signal.

APPLICATIONS INFORMATION

Double sideband suppressed carrier modulation is the basic application of the MC1496. The suggested circuit for this application is shown on the front page of this data sheet.

In some applications, it may be necessary to operate the MC1496 with a single dc supply voltage instead of dual supplies. Figure 26 shows a balanced modulator designed for operation with a single 12 Vdc supply. Performance of this circuit is similar to that of the dual supply modulator.

AM Modulator

The circuit shown in Figure 27 may be used as an amplitude modulator with a minor modification.

All that is required to shift from suppressed carrier to AM operation is to adjust the carrier null potentiometer for the proper amount of carrier insertion in the output signal.

However, the suppressed carrier null circuitry as shown in Figure 27 does not have sufficient adjustment range. Therefore, the modulator may be modified for AM operation by changing two resistor values in the null circuit as shown in Figure 28.

Product Detector

The MC1496 makes an excellent SSB product detector (see Figure 29).

This product detector has a sensitivity of 3.0 microvolts and a dynamic range of 90 dB when operating at an intermediate frequency of 9.0 MHz.

The detector is broadband for the entire high frequency range. For operation at very low intermediate frequencies down to 50 kHz the 0.1 μF capacitors on Pins 8 and 10 should be increased to 1.0 μF. Also, the output filter at Pin 12 can be tailored to a specific intermediate frequency and audio amplifier input impedance.

As in all applications of the MC1496, the emitter resistance between Pins 2 and 3 may be increased or decreased to adjust circuit gain, sensitivity, and dynamic range.

This circuit may also be used as an AM detector by introducing carrier signal at the carrier input and an AM signal at the SSB input.

The carrier signal may be derived from the intermediate frequency signal or generated locally. The carrier signal may be introduced with or without modulation, provided its level is sufficiently high to saturate the upper quad differential

MC1496, B

amplifier. If the carrier signal is modulated, a 300 mVrms input level is recommended.

Doubly Balanced Mixer

The MC1496 may be used as a doubly balanced mixer with either broadband or tuned narrow band input and output networks.

The local oscillator signal is introduced at the carrier input port with a recommended amplitude of 100 mVrms.

Figure 30 shows a mixer with a broadband input and a tuned output.

Frequency Doubler

The MC1496 will operate as a frequency doubler by introducing the same frequency at both input ports.

Figures 31 and 32 show a broadband frequency doubler and a tuned output very high frequency (VHF) doubler, respectively.

Phase Detection and FM Detection

The MC1496 will function as a phase detector. High–level input signals are introduced at both inputs. When both inputs are at the same frequency the MC1496 will deliver an output which is a function of the phase difference between the two input signals.

An FM detector may be constructed by using the phase detector principle. A tuned circuit is added at one of the inputs to cause the two input signals to vary in phase as a function of frequency. The MC1496 will then provide an output which is a function of the input signal frequency.

TYPICAL APPLICATIONS

Figure 26. Balanced Modulator (12 Vdc Single Supply)

Figure 27. Balanced Modulator–Demodulator

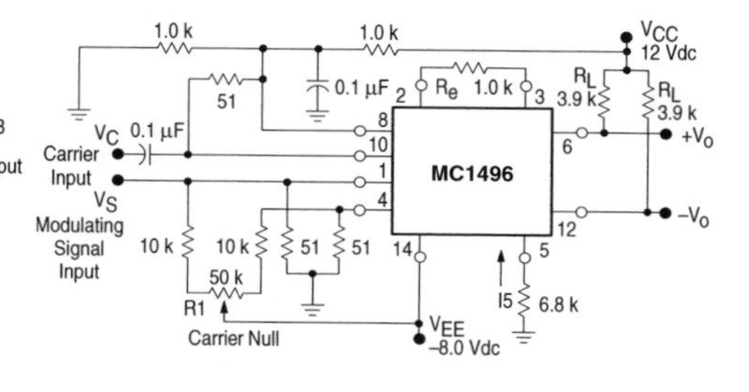

Figure 28. AM Modulator Circuit

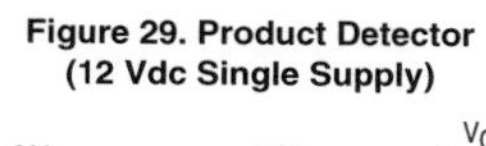

Figure 29. Product Detector (12 Vdc Single Supply)

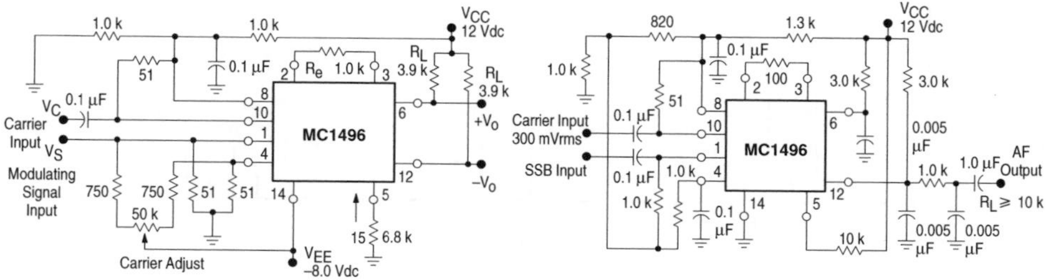

MC1496, B

Figure 30. Doubly Balanced Mixer (Broadband Inputs, 9.0 MHz Tuned Output)

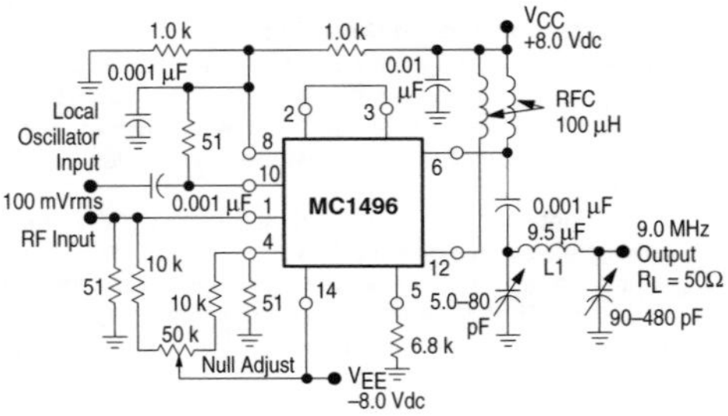

L1 = 44 Turns AWG No. 28 Enameled Wire, Wound on Micrometals Type 44–6 Toroid Core.

Figure 31. Low–Frequency Doubler

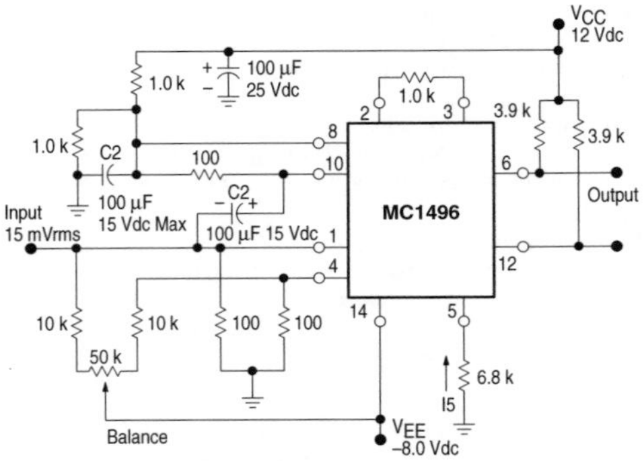

Figure 32. 150 to 300 MHz Doubler

L1 = 1 Turn AWG No. 18 Wire, 7/32″ ID

Balanced Modulator Spectrum

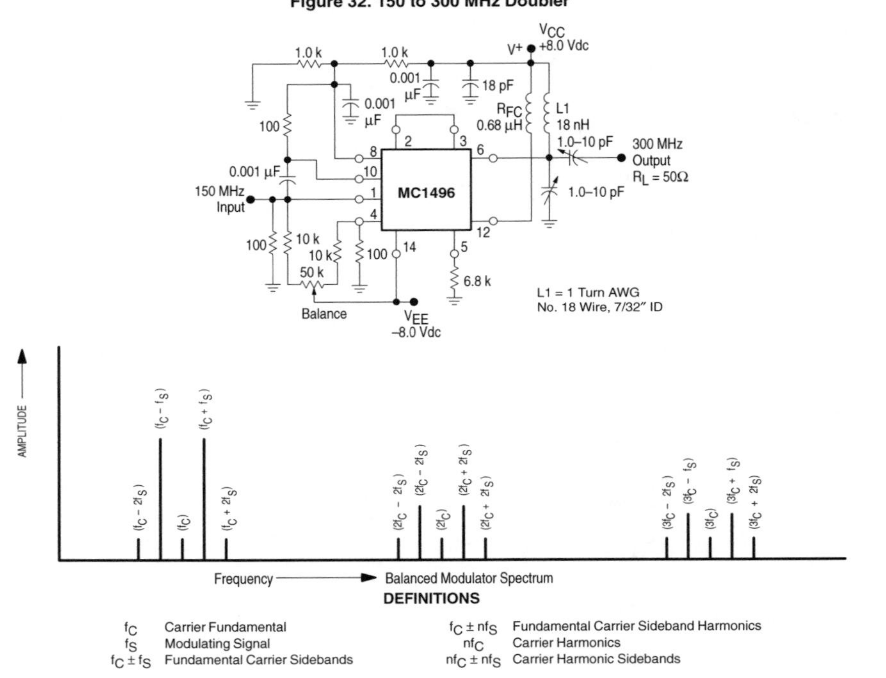

DEFINITIONS

f_C	Carrier Fundamental	$f_C \pm nf_S$	Fundamental Carrier Sideband Harmonics
f_S	Modulating Signal	nf_C	Carrier Harmonics
$f_C \pm f_S$	Fundamental Carrier Sidebands	$nf_C \pm nf_S$	Carrier Harmonic Sidebands

MC1496, B

OUTLINE DIMENSIONS

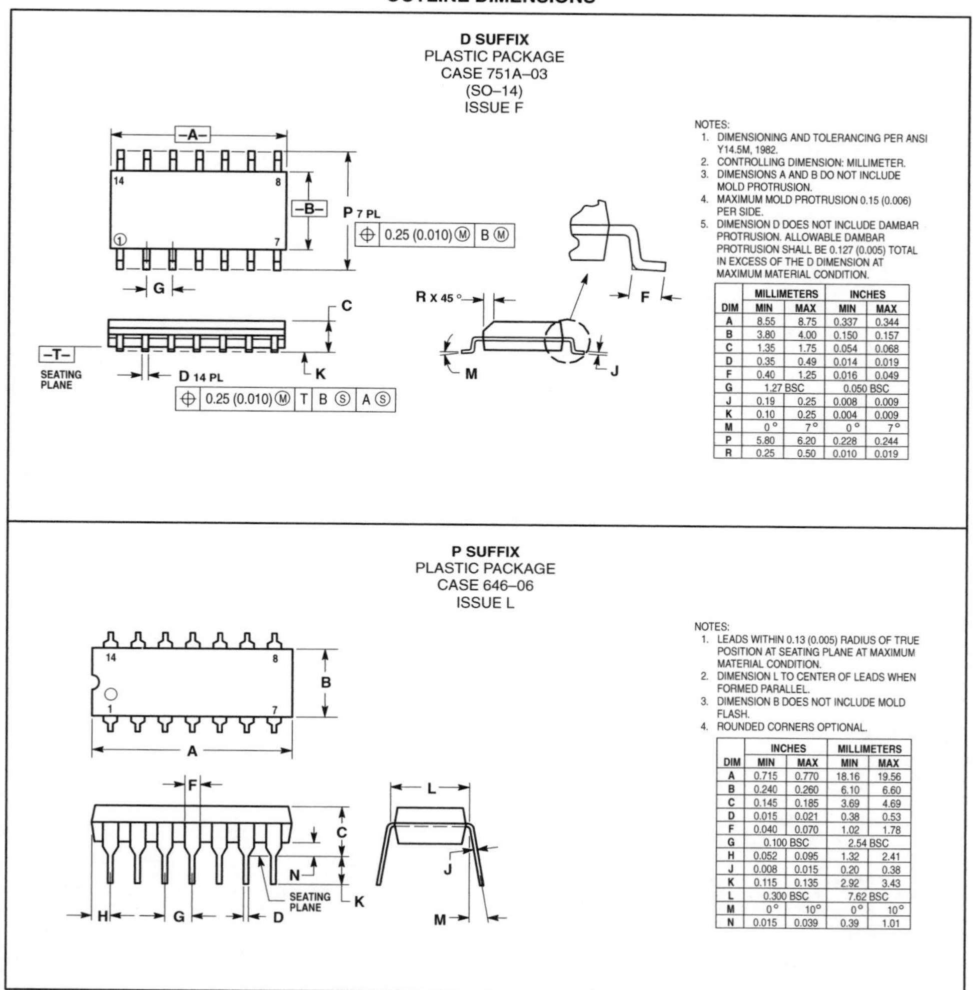

D SUFFIX
PLASTIC PACKAGE
CASE 751A–03
(SO–14)
ISSUE F

NOTES:
1. DIMENSIONING AND TOLERANCING PER ANSI Y14.5M, 1982.
2. CONTROLLING DIMENSION: MILLIMETER.
3. DIMENSIONS A AND B DO NOT INCLUDE MOLD PROTRUSION.
4. MAXIMUM MOLD PROTRUSION 0.15 (0.006) PER SIDE.
5. DIMENSION D DOES NOT INCLUDE DAMBAR PROTRUSION. ALLOWABLE DAMBAR PROTRUSION SHALL BE 0.127 (0.005) TOTAL IN EXCESS OF THE D DIMENSION AT MAXIMUM MATERIAL CONDITION.

	MILLIMETERS		INCHES	
DIM	MIN	MAX	MIN	MAX
A	8.55	8.75	0.337	0.344
B	3.80	4.00	0.150	0.157
C	1.35	1.75	0.054	0.068
D	0.35	0.49	0.014	0.019
F	0.40	1.25	0.016	0.049
G	1.27 BSC		0.050 BSC	
J	0.19	0.25	0.008	0.009
K	0.10	0.25	0.004	0.009
M	0°	7°	0°	7°
P	5.80	6.20	0.228	0.244
R	0.25	0.50	0.010	0.019

P SUFFIX
PLASTIC PACKAGE
CASE 646–06
ISSUE L

NOTES:
1. LEADS WITHIN 0.13 (0.005) RADIUS OF TRUE POSITION AT SEATING PLANE AT MAXIMUM MATERIAL CONDITION.
2. DIMENSION L TO CENTER OF LEADS WHEN FORMED PARALLEL.
3. DIMENSION B DOES NOT INCLUDE MOLD FLASH.
4. ROUNDED CORNERS OPTIONAL.

	INCHES		MILLIMETERS	
DIM	MIN	MAX	MIN	MAX
A	0.715	0.770	18.16	19.56
B	0.240	0.260	6.10	6.60
C	0.145	0.185	3.69	4.69
D	0.015	0.021	0.38	0.53
F	0.040	0.070	1.02	1.78
G	0.100 BSC		2.54 BSC	
H	0.052	0.095	1.32	2.41
J	0.008	0.015	0.20	0.38
K	0.115	0.135	2.92	3.43
L	0.300 BSC		7.62 BSC	
M	0°	10°	0°	10°
N	0.015	0.039	0.39	1.01

MC1496, B

How to reach us:
USA/EUROPE/Locations Not Listed: Motorola Literature Distribution; P.O. Box 20912; Phoenix, Arizona 85036. 1–800–441–2447 or 602–303–5454

MFAX: RMFAX0@email.sps.mot.com – TOUCHTONE 602–244–6609
INTERNET: http://Design–NET.com

JAPAN: Nippon Motorola Ltd.; Tatsumi–SPD–JLDC, 6F Seibu–Butsuryu–Center, 3–14–2 Tatsumi Koto–Ku, Tokyo 135, Japan. 03–81–3521–8315

ASIA/PACIFIC: Motorola Semiconductors H.K. Ltd.; 8B Tai Ping Industrial Park, 51 Ting Kok Road, Tai Po, N.T., Hong Kong. 852–26629298

◊

3

Filter applications

3.1 Introduction

Electronic filters have many applications in the telecommunications and data communications industry. One such application, which involves a multiple channel communications system employing a technique known as time-division multiplexing (TDM), is shown in Fig. 3.1. In this system several channels are transmitted through a medium such as an optical fibre, as shown here, or through a co-axial cable or waveguide. Multiplexing means combining several signals into one, and this is accomplished in TDM by allocating time slots for each channel so that each channel is transmitted at a particular time. If the signals are synchronized correctly there will be no interference between them. At the transmitter end a multiplexer is used to combine the signals, while at the receiver end a demultiplexer is used to separate the original channels.

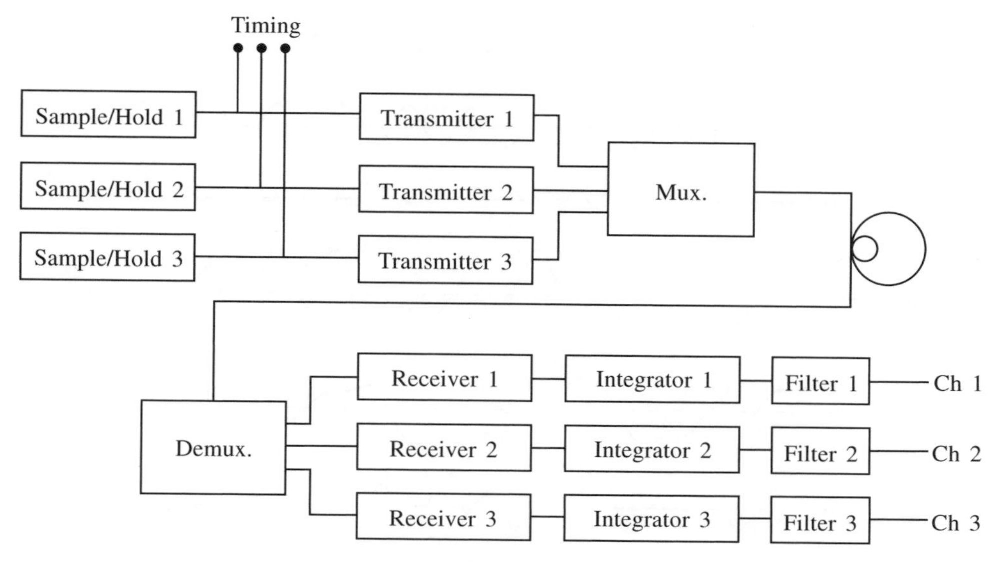

Fig. 3.1

However, when the channel signals arrive at the receivers they have deteriorated in shape and amplitude. In order to clean them up they are reconstructed by an integrator which sums up the incoming signal very much as in mathematical integration. Once this has been done a filter is used to pass the wanted channel frequencies while attenuating the unwanted signals such as noise.

The combined functions of the integrator and filter cause the transmitted channels to be reproduced. In this case, where three channels are involved, each filter will be designed to pass the particular channel frequency and its related information, hence a band of frequencies is passed by each filter.

This is an example of where filters are used to pass bands of frequencies such as the voice band (300–3400 Hz). However, filters can also be used to pass frequencies below a certain frequency while attenuating all frequencies above it. Similarly, it is possible to construct a filter which passes all frequencies above a certain frequency while attenuating all frequencies below it.

Other applications are the following: noise filtering; guard band separation of channels; bandpass selection; boosting and cutting certain bands in the frequency spectrum; and harmonic reduction. Some of these will be investigated later.

Sine waves of different amplitudes and frequencies are shown in Fig. 3.2(a)–(d). It should be appreciated that the majority of filters have to be capable of handling a mixture of such sine waves, as shown in Fig. 3.2(e); the effect of reducing the amplitudes of the signals in Fig. 3.2(d)–(e) is shown in Fig. 3.2(f). Figure 3.2(g) shows what happens when the signal in Fig. 3.2(b) is reduced and that in Fig. 3.2(a) is eliminated. It is therefore possible to use filters to alter amplitudes and frequencies, depending on the requirements of the system.

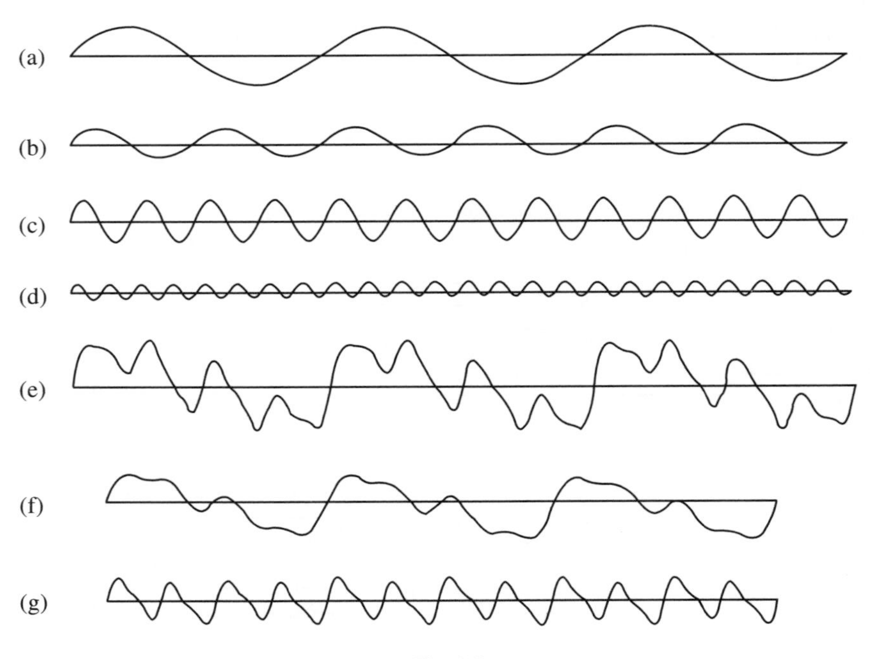

Fig. 3.2

Finally, the filters discussed in this chapter are used in sine or continuous wave circuits. However, certain circuits such as integrators and differentiators utilize passive high-pass and low-pass networks to process square waves and produce wave shaping. When fed through a filter the square wave is modified: the high-frequency edges are rounded when passing through a low-pass filter, while the flat top and bottom are distorted when passing through a high-pass filter.

3.2 Passive filters

The most elementary types of filters are constructed from *RC* networks and are known as passive filters as they dissipate part of the signal power and pass the rest. Figure 3.3(a) shows a passive low-pass filter, while Fig. 3.3(b) shows a passive high-pass filter. These form the basis of more sophisticated filters. Each has a cut-off frequency, which may be derived by considering the high-pass filter as a voltage divider. From Fig. 3.3(b) we have

$$\frac{V_o}{V_i} = \frac{R}{\sqrt{R^2 + X_c^2}} \tag{3.1}$$

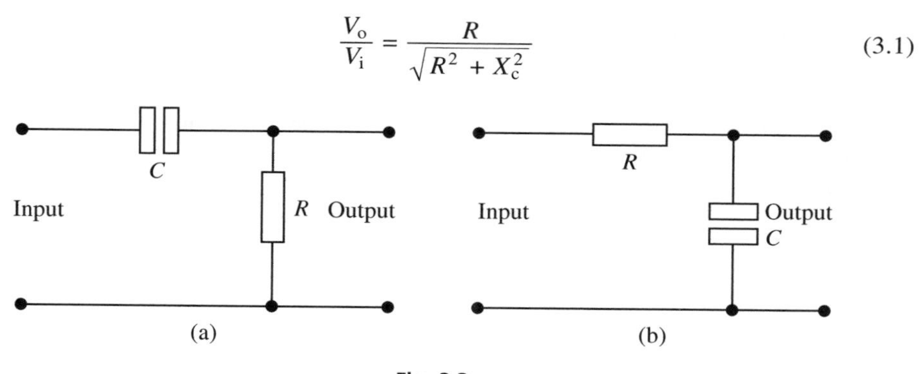

Fig. 3.3

and at the cut-off frequency the gain falls by 3 dB or $1/\sqrt{2}$. Also at this frequency $R = X_c$, which gives

$$R = X_c = \frac{1}{2\pi f_c C}$$

hence

$$f_c = \frac{1}{2\pi RC} \tag{3.2}$$

A similar result can be derived for the low pass filter, but for both first-order filters the following points should be considered.

(a) Cascading or connecting these networks in series causes the roll-off of the frequency response to increase by 20 dB/decade for each filter, where 'decade' refers to a one-to-ten range of frequencies, such as 1–10 Hz, 10–100 Hz, etc.: observe that on a logarithmic scale, such ranges span an equal distance (see Figs 3.42 and 3.43).

(b) A low-pass filter causes a phase lag between the output and input voltages, while a high-pass filter causes a phase lead between the output and input voltages. This has an important bearing on filters used in certain oscillators.

3.3 Active filters

The use of operational amplifiers in active filter devices is now well established in communications systems. Their main advantages over passive filters are:

(a) flexibility in design and construction;

(b) the absence of inductors, which at low frequencies is useful due to their large size and cost;

(c) low-frequency applications down to 1 Hz;

(d) the buffering effect due to the high input impedance and the low output impedance;

(e) with gain setting resistors the op-amp is capable of providing gain, hence the input signal is not attenuated as it is in passive filters;

(f) they are easier to tune than passive filters.

It is as well at this stage to appreciate that there are many types of filter, such as crystal, acoustical and digital filters, all of which have a specific application. In this chapter we will investigate active filters which are of the analogue type but can be used in either digital or analogue system applications.

Filter response

Associated with a filter's performance is the frequency response, which involves a plot of frequency against gain or against attenuation. This graph involves a response for all frequencies which the filter is designed to pass. At a particular frequency, known as the cut-off frequency, the response starts to decrease in amplitude. This is known as the roll-off and is a measure of how sharply the filter responds to attenuate frequencies above or below the cut-off frequency.

The filters in this chapter will have input *RC* networks, and as the signal frequency decreases the capacitive reactance X_c increases. This causes less voltage to be applied across the input impedance of the amplifier because more is dropped across X_c. This reduces the overall gain of the filter, and a critical point is reached when the output voltage is 0.707, i.e. $1/\sqrt{2}$, of the input ($V_o = 0.707V_i$). This condition occurs when $X_c = R$ and is called the –3 dB point of the response as the overall gain is 3 dB down on the pass-band gain. The frequency at which this occurs is the cut-off frequency. This discussion applies to all filter types.

All filters have four basic applications which can be easily understood from the ideal responses shown below. Note that an ideal response is one which has a vertical roll-off at the cut-off frequency. In practice this is not possible, but certain sophisticated filters tend to approach it. The four ideal configurations are shown in Fig. 3.4, in which the pass and stop bands are shown.

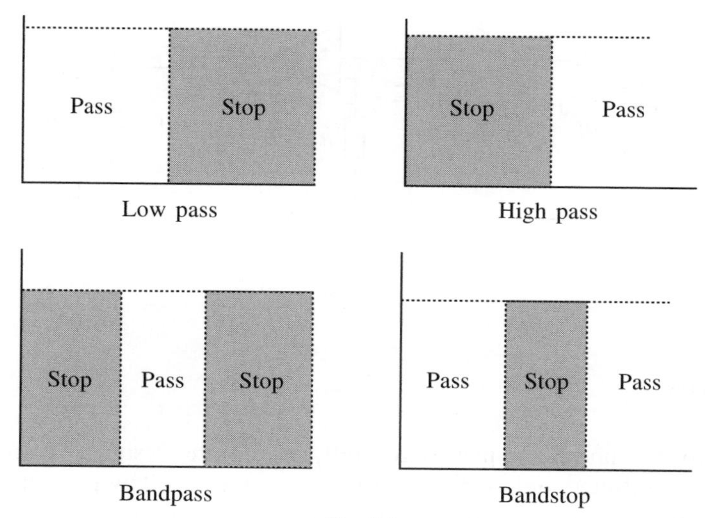

Fig. 3.4

Cut-off frequency and roll-off rate

As has been mentioned, no filter achieves the ideal response shown in Fig. 3.4, but the higher the order of the filter the closer it approaches the ideal case. This is shown in Fig. 3.5, which shows a multiple response diagram. It can be seen from this diagram that the roll-off rate increases with the order of the filter. This filter order is dependent on the number of *RC* networks (number of poles) included in the filter design. For example, if a single *RC* network is used with a filter it is referred to as a single-pole filter, while two *RC* networks produce a two-pole filter. Correspondingly the roll-off would be 20 dB/decade and 40 dB/decade, respectively. Hence increasing the number of *RC* networks increases the order of the filter. A three-pole or third-order filter is shown in Fig. 3.6.

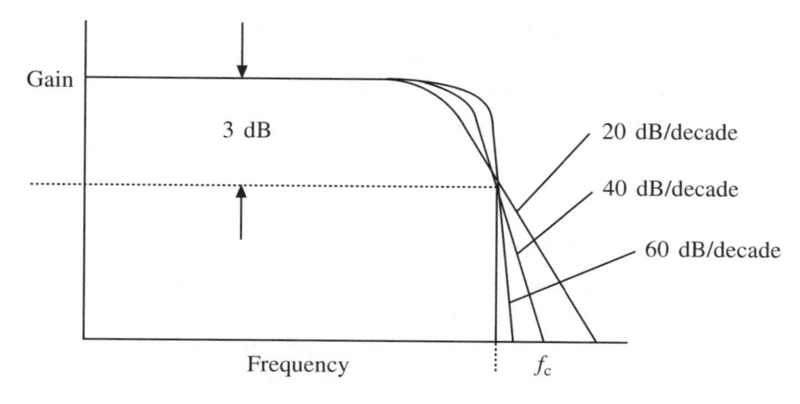

Fig. 3.5

It is normally not necessary to go beyond a fourth-order filter, but if this situation arises then it is a simple matter of cascading first and second-order filters to achieve higher orders. We will now examine these two important filters in detail and see how they can be realized in a practical way.

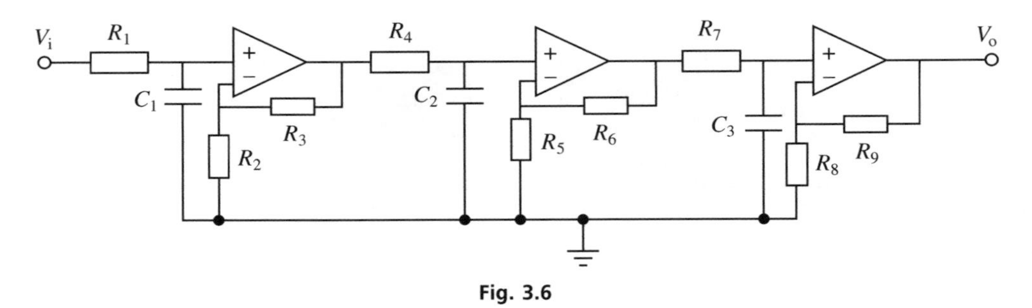

Fig. 3.6

Filter types

There are two fundamental responses generally used in the design of filters; these are referred to as the Butterworth and Chebyshev responses. The low-pass filter responses for these types are shown in Fig. 3.7. As can be seen, the two responses are quite different. The Butterworth type has what is called a maximally flat response in the pass band. Hence there is no ripple in this type of filter and the cut-off frequency is generally taken at the 3 dB level as shown. Note that in Fig. 3.7(a) the stop band lies between 0 Hz and f_c. In practice this may not be the case, and a minimum gain may be stipulated (say) between point A and f_c.

The maximally flat response of the Butterworth is good at frequencies around about zero hertz, but the response is poorer near the edge of the pass band. The Chebyshev filter can solve this problem. The Chebyshev response shown in Fig. 3.7(b) contains a ripple in the pass band. However, the attenuation increases more rapidly outside the pass band than the Butterworth. The greater the ripple, the more selective is the filter. The pass band is not so easily defined but is usually taken from the point where the highest-frequency peak ripple occurs. If, for example, the Chebyshev high-pass filter in Fig. 3.7(b) has a 0.5 dB ripple as shown and $f_r = 1$ kHz, then its response would be given as ± 0.5 dB from 1 kHz onwards with a rapidly increasing attenuation for frequencies less than 1 kHz. In some applications, however, the 3 dB bandwidth is required as shown at point C on Fig. 3.7(b), and this may be calculated using what are called transfer functions. These will be discussed later.

Filter orders

Filter orders have already been mentioned, and it can be seen from Fig. 3.4 that the orders would have to be infinitely high in order to achieve ideal responses.

The order of a Chebyshev or Butterworth filter determines the sharpness or roll-off of the response, but the interpretation of order is slightly different because of the ripple pass band in the Chebyshev filter. In this case the number of ripple peaks in the pass band determines the order (n) of the filter. This is shown in Fig. 3.8. For example, in Fig. 3.8(a) $n = 2$ and in Fig. 3.8(c) $n = 4$. Note that unlike the Chebyshev filter, the Butterworth low-pass filter will be 3 dB down on its maximum value no matter what the order is. The same points apply to the high-pass filter responses.

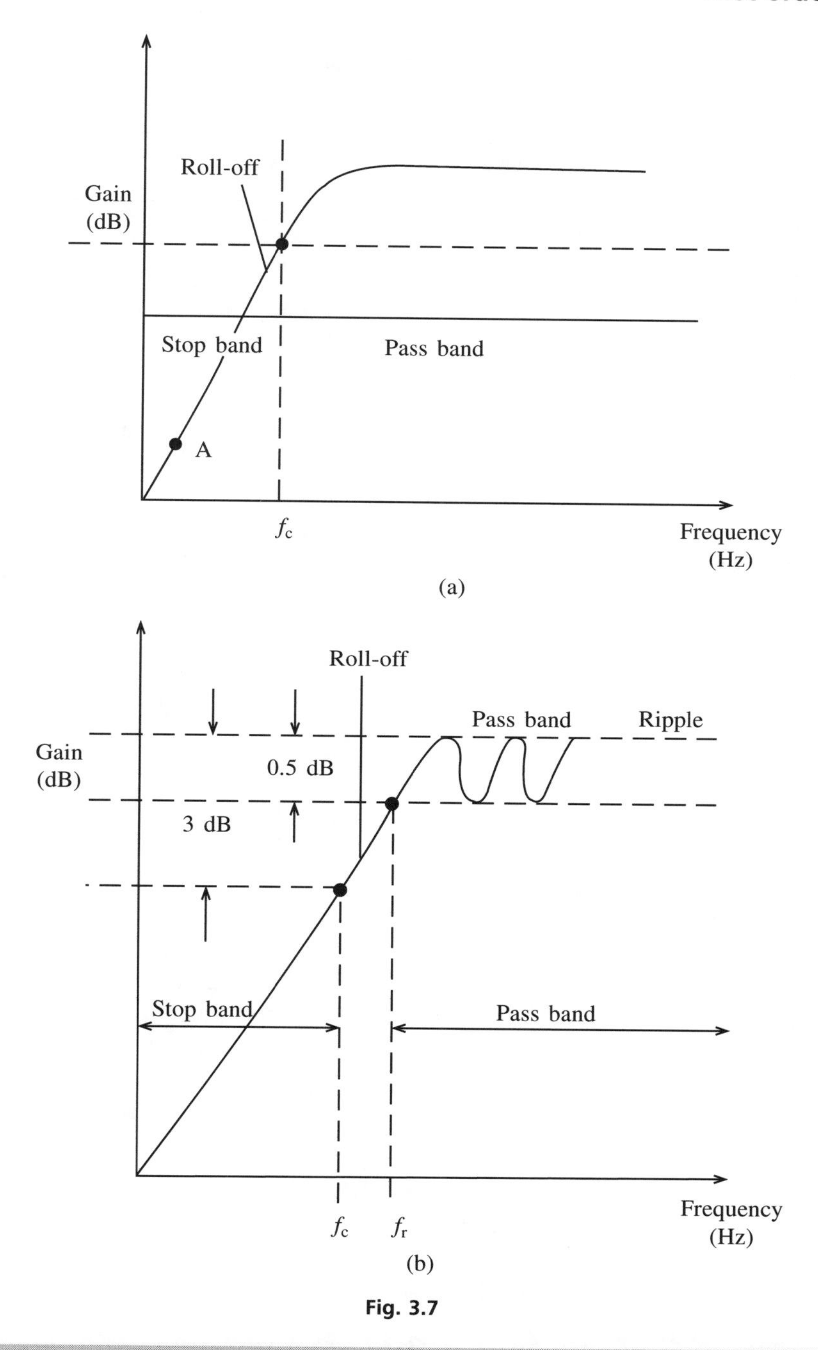

Fig. 3.7

3.4 First-order filters

The first-order filter is the simplest type and forms the basis of all other filters. Normally, what is called the Butterworth type is analysed. We will look at the low-pass filter first, a circuit for which is shown in Fig. 3.9.

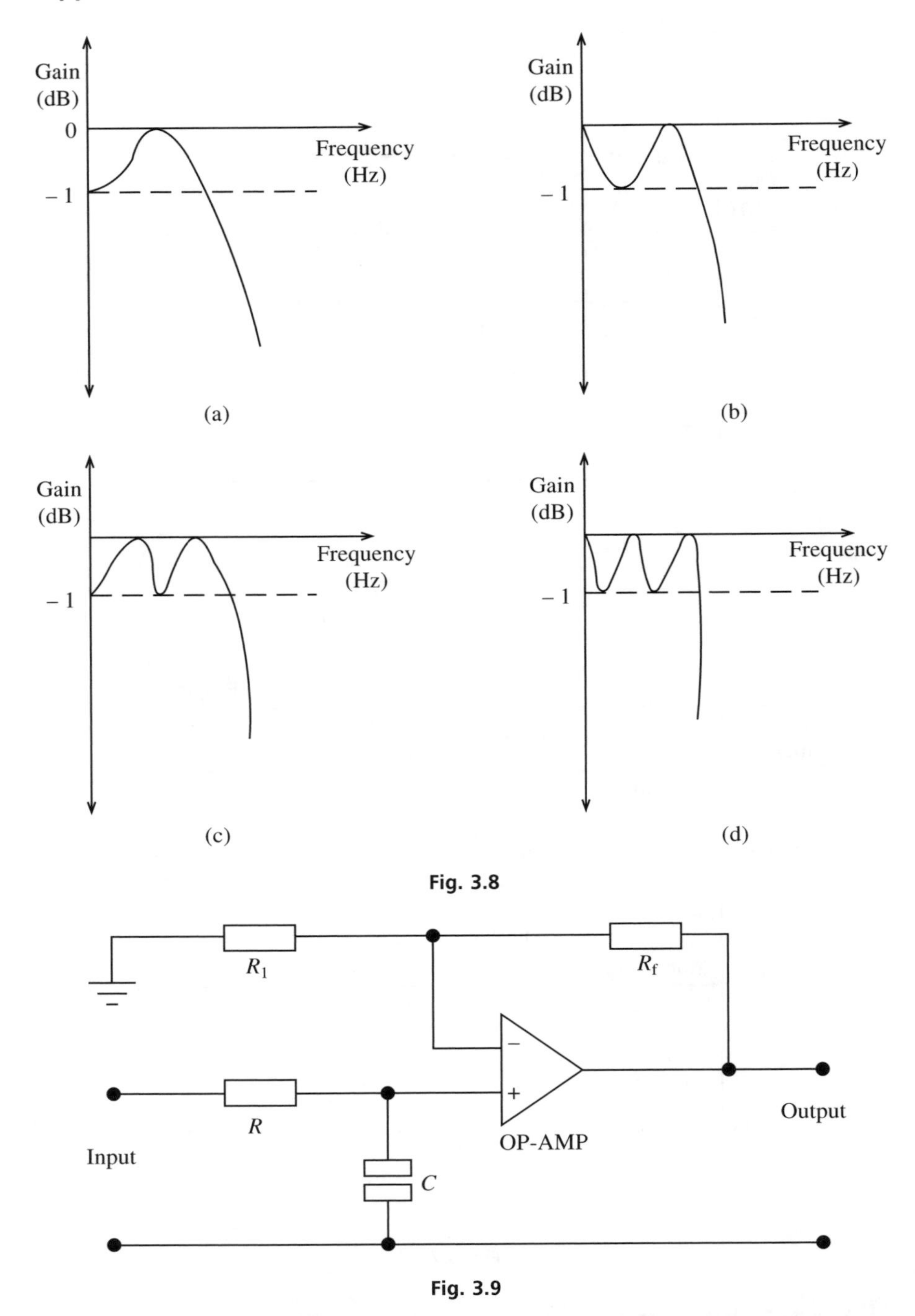

Fig. 3.8

Fig. 3.9

In this circuit note that the op-amp is ideal, i.e. it draws no current, and also it is used in the non-inverting mode in order to prevent loading down of the *RC* network. *R* and *C* act as a voltage-dividing network, and hence we have that

$$V = -\frac{jX_c}{R - jX_c} = V_i$$

Simplifying this expression gives

$$V = \frac{V_i}{1 + j2\pi RC}$$

The output voltage is given as

$$V_o = \left(1 + \frac{R_f}{R_i}\right) V$$

Hence

$$V_o = \left(1 + \frac{R_f}{R_i}\right) \frac{V_i}{1 + j2\pi RC}$$

or

$$\frac{V_o}{V_i} = \frac{A}{1 + j(f/f_{3dB})} \qquad (3.3)$$

Note that

$$f_{L(3dB)} = \frac{1}{2\pi RC} \qquad (3.4)$$

This has the characteristics of a first-order low-pass filter. When $\omega = 0$ then the pass-band gain is

$$\frac{V_o}{V_i} = \frac{R_2}{R_1} = K \qquad (3.5)$$

This is simply the amplifier gain. Note also that when

$$\omega = \frac{1}{RC}$$

the gain has dropped by 3 dB after which the gain falls off at the rate of 20 dB/decade. A typical response for this filter is shown in Fig. 3.10.

A similar analysis may be carried out for the first-order high-pass filter, which is shown in Fig. 3.11. Note that these two filters are identical except that R and C have been interchanged. The output voltage is given by

$$V_o = \left(1 + \frac{R_f}{R_1}\right)\left(\frac{2j\pi fRC}{1 + j2\pi RC}\right) V_i$$

or

$$\frac{V_o}{V_i} = A\,\frac{j(f/f_{3dB})}{1 + j(f/f_{3dB})}$$

Note that

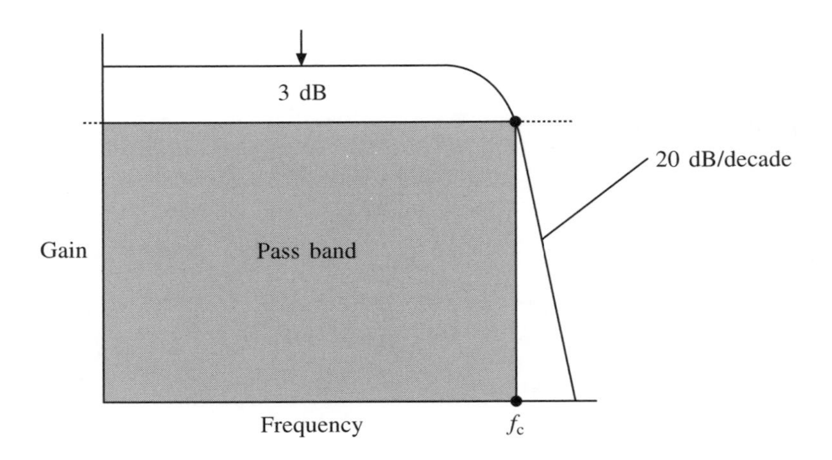

Fig. 3.10

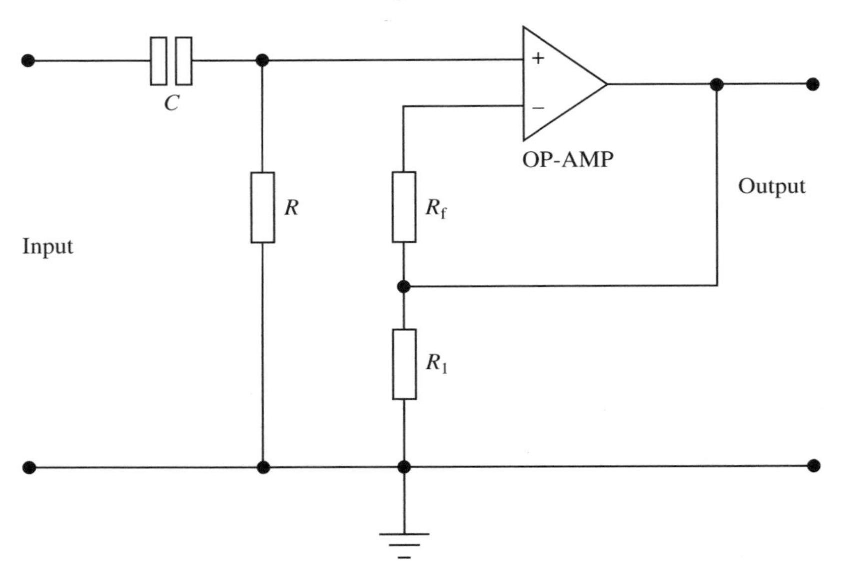

Fig. 3.11

$$f_{H(3dB)} = \frac{1}{2\pi RC} \tag{3.6}$$

The response for this filter is shown below in Fig. 3.12.

3.5 Design of first-order filters

Low- and high-pass first-order filters may be designed very easily if certain steps are followed:

1. The cut-off frequency must be known.

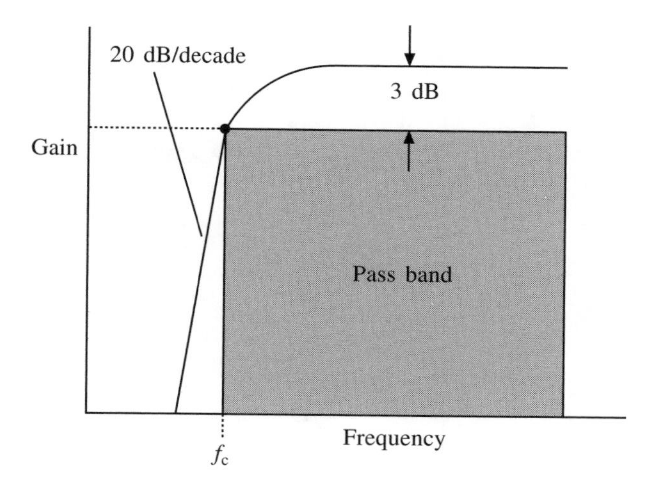

Fig. 3.12

2. A value of C less than 1 μF (say) should be chosen.
3. Then calculate the value of R from equation (3.4) or (3.6), depending on the filter being designed.
4. Determine a value of A and calculate R_f and R_1.

Example 3.1
Design a low-pass filter at a cut-off frequency of 2.4 kHz with a pass-band gain of 3.

Solution
Select a value of $C = 0.025\ \mu F$. This will give

$$R = \frac{1}{2\pi \times 2.4 \times 10^3 \times 0.025 \times 10^{-6}} = 2.7\ k\Omega$$

Since the pass-band gain is 3 then

$$3 = 1 + \frac{R_f}{R_i}$$

Hence $R_f = 2R_i$ and so various values are possible. If an unusual value is calculated then a potentiometer may be used to set the values. It should also be mentioned at this point that with advanced semiconductor technology a selection of very low values of capacitance in the nanofarad range is available from many manufacturers in chip form.

In order to complete the exercise the practical circuit is shown in Fig. 3.13 and this can now be set up on a printed circuit board.

Example 3.2
Design a high-pass filter at a cut-off frequency of 1 kHz with a passband gain of 2.

Solution
Once again select a suitable value of C, such as 0.01 μF. Hence, since the cut-off frequency

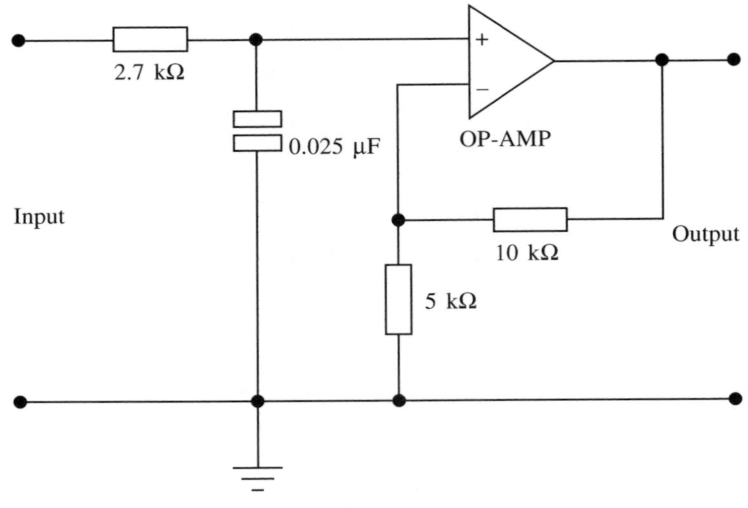

Fig. 3.13

is 1 kHz, $R = 15.9$ kΩ. Since $A = 2$, the two feedback resistors are equal. Several solutions are possible, such as 10 kΩ.

3.6 Second-order filters

As has already been mentioned, the higher the order of filter the sharper the cut-off. For certain applications, such as radio relay applications and channel separation, it is necessary to have higher-order filters. This chapter only looks at first and second-order filters but many higher orders can be designed by simply cascading these two types; indeed, this is one of the big advantages of using the active filter.

Low-pass second-order filters

Consider two low-pass first-order filters with the same cut-off frequencies but different pass-band gains

$$\frac{K_1}{1 + \mathrm{j}f/f_{3\mathrm{dB}}} \qquad\qquad \frac{K_2}{1 + \mathrm{j}f/f_{3\mathrm{dB}}}$$

If these filters are now cascaded, then the overall function will appear as follows,

$$\frac{K_1 K_2}{(1 + \mathrm{j}af)^2} = \frac{K}{(1 + \mathrm{j}af)^2}$$

where $a = 1/f_{3\mathrm{dB}}$ and $K = K_1 K_2$.

Expanding the above expression will give

$$\frac{V_o}{V_i} = \frac{K}{a^2(\mathrm{j}\omega)^2 + 2a(\mathrm{j}\omega) + 1}$$

and in general terms this is stated as

$$\frac{V_o}{V_i} = \frac{K}{a_2\,(j\omega)^2 + a_1(j\omega) + 1} \tag{3.7}$$

where a_1 and a_2 are constants. This expression is the characteristic of a second-order filter, and from it two basic types of filter may be deduced, depending on the values of a_1 and a_2: Butterworth flat response, where $a_1^2 = 2a_2$; and Chebyshev ripple response, where $a_1^2 < 2a_2$. The responses of both these filters has already been given, but they are combined in Fig. 3.14.

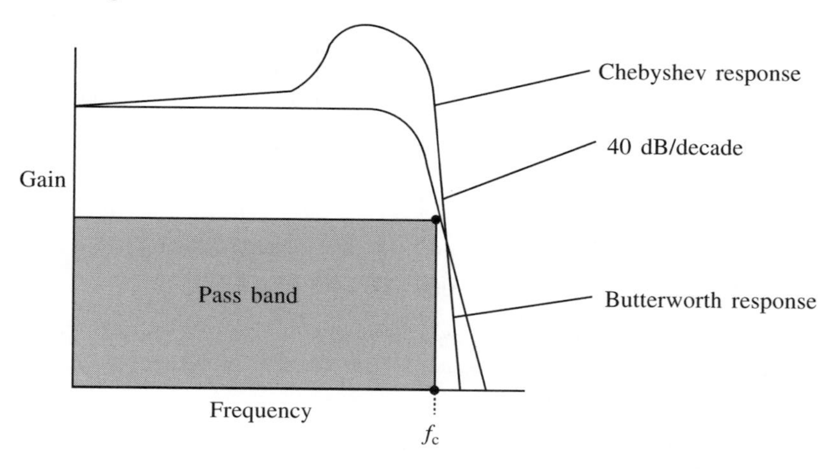

Fig. 3.14

The Butterworth response is generally a flatter response than the Chebyshev, but the Chebyshev filter has a faster rate of cut-off immediately after the cut-off frequency. Because of its flat response the Butterworth filter is more popular, but the ripple response of the Chebyshev has applications in satellite transponders where channel separation is tight.

Both these filters can be represented by many circuits, but the easiest configuration is known as the Sallen–Key circuit from which most filters may be designed, provided the pass-band gain and cut-off frequency are known. The typical circuit configuration is shown in Fig. 3.15.

By using circuit analysis the general transfer function for the circuit in Fig. 3.15 may be determined as follows:

$$\frac{V_o}{V_i} = \frac{K/R_1 R_2 C_1 C_2}{s^2 + s\,\{1/R_1 C_1 + 1/R_2 C_1 + (1 - K)/R_2 C_2\} + 1/R_1 R_2 C_1 C_2} \tag{3.8}$$

where $K = 1 + R_a/R_b$ (the d.c. gain) and $s = j\omega$. A full analysis of the transfer function can be found in standard texts on filters.

The denominator term in equation (3.7) is known as the polynomial for the nth-order filter. These polynomials may be derived for any filter type or order, but it is more convenient to use polynomial tables. Examples given in this text will use polynomials which are shown in Table 3.1.

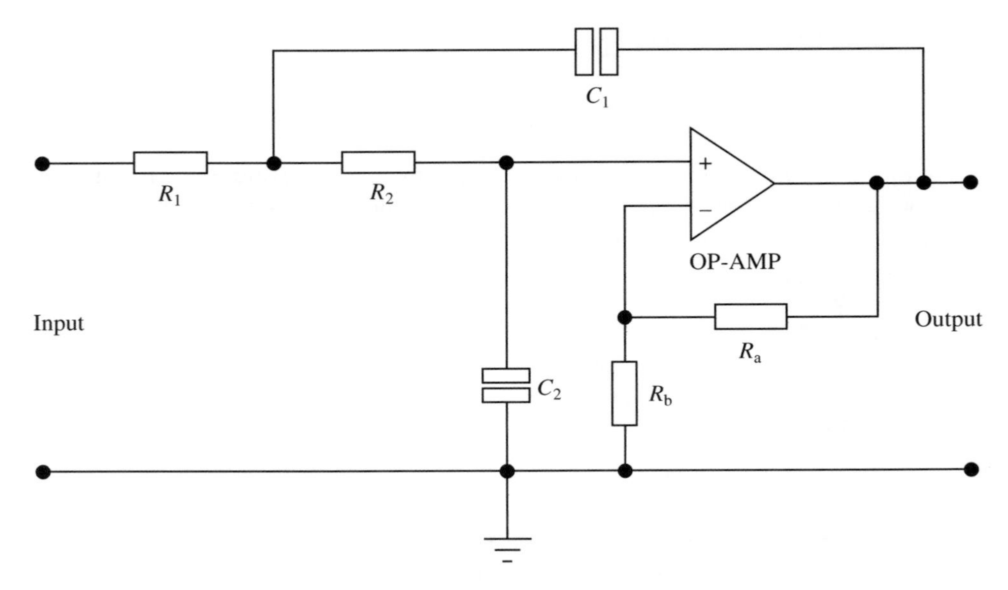

Fig. 3.15

Table 3.1

n	Butterworth polynomials
1	$S + 1$
2	$S^2 + 1.414 + 1$
3	$(S + 1)(S^2 + 1.414 + 1)$
4	$(S^2 + 0.765S + 1)(S^2 + 1.848S + 1)$
5	$(S + 1)(S^2 + 0.765S + 1)(S^2 + 1.848S + 1)$
6	$(S^2 + 0.518S + 1)(S^2 + 1.414S + 1)(S^2 + 1.932S + 1)$
	Chebyshev polynomials (0.5 dB ripple)
1	$S + 2.863$
2	$S^2 + 1.425S + 1.516$
3	$(S + 0.626)(S^2 + 0.626S + 1.142)$
4	$(S^2 + 0.351S + 1.064)(S^2 + 0.845S + 0.356)$
5	$(S + 0.362)(S^2 + 0.224S + 1.036)(S^2 + 0.586S + 0.477)$
6	$(S^2 + 0.155S + 1.024)(S^2 + 0.414S + 0.548)(S^2 + 0.580S + 0.157)$
	Chebyshev polynomials (1 dB ripple)
1	$S + 1.965$
2	$S^2 + 1.098S + 1.103$
3	$(S + 0.494)(S^2 + 0.494S + 0.994)$
4	$(S^2 + 0.279S + 0.987)(S^2 + 0.674S + 0.279)$
5	$(S + 0.289)(S^2 + 0.179S + 0.989)(S^2 + 0.469S + 0.429)$
6	$(S^2 + 0.124S + 0.991)(S^2 + 0.340S + 0.558)(S^2 + 0.464S + 0.125)$

This form of the general transfer function is related to Fig. 3.15 where R_1, C_1 etc. are the components used in the Sallen–Key circuit after a multiplication factor has been applied. This is called denormalization.

Later in this chapter normalized filter tables, given in Table 3.2, will be used. The term 'normalization' is defined usually as the scaling or standardization of a certain parameter. In the case of the filter tables the values are normalized to an angular cut-off frequency of 1 rad/s or 1 Hz. A multiplier is used in order to calculate the actual values of the components which will be used in the printed circuit board design. The operation of this multiplier is known as denormalization and will be fully demonstrated by the examples given later.

Table 3.2

Normalized tables for second-order filters

(a) Low-pass normalized filter (second order) with cut-off frequency of 1 rad/s

Filter type	R_1	R_2	C_1	C_2	Gain K
Butterworth	1.000	1.000	1.000	1.000	1.585
	1.000	1.000	1.414	0.707	1.000
	1.000	1.000	0.874	1.144	2.000
	0.707	1.414	1.000	1.000	2.000
0.5 dB ripple Chebyshev	0.812	0.812	1.000	1.000	1.842
	1.000	1.000	1.403	0.470	1.000
	1.000	1.000	0.771	0.856	2.000
	0.701	0.940	1.000	1.000	2.000
1 dB ripple Chebyshev	0.952	0.952	1.000	1.000	1.954
	1.000	1.000	1.822	0.498	1.000
	1.000	1.000	0.938	0.967	2.000
	0.911	0.996	1.000	1.000	2.000

(b) High-pass normalized filter (second order) with cut-off frequency of 1 rad/s

Filter type	R_1	R_2	C_1	C_2	Gain K
Butterworth	1.000	1.000	1.000	1.000	1.585
	0.707	1.414	1.000	1.000	1.000
	1.000	1.000	1.414	0.707	2.000
	1.144	0.874	1.000	1.000	2.000
0.5 dB ripple Chebyshev	1.231	1.231	1.000	1.000	1.842
	0.713	2.127	1.000	1.000	1.000
	1.000	1.000	1.426	1.064	2.000
	1.247	1.169	1.000	1.000	2.000
1 dB ripple Chebyshev	1.050	1.050	1.000	1.000	1.954
	0.549	2.009	1.000	1.000	1.000
	1.000	1.000	1.097	1.004	2.000
	1.066	1.034	1.000	1.000	2.000

It is as well to appreciate at this point that filter problems may be solved by using four main methods: the transfer function; normalized tables; identical components; and software. The use of software is widespread and there are many software packages which can be easily used by the novice. The suitability of these packages is a personal matter, but the author has found that the use of spreadsheets gives excellent results. The other three methods of solving active filter problems will be demonstrated by example.

3.7 Using the transfer function

Example 3.3

Determine suitable values for R_1, R_2, C_1 and C_2 for a second-order Butterworth filter with an upper cut-off frequency of 4 kHz and a pass-band gain of 20.

Solution

A problem of this nature requires a normalized response before it can be solved. The second-order Butterworth normalized response in this case will be given as

$$\frac{H}{(j\omega)^2 + 1.414(j\omega) + 1} \tag{3.9}$$

and the transfer function will be as stated previously in equation (3.8). If we multiply top and bottom of the right-hand side this equation by $R_1R_2C_1C_2$ and substitute $K = 20$, we obtain

$$\frac{V_o}{V_i} = \frac{20}{R_1R_2C_1C_2s^2 + s\{R_2C_2 + (1-K)R_1C_1 + R_1C_2\} + 1/R_1R_2C_1C_2} \tag{3.10}$$

The next step is to equate the coefficients of equations (3.9) and (3.10): for the s^2 terms

$$R_1R_2C_1C_2 = 1 \tag{3.11}$$

and for the s terms

$$R_2C_2 + (1-20)R_1C_1 + R_1C_2 = 1.414 \tag{3.12}$$

From (3.11) we may write

$$R_2C_2 = 1 \text{ and } R_1C_1 = 1$$

as this will satisfy the right-hand side of the equation. Substituting in (3.12) will give

$$1 - 19 + R_1C_2 = 1.414$$

$\therefore$

$$R_1C_2 = 19.414$$

Letting $R_1 = 1\ \Omega$ gives $C_2 = 19.414$ F. Since $R_2C_2 = 1$, we have $R_2 = 1/19.414 = 0.052\ \Omega$. Finally $R_1C_1 = 1$, hence $C_1 = 1$F. We now have all the values which will enable us to build the filter, but remember these are normalized values and they have to be denormalized. The method of doing this is shown below.

We will assume a denormalization factor of 10^4. Note that 10^3 or 10^5 could have been used: this is purely arbitrary. Then

$$R_1' = 1 \times 10^4 = 10\ \text{k}\Omega \tag{3.13}$$

Similarly,

$$R_2' = 1 \times 10^4/19.414 = 515\ \Omega \tag{3.14}$$

The capacitors are treated in a different way, but all you need to know is that the normalized values are divided by the cut-off frequency and the denormalization factor 10^4 as before:

$$C_1' = \frac{1}{10^4 \times 2\pi \times 5 \times 10^3} = 3.18 \text{ nF} \tag{3.15}$$

$$C_2' = \frac{19.44}{10^4 \times 2\pi \times 5 \times 10^3} = 65 \text{ nF} \tag{3.16}$$

The filter can now be built using the Sellen–Key circuit in Fig. 3.16.

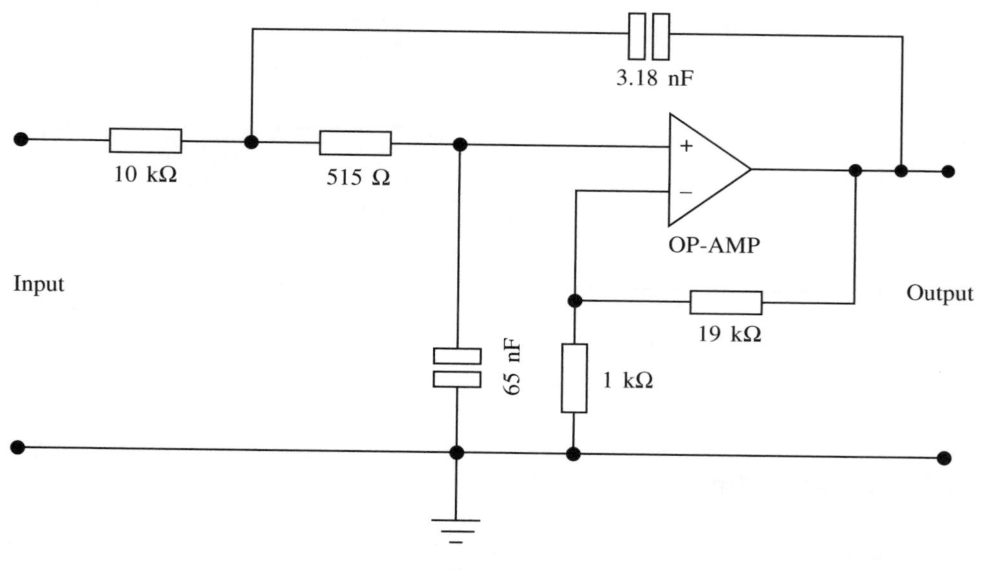

Fig. 3.16

Example 3.4

Design the same filter as in Example 3.3, but with a Chebyshev response given by the following normalized transfer function:

$$\frac{V_o}{V_i} = \frac{H}{1.4125(j\omega)^2 + 0.9109(j\omega) + 1}$$

Solution

Once again using the procedure adopted in the previous example and equating the coefficients,

$$R_1 = 1\ \Omega,\ R_2 = \frac{1}{26.74} = 0.0374\ \Omega$$

$$C_1 = 1.4125 \text{ F},\ C_2 = 26.74 \text{ F}$$

Denormalizing these values as before gives

$$C_1' = \frac{1.4125}{10^4 \times 2\pi \times 5 \times 10^3} = 4.49 \text{ nF}$$

$$C_2' = \frac{26.74}{10^4 \times 2\pi \times 5 \times 10^3} = 85.2 \text{ nF}$$

Also $R_1' = 10\ \text{k}\Omega$ and $R_2' = 374\ \Omega$. The circuit is shown in Fig. 3.17.

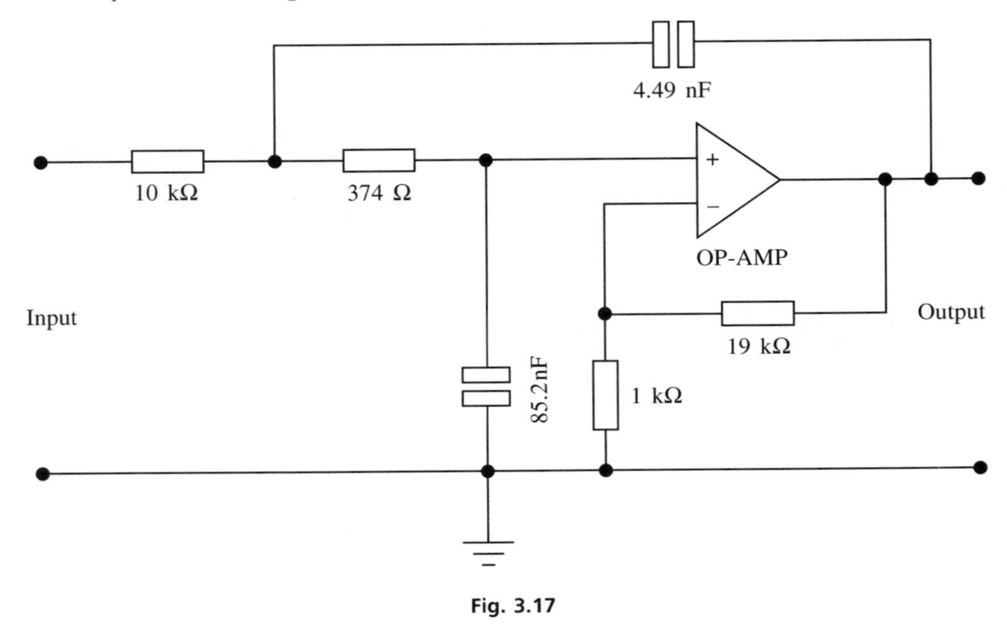

Fig. 3.17

3.8 Using normalized tables

If normalized tables are available these can be easily used without much calculation. A set of these tables is shown in Table 3.2. As can be seen, if the pass-band gain (K) is known it is simply a matter of selecting the appropriate values. Note that several combinations may be possible, as was the case with the previous method.

Remember these are normalized values, and they have to be denormalized as before. This method is a lot easier than the analytical method discussed previously and where tables are available for a certain pass-band gain this method is by far the easiest to apply.

Example 3.5
It is required to design a low-pass Butterworth filter with a pass-band gain of 2 and 3 dB cut-off frequency of 3.2 kHz.

Solution
Consulting the table gives a choice of components in this case, but we will select the following (the choice is purely arbitrary):

$$R_1 = 1.000\ \Omega \quad R_2 = 1.000\ \Omega$$

$$C_1 = 0.874\ \text{F} \quad C_2 = 1.144\ \text{F}$$

These are normalized as usual:

$$C_1' = \frac{0.874}{2\pi \times 3.2 \times 10^3 \times 10^4} = 4.35\ \text{nF}$$

$$C_2' = \frac{1.144}{2\pi \times 3.2 \times 10^3 \times 10^4} = 5.7 \text{ nF}$$

Also $R_1' = R_2' = 10\ \text{k}\Omega$. The gain setting resistors are chosen in the usual way.

3.9 Using identical components

It is simpler sometimes to use equal components, but it is necessary to adhere to the particular pass-band gain on the normalized tables. In many applications this method should be considered first.

Select a cut-off frequency value and then choose a common value for $C = C_1 = C_2$ – some value less than 1 mF, say. Since $R = R_1 = R_2$, R can now be calculated as follows.

$$R = \frac{1}{2\pi f C} \tag{3.17}$$

Note also that the pass-band gain has to be 1.585, this being obtained from the normalized tables.

Example 3.6
It is required to design a second-order low-pass filter with a cut-off frequency of 3 kHz.

Solution
Let $C_1 = C_2 = 0.047\ \mu\text{F}$. Hence

$$R_1 = R_2 = \frac{10^6}{2\pi \times 3 \times 10^3 \times 0.047} = 1128.76\ \Omega$$

Selecting the gain setting resistors is once again achieved by using the fact that these have to satisfy the equation

$$A = 1 + \frac{R_a}{R_b}$$

Hence $R_a = 0.586 R_b$ and several combinations are possible.

3.10 Second-order high-pass filters

High-pass filters may be designed in a similar manner to low-pass second-order filters, but in this case the normalized response is slightly different. The response for such a filter may be given as

$$\frac{Hs^2}{a_2 + a_1 s + s^2} \tag{3.18}$$

As before two cases are deduced: the Chebyshev response, where $a_1^2 < 2a_2$; and the Butterworth response, where $a_1^2 < 2a_2$. These responses are shown in Figure 3.18.

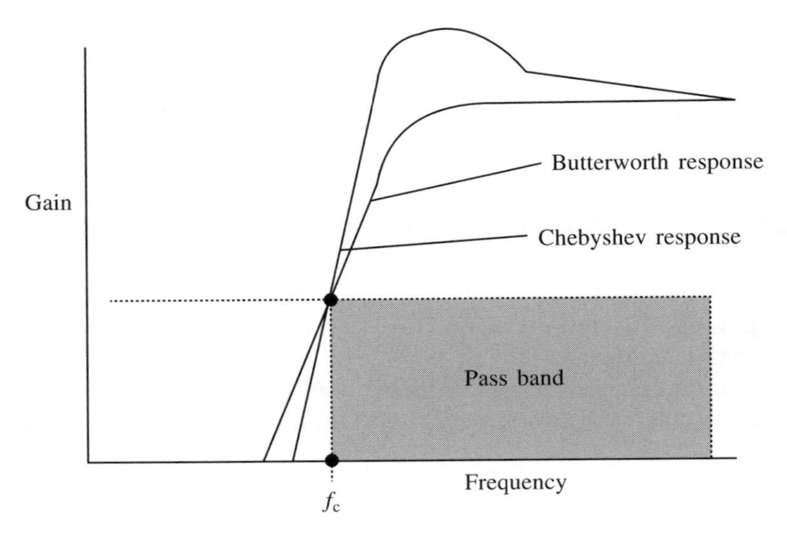

Fig. 3.18

As before, a Sallen and Key circuit can be drawn, and this is almost identical to the low-pass circuit except that the components are interchanged. Such a circuit is shown in Fig. 3.19.

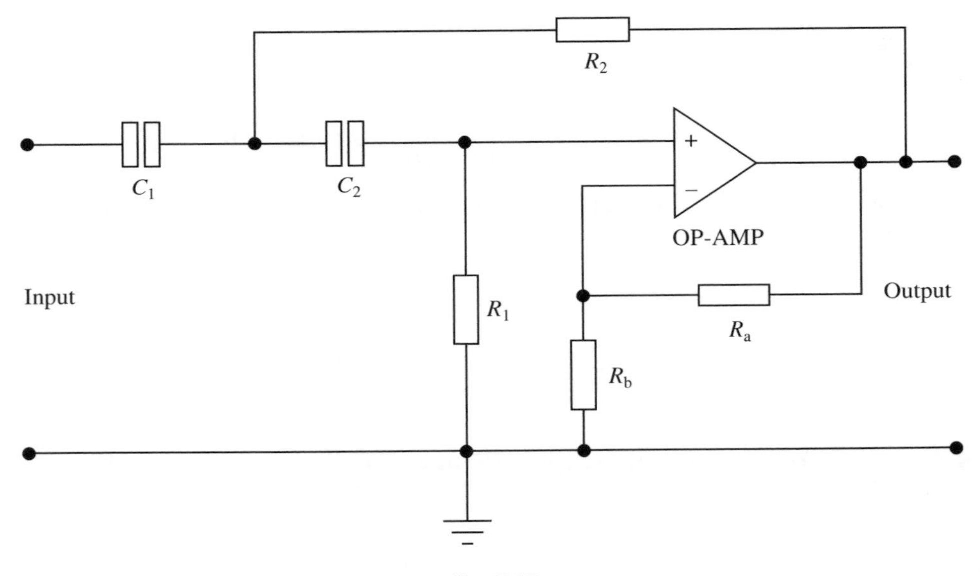

Fig. 3.19

The transfer function is the same as for the low-pass filter, but it should be remembered that the components have been interchanged and because of this it will now take the form

$$\frac{V_o}{V_i} = \frac{Ks^2}{s^2 + s\{1/R_2C_1 + 1/R_2C_2 + (1 - K)/R_1C_1\} + 1/R_1C_1R_2C_2} \tag{3.19}$$

which is in the form

$$\frac{Ks^2}{a_2 + sa_1 + s^2}$$

Problems are tackled in exactly the same way as for the low-pass case, and normalized tables may be used in a similar fashion. The following worked examples will now clarify the principles discussed so far.

Example 3.7
Draw the circuit of a first-order low-pass Butterworth filter having a cut-off frequency of 10 kHz and a pass-band gain of unity.

Solution
Choose a value $C = 0.001$ μF. Hence

$$R = \frac{10^6}{2\pi \times 10^4 \times 0.001} = 15.9 \text{ k}\Omega$$

The circuit for this solution is shown in Fig. 3.20.

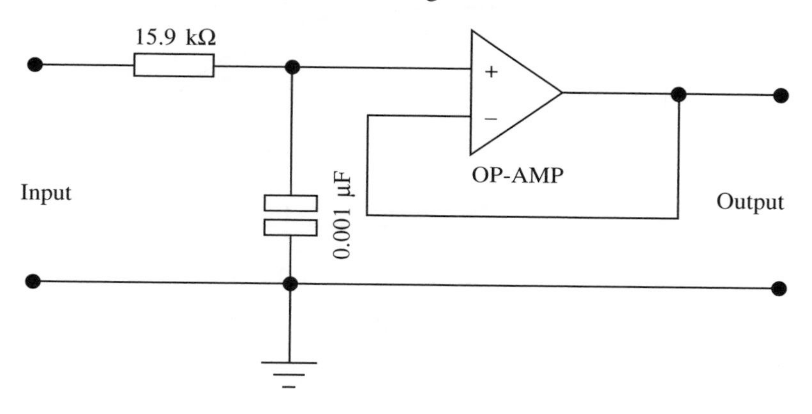

Fig. 3.20

Example 3.8
Figure 3.21 represents a first-order filter. Draw the response for this filter showing scaling and relevant points.

Solution
Gain is given by

$$\frac{V_o}{V_i} = 1 + \frac{R_2}{R_1} = 1 + \frac{10}{10} = 2$$

Since $R = 15.6$ kΩ and $C = 0.01$ μF,

$$f = \frac{1}{2\pi RC} = \frac{1}{2\pi \times 15.6 \times 10^3 \times 0.01 \times 10^{-6}} = 1.020 \text{ kHz}$$

The response for this problem is shown in Fig. 3.22.

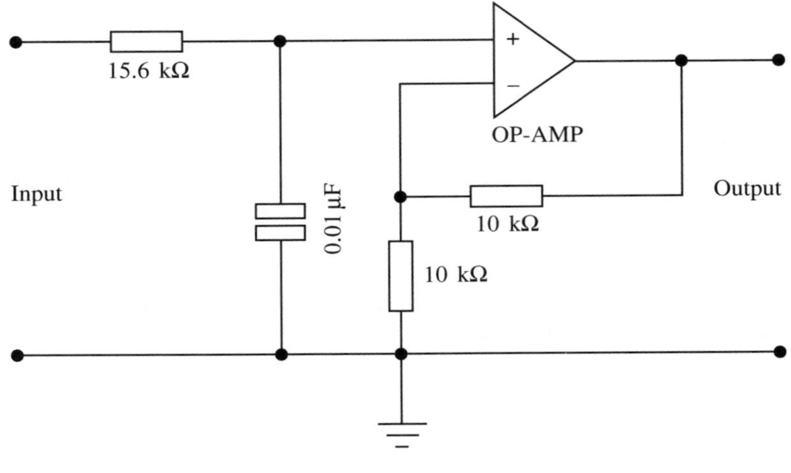

Fig. 3.21

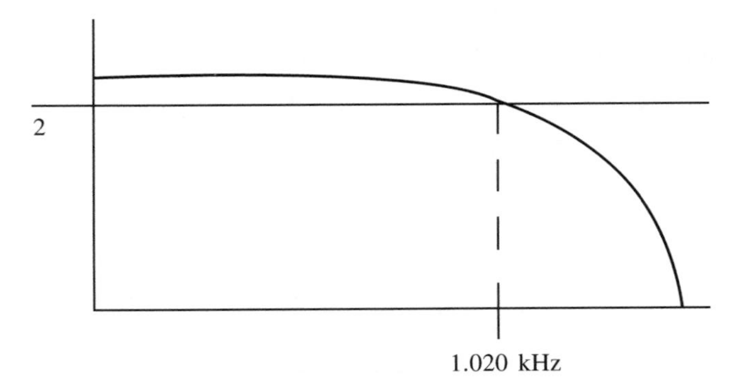

Fig. 3.22

Example 3.9
Design a –40 dB/decade low pass filter at a cut-off frequency of 10 krad/s, assuming equal value components.

Solution
As equal value components are used, from the normalized tables the gain must be 1.585. Hence, as the angular frequency is 10 krad/s,

$$C = \frac{1}{2\pi f R}$$

and selecting a value for R at random, say 36 kΩ, then we simply apply this to the formula as follows:

$$C = \frac{1}{10^4 \times 36 \times 10^3} = 2.8 \text{ nF}$$

The circuit is shown in Fig. 3.23.

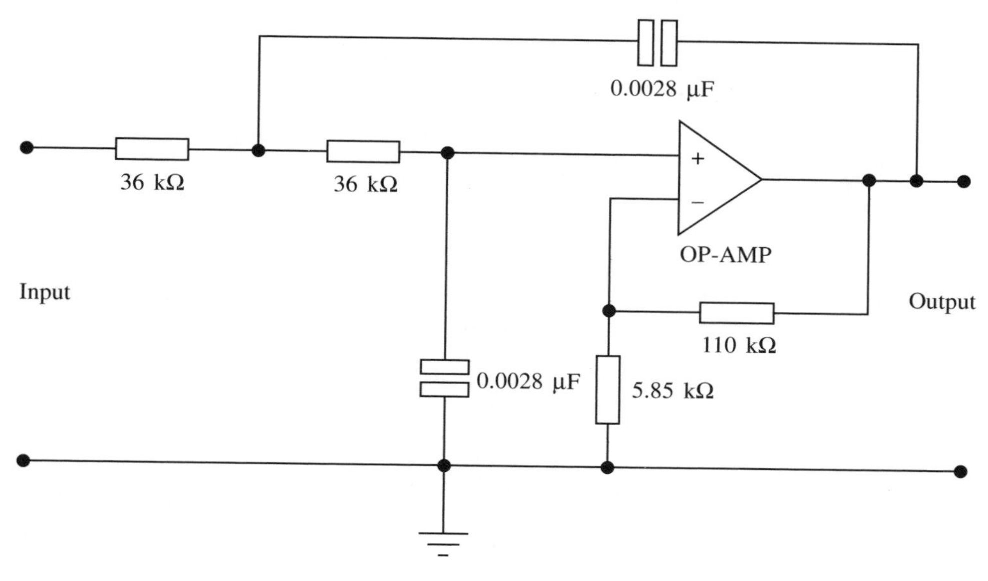

Fig. 3.23

Example 3.10

Design a second-order high-pass filter which has a Butterworth response with a pass--band gain of 25 and a 3 dB cut-off frequency of 20 kHz. Note the second-order Butterworth coefficients are $a_2 = 1$ and $a_1 = 1.414$.

Solution

This type of problem unfortunately cannot be solved by the normalized tables, hence the analytical method will be used.

The second-order Butterworth response is given by

$$\frac{V_o}{V_i} = \frac{Ks^2}{1 + 1.414s + s^2}$$

$$= \left[\frac{Ks^2}{1/R_1R_2C_1C_2 + \{(1/R_2C_2) + (1/R_2C_1) - (24/R_1C_1)\}\, s + s^2}\right]$$

Equating as usual gives

$$\frac{1}{R_1R_2C_1C_2} = 1 \qquad (3.20)$$

$$\frac{1}{R_2C_2} + \frac{1}{R_2C_1} - \frac{24}{R_1C_1} = 1.414 \qquad (3.21)$$

Let

$$\frac{1}{R_2C_2} = 1$$

Therefore, from (3.20)

$$\frac{1}{R_1 C_1} = 1$$

Hence substituting in (3.21) gives,

$$1 + \frac{1}{R_2 C_1} - 24 = 1.414$$

i.e.

$$\frac{1}{R_2 C_1} = 24.414$$

Letting $C_1 = 1$ F gives $R_2 = 1/24.414 = 0.0410\ \Omega$; thus $C_2 = 24.414$ F. Also $C_1 = 1/R_1$, therefore $R_1 = 1\ \Omega$. Assuming a denormalizing factor of 10^4, we have

$$C_1' = \frac{1}{2\pi \times 2 \times 10^4 \times 10^4} = 0.79 \text{ nF}$$

$$C_2' = \frac{24.414}{2\pi \times 2 \times 10^4 \times 10^4}$$

$$R_2' = \frac{10^4}{24.414} = 410\ \Omega$$

$$R_1' = 10 \text{ k}\Omega$$

Also since

$$25 = 1 + \frac{R_a}{R_b}$$

we have $R_a = 1$ kΩ and $R_b = 24$ kΩ. The circuit is shown in Fig. 3.24.

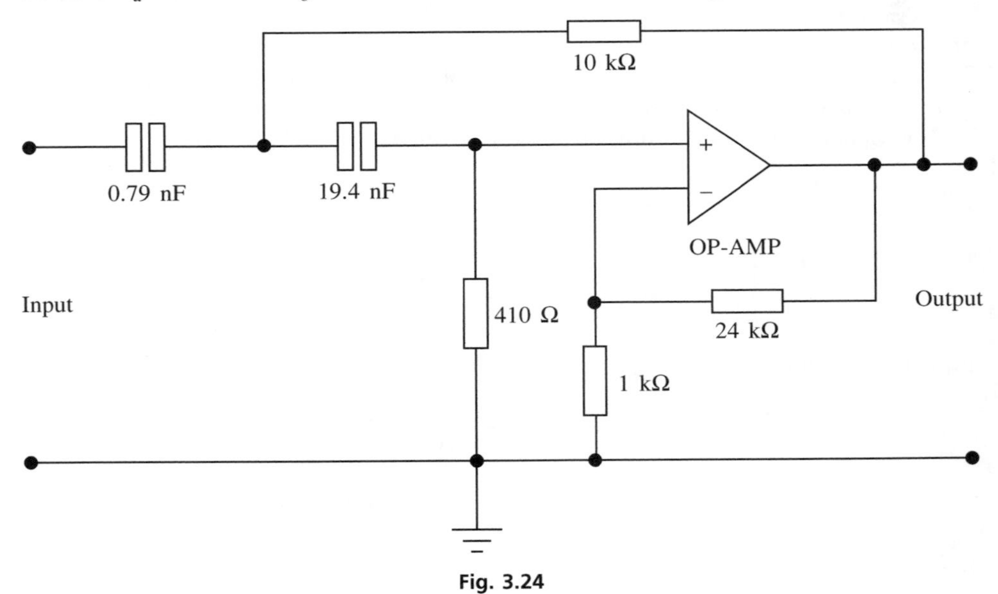

Fig. 3.24

Example 3.11

Show how a third-order low-pass filter may be designed using a first- and second-order combination in order to achieve a pass-band gain of 2 and a cut-off frequency of 5 kHz.

Solution
For the first-order stage we have

$$R = \frac{1}{2\pi fC}$$

Choosing a value for $C = 0.01\ \mu\text{F}$,

$$R = \frac{10^6}{2\pi \times 5 \times 10^3 \times 0.01} = 3.18\ \text{k}\Omega \text{ (use a 5 k}\Omega\text{ pot)}$$

For the second-order stage the normalized tables are used for a pass-band gain of 2. Select $R_1 = R_2 = 1$, $C_1 = 0.874$ and $C_2 = 1.414$. Using a denormalizing factor of 10^4 gives the following values:

$$R_1' = R_2' = 10\ \text{k}\Omega$$

$$R_1' = \frac{0.874}{6.28 \times 10^4 \times 5 \times 10^3} = 2.78\ \text{nF}$$

$$C_2' = \frac{1.414}{6.28 \times 10^4 \times 5 \times 10^3} = 4.50\ \text{nF}$$

$R_a/R_b = 1$, hence let $R_a = R_b = 10\ \text{k}\Omega$. The circuit is shown in Fig. 3.25.

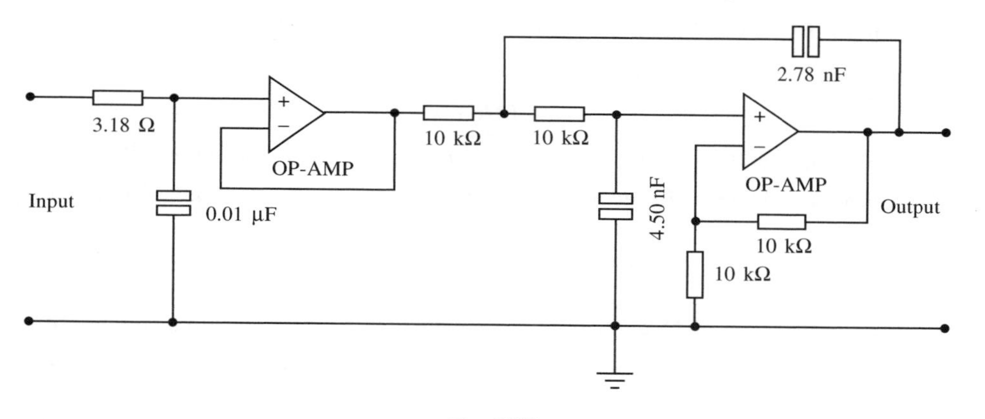

Fig. 3.25

3.11 Additional problems

1. Design a first-order filter with a gain of 10 and a cut-off frequency of 200 Hz. Use a 0.05 μF capacitor.
2. Design a second-order 1 dB ripple high-pass Chebyshev filter with a gain of 2 and a cut-off frequency of 500 Hz.
3. Draw the response of a low-pass second-order Butterworth filter having a gain of unity and a cut-off frequency of 1.5 kHz. Also on the same sketch estimate the response of a fourth-order low pass filter of the Butterworth type having the same gain and cut-off frequency.

4. Design a first-order low-pass filter with a gain of 1 and a cut-off frequency of 3 kHz.
5. By the use of a schematic diagram show how a fifth-order filter having unity gain might be designed.
6. A low-pass second-order Chebyshev filter has a normalized response given by

$$\frac{V_o}{V_i} = \frac{K}{s^2 + 0.803\,816s + 0.823\,060}$$

This is the response for a 2 dB ripple filter. Assuming a gain of 12 and a cut-off frequency of 3.4 kHz, design a suitable filter.

7. Design a third-order low pass Butterworth filter using a second-order and first-order filter. The gain per stage has to be 1.585 and the cut-off frequency 1 kHz.
8. State four reasons why operational amplifiers are used in active filters.
9. In a communications channel it is desirable to pass all frequencies above 10 kHz and all frequencies below 5 kHz. A flat response is desired over a gain of unity with a roll-off of 40 dB/decade. Show how this may be achieved.
10. Design a second-order filter having a gain of 2 and a cut-off frequency of 2.5 kHz. Use two first-order filters to accomplish this.
11. A high-pass second-order Butterworth filter has a normalized response given by

$$\frac{V_o}{V_i} = \frac{Ks^2}{s^2 + 0.7632s + 0.822}$$

Assuming a gain of 10 and a cut-off frequency of 4.3 kHz, design a suitable filter.

12. A low-pass Chebyshev second-order filter has a normalized response given by

$$\frac{V_o}{V_i} = \frac{K}{1.365s^2 + 0.968s + 1}$$

Assuming a gain of 15 and a cut-off frequency of 1.2 kHz, design a suitable filter.

13. Design a first-order low-pass filter having a cut-off frequency of 3.4 kHz and a gain of 4.
14. Design a high-pass filter having a cut-off frequency of 8.5 kHz and a gain of 8.
15. A low-pass second-order Butterworth filter has a normalized response given by

$$\frac{V_o}{V_i} = \frac{Ks^2}{s^2 + 0.998s + 1.22}$$

Assuming a gain of 6 and a cut-off frequency of 600 Hz, design a suitable filter.

3.12 Bandpass filters

Previously we have looked at single low- or high-pass filters, but a common application of filters is where a band of frequencies has to be passed while all other frequencies are stopped. This is called a bandpass filter. Such a filter may be formed from a low- and a high-pass filter in cascade. Generally the low pass is followed by the high pass, but the order of cascade is not important as the same result will be produced.

Consider Fig. 3.26. The following points should be noted from this diagram.

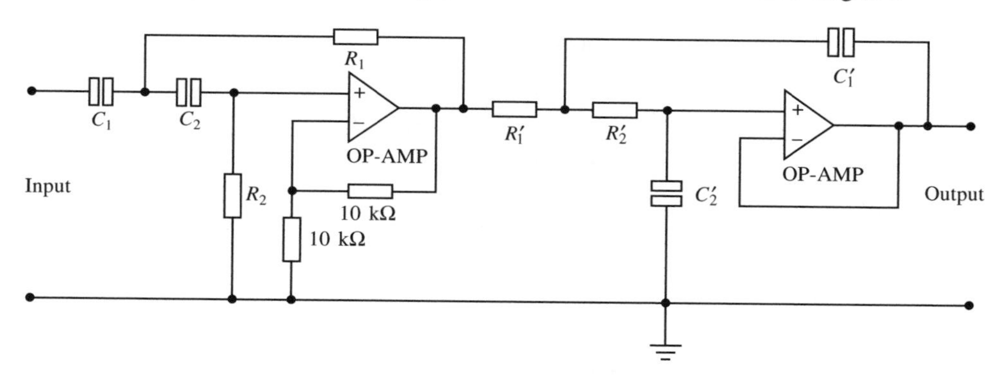

Fig. 3.26

1. A second-order low-pass filter is cascaded with a second-order high-pass filter. Note that the labelling of the components should correspond with the normalized tables.
2. The gain of the low-pass filter is unity, while that of the high pass filter is 2. This gives an overall gain of 2.
3. The overall response will give two cut-off frequencies.
4. No buffering is required as op-amps are used.

Example 3.12

Design a first-order bandpass filter which has a pass-band gain of 4, a lower cut-off frequency f_l = 200 Hz and an upper cut-off frequency of f_h = 1 kHz. Draw the frequency response of this filter.

Solution

As the gain has to be 4 overall then each filter should have a gain of 2. Hence, if the filter uses op-amps in the non-inverting mode then R_a and R_b are calculated by using:

$$K = 1 + \frac{R_a}{R_b}$$

Let R_1 = 10 kΩ. So both filter sections will have gain setting resistors of 10 kΩ. The values for both sections of the filter are calculated as follows. For the high-pass section,

$$f_l = \frac{1}{2\pi RC}$$

Let C = 0.05 μF. Then

$$R = \frac{1}{200 \times 6.28 \times 0.05 \times 10^{-6}} = 15.9 \text{ k}\Omega$$

For the low-pass section,

$$f_h = \frac{1}{2\pi RC}$$

$$R = \frac{1}{6.28 \times 10^3 \times 0.01 \times 10^{-6}}$$

$$= 15.9 \text{ k}\Omega$$

The response for this filter is shown in Fig. 3.27.

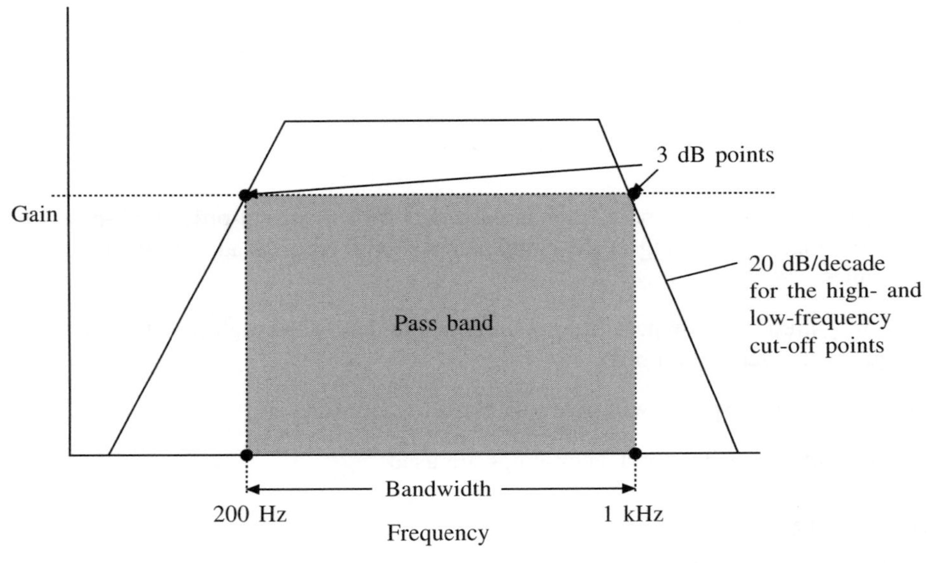

Fig. 3.27

Example 3.13

Design a filter which when cascaded with the high-pass filter in Fig. 3.28 will give an overall bandwidth of 35 krad/s and an overall maximum gain of 3.17 at the centre frequency. The response should be flat and the roll-off 40 dB/decade.

Solution

For the high-pass filter in Fig. 3.28, normalized values can be calculated by noting that $R_1/R_2 = 1$ and $C_1/C_2 = 2$. Hence $R_1 = R_2 = 1\ \Omega$, $C_1 = 1.414$ F and $C_2 = 0.707$ F. So from the tables the pass-band gain is 2 for these normalized values. Also

$$\omega = \frac{1.414}{10^4 \times 28.3 \times 10^{-9}} = 5 \text{ krad/s}$$

Hence a low-pass filter is required with a cut-off frequency of

$$5 \times 10^3 + 35 \times 10^3 = 40 \text{ krad/s}$$

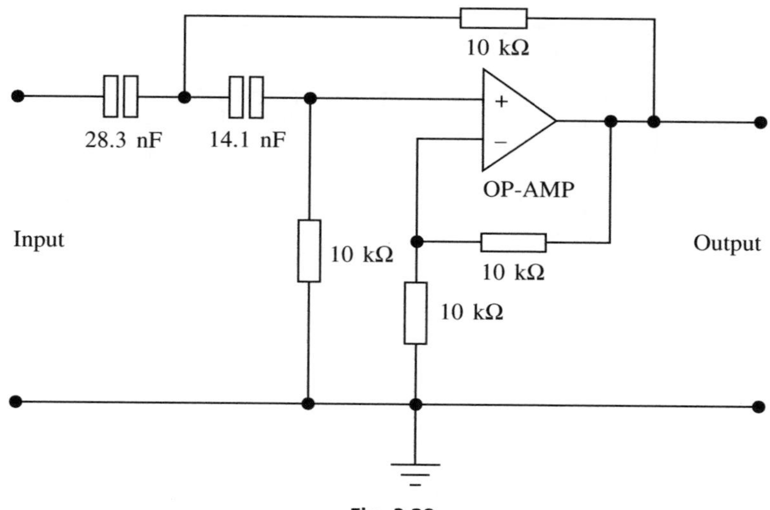

Fig. 3.28

(this is the upper cut-off frequency). Since the maximum gain at the centre frequency has to be 3.17 = 10 dB then the gain of the second filter is

$$3.17/2 = 1.585$$

So the gain of the second filter has to be 1.585, and from the normalized tables for a low-pass Butterworth we have

$$R_1 = R_2 = 1\ \Omega, \qquad C_1 = C_2 = 1\ \text{F}$$

If a denormalization factor of 10^4 is used and $\omega = 40$ krad/s, then

$$R_1' = R_2' = 10\ \text{k}\Omega \qquad C_1' = C_2' = \frac{1}{10^4 \times 4 \times 10^4} = 2.5\ \text{nF}$$

Finally,

$$1 + R_a/R_b = 1.585$$

$$\therefore \qquad R_a/R_b = 0.585$$

Select $R_a = 10$ kΩ and $R_b = 17$ kΩ. The complete filter is shown in Fig. 3.29.

Example 3.14

It is required to build a third-order low-pass filter with a cut-off frequency of 1 kHz and a pass-band gain of 2. Design such a filter.

Solution

A first-order and second-order filter can be connected in series to satisfy this circuit. In order to guarantee a Butterworth response the gain values of both circuits must be adhered to so for the first order a pass-band gain of 1 will be set, while the second order will have a pass-band gain of 2. The usual calculations are carried out using the normalized tables and the Butterworth low-pass normalized values. The full circuit is given in Fig. 3.30.

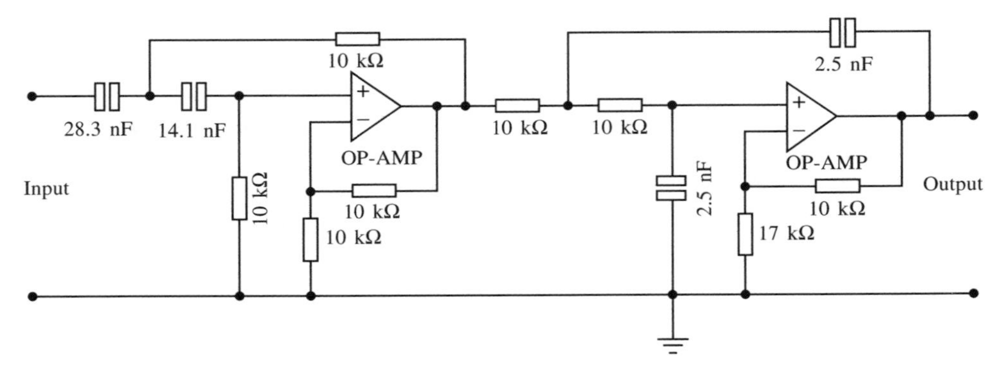

Fig. 3.29

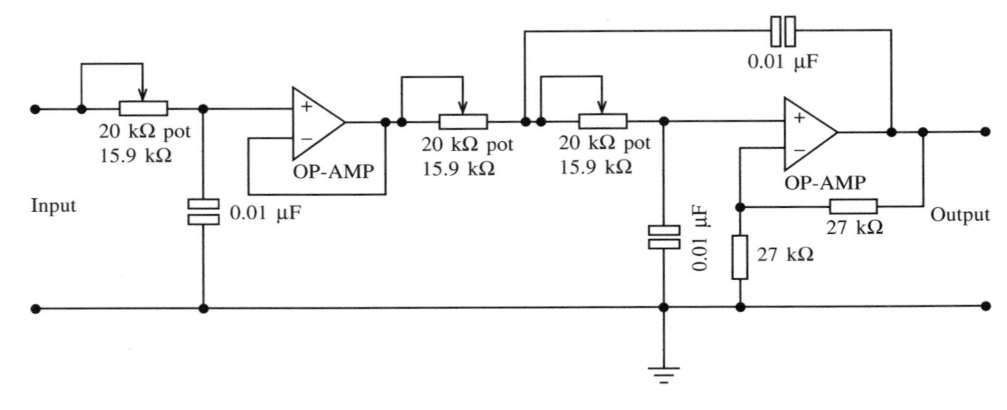

Fig. 3.30

3.13 Additional problems

1. Design a fourth-order high-pass Butterworth filter with a pass-band gain of 4 and a cut-off frequency of 2.5 kHz.
2. Design a low-pass fourth-order Chebyshev 0.5 dB ripple filter with a cut-off frequency of 5 kHz and a pass-band gain of 2.
3. Design a Butterworth band-pass filter with $f_l = 400$ Hz and $f_h = 2$ kHz. The pass-band gain should be 4. Draw the frequency response.
4. A communications system requires a wide-band filter having a band-pass centre frequency of 3.2 kHz and a bandwidth of 800 Hz. If the filter requirements are a roll-off of 40 dB/decade and an overall pass-band gain $K = 1$, design an appropriate filter.

3.14 Switched capacitor filter

Switched capacitor filters have become popular mainly because they require no external components such as capacitors or inductors. Besides offering a very sharp cut-off frequency, these filters have the following advantages: low cost; high accuracy; good temperature stability; and few external components are required. The main disadvantage is that they generate more noise than standard active filters.

The operation of any RC filter depends on the value of the selected resistors and capacitors. Briefly, the switched capacitor filter simulates the resistance by using a capacitor and a few switches.

In Figure 3.31(a) the value of the simulated resistor is proportional to the rate at which the switches are opened and closed in Fig. 3.31(b). If a voltage V_{in} is applied to the resistor then the current through it is given by

$$I = \frac{V_{in}}{R} \tag{3.22}$$

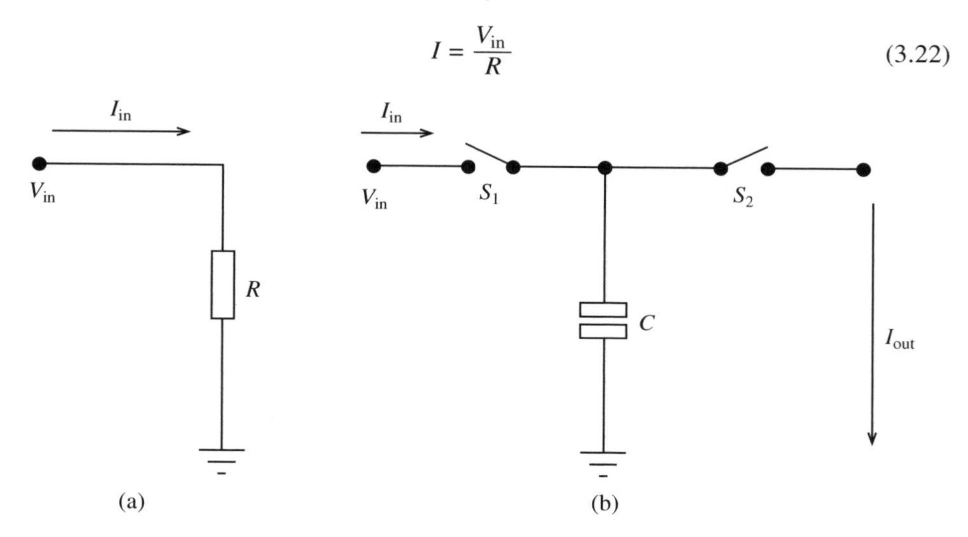

Fig. 3.31

Figure 3.31(b) consists of a capacitor and two switches, which, in practice, would be MOS transistors etched on the integrated circuit. When S_1 is open V_{in} is applied to the capacitor C and hence the total charge on the capacitor is

$$Q = V_{in}\, C \tag{3.23}$$

When S_1 is open and S_2 closed, the charge Q flows to ground. Furthermore, if the switches have no resistance, i.e. they are ideal switches, C will charge and discharge instantly.

Figure 3.32 shows the current into and out of the switched capacitor filter as a function of time. If the switches are opened and closed at a faster rate, the bursts of current will have the same amplitude but will occur more often. Hence the average current will be greater for a higher switching rate. The average current flowing through the capacitor is

$$I_{ave} = \frac{Q}{T} = \frac{V_{in}\, C}{T} = V_{in}\, C f_{clk} \tag{3.24}$$

where T is the time between S_1 and S_2 closing. The equivalent resistance can now be given by

$$R = \frac{V_{in}}{I_{ave}} = \frac{V_{in}}{V_{in}\, C f_{clk}} = \frac{1}{C f_{clk}} \tag{3.25}$$

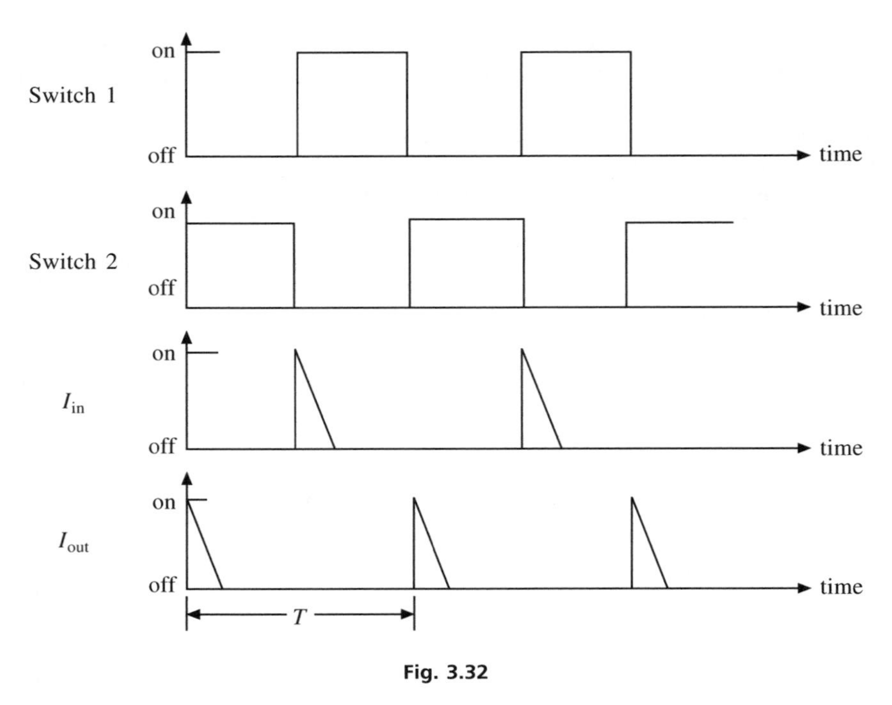

Fig. 3.32

This expression indicates that R is dependent on the clock frequency as C is constant. It should be noted that V_{in} must change at a rate much slower than f_{clk} especially when V_{in} is an a.c. signal.

3.15 Monolithic switched capacitor filter

There are many types of switched capacitor chip on the market, and one of the most common is the MF100 universal switched capacitor filter manufactured by National Semiconductor. It can be used as a bandpass, low-pass, high-pass or notch filter simply by connecting the appropriate resistors externally. The values of these resistors determine the shape of the amplitude and phase responses, while the centre frequency is set by the external clock. The following points should be noted about the MF100:

1. It is a second-order filter.
2. The maximum recommended clock frequency is 1 MHz.
3. Eight different connecting modes are shown in the data sheets, but for most applications mode 3 is used. This will give low-pass, high-pass and bandpass responses.
4. Mode 3 also allows independent adjustment of gain, Q factor and the clock-to-centre frequency ratio. This last feature is particularly advantageous if the only available clock has a frequency other than 50 or 100 times the desired centre frequency or if an application requires two or more filters, each with different centre or cut-off frequencies.

5. The MF10 chip is a dual version of the MF100.
6. The MF100 can operate with a single or split power supply, but the total supply must be between 8 and 14 V.
7. The f_{clk}/f_o ratio affects the performance of the filter. A ratio of 100 : 1 reduces aliasing and is recommended for wide-band input signals. For noise-sensitive applications a ratio of 50 : 1 is better.

Example 3.15
It is required to design a second-order Butterworth low-pass filter with a cut-off frequency of 500 Hz and a pass-band gain of –2.

Solution
Mode 1 is selected as it inverts the signal polarity and also configures for low-pass, band-pass and notch filters. For Mode 1 the following relationships hold:

$$H_{OLP} = -\frac{R_2}{R_1}$$

$$\therefore \qquad R_2 = -R_1 H_{OLP}$$

Let $R_1 = 10$ kΩ. Hence $R_2 = 20$ kΩ. Also

$$Q = \frac{f_o}{BW} = \frac{R_3}{R_2}$$

Since $Q = 0.707$ for a second-order Butterworth low-pass filter,

$$R_3 = QR_2 = 0.707 \times 20 \times 10^3 = 14.14 \text{ k}\Omega$$

(Use 15 kΩ.) Since the cut-off frequency is 500 Hz and $f_{clk}/f_o = 50 : 1$, the external clock frequency is

$$50 \times 500 = 25 \text{ kHz}$$

L. sh (pin 7) should be connected to ground (pin 11) since the clock is CMOS. Finally, pin 5 should be connected to pin 6. The complete circuit is shown in Fig. 3.33.

3.16 The notch filter

A notch filter is sometimes referred to as a frequency rejection circuit as it functions as a bandstop filter passing all frequencies on either side with a flat response, while filtering out a narrow band of frequencies between these two states. Such filters are commonly used for guard bands in multi-channel systems and to remove mains interference from audio circuits. A typical response for a notch filter is shown in Fig. 3.34.

There are two common methods of producing such a filter: using a twin-T network and using a state variable filter. Both methods may be incorporated in an integrated circuit, but a discrete method will be discussed here for the sake of understanding the principles involved.

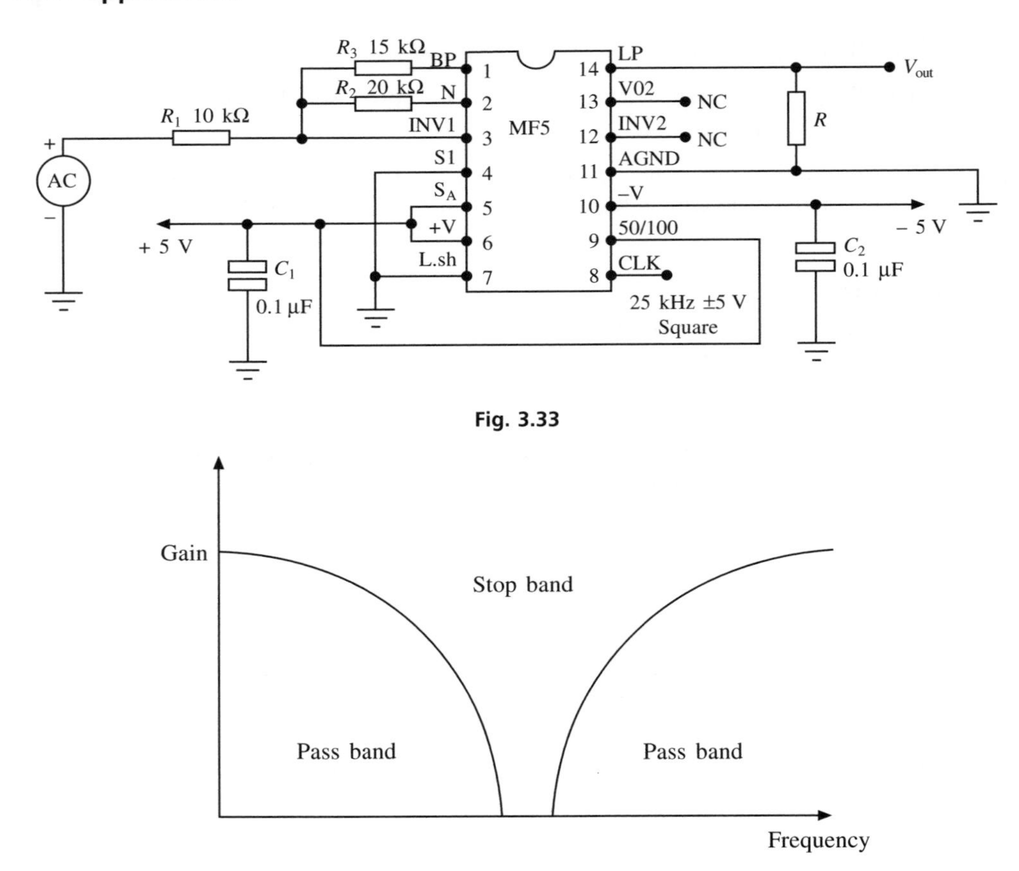

Fig. 3.33

Fig. 3.34

Twin-T network

Figure 3.35 shows a passive twin-T network. Note the values of the components and their configuration. Frequent problems arise with this circuit because of lack of precision when choosing the components. Also the bandwidth of the notch can be wide. In other words,

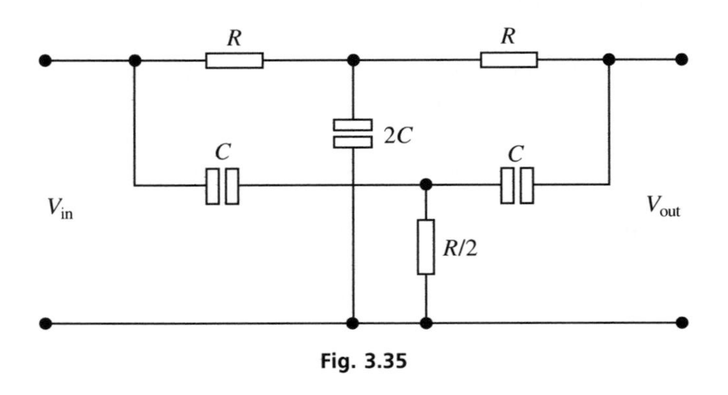

Fig. 3.35

the Q factor is low. This can easily be improved by using an active circuit such as the one shown in Fig. 3.36.

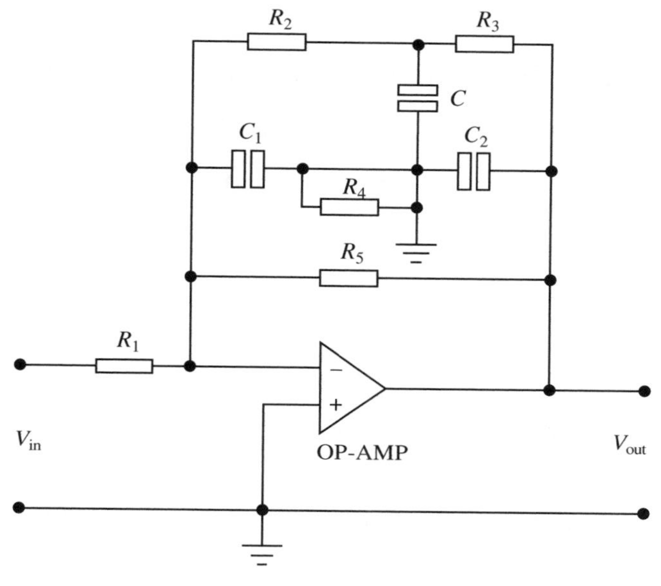

Fig. 3.36

The centre frequency of the twin-T network may be calculated by using the characteristic expression

$$f = \frac{1}{2\pi RC} \tag{3.26}$$

This is the frequency at which the signals passing along the two branches appear to be in antiphase and hence cancel. This cancellation effect causes a sharp dip in the response at and close to the resonant frequency.

This filter is useful but only for a fixed frequency. A higher Q value with frequency tuning may be achieved by using a state variable filter.

The state variable filter

This filter is widely used in bandpass applications and usually comes in integrated circuit form. However, it can be constructed using a summing amplifier and two integrators as shown in Fig. 3.37.

Note that this filter can be configured as a low-pass and a high-pass circuit as well as a bandpass filter. The centre frequency is set by the values of R and C in both the integrators, and when used as a bandpass filter the critical frequencies (f_c) of the integrators are usually equal.

At frequencies below the critical frequency the input signal passes through the summing amplifier and integrators and, as can be seen from Fig. 3.34, is fed back to the summer amplifiers in antiphase. Hence the feedback and input signals cancel for all frequencies

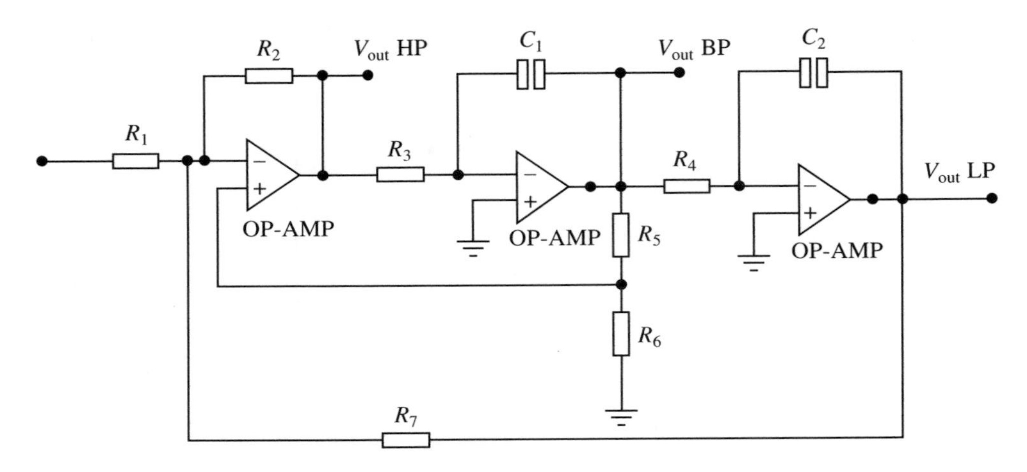

Fig. 3.37

below the critical frequency. This is ideal due to capacitor and resistor tolerances, but the cut-off is sharp in practice. As the low-pass response of the integrators rolls off, the feedback voltage reduces and the input passes through the bandpass output. For signals above the critical frequency the low-pass response disappears and prevents the input signal from passing through the integrators. This results in the bandpass output peaking sharply at the critical frequency.

The Q factor or selectivity of the filter is determined by R_5 and R_6 in Fig. 3.37 and may be calculated from the expression

$$Q = \frac{1}{3}\left(\frac{R_5}{R_6} + 1\right) \tag{3.27}$$

The filter is normally set for a high Q factor, but the high-pass and low-pass filters cannot be simultaneously set for optimum conditions. This is not important, however, when the state variable filter is being used as a notch filter. Figure 3.38 shows how the state variable filter can be used as a notch filter by connecting the high and low-pass outputs to a summer amplifier.

This type of filter can be tuned manually by switching in capacitors or including variable capacitors in the integrator circuits. RV_1 may also be included to alter the gain of

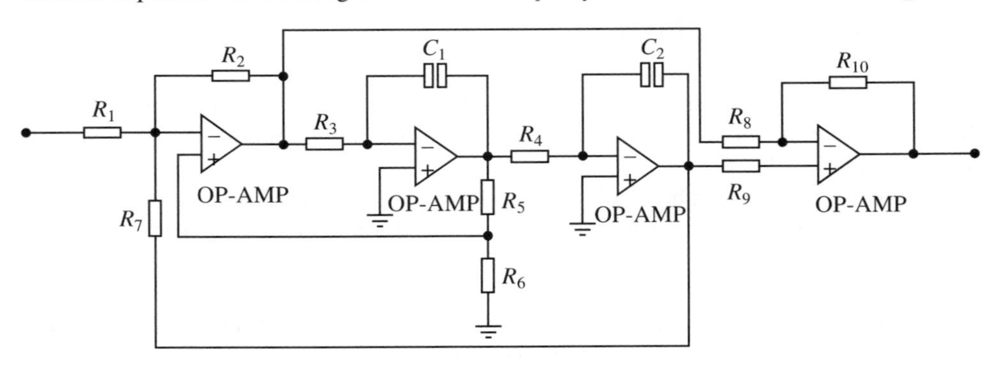

Fig. 3.38

the filter output, while RV_2 and RV_3 are usually ganged variable resistors used to vary the frequency as they are varied from 1 kΩ to 10 kΩ. A practical filter using these techniques is shown in Fig. 3.39. Note that in order to optimize the low and high-pass outputs a damping circuit would normally be connected between the bandpass output and the input of the filter. However, as this configuration is being used as a notch filter this is not necessary. It should be appreciated that this filter is manually tuned, but where electronic tuning is required the switched capacitor filter already mentioned is used.

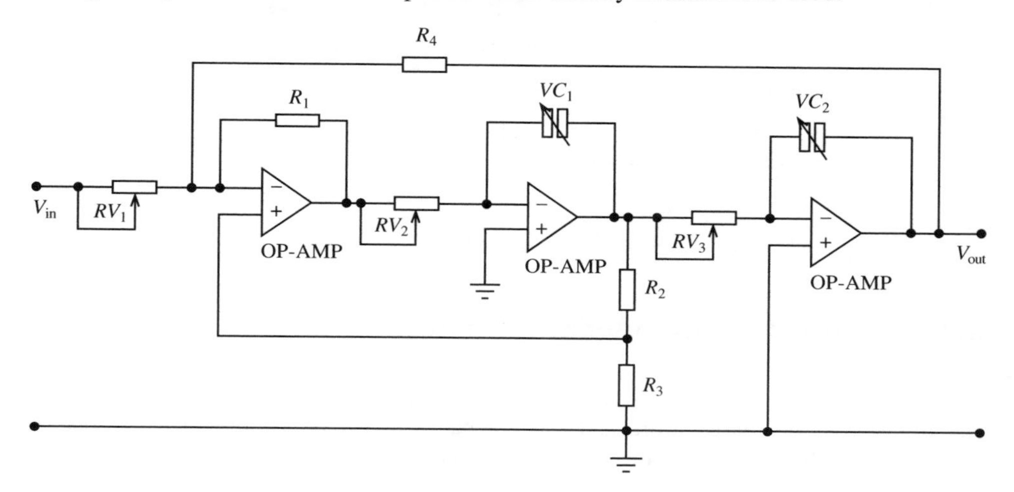

Fig. 3.39

Example 3.16

A notch filter has to be designed in such a way as to eliminate a 50 Hz hum on a data communications line. In order to achieve this a Q factor of 40 is required. Design a suitable circuit which would practically achieve this.

Solution

The best design for this type of application would be a state variable filter using the summer amplifier. Select a capacitor value of 0.2 μF and determine the integrator resistor values.

$$R = \frac{1}{2\pi fC} = \frac{10^6}{2\pi \times 50 \times 0.2} = 15.9\ \text{k}\Omega$$

Also

$$Q = \frac{1}{3}\left(\frac{R_5}{R_6} + 1\right)$$

$$R_5 = (3Q - 1)\ R_6$$

Select $R_6 = 1$ kΩ. Then

$$R_5 = \{(3 \times 40) - 1\}1$$
$$= 119\ \text{k}\Omega$$

The complete circuit may now be drawn with a unity gain summer amplifier using 1 kΩ resistors. This is shown in Fig. 3.40.

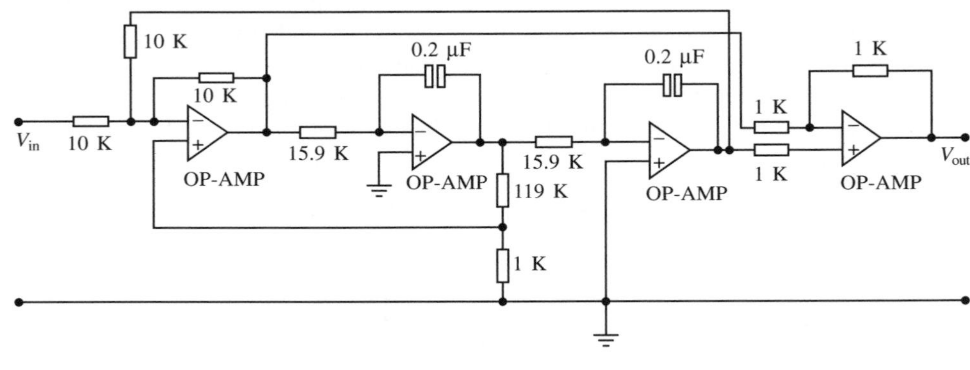

Fig. 3.40

3.17 Choosing components for filters

The selection of components in the construction of filters is more precise than in many electronic circuits as sharp cut-offs and selection bands have to be accommodated. Capacitor selection is perhaps more important as these encompass a large range of materials and tolerances.

Resistor selection

Generally fixed resistors should have tolerances of ± 1% or ± 2%, but ± 5% is adequate for less critical circuit design. Tolerances less than this may be required for notch filters. Carbon track resistors may be suitable if they are properly calibrated on a bridge such as a Wayne–Kerr bridge. An alternative to this would be Cermet track variable resistors, which would give better reliability. However, for greater accuracy a bridge should always be used if available.

Capacitor selection

Silvered mica

These capacitors have the highest tolerance (±1%) but the maximum value commonly available may only be 4.7 nF. They have good temperature stability, and this is important if the filter has to operate over a wide range of temperatures.

Polystyrene

These capacitors are most suitable for filters because of their close tolerance and large capacitance range. They also have excellent temperature stability.

Ceramic

These come in three types, namely metallized, resin-dipped and disc. The metallized type has good tolerance (± 2%) and temperature stability. The resin-dipped type has tolerances of ±5%. Disc types have very poor tolerance, making them unsuitable for filter design.

Polyester

When a capacitor of larger value is required this may be the choice. Their tolerance is between ± 5% and ±10% and their temperature stability is poor.

Electrolytic

These capacitors have a tolerance of ±20% or more and their capacitance is likely to change more quickly with use. This, together with the fact that they are polarized, makes them unsuitable for filter circuits.

Tantalum

These capacitors are also unsuitable for filter design for the same reasons as electrolytic capacitors.

3.18 Testing filter response

There are two basic methods of measuring filter response: the signal generator and oscilloscope method, and the sweep frequency method.

Signal generator and oscilloscope method

This method is the one frequently adopted due to the availability of equipment. The test set-up is shown in Fig. 3.41. The procedure is as follows

1. The amplitude of the signal generator is set to a suitable voltage level with no distortion showing on the oscilloscope.

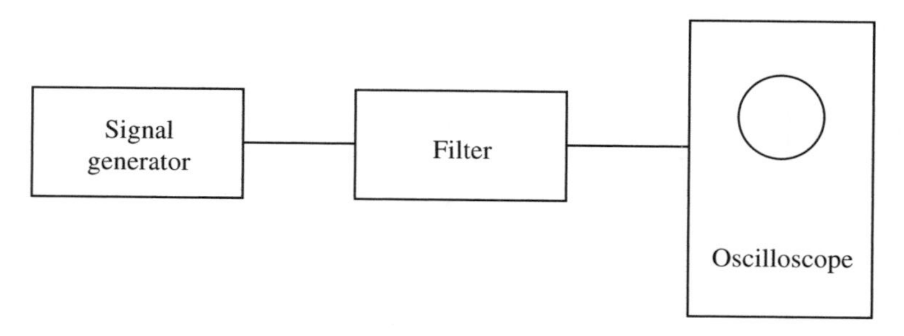

Fig. 3.41

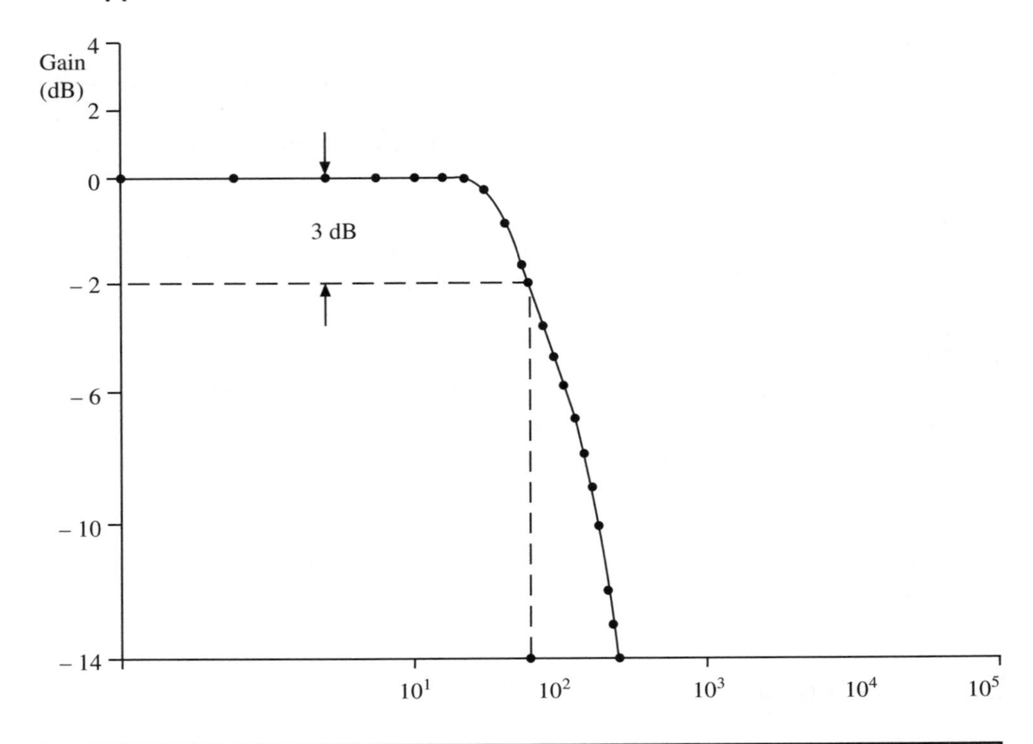

Frequency	V_{in}	V_{out}	$G = V_{out}/V_{in}$	dB = 20 log G
100	20	20	1	0
250	20	20	1	0
500	20	20	1	0
750	20	20	1	0
1000	20	20	1	0
1250	20	20	1	0
1500	20	20	1	0
1750	20	19	0.95	–0.44
2000	20	17	0.85	–1.41
2250	20	15	0.75	–2.5
2500	20	14	0.7	–3
2750	20	12	0.6	–4.4
3000	20	11	0.55	–5.2
3250	20	10	0.5	–6
3500	20	9	0.45	–7
3750	20	8	0.4	–8
4000	20	7	0.35	–9
4250	20	6	0.3	–10.5
4500	20	5	0.25	–12
4750	20	5	0.25	–13
5000	20	4	0.2	–14

Fig. 3.42

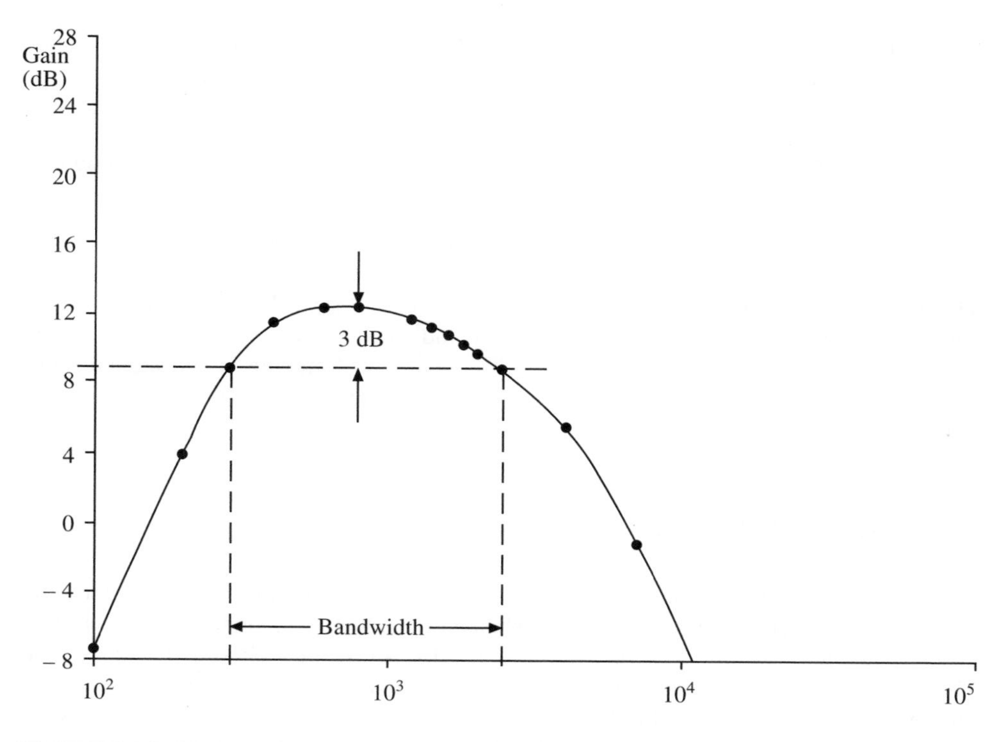

Frequency (Hz)	V_{in}	V_{out}	$G = V_{out}/V_{in}$	dB = 20 log G
100	3.6	1.5	0.42	– 7.6
200	3.6	5.8	1.61	4.14
400	3.6	13	3.61	11.5
600	3.6	14.4	4	12.04
800	3.6	14.4	4	12.04
1000	3.6	14.4	4	12.04
1200	3.6	14.2	3.94	11.91
1400	3.6	14	3.89	11.8
1600	3.6	13.6	3.78	11.5
1800	3.6	13.3	3.74	11.05
2000	3.6	12.8	3.56	11
2200	3.6	12.4	3.44	10.73
2400	3.6	11.8	3.28	10.32
2600	3.6	11.2	3.11	9.86
2800	3.6	10.6	2.94	9.37
3000	3.6	10	2.78	8.88
3200	3.6	9.4	2.61	8.33
3400	3.6	8.4	2.33	7.35
3600	3.6	8.2	2.28	7.16
3800	3.6	7.6	2.11	6.49
4000	3.6	7	1.99	5.76
7000	3.6	0.18	0.05	– 1

Fig. 3.43

2. The frequency of the signal generator is increased in predetermined steps. Sufficient steps should be selected to give an accurate response when plotted.
3. The input voltage should remain constant for each output.
4. After sufficient points have been recorded, a table similar to the one shown in Fig. 3.42 should be prepared.
5. The graph is then plotted on log-linear graph paper as shown in Fig. 3.42.

This is a typical response for a second order low-pass Butterworth filter. Note the cut-off frequency, 3 dB point and roll-off which are indicated. Also it is customary to plot decibels vertically on the linear scale while frequency is plotted on the horizontal scale.

A second example is shown in Fig. 3.43. In this example a bandpass filter has been used.

The sweep frequency method

This method involves more sophisticated equipment available at the larger telecommunications companies and more sophisticated teaching laboratories. It is a more efficient method and produces very accurate results. A test set-up is shown in Fig. 3.44. The sweep frequency generator uses two preset limits sometimes called markers; depending on the expected response of the filter, the generator is set between these limits. As the input frequency sweeps through the required range, a response curve is traced out on the spectrum analyser as shown in Fig. 3.44.

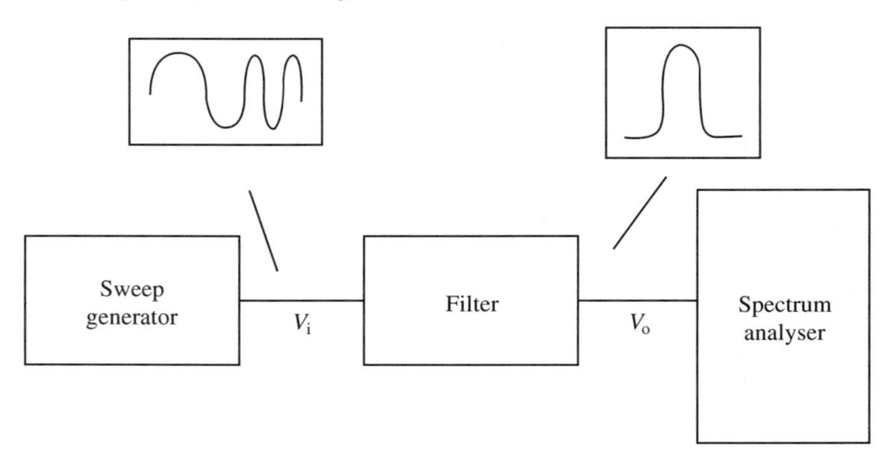

Fig. 3.44

Data sheets

January 1995

LMF100 High Performance Dual Switched Capacitor Filter

General Description

The LMF100 consists of two independent general purpose high performance switched capacitor filters. With an external clock and 2 to 4 resistors, various second-order and first-order filtering functions can be realized by each filter block. Each block has 3 outputs. One output can be configured to perform either an allpass, highpass, or notch function. The other two outputs perform bandpass and lowpass functions. The center frequency of each filter stage is tuned by using an external clock or a combination of a clock and resistor ratio. Up to a 4th-order biquadratic function can be realized with a single LMF100. Higher order filters are implemented by simply cascading additional packages, and all the classical filters (such as Butterworth, Bessel, Elliptic, and Chebyshev) can be realized.

The LMF100 is fabricated on National Semiconductor's high performance analog silicon gate CMOS process, LMCMOS™. This allows for the production of a very low offset, high frequency filter building block. The LMF100 is pin-compatible with the industry standard MF10, but provides greatly improved performance.

Features

- Wide 4V to 15V power supply range
- Operation up to 100 kHz
- Low offset voltage (50:1 or 100:1 mode) typically Vos1 = ±5 mV, Vos2 = ±15 mV, Vos3 = ±15 mV
- Low crosstalk −60 dB
- Clock to center frequency ratio accuracy ±0.2% typical
- $f_0 \times Q$ range up to 1.8 MHz
- Pin-compatible with MF10

4th Order 100 kHz Butterworth Lowpass Filter

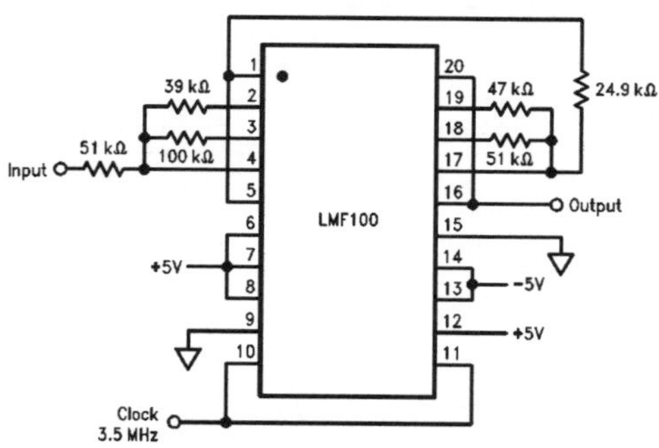

TL/H/5645–2

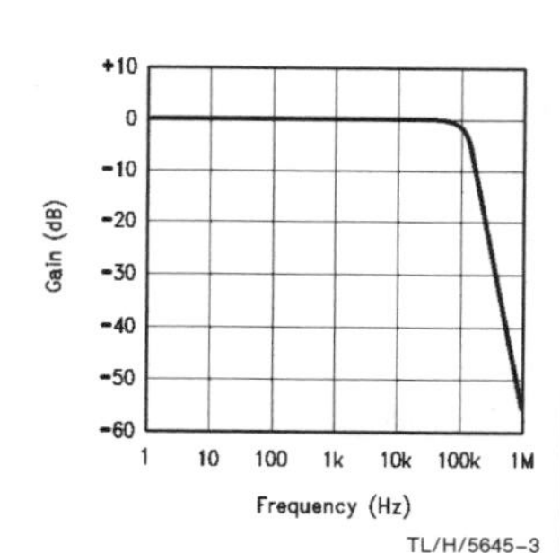

TL/H/5645–3

Connection Diagram

Surface Mount and Dual-In-Line Package

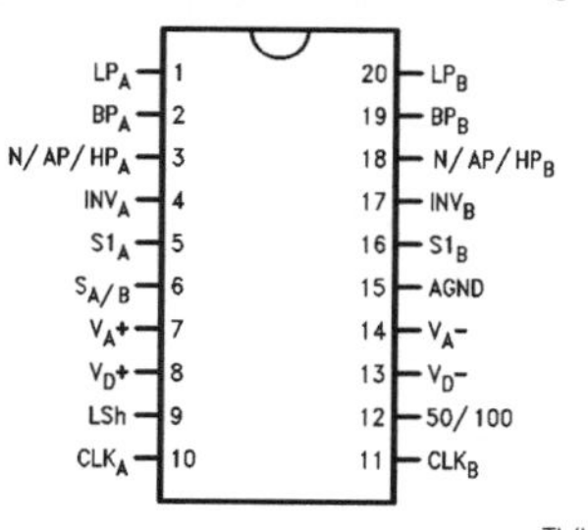

TL/H/5645–18

Top View

Order Number LMF100AE/883 or 5962-9153301M2A, LMF100AJ, LMF100AJ/883 or 5962-9153301MRA, LMF100CIJ, LMF100ACN, LMF100CCN, LMF100CIN or LMF100CIWM
See NS Package Number J20A, N20A or M20B

RRD-B30M115/Printed in U. S. A.

Absolute Maximum Ratings (Note 1)

If Military/Aerospace specified devices are required, please contact the National Semiconductor Sales Office/Distributors for availability and specifications. (Note 14)

Supply Voltage ($V^+ - V^-$)	16V
Voltage at Any Pin	$V^+ + 0.3V$ $V^- - 0.3V$
Input Current at Any Pin (Note 2)	5 mA
Package Input Current (Note 2)	20 mA
Power Dissipation (Note 3)	500 mW
Storage Temperature	150°C
ESD Susceptability (Note 11)	2000V

Soldering Information

N Package: 10 sec.	260°C
J Package: 10 sec.	300°C
SO Package: Vapor Phase (60 sec.)	215°C
Infrared (15 sec.)	220°C

See AN-450 "Surface Mounting Methods and Their Effect on Product Reliability" (Appendix D) for other methods of soldering surface mount devices.

Operating Ratings (Note 1)

Temperature Range	$T_{MIN} \le T_A \le T_{MAX}$
LMF100ACN, LMF100CCN	$0°C \le T_A \le +70°C$
LMF100CIJ, LMF100CIN, LMF100CIWM	$-40°C \le T_A \le +85°C$
LMF100AJ, MF100AJ/883, LMF100AE/883	$-55°C \le T_A \le +125°C$
Supply Voltage	$4V \le V^+ - V^- \le 15V$

Electrical Characteristics

The following specifications apply for Mode 1, Q = 10 ($R_1 = R_3 = 100k$, $R_2 = 10k$), $V^+ = +5V$ and $V^- = -5V$ unless otherwise specified. **Boldface limits apply for T_{MIN} to T_{MAX}**; all other limits $T_A = T_J = 25°C$.

Symbol	Parameter		Conditions		LMF100ACN, LMF100CCN			LMF100AJ, LMF100CIN, LMF100CIWM, LMF100CIJ			Units
					Typical (Note 8)	Tested Limit (Note 9)	Design Limit (Note 10)	Typical (Note 8)	Tested Limit (Note 9)	Design Limit (Note 10)	
I_s	Maximum Supply Current		f_{CLK} = 250 kHz No Input Signal		9	13	**13**	9	**13**		mA
f_0	Center Frequency Range	MIN			0.1			0.1			Hz
		MAX			100			100			kHz
f_{CLK}	Clock Frequency Range	MIN			5.0			5.0			Hz
		MAX			3.5			3.5			MHz
f_{CLK}/f_0	Clock to Center Frequency Ratio Deviation		V_{Pin12} = 5V or 0V f_{CLK} = 1 MHz	LMF100A	±0.2	±0.6	**±0.6**	±0.2	**±0.6**		%
				LMF100C	±0.2	±0.8	**±0.8**	±0.2	**±0.8**		%
$\frac{\Delta Q}{Q}$	Q Error (MAX) (Note 4)		Q = 10, Mode 1 V_{Pin12} = 5V or 0V f_{CLK} = 1 MHz	LMF100A	±0.5	±4	**±5**	±0.5	**±5**		%
				LMF100C	±0.5	±5	**±6**	±0.5	**±6**		%
H_{OBP}	Bandpass Gain at f_0		f_{CLK} = 1 MHz		0	±0.4	**±0.4**	0	**±0.4**		dB
H_{OLP}	DC Lowpass Gain		$R_1 = R_2 = 10k$ f_{CLK} = 250 kHz		0	±0.2	**±0.2**	0	**±0.2**		dB
V_{OS1}	DC Offset Voltage (Note 5)		f_{CLK} = 250 kHz		±5.0	±15	**±15**	±5.0	**±15**		mV
V_{OS2}	DC Offset Voltage (Note 5)		f_{CLK} = 250 kHz	$S_{A/B} = V^+$	±30	±80	**±80**	±30	**±80**		mV
				$S_{A/B} = V^-$	±15	±70	**±70**	±15	**±70**		mV
V_{OS3}	DC Offset Voltage (Note 5)		f_{CLK} = 250 kHz		±15	±40	**±60**	±15	**±60**		mV
	Crosstalk (Note 6)		A Side to B Side or B Side to A Side		−60			−60			dB
	Output Noise (Note 12)		f_{CLK} = 250 kHz 20 kHz Bandwidth 100:1 Mode	N	40			40			μV
				BP	320			320			
				LP	300			300			
	Clock Feedthrough (Note 13)		f_{CLK} = 250 kHz 100:1 Mode		6			6			mV
V_{OUT}	Minimum Output Voltage Swing		R_L = 5k (All Outputs)		+4.0 −4.7	±3.8	**±3.7**	+4.0 −4.7	**±3.7**		V
			R_L = 3.5k (All Outputs)		+3.9 −4.6			+3.9 −4.6			V
GBW	Op Amp Gain BW Product				5			5			MHz
SR	Op Amp Slew Rate				20			20			V/μs

Electrical Characteristics

The following specifications apply for Mode 1, Q = 10 (R_1 = R_3 = 100k, R_2 = 10k), V^+ = +5V and V^- = −5V unless otherwise specified. **Boldface limits apply for T_{MIN} to T_{MAX}**; all other limits T_A = T_J = 25°C. (Continued)

Symbol	Parameter		Conditions	LMF100ACN, LMF100CCN			LMF100AJ, LMF100CIN, LMF100CIWM, LMF100CIJ			Units
				Typical (Note 8)	Tested Limit (Note 9)	Design Limit (Note 10)	Typical (Note 8)	Tested Limit (Note 9)	Design Limit (Note 10)	
I_{SC}	Maximum Output Short Circuit Current (Note 7)	Source	(All Outputs)	12			12			mA
		Sink		45			45			mA
I_{IN}	Input Current on Pins: 4, 5, 6, 9, 10, 11, 12, 16, 17				10			**10**		μA

Electrical Characteristics

The following specifications apply for Mode 1, Q = 10 (R_1 = R_3 = 100k, R_2 = 10k), V^+ = +2.50V and V^- = −2.50V unless otherwise specified. **Boldface limits apply for T_{MIN} to T_{MAX}**; all other limits T_A = T_J = 25°C.

Symbol	Parameter		Conditions		LMF100ACN, LMF100CCN			LMF100AJ, LMF100CIN, LMF100CIWM, LMF100CIJ			Units
					Typical (Note 8)	Tested Limit (Note 9)	Design Limit (Note 10)	Typical (Note 8)	Tested Limit (Note 9)	Design Limit (Note 10)	
I_s	Maximum Supply Current		f_{CLK} = 250 kHz No Input Signal		8	12	**12**	8	**12**		mA
f_0	Center Frequency Range	MIN			0.1			0.1			Hz
		MAX			50			50			kHz
f_{CLK}	Clock Frequency Range	MIN			5.0			5.0			Hz
		MAX			1.5			1.5			MHz
f_{CLK}/f_0	Clock to Center Frequency Ratio Deviation		V_{Pin12} = 2.5V or 0V f_{CLK} = 1 MHz	LMF100A	±0.2	±0.6	**±0.8**	±0.2	**±0.8**		%
				LMF100C	±0.2	±1	**±1**	±0.2	**±1**		%
$\frac{\Delta Q}{Q}$	Q Error (MAX) (Note 4)		Q = 10, Mode 1 V_{Pin12} = 5V or 0V f_{CLK} = 1 MHz	LMF100A	±0.5	±4	**±6**	±0.5	**±6**		%
				LMF100C	±0.5	±5	**±8**	±0.5	**±8**		%
H_{OBP}	Bandpass Gain at f_0		f_{CLK} = 1 MHz		0	±0.4	**±0.5**	0	**±0.5**		dB
H_{OLP}	DC Lowpass Gain		R_1 = R_2 = 10k f_{CLK} = 250 kHz		0	±0.2	**±0.2**	0	**±0.2**		dB
V_{OS1}	DC Offset Voltage (Note 5)		f_{CLK} = 250 kHz		±5.0	±15	**±15**	±5.0	**±15**		mV
V_{OS2}	DC Offset Voltage (Note 5)		f_{CLK} = 250 kHz	$S_{A/B}$ = V^+	±20	±60	**±60**	±20	**±60**		mV
				$S_{A/B}$ = V^-	±10	±50	**±60**	±10	**±60**		mV
V_{OS3}	DC Offset Voltage (Note 5)		f_{CLK} = 250 kHz		±10	±25	**±30**	±10	**±30**		mV
	Crosstalk (Note 6)		A Side to B Side or B Side to A Side		−65			−65			dB
	Output Noise (Note 12)		f_{CLK} = 250 kHz 20 kHz Bandwidth 100:1 Mode	N	25			25			μV
				BP	250			250			μV
				LP	220			220			μV
	Clock Feedthrough (Note 13)		f_{CLK} = 250 kHz 100:1 Mode		2			2			mV
V_{OUT}	Minimum Output Voltage Swing		R_L = 5k (All Outputs)		+1.6 −2.2	±1.5	**±1.4**	+1.6 −2.2	**±1.4**		V
			R_L = 3.5k (All outputs)		+1.5 −2.1			+1.5 −2.1			V
GBW	Op Amp Gain BW Product				5			5			MHz
SR	Op Amp Slew Rate				18			18			V/μs
I_{SC}	Maximum Output Short Circuit Current (Note 7)	Source	(All Outputs)		10			10			mA
		Sink			20			20			mA

Logic Input Characteristics **Boldface limits apply for T_{MIN} to T_{MAX};** all other limits $T_A = T_J = 25°C$.

Parameter		Conditions	LMF100ACN, LMF100CCN			LMF100AJ, LMF100CIN, LMF100CIWM, LMF100CIJ			Units
			Typical (Note 8)	Tested Limit (Note 9)	Design Limit (Note 10)	Typical (Note 8)	Tested Limit (Note 9)	Design Limit (Note 10)	
CMOS Clock Input Voltage	MIN Logical "1"	$V^+ = +5V$, $V^- = -5V$, $V_{LSh} = 0V$		+3.0	**+3.0**		**+3.0**		V
	MAX Logical "0"			−3.0	**−3.0**		**−3.0**		V
	MIN Logical "1"	$V^+ = +10V$, $V^- = 0V$, $V_{LSh} = +5V$		+8.0	**+8.0**		**+8.0**		V
	MAX Logical "0"			+2.0	**+2.0**		**+2.0**		V
TTL Clock Input Voltage	MIN Logical "1"	$V^+ = +5V$, $V^- = -5V$, $V_{LSh} = 0V$		+2.0	**+2.0**		**+2.0**		V
	MAX Logical "0"			+0.8	**+0.8**		**+0.8**		V
	MIN Logical "1"	$V^+ = +10V$, $V^- = 0V$, $V_{LSh} = 0V$		+2.0	**+2.0**		**+2.0**		V
	MAX Logical "0"			+0.8	**+0.8**		**+0.8**		V
CMOS Clock Input Voltage	MIN Logical "1"	$V^+ = +2.5V$, $V^- = -2.5V$, $V_{LSh} = 0V$		+1.5	**+1.5**		**+1.5**		V
	MAX Logical "0"			−1.5	**−1.5**		**−1.5**		V
	MIN Logical "1"	$V^+ = +5V$, $V^- = 0V$, $V_{LSh} = +2.5V$		+4.0	**+4.0**		**+4.0**		V
	MAX Logical "0"			+1.0	**+1.0**		**+1.0**		V
TTL Clock Input Voltage	MIN Logical "1"	$V^+ = +5V$, $V^- = 0V$, $V_{LSh} = 0V$, $V_D^+ = 0V$		+2.0	**+2.0**		**+2.0**		V
	MAX Logical "0"			+0.8	**+0.8**		**+0.8**		V

Note 1: Absolute Maximum Ratings indicate limits beyond which damage to the device may occur. Operating Ratings indicate conditions for which the device is intended to be functional. These ratings do not guarantee specific performance limits, however. For guaranteed specifications and test conditions, see the Electrical Characteristics. The guaranteed specifications apply only for the test conditions listed. Some performance characteristics may degrade when the device is not operated under the listed test conditions.

Note 2: When the input voltage (V_{IN}) at any pin exceeds the power supply rails ($V_{IN} < V^-$ or $V_{IN} > V^+$) the absolute value of current at that pin should be limited to 5 mA or less. The sum of the currents at all pins that are driven beyond the power supply voltages should not exceed 20 mA.

Note 3: The maximum power dissipation must be derated at elevated temperatures and is dictated by T_{JMAX}, θ_{JA}, and the ambient temperature, T_A. The maximum allowable power dissipation at any temperature is $P_D = (T_{JMAX} - T_A)/\theta_{JA}$ or the number given in the Absolute Maximum Ratings, whichever is lower. For this device, $T_{JMAX} = 125°C$, and the typical junction-to-ambient thermal resistance of the LMF100ACN/CCN/CIN when board mounted is 55°C/W. For the LMF100AJ/CIJ, this number increases to 95°C/W and for the LMF100CIWM this number is 66°C/W.

Note 4: The accuracy of the Q value is a function of the center frequency (f_0). This is illustrated in the curves under the heading "Typical Peformance Characteristics".

Note 5: V_{os1}, V_{os2}, and V_{os3} refer to the internal offsets as discussed in the Applications Information section 3.4.

Note 6: Crosstalk between the internal filter sections is measured by applying a 1 V_{RMS} 10 kHz signal to one bandpass filter section input and grounding the input of the other bandpass filter section. The crosstalk is the ratio between the output of the grounded filter section and the 1 V_{RMS} input signal of the other section.

Note 7: The short circuit source current is measured by forcing the output that is being tested to its maximum positive voltage swing and then shorting that output to the negative supply. The short circuit sink current is measured by forcing the output that is being tested to its maximum negative voltage swing and then shorting that output to the positive supply. These are the worst case conditions.

Note 8: Typicals are at 25°C and represent most likely parametric norm.

Note 9: Tested limits are guaranteed to National's AOQL (Average Outgoing Quality Level).

Note 10: Design limits are guaranteed to National's AOQL (Average Outgoing Quality Level) but are not 100% tested.

Note 11: Human body model, 100 pF discharged through a 1.5 kΩ resistor.

Note 12: In 50:1 mode the output noise is 3 dB higher.

Note 13: In 50:1 mode the clock feedthrough is 6 dB higher.

Note 14: A military RETS specification is available upon request.

Typical Performance Characteristics

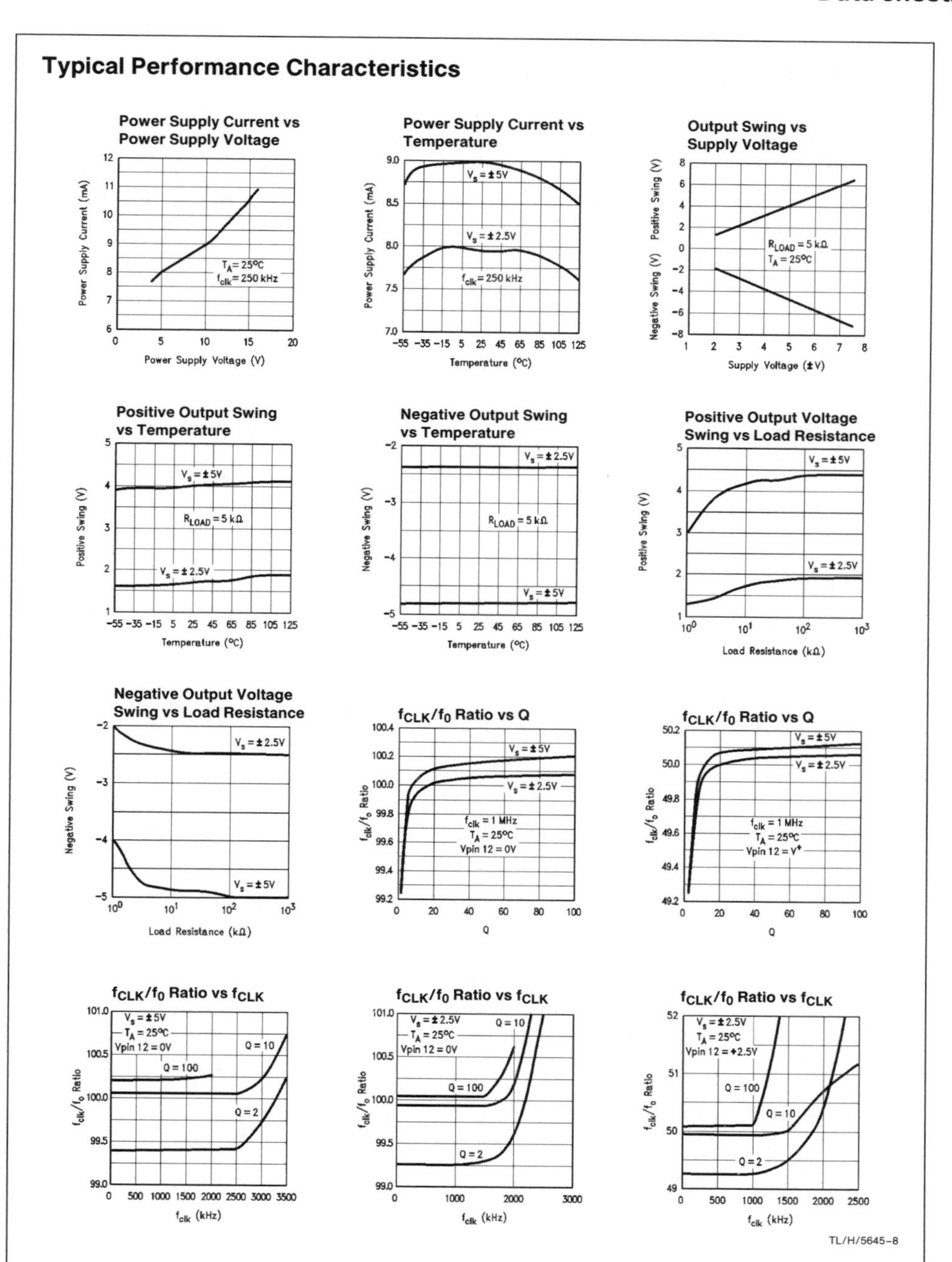

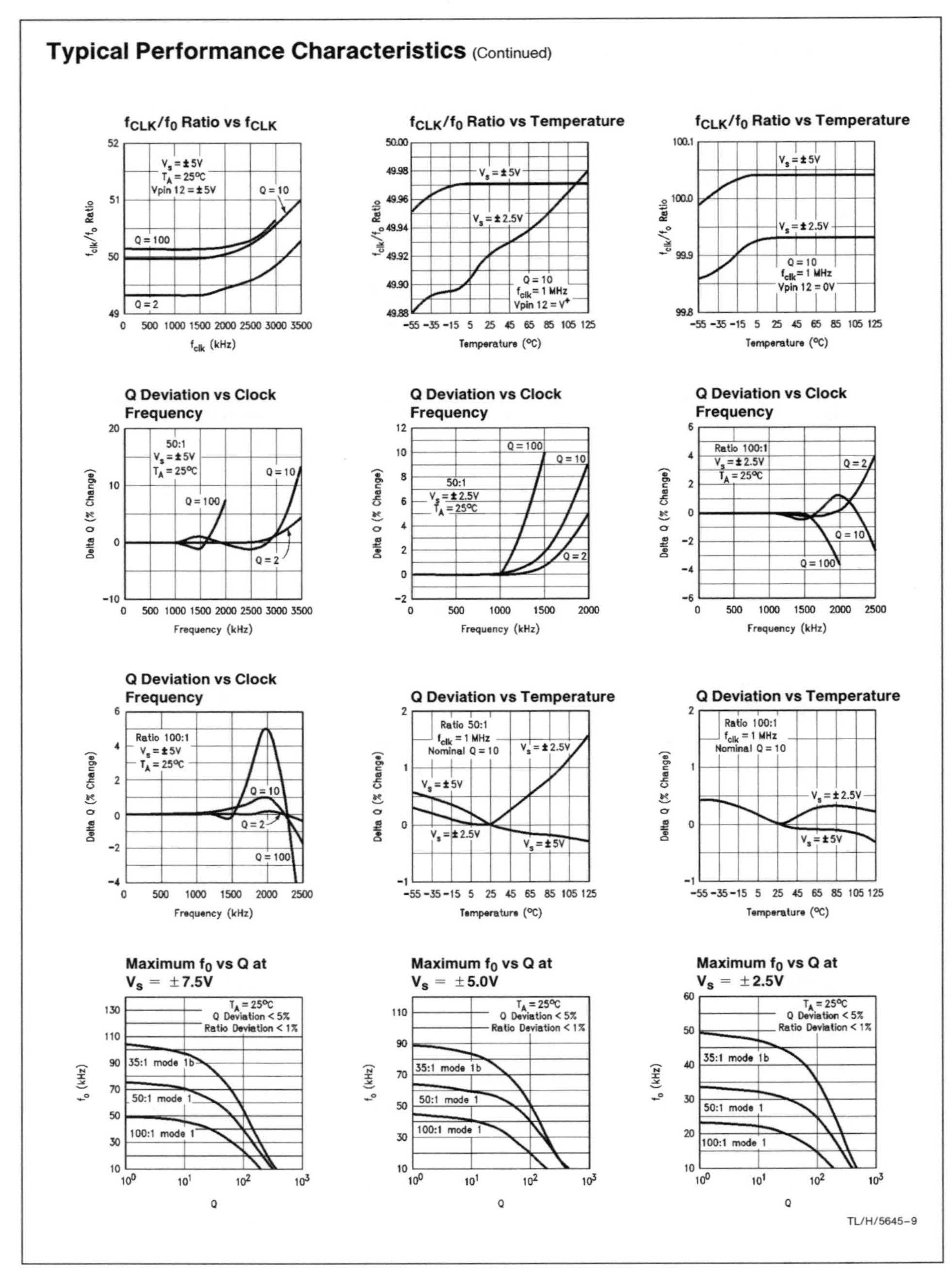
Typical Performance Characteristics (Continued)
fCLK/f0 Ratio vs fCLK
fCLK/f0 Ratio vs Temperature
fCLK/f0 Ratio vs Temperature
Q Deviation vs Clock Frequency
Q Deviation vs Clock Frequency
Q Deviation vs Clock Frequency
Q Deviation vs Clock Frequency
Q Deviation vs Temperature
Q Deviation vs Temperature
Maximum f0 vs Q at Vs = ±7.5V
Maximum f0 vs Q at Vs = ±5.0V
Maximum f0 vs Q at Vs = ±2.5V
TL/H/5645-9

LMF100 System Block Diagram

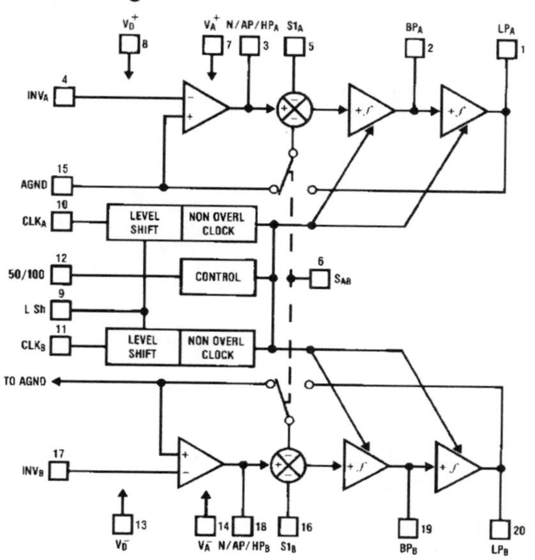

TL/H/5645-1

Pin Descriptions

LP(1,20), BP(2,19), N/AP/HP(3,18) The second order lowpass, bandpass and notch/allpass/highpass outputs. These outputs can typically swing to within 1V of each supply when driving a 5 kΩ load. For optimum performance, capacitive loading on these outputs should be minimized. For signal frequencies above 15 kHz the capacitance loading should be kept below 30 pF.

INV(4,17) The inverting input of the summing opamp of each filter. These are high impedance inputs. The non-inverting input is internally tied to AGND so the opamp can be used only as an inverting amplifier.

S1(5,16) S1 is a signal input pin used in modes 1b, 4, and 5. The input impedance is $1/f_{CLK} \times 1$ pF. The pin should be driven with a source impedance of less than 1 kΩ. If S1 is not driven with a signal it should be tied to AGND (mid-supply).

$S_{A/B}$(6) This pin activates a switch that connects one of the inputs of each filter's second summer either to AGND ($S_{A/B}$ tied to V^-) or to the lowpass (LP) output ($S_{A/B}$ tied to V^+). This offers the flexibility needed for configuring the filter in its various modes of operation.

V_A^+(7)* This is both the analog and digital positive supply.

V_D^+(8)* This pin needs to be tied to V^+ except when the device is to operate on a single 5V supply and a TTL level clock is applied. For 5V, TTL operation, V_D^+ should be tied to ground (0V).

V_A^-(14), V_D^-(13) Analog and digital negative supplies. V_A^- and V_D^- should be derived from the same source. They have been brought out separately so they can be bypassed by separate capacitors, if desired. They can also be tied together externally and bypassed with a single capacitor.

Pin Descriptions (Continued)

LSh(9) — Level shift pin. This is used to accommodate various clock levels with dual or single supply operation. With dual ±5V supplies and CMOS (±5V) or TTL (0V–5V) clock levels, LSh should be tied to system ground.

For 0V–10V single supply operation the AGND pin should be biased at +5V and the LSh pin should be tied to the system ground for TTL clock levels. LSh should be biased at +5V for ±5V CMOS clock levels.

The LSh pin is tied to system ground for ±2.5V operation. For single 5V operation the LSh and V_D+ pins are tied to system ground for TTL clock levels.

CLK(10,11) — Clock inputs for the two switched capacitor filter sections. Unipolar or bipolar clock levels may be applied to the CLK inputs according to the programming voltage applied to the LSh pin. The duty cycle of the clock should be close to 50%, especially when clock frequencies above 200 kHz are used. This allows the maximum time for the internal opamps to settle, which yields optimum filter performance.

50/100(12)* — By tying this pin to V^+ a 50:1 clock to filter center frequency ratio is obtained. Tying this pin at mid-supply (i.e., system ground with dual supplies) or to V^- allows the filter to operate at a 100:1 clock to center frequency ratio.

AGND(15) — This is the analog ground pin. This pin should be connected to the system ground for dual supply operation or biased to mid-supply for single supply operation. For a further discussion of mid-supply biasing techniques see the Applications Information (Section 3.2). For optimum filter performance a "clean" ground must be provided.

*This device is pin-for-pin compatible with the MF10 except for the following changes:

1. Unlike the MF10, the LMF100 has a single positive supply pin (V_A+).

2. On the LMF100 V_D+ is a control pin and is not the digital positive supply as on the MF10.

3. Unlike the MF10, the LMF100 does not support the current limiting mode. When the 50/100 pin is tied to V^- the LMF100 will remain in the 100:1 mode.

1.0 Definitions of Terms

f_{CLK}: the frequency of the external clock signal applied to pin 10 or 11.

f_0: center frequency of the second order function complex pole pair. f_0 is measured at the bandpass outputs of the LMF100, and is the frequency of maximum bandpass gain. *(Figure 1).*

f_{notch}: the frequency of minimum (ideally zero) gain at the notch outputs.

f_z: the center frequency of the second order complex zero pair, if any. If f_z is different from f_0 and if Q_z is high, it can be observed as the frequency of a notch at the allpass output. *(Figure 13).*

Q: "quality factor" of the 2nd order filter. Q is measured at the bandpass outputs of the LMF100 and is equal to f_0 divided by the −3 dB bandwidth of the 2nd order bandpass filter *(Figure 1)*. The value of Q determines the shape of the 2nd order filter responses as shown in *Figure 6*.

Q_z: the quality factor of the second order complex zero pair, if any. Q_Z is related to the allpass characteristic, which is written:

$$H_{AP}(s) = \frac{H_{OAP}\left(s^2 - \frac{s\omega_o}{Q_z} + \omega_o^2\right)}{s^2 + \frac{s\omega_o}{Q} + \omega_o^2}$$

where $Q_Z = Q$ for an all-pass response.

H_{OBP}: the gain (in V/V) of the bandpass output at $f = f_0$.

H_{OLP}: the gain (in V/V) of the lowpass output as $f \rightarrow 0$ Hz *(Figure 2)*.

H_{OHP}: the gain (in V/V) of the highpass output as $f \rightarrow f_{CLK}/2$ *(Figure 3)*.

H_{ON}: the gain (in V/V) of the notch output as $f \rightarrow 0$ Hz and as $f \rightarrow f_{CLK}/2$, when the notch filter has equal gain above and below the center frequency *(Figure 4)*. When the low-frequency gain differs from the high-frequency gain, as in modes 2 and 3a *(Figures 10* and *12)*, the two quantities below are used in place of H_{ON}.

H_{ON1}: the gain (in V/V) of the notch output as $f \rightarrow 0$ Hz.

H_{ON2}: the gain (in V/V) of the notch output as $f \rightarrow f_{CLK}/2$.

1.0 Definitions of Terms (Continued)

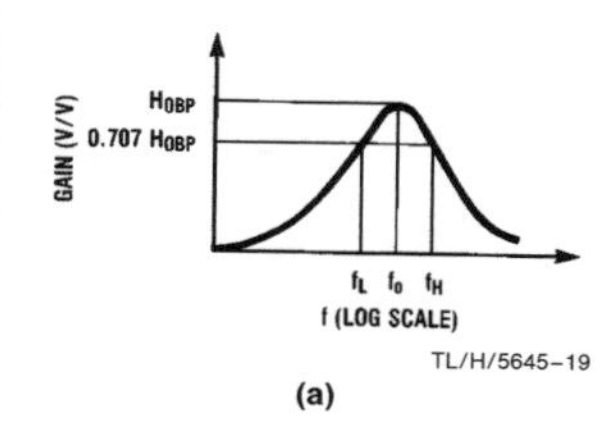

TL/H/5645–19

(a)

TL/H/5645–20

(b)

$$H_{BP}(s) = \frac{H_{OBP}\,\frac{\omega_O}{Q}\,s}{s^2 + \frac{s\omega_o}{Q} + \omega_o^2}$$

$$Q = \frac{f_0}{f_H - f_L};\ f_0 = \sqrt{f_L f_H}$$

$$f_L = f_0\left(\frac{-1}{2Q} + \sqrt{\left(\frac{1}{2Q}\right)^2 + 1}\right)$$

$$f_H = f_0\left(\frac{1}{2Q} + \sqrt{\left(\frac{1}{2Q}\right)^2 + 1}\right)$$

$$\omega_o = 2\pi f_0$$

FIGURE 1. 2nd-Order Bandpass Response

TL/H/5645–21

(a)

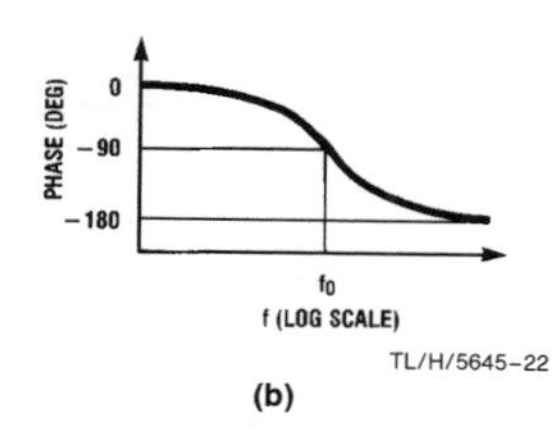

TL/H/5645–22

(b)

$$H_{LP}(s) = \frac{H_{OLP}\,\omega_O^2}{s^2 + \frac{s\omega_o}{Q} + \omega_o^2}$$

$$f_c = f_0 \times \sqrt{\left(1 - \frac{1}{2Q^2}\right) + \sqrt{\left(1 - \frac{1}{2Q^2}\right)^2 + 1}}$$

$$f_p = f_0\sqrt{1 - \frac{1}{2Q^2}}$$

$$H_{OP} = H_{OLP} \times \frac{1}{\frac{1}{Q}\sqrt{1 - \frac{1}{4Q^2}}}$$

FIGURE 2. 2nd-Order Low-Pass Response

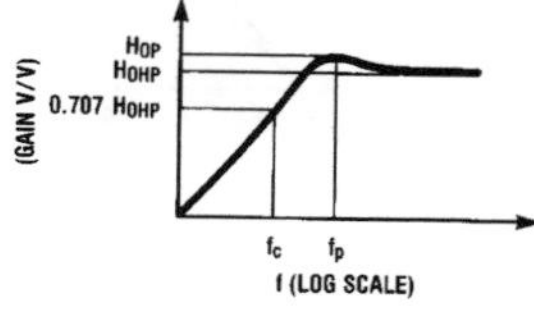

TL/H/5645–23

(a)

TL/H/5645–24

(b)

$$H_{HP}(s) = \frac{H_{OHP}\,s^2}{s^2 + \frac{s\omega_o}{Q} + \omega_o^2}$$

$$f_c = f_0 \times \left[\sqrt{\left(1 - \frac{1}{2Q^2}\right) + \sqrt{\left(1 - \frac{1}{2Q^2}\right)^2 + 1}}\right]^{-1}$$

$$f_p = f_0 \times \left[\sqrt{1 - \frac{1}{2Q^2}}\right]^{-1}$$

$$H_{OP} = H_{OHP} \times \frac{1}{\frac{1}{Q}\sqrt{1 - \frac{1}{4Q^2}}}$$

FIGURE 3. 2nd-Order High-Pass Response

1.0 Definitions of Terms (Continued)

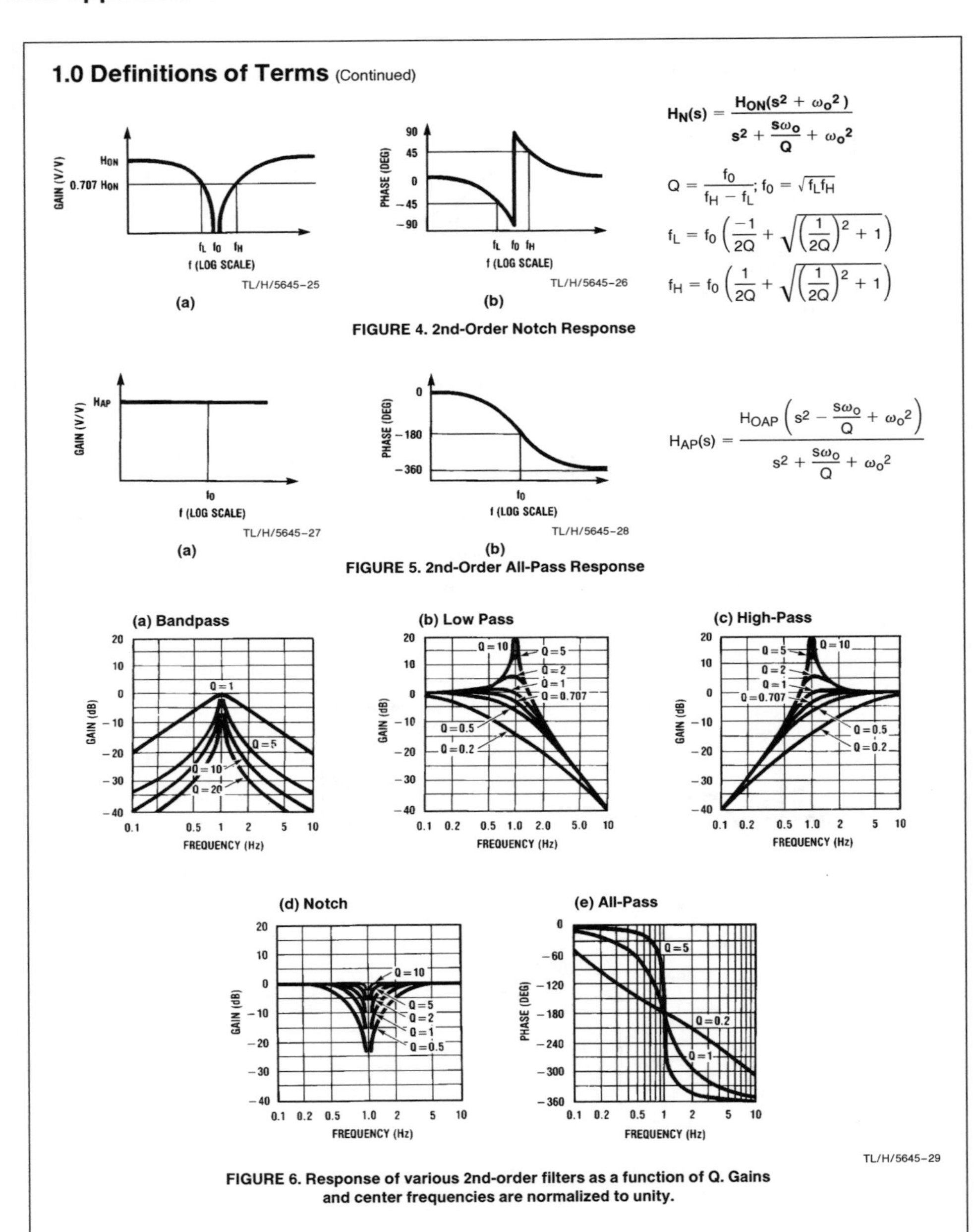

$$H_N(s) = \frac{H_{ON}(s^2 + \omega_o^2)}{s^2 + \frac{s\omega_o}{Q} + \omega_o^2}$$

$$Q = \frac{f_0}{f_H - f_L};\ f_0 = \sqrt{f_L f_H}$$

$$f_L = f_0\left(\frac{-1}{2Q} + \sqrt{\left(\frac{1}{2Q}\right)^2 + 1}\right)$$

$$f_H = f_0\left(\frac{1}{2Q} + \sqrt{\left(\frac{1}{2Q}\right)^2 + 1}\right)$$

FIGURE 4. 2nd-Order Notch Response

$$H_{AP}(s) = \frac{H_{OAP}\left(s^2 - \frac{s\omega_o}{Q} + \omega_o^2\right)}{s^2 + \frac{s\omega_o}{Q} + \omega_o^2}$$

FIGURE 5. 2nd-Order All-Pass Response

FIGURE 6. Response of various 2nd-order filters as a function of Q. Gains and center frequencies are normalized to unity.

2.0 Modes of Operation

The LMF100 is a switched capacitor (sampled data) filter. To fully describe its transfer functions, a time domain analysis is appropriate. Since this is cumbersome, and since the LMF100 closely approximates continuous filters, the following discussion is based on the well-known frequency domain. Each LMF100 can produce two full 2nd order functions. See Table I for a summary of the characteristics of the various modes.

MODE 1: Notch 1, Bandpass, Lowpass Outputs:

$f_{notch} = f_0$ (See *Figure 7*)

f_0 = center frequency of the complex pole pair

$= \frac{f_{CLK}}{100}$ or $\frac{f_{CLK}}{50}$

f_{notch} = center frequency of the imaginary zero pair = f_0.

H_{OLP} = Lowpass gain (as $f \rightarrow 0$) $= -\frac{R2}{R1}$

H_{OBP} = Bandpass gain (at $f = f_0$) $= -\frac{R3}{R1}$

H_{ON} = Notch output gain as $\left.\begin{matrix} f \rightarrow 0 \\ f \rightarrow f_{CLK}/2 \end{matrix}\right\} = \frac{-R_2}{R_1}$

$Q = \frac{f_0}{BW} = \frac{R3}{R2}$

= quality factor of the complex pole pair

BW = the −3 dB bandwidth of the bandpass output.

Circuit dynamics:

$$H_{OLP} = \frac{H_{OBP}}{Q} \text{ or } H_{OBP} = H_{OLP} \times Q$$

$$= H_{ON} \times Q.$$

$$H_{OLP(peak)} \cong Q \times H_{OLP} \text{ (for high Q's)}$$

MODE 1a: Non-Inverting BP, LP (See *Figure 8*)

$f_0 = \frac{f_{CLK}}{100}$ or $\frac{f_{CLK}}{50}$

$Q = \frac{R3}{R2}$

$H_{OLP} = -1$; $H_{OLP(peak)} \cong Q \times H_{OLP}$ (for high Q's)

$H_{OBP_1} = -\frac{R3}{R2}$

$H_{OBP_2} = 1$ (non-inverting)

Circuit dynamics: $H_{OBP_1} = Q$

Note: V_{IN} should be driven from a low impedance (<1 kΩ) source.

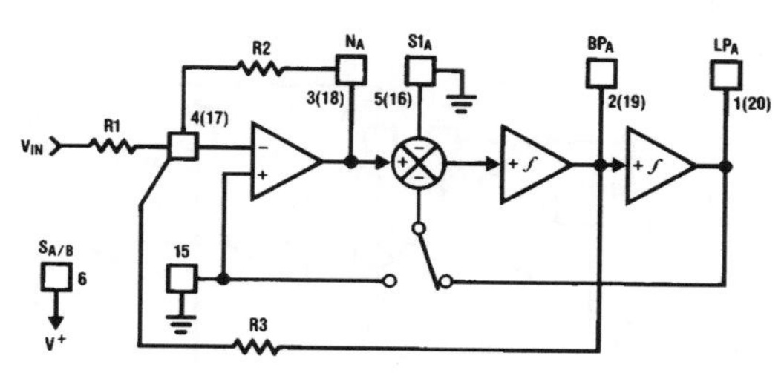

FIGURE 7. MODE 1

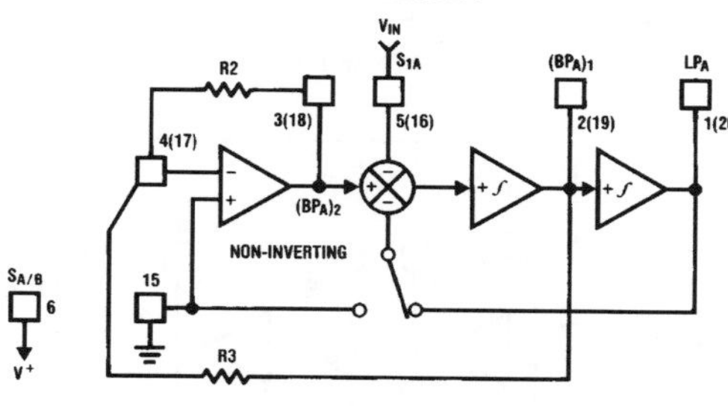

FIGURE 8. MODE 1a

2.0 Modes of Operation (Continued)

MODE 1b: Notch 1, Bandpass, Lowpass Outputs: $f_{notch} = f_0$ (See *Figure 9*)

f_0 = center frequency of the complex pole pair

$$= \frac{f_{CLK}}{100} \times \sqrt{2} \text{ or } \frac{f_{CLK}}{50} \times \sqrt{2}$$

f_{notch} = center frequency of the imaginary zero pair = f_0.

H_{OLP} = Lowpass gain (as $f \rightarrow 0$) $= -\frac{R2}{2R1}$

H_{OBP} = Bandpass gain (at $f = f_0$) $= -\frac{R3}{R1}$

H_{ON} = Notch output gain as $\left.\begin{matrix} f \rightarrow 0 \\ f \rightarrow f_{CLK}/2 \end{matrix}\right\} = \frac{-R_2}{R_1}$

$$Q = \frac{f_0}{BW} = \frac{R3}{R2} \times \sqrt{2}$$

= quality factor of the complex pole pair

BW = the −3 dB bandwidth of the bandpass output.

Circuit dynamics:

$$H_{OLP} = \frac{H_{OBP}}{\sqrt{2}\,Q} \text{ or } H_{OBP} = H_{OLP} \times Q \times \sqrt{2}$$

$$H_{OBP} = \frac{H_{ON} \times Q}{\sqrt{2}}$$

$H_{OLP(peak)} \cong Q \times H_{OLP}$ (for high Q's)

MODE 2: Notch 2, Bandpass, Lowpass: $f_{notch} < f_0$ (See *Figure 10*)

f_0 = center frequency

$$= \frac{f_{CLK}}{100}\sqrt{\frac{R2}{R4} + 1} \text{ or } \frac{f_{CLK}}{50}\sqrt{\frac{R2}{R4} + 1}$$

$$f_{notch} = \frac{f_{CLK}}{100} \text{ or } \frac{f_{CLK}}{50}$$

Q = quality factor of the complex pole pair

$$= \frac{\sqrt{R2/R4 + 1}}{R2/R3}$$

H_{OLP} = Lowpass output gain (as $f \rightarrow 0$)

$$= -\frac{R2/R1}{R2/R4 + 1}$$

H_{OBP} = Bandpass output gain (at $f = f_0$) $= -R3/R1$

H_{ON_1} = Notch output gain (as $f \rightarrow 0$)

$$= -\frac{R2/R1}{R2/R4 + 1}$$

H_{ON_2} = Notch output gain $\left(\text{as } f \rightarrow \frac{f_{CLK}}{2}\right) = -R2/R1$

Filter dynamics: $H_{OBP} = Q\sqrt{H_{OLP} H_{ON_2}} = \sqrt{H_{ON_1} H_{ON_2}}$

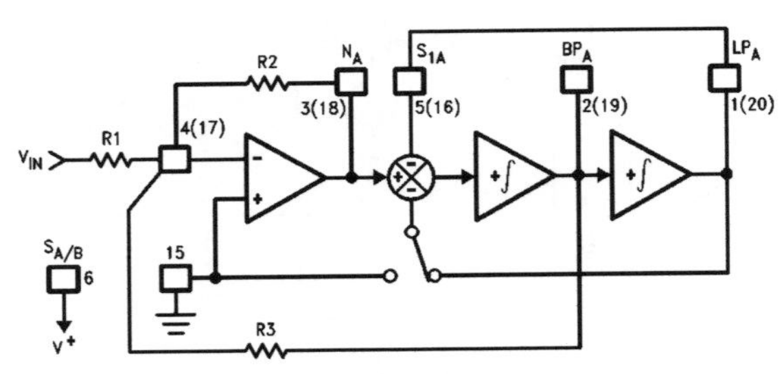

FIGURE 9. MODE 1b

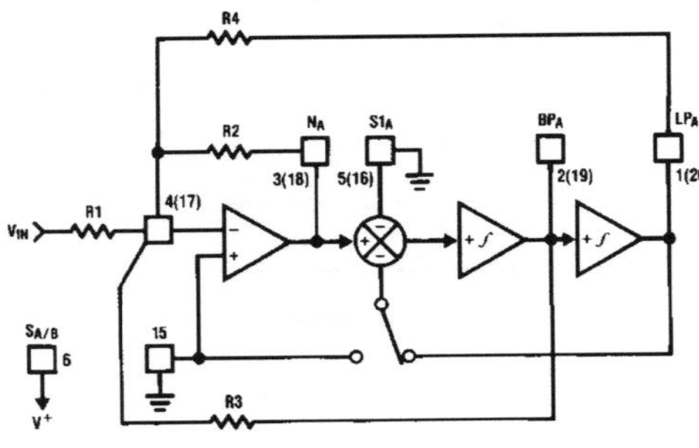

FIGURE 10. MODE 2

2.0 Modes of Operation (Continued)

MODE 3: Highpass, Bandpass, Lowpass Outputs (See *Figure 11*)

$f_0 = \frac{f_{CLK}}{100} \times \sqrt{\frac{R2}{R4}}$ or $\frac{f_{CLK}}{50} \times \sqrt{\frac{R2}{R4}}$

Q = quality factor of the complex pole pair

$= \sqrt{\frac{R2}{R4}} \times \frac{R3}{R2}$

H_{OHP} = Highpass gain $\left(\text{at f} \rightarrow \frac{f_{CLK}}{2}\right) = -\frac{R2}{R1}$

H_{OBP} = Bandpass gain (at $f = f_0$) $= -\frac{R3}{R1}$

H_{OLP} = Lowpass gain (as $f \rightarrow 0$) $= -\frac{R4}{R1}$

Circuit dynamics: $\frac{R2}{R4} = \frac{H_{OHP}}{H_{OLP}}$; $H_{OBP} = \sqrt{H_{OHP} \times H_{OLP}} \times Q$

$H_{OLP(peak)} \cong Q \times H_{OLP}$ (for high Q's)

$H_{OHP(peak)} \cong Q \times H_{OHP}$ (for high Q's)

MODE 3a: HP, BP, LP and Notch with External Op Amp (See *Figure 12*)

$f_0 = \frac{f_{CLK}}{100} \times \sqrt{\frac{R2}{R4}}$ or $\frac{f_{CLK}}{50} \times \sqrt{\frac{R2}{R4}}$

$Q = \sqrt{\frac{R2}{R4}} \times \frac{R3}{R2}$

$H_{OHP} = -\frac{R2}{R1}$

$H_{OBP} = -\frac{R3}{R1}$

$H_{OLP} = -\frac{R4}{R1}$

f_n = notch frequency $= \frac{f_{CLK}}{100}\sqrt{\frac{R_h}{R_l}}$ or $\frac{f_{CLK}}{50}\sqrt{\frac{R_h}{R_l}}$

H_{ON} = gain of notch at

$f = f_0 = \left\| Q\left(\frac{R_g}{R_l} H_{OLP} - \frac{R_g}{R_h} H_{OHP}\right) \right\|$

H_{n1} = gain of notch (as $f \rightarrow 0$) $= \frac{R_g}{R_l} \times H_{OLP}$

H_{n2} = gain of notch $\left(\text{as f} \rightarrow \frac{f_{CLK}}{2}\right)$

$= -\frac{R_g}{R_h} \times H_{OHP}$

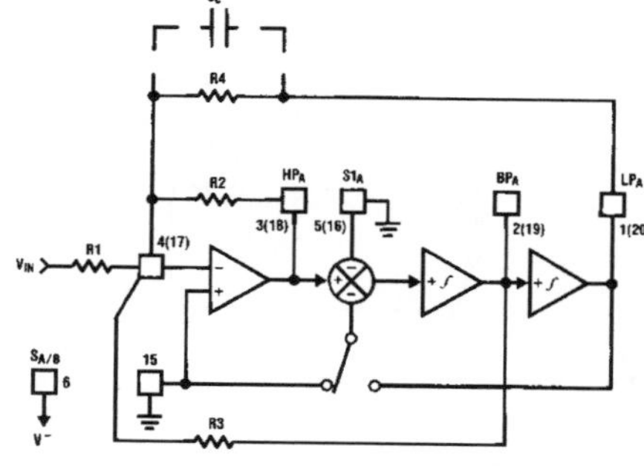

*In Mode 3, the feedback loop is closed around the input summing amplifier; the finite GBW product of this op amp causes a slight Q enhancement. If this is a problem, connect a small capacitor (10 pF – 100 pF) across R4 to provide some phase lead.

TL/H/5645-5

FIGURE 11. MODE 3

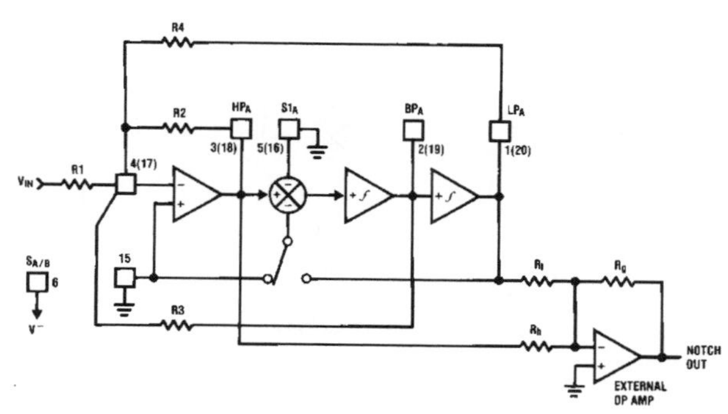

TL/H/5645-10

FIGURE 12. MODE 3a

2.0 Modes of Operation (Continued)

MODE 4: Allpass, Bandpass, Lowpass Outputs
(See *Figure 13*)

f_0 = center frequency

$= \frac{f_{CLK}}{100}$ or $\frac{f_{CLK}}{50}$;

f_z* = center frequency of the complex zero ≈ f_0

$Q = \frac{f_0}{BW} = \frac{R3}{R2}$;

Q_z = quality factor of complex zero pair $= \frac{R3}{R1}$

For AP output make R1 = R2

H_{OAP}* = Allpass gain $\left(\text{at } 0 < f < \frac{f_{CLK}}{2}\right) = -\frac{R2}{R1} = -1$

H_{OLP} = Lowpass gain (as f → 0)

$= -\left(\frac{R2}{R1} + 1\right) = -2$

H_{OBP} = Bandpass gain (at f = f_0)

$= -\frac{R3}{R2}\left(1 + \frac{R2}{R1}\right) = -2\left(\frac{R3}{R2}\right)$

Circuit dynamics: $H_{OBP} = (H_{OLP}) \times Q = (H_{OAP} + 1)Q$

*Due to the sampled data nature of the filter, a slight mismatch of f_z and f_0 occurs causing a 0.4 dB peaking around f_0 of the allpass filter amplitude response (which theoretically should be a straight line). If this is unacceptable, Mode 5 is recommended.

MODE 5: Numerator Complex Zeros, BP, LP
(See *Figure 14*)

$f_0 = \sqrt{1 + \frac{R2}{R4}} \times \frac{f_{CLK}}{100}$ or $\sqrt{1 + \frac{R2}{R4}} \times \frac{f_{CLK}}{50}$

$f_z = \sqrt{1 - \frac{R1}{R4}} \times \frac{f_{CLK}}{100}$ or $\sqrt{1 - \frac{R1}{R4}} \times \frac{f_{CLK}}{50}$

$Q = \sqrt{1 + R2/R4} \times \frac{R3}{R2}$

$Q_Z = \sqrt{1 - R1/R4} \times \frac{R3}{R1}$

$H_{0_{z1}}$ = gain at C.Z. output (as f → 0 Hz)
$= \frac{-R2(R4 - R1)}{R1(R2 + R4)}$

$H_{0_{z2}}$ = gain at C.Z. output $\left(\text{as } f \to \frac{f_{CLK}}{2}\right) = \frac{-R2}{R1}$

$H_{OBP} = -\left(\frac{R2}{R1} + 1\right) \times \frac{R3}{R2}$

$H_{OLP} = -\left(\frac{R2 + R1}{R2 + R4}\right) \times \frac{R4}{R1}$

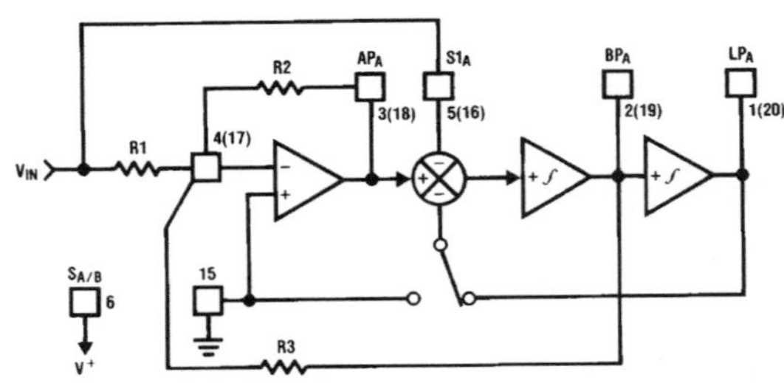

FIGURE 13. MODE 4

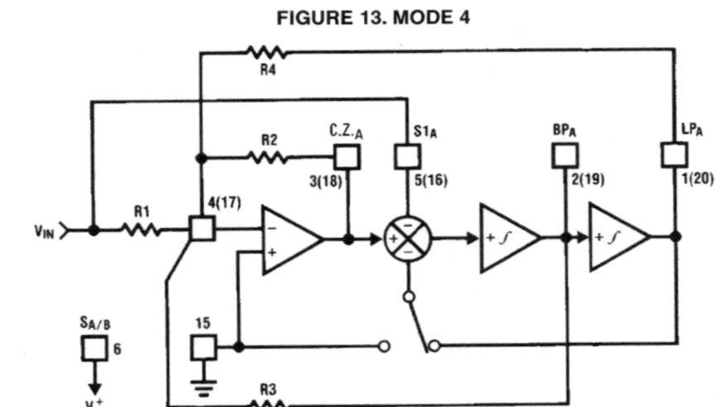

FIGURE 14. MODE 5

2.0 Modes of Operation (Continued)

MODE 6a: Single Pole, HP, LP Filter (See *Figure 15*)

f_c = cutoff frequency of LP or HP output

$= \frac{R2}{R3}\frac{f_{CLK}}{100}$ or $\frac{R2}{R3}\frac{f_{CLK}}{50}$

$H_{OLP} = -\frac{R3}{R1}$

$H_{OHP} = -\frac{R2}{R1}$

MODE 6b: Single Pole LP Filter (Inverting and Non-Inverting) (See *Figure 16*)

f_c = cutoff frequency of LP outputs

$\cong \frac{R2}{R3}\frac{f_{CLK}}{100}$ or $\frac{R2}{R3}\frac{f_{CLK}}{50}$

H_{OLP_1} = 1 (non-inverting)

$H_{OLP_2} = -\frac{R3}{R2}$

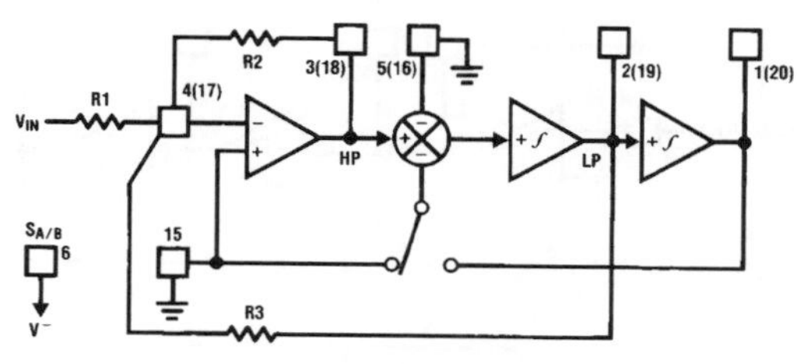

TL/H/5645-16

FIGURE 15. MODE 6a

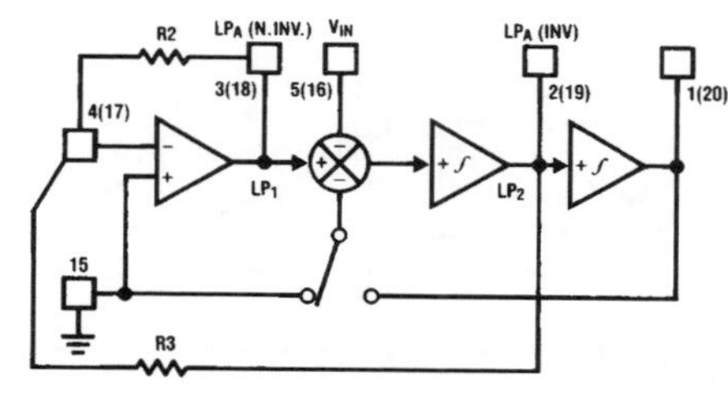

TL/H/5645-7

FIGURE 16. MODE 6b

2.0 Modes of Operation (Continued)

MODE 6c: Single Pole, AP, LP Filter (See *Figure 17*)

$f_c = \frac{f_{CLK}}{50}$ or $\frac{f_{CLK}}{100}$

$H_{OAP} = 1$ (as $f \rightarrow 0$)

$H_{OAP} = -1$ (as $f \rightarrow f_{CLK}/2$)

$H_{OLP} = -2$

$R_1 = R_2 = R_3$

MODE 7: Summing Integrator (See *Figure 18*)

τ = integrator time constant

$= \frac{16}{f_{CLK}}$ or $\frac{8}{f_{CLK}}$

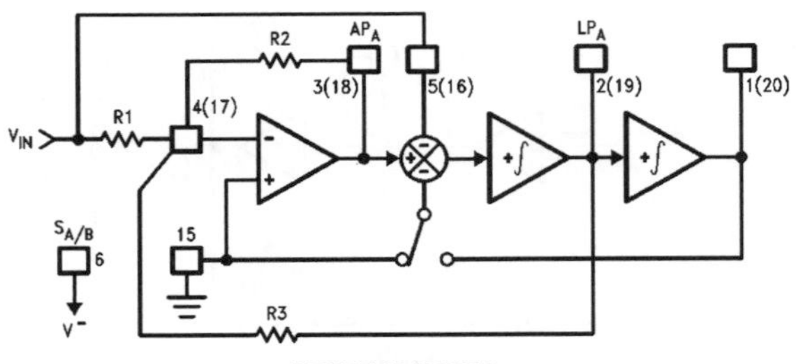

FIGURE 17. MODE 6c

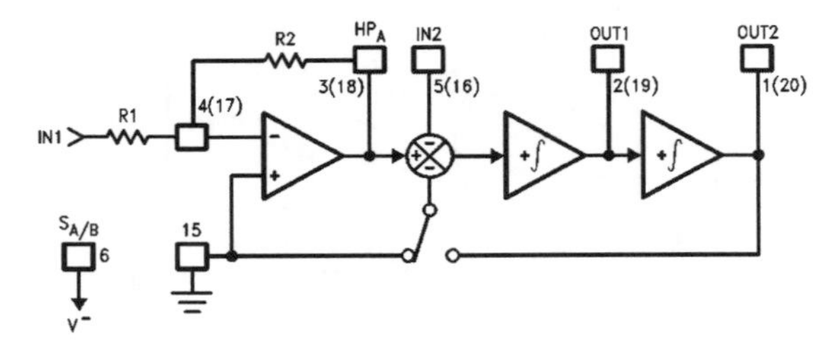

Equivalent Circuit

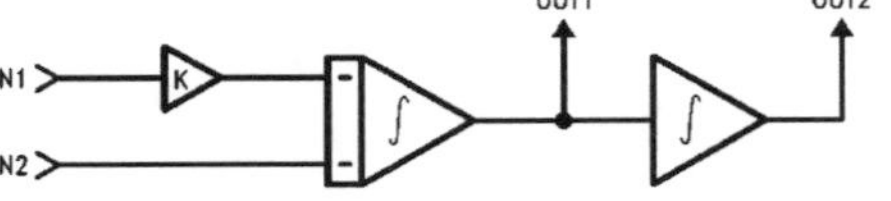

$K = \frac{R_2}{R_1}$

$OUT1 = -\frac{k}{\tau}\int IN1\,dt - \frac{1}{\tau}\int IN2\,dt$

$OUT2 = \frac{1}{\tau}\int OUT1\,dt$

FIGURE 18. MODE 7

2.0 Modes of Operation (Continued)

TABLE I. Summary of Modes. Realizable filter types (e.g. low-pass) denoted by asterisks. Unless otherwise noted, gains of various filter outputs are inverting and adjustable by resistor ratios.

Mode	BP	LP	HP	N	AP	Number of Resistors	Adjustable f_{CLK}/f_0	Notes
1	*	*		*		3	No	
1a	(2) $H_{OBP1} = -Q$ $H_{OBP2} = +1$	$H_{OLP} = +1$				2	No	May need input buffer. Poor dynamics for high Q.
1b	*	*		*		3	No	Useful for high frequency applications.
2	*	*		*		3	Yes (above $f_{CLK}/50$ or $f_{CLK}/100$)	
3	*	*	*			4	Yes	Universal State-Variable Filter. Best general-purpose mode.
3a	*	*	*	*		7	Yes	As above, but also includes resistor-tuneable notch.
4	*	*			*	3	No	Gives Allpass response with $H_{OAP} = -1$ and $H_{OLP} = -2$.
5	*	*			*	4	Yes	Gives flatter allpass response than above if $R_1 = R_2 = 0.02R_4$.
6a		*	*			3	Yes	Single pole.
6b		(2) $H_{OLP1} = +1$ $H_{OLP2} = \frac{-R3}{R2}$				2	Yes	Single pole.
6c		*			*	3	No	Single pole.
7						2	Yes	Summing integrator with adjustable time constant.

3.0 Applications Information

The LMF100 is a general purpose dual second-order state variable filter whose center frequency is proportional to the frequency of the square wave applied to the clock input (f_{CLK}). The various clocking options are summarized in the following table.

Clocking Options

Power Supply	Clock Levels	LSh	V_D+
−5V and +5V	TTL (0V to +5V)	0V	+5V
−5V and +5V	CMOS (−5V to +5V)	0V	+5V
0V and 10V	TTL (0V to 5V)	0V	+10V
0V and 10V	CMOS (0V to +10V)	+5V	+10V
−2.5V and +2.5V	CMOS (−2.5V to +2.5V)	0V	+2.5V
0V and 5V	TTL (0V to +5V)	0V	0V
0V and 5V	CMOS (0V to +5V)	+2.5V	+5V

By connecting pin 12 to the appropriate dc voltage, the filter center frequency, f_0, can be made equal to either $f_{CLK}/100$ or $f_{CLK}/50$. f_0 can be very accurately set (within ±0.6%) by using a crystal clock oscillator, or can be easily varied over a wide frequency range by adjusting the clock frequency. If desired, the f_{CLK}/f_0 ratio can be altered by external resistors as in *Figures 10, 11, 12, 13, 14, 15* and *16*. This is useful when high-order filters (greater than two) are to be realized by cascading the second-order sections. This allows each stage to be stagger tuned while using only one clock. The filter Q and gain are set by external resistor ratios.

All of the five second-order filter types can be built using either section of the LMF100. These are illustrated in *Figures 1* through *5* along with their transfer functions and some related equations. *Figure 6* shows the effect of Q on the shapes of these curves.

3.0 Applications Information (Continued)

3.1 DESIGN EXAMPLE

In order to design a filter using the LMF100, we must define the necessary values of three parameters for each second-order section: f_0, the filter section's center frequency; H_0, the passband gain; and the filter's Q. These are determined by the characteristics required of the filter being designed.

As an example, let's assume that a system requires a fourth-order Chebyshev low-pass filter with 1 dB ripple, unity gain at dc, and 1000 Hz cutoff frequency. As the system order is four, it is realizable using both second-order sections of an LMF100. Many filter design texts (and National's Switched Capacitor Filter Handbook) include tables that list the characteristics (f_0 and Q) of each of the second-order filter sections needed to synthesize a given higher-order filter. For the Chebyshev filter defined above, such a table yields the following characteristics:

$f_{0A} = 529$ Hz $\quad Q_A = 0.785$

$f_{0B} = 993$ Hz $\quad Q_B = 3.559$

For unity gain at dc, we also specify:

$H_{0A} = 1$

$H_{0B} = 1$

The desired clock-to-cutoff-frequency ratio for the overall filter of this example is 100 and a 100 kHz clock signal is available. Note that the required center frequencies for the two second-order sections will not be obtainable with clock-to-center-frequency ratios of 50 or 100. It will be necessary to adjust $\frac{f_{CLK}}{f_0}$ externally. From Table I, we see that Mode 3 can be used to produce a low-pass filter with resistor-adjustable center frequency.

In most filter designs involving multiple second-order stages, it is best to place the stages with lower Q values ahead of stages with higher Q, especially when the higher Q is greater than 0.707. This is due to the higher relative gain at the center frequency of a higher-Q stage. Placing a stage with lower Q ahead of a higher-Q stage will provide some attenuation at the center frequency and thus help avoid clipping of signals near this frequency. For this example, stage A has the lower Q (0.785) so it will be placed ahead of the other stage.

For the first section, we begin the design by choosing a convenient value for the input resistance: $R_{1A} = 20k$. The absolute value of the passband gain H_{OLPA} is made equal to 1 by choosing R_{4A} such that: $R_{4A} = -H_{OLPA}R_{1A} = R_{1A} = 20k$. If the 50/100/CL pin is connected to mid-supply for nominal 100:1 clock-to-center-frequency ratio, we find R_{2A} by:

$$R_{2A} = R_{4A}\frac{f_{0A}^2}{(f_{CLK}/100)^2} = 2 \times 10^4 \times \frac{(529)^2}{(1000)^2} = 5.6\text{k and}$$

$$R_{3A} = Q_A\sqrt{R_{2A}R_{4A}} = 0.785\sqrt{5.6 \times 10^3 \times 2 \times 10^4} = 8.3\text{k}$$

The resistors for the second section are found in a similar fashion:

$R_{1B} = 20k$

$R_{4B} = R_{1B} = 20k$

$$R_{2B} = R_{4B}\frac{f_{0B}^2}{(f_{CLK}/100)^2} = 20\text{k}\frac{(993)^2}{(1000)^2} = 19.7\text{k}$$

$$R_{3B} = Q_B\sqrt{R_{2B}R_{4B}} = 3.559\sqrt{1.97 \times 10^4 \times 2 \times 10^4} = 70.6\text{k}$$

The complete circuit is shown in *Figure 19* for split ±5V power supplies. Supply bypass capacitors are highly recommended.

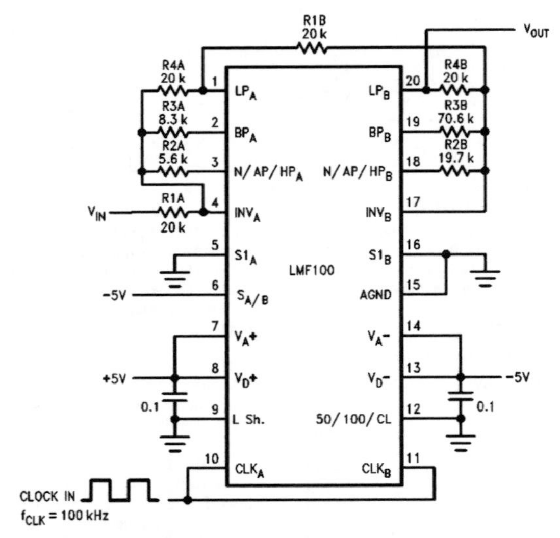

FIGURE 19. Fourth-order Chebyshev low-pass filter from example in 3.1. ±5V power supply. 0V–5V TTL or ±5V CMOS logic levels.

3.0 Applications Information (Continued)

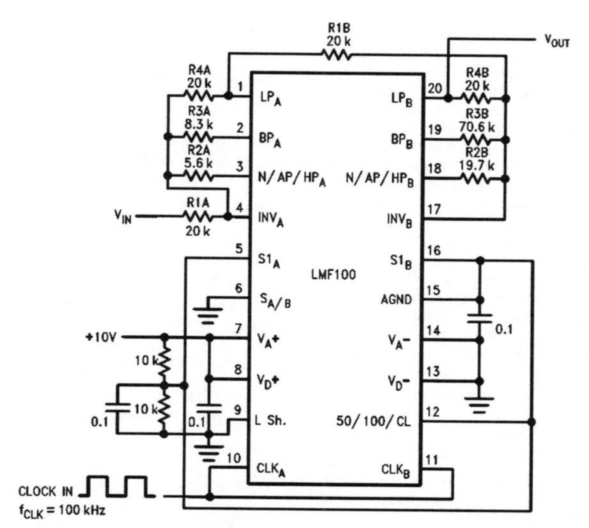

FIGURE 20. Fourth-order Chebyshev low-pass filter from example in 3.1. Single + 10V power supply. 0V–5V TTL logic levels. Input signals should be referred to half-supply or applied through a coupling capacitor.

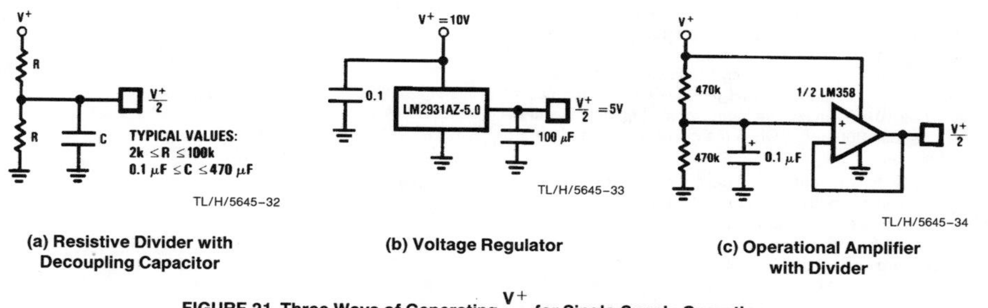

(a) Resistive Divider with Decoupling Capacitor

(b) Voltage Regulator

(c) Operational Amplifier with Divider

FIGURE 21. Three Ways of Generating $\frac{V^+}{2}$ for Single-Supply Operation

3.0 Applications Information (Continued)

3.2 SINGLE SUPPLY OPERATION

The LMF100 can also operate with a single-ended power supply. *Figure 20* shows the example filter with a single-ended power supply. $V_A{}^+$ and $V_D{}^+$ are again connected to the positive power supply (4 to 15 volts), and $V_A{}^-$ and $V_D{}^-$ are connected to ground. The A_{GND} pin must be tied to $V^+/2$ for single supply operation. This half-supply point should be very "clean", as any noise appearing on it will be treated as an input to the filter. It can be derived from the supply voltage with a pair of resistors and a bypass capacitor *(Figure 21a)*, or a low-impedance half-supply voltage can be made using a three-terminal voltage regulator or an operational amplifier *(Figures 21b* and *21c)*. The passive resistor divider with a bypass capacitor is sufficient for many applications, provided that the time constant is long enough to reject any power supply noise. It is also important that the half-supply reference present a low impedance to the clock frequency, so at very low clock frequencies the regulator or op-amp approaches may be preferable because they will require smaller capacitors to filter the clock frequency. The main power supply voltage should be clean (preferably regulated) and bypassed with 0.1 μF.

3.3 DYNAMIC CONSIDERATIONS

The maximum signal handling capability of the LMF100, like that of any active filter, is limited by the power supply voltages used. The amplifiers in the LMF100 are able to swing to within about 1 volt of the supplies, so the input signals must be kept small enough that none of the outputs will exceed these limits. If the LMF100 is operating on ±5 volts, for example, the outputs will clip at about $8V_{p\text{-}p}$. The maximum input voltage multiplied by the filter gain should therefore be less than $8V_{p\text{-}p}$.

Note that if the filter Q is high, the gain at the lowpass or highpass outputs will be much greater than the nominal filter gain *(Figure 6)*. As an example, a lowpass filter with a Q of 10 will have a 20 dB peak in its amplitude response at f_0. If the nominal gain of the filter (H_{OLP}) is equal to 1, the gain at f_0 will be 10. The maximum input signal at f_0 must therefore be less than 800 $mV_{p\text{-}p}$ when the circuit is operated on ±5 volt supplies.

Also note that one output can have a reasonable small voltage on it while another is saturated. This is most likely for a circuit such as the notch in Mode 1 *(Figure 7)*. The notch output will be very small at f_0, so it might appear safe to apply a large signal to the input. However, the bandpass will have its maximum gain at f_0 and can clip if overdriven. If one output clips, the performance at the other outputs will be degraded, so avoid overdriving any filter section, even ones whose outputs are not being directly used. Accompanying *Figures 7* through *17* are equations labeled "circuit dynamics", which relate the Q and the gains at the various outputs. These should be consulted to determine peak circuit gains and maximum allowable signals for a given application.

3.4 OFFSET VOLTAGE

The LMF100's switched capacitor integrators have a slightly higher input offset voltage than found in a typical continuous time active filter integrator. Because of National's new LMCMOS process and new design techniques the internal offsets have been minimized, compared to the industry standard MF10. *Figure 22* shows an equivalent circuit of the LMF100 from which the output dc offsets can be calculated.

Typical values for these offsets with $S_{A/B}$ tied to V^+ are:

V_{OS1} = opamp offset = ±5 mV

V_{OS2} = ±30 mV at 50:1 or 100:1

V_{OS3} = ±15 mV at 50:1 or 100:1

When $S_{A/B}$ is tied to V^-, V_{OS2} will approximately halve. The dc offset at the BP output is equal to the input offset of the lowpass integrator (V_{OS3}). The offsets at the other outputs depend on the mode of operation and the resistor ratios, as described in the following expressions.

Mode 1 and Mode 4

$$V_{OS(N)} = V_{OS1}\left(\frac{1}{Q} + 1 + \|H_{OLP}\|\right) - \frac{V_{OS3}}{Q}$$

$$V_{OS(BP)} = V_{OS3}$$

$$V_{OS(LP)} = V_{OS(N)} - V_{OS2}$$

Mode 1a

$$V_{OS(N.INV.BP)} = \left(1 + \frac{1}{Q}\right)V_{OS1} - \frac{V_{OS3}}{Q}$$

$$V_{OS(INV.BP)} = V_{OS3}$$

$$V_{OS(LP)} = V_{OS(N.INV.BP)} - V_{OS2}$$

Mode 1b

$$V_{OS(N)} = V_{OS1}\left(1 + \frac{R2}{R3} + \frac{R2}{R1}\right) - \frac{R2}{R3}V_{OS3}$$

$$V_{OS(BP)} = V_{OS3}$$

$$V_{OS(LP)} = \frac{V_{OS(N)}}{2} - \frac{V_{OS2}}{2}$$

Mode 2 and Mode 5

$$V_{OS(N)} = \left(\frac{R2}{Rp} + 1\right)V_{OS1} \times \frac{1}{1 + R2/R4} + V_{OS2}\frac{1}{1 + R4/R2} - \frac{V_{OS3}}{Q\sqrt{1 + R2/R4}};$$

$$R_p = R1\|R3\|R4$$

$$V_{OS(BP)} = V_{OS3}$$

$$V_{OS(LP)} = V_{OS(N)} - V_{OS2}$$

Mode 3

$$V_{OS(HP)} = V_{OS2}$$

$$V_{OS(BP)} = V_{OS3}$$

$$V_{OS(LP)} = V_{OS1}\left[1 + \frac{R4}{R_p}\right] - V_{OS2}\left(\frac{R4}{R2}\right) - V_{OS3}\left(\frac{R4}{R3}\right)$$

$$R_p = R1\|R2\|R3$$

Mode 6a and 6c

$$V_{OS(HP)} = V_{OS2}$$

$$V_{OS(LP)} = V_{OS1}\left(1 + \frac{R_3}{R_2} + \frac{R_3}{R_1}\right) - \frac{R_3}{R_2}V_{OS2}$$

Mode 6b

$$V_{OS(LP\ (N.INV))} = V_{OS2}$$

$$V_{OS(LP\ (INV))} = V_{OS1}\left(1 + \frac{R_3}{R_2}\right) - \frac{R_3}{R_2}V_{OS2}$$

3.0 Applications Information (Continued)

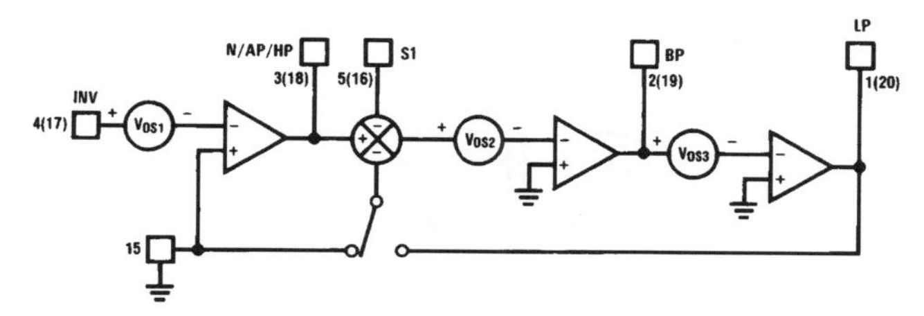

FIGURE 22. Offset Voltage Sources

In many applications, the outputs are ac coupled and dc offsets are not bothersome unless large signals are applied to the filter input. However, larger offset voltages will cause clipping to occur at lower ac signal levels, and clipping at any of the outputs will cause gain nonlinearities and will change f_0 and Q. When operating in Mode 3, offsets can become excessively large if R_2 and R_4 are used to make f_{CLK}/f_0 significantly higher than the nominal value, especially if Q is also high.

For example, *Figure 23* shows a second-order 60 Hz notch filter. This circuit yields a notch with about 40 dB of attenuation at 60 Hz. A notch is formed by subtracting the bandpass output of a mode 3 configuration from the input using the unused side B opamp. The Q is 10 and the gain is 1 V/V in the passband. However, f_{CLK}/f_0 = 1000 to allow for a wide input spectrum. This means that for pin 12 tied to ground (100:1 mode), R4/R2 = 100. The offset voltage at the lowpass output (LP) will be about 3V. However, this is an extreme case and the resistor ratio is usually much smaller. Where necessary, the offset voltage can be adjusted by using the circuit of *Figure 24*. This allows adjustment of V_{OS1}, which will have varying effects on the different outputs as described in the above equations. Some outputs cannot be adjusted this way in some modes, however ($V_{OS(BP)}$ in modes 1a and 3, for example).

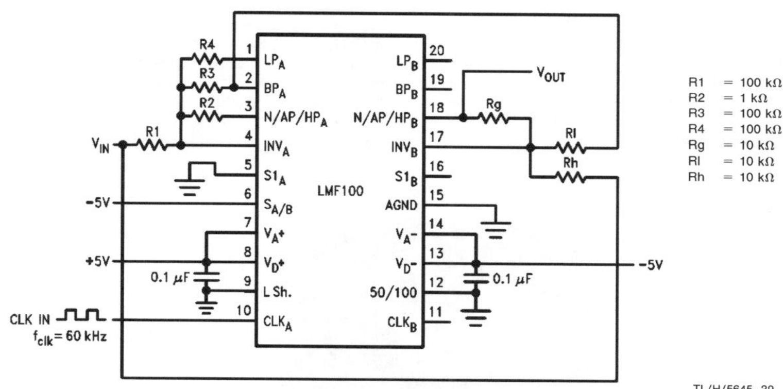

FIGURE 23. Second-Order Notch Filter

3.0 Applications Information (Continued)

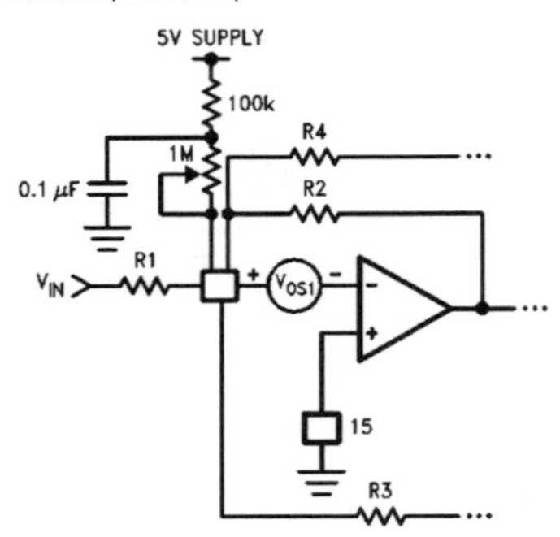

TL/H/5645–13

FIGURE 24. Method for Trimming V_{OS}

3.5 SAMPLED DATA SYSTEM CONSIDERATIONS

The LMF100 is a sampled data filter, and as such, differs in many ways from conventional continuous-time filters. An important characteristic of sampled-data systems is their effect on signals at frequencies greater than one-half the sampling frequency. (The LMF100's sampling frequency is the same as its clock frequency.) If a signal with a frequency greater than one-half the sampling frequency is applied to the input of a sampled data system, it will be "reflected" to a frequency less than one-half the sampling frequency. Thus, an input signal whose frequency is $f_s/2 + 100$ Hz will cause the system to respond as though the input frequency was $f_s/2 - 100$ Hz. This phenomenon is known as "aliasing", and can be reduced or eliminated by limiting the input signal spectrum to less than $f_s/2$. This may in some cases require the use of a bandwidth-limiting filter ahead of the LMF100 to limit the input spectrum. However, since the clock frequency is much higher than the center frequency, this will often not be necessary.

Another characteristic of sampled-data circuits is that the output signal changes amplitude once every sampling period, resulting in "steps" in the output voltage which occur at the clock rate *(Figure 25)*. If necessary, these can be "smoothed" with a simple R-C low-pass filter at the LMF100 output.

The ratio of f_{CLK} to f_c (normally either 50:1 or 100:1) will also affect performance. A ratio of 100:1 will reduce any aliasing problems and is usually recommended for wide-band input signals. In noise-sensitive applications, a ratio of 100:1 will result in 3 dB lower output noise for the same filter configuration.

The accuracy of the f_{CLK}/f_0 ratio is dependent on the value of Q. This is illustrated in the curves under the heading "Typical Performance Characteristics". As Q is changed, the true value of the ratio changes as well. Unless the Q is low, the error in f_{CLK}/f_0 will be small. If the error is too large for a specific application, use a mode that allows adjustment of the ratio with external resistors.

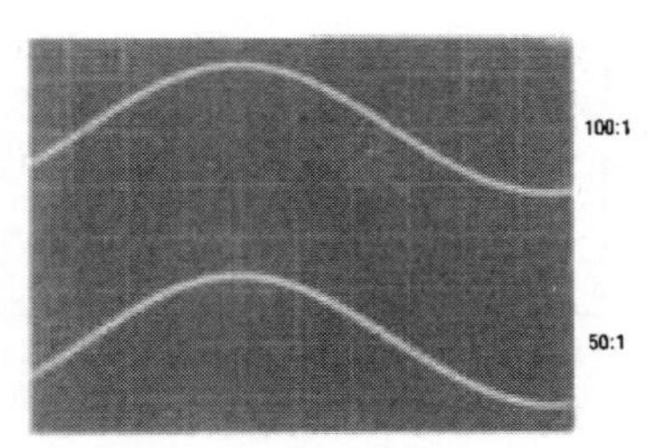

TL/H/5645–35

FIGURE 25. The Sampled-Data Output Waveform

Physical Dimensions inches (millimeters)

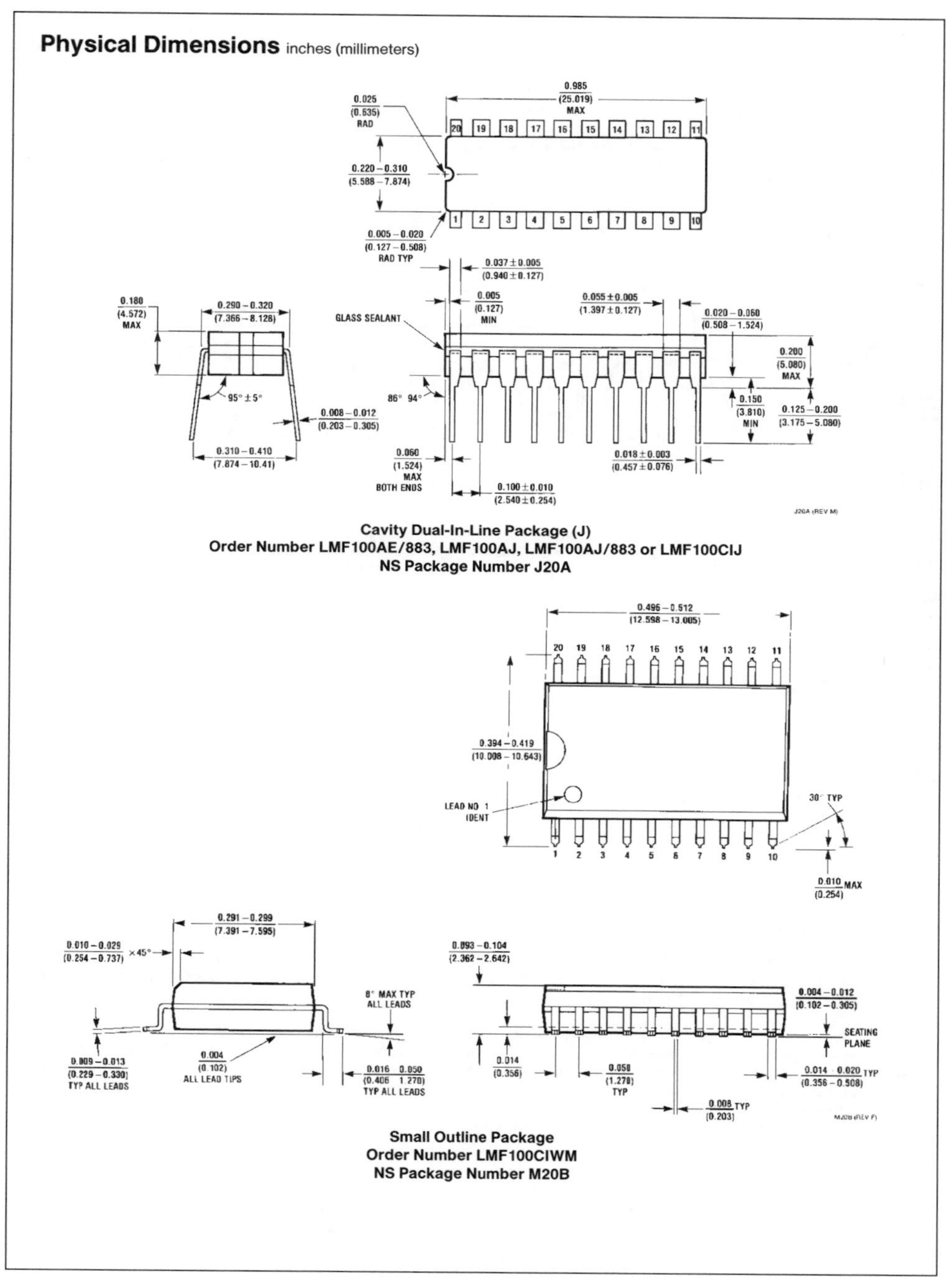

Cavity Dual-In-Line Package (J)
Order Number LMF100AE/883, LMF100AJ, LMF100AJ/883 or LMF100CIJ
NS Package Number J20A

Small Outline Package
Order Number LMF100CIWM
NS Package Number M20B

Physical Dimensions inches (millimeters) (Continued)

Lit. # 108800

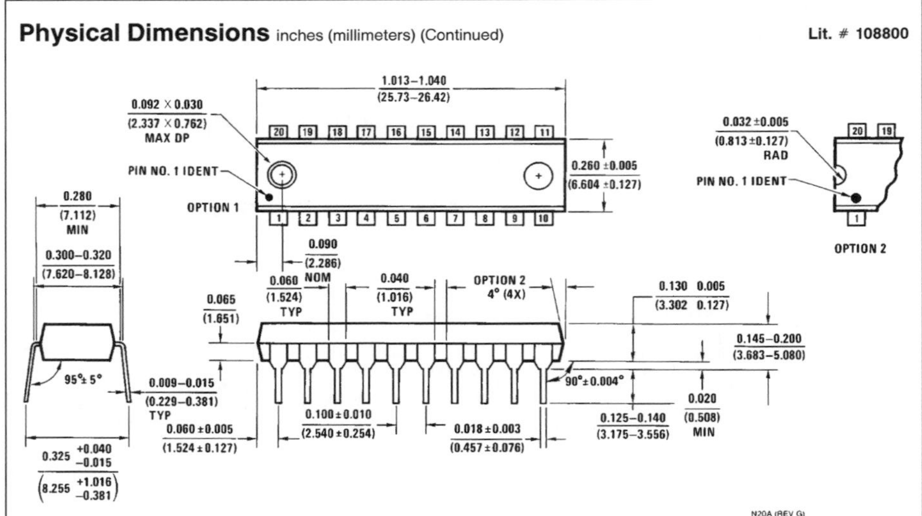

Molded Dual-In-Line Package (N)
Order Number LMF100ACN, LMF100CCN or LMF100CIN
NS Package Number N20A

National Semiconductor Corporation
1111 West Bardin Road
Arlington, TX 76017
Tel: 1(800) 272-9959
Fax: 1(800) 737-7018

National Semiconductor Europe
Fax: (+49) 0-180-530 85 86
Email: cnjwge@tevm2.nsc.com
Deutsch Tel: (+49) 0-180-530 85 85
English Tel: (+49) 0-180-532 78 32
Français Tel: (+49) 0-180-532 93 58
Italiano Tel: (+49) 0-180-534 16 80

National Semiconductor Hong Kong Ltd.
13th Floor, Straight Block,
Ocean Centre, 5 Canton Rd.
Tsimshatsui, Kowloon
Hong Kong
Tel: (852) 2737-1600
Fax: (852) 2736-9960

National Semiconductor Japan Ltd.
Tel: 81-043-299-2309
Fax: 81-043-299-2408

4

Tuned amplifier applications

4.1 Introduction

A tuned amplifier is one which operates over a band of frequencies centred on a resonant frequency. Its two main requirements are to provide high gain and good selectivity, both of which can be achieved by techniques which will be mentioned in this chapter.

There are many applications in telecommunications work where it is necessary to amplify a narrow band of frequencies centred on one frequency. One such example is shown in the frequency-modulated transmitter of Fig. 4.1. In this case the radio frequency (RF) amplifier is used to increase the amplitude of the oscillator output, while the frequency tripler stages are used to increase the transmitted frequency to 104.5 MHz. Both these stages are forms of tuned amplifier.

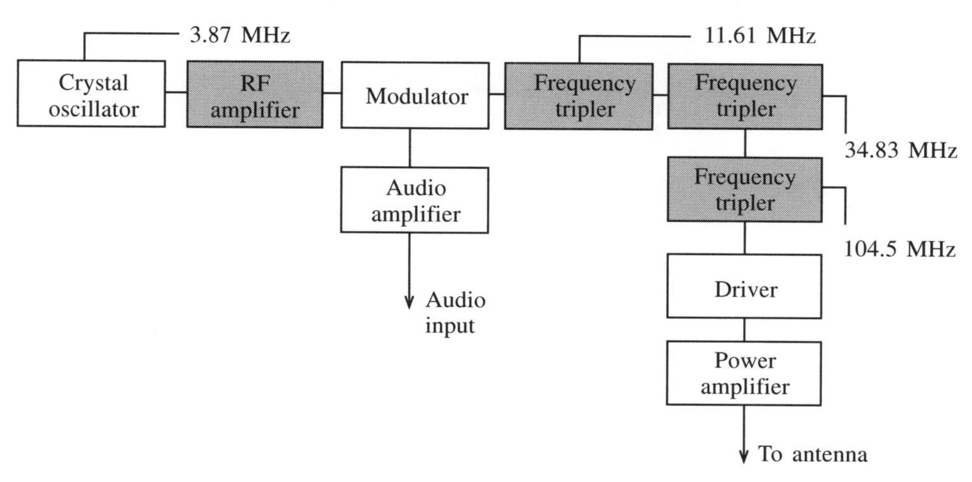

Fig. 4.1

The following are among the applications of tuned amplifiers:

(a) intermediate frequency (IF) amplifier in a superheterodyne receiver;

(b) very narrow-band IF amplifier in a spectrum analyser;

(c) IF amplifier in a satellite transponder;

(d) RF amplifiers in receivers;

(e) wide-band tuned amplifiers for video amplification;

(f) wide-band tuned amplifiers for Y-amplifiers in oscilloscopes;

(g) UHF radio relay systems.

It is as well to note at this point that it is virtually impossible to tell the difference between IF and RF amplifiers from a schematic diagram. The IF amplifier takes its name from its function within a receiver and not from the range of frequencies over which it works. It is possible for both IF and RF amplifiers to work over the same frequency ranges, but from the applications given above it can be seen that tuned amplifiers fall into two specific types, namely narrow-band and wide-band.

4.2 Tuned circuits

In order for a tuned amplifier to work it must have a tuned load. Tuned circuits may be parallel or series, but normally a parallel tuned load is used in a tuned amplifier. Figure 4.2 shows such a circuit with the d.c. resistance (r) of the coil also included.

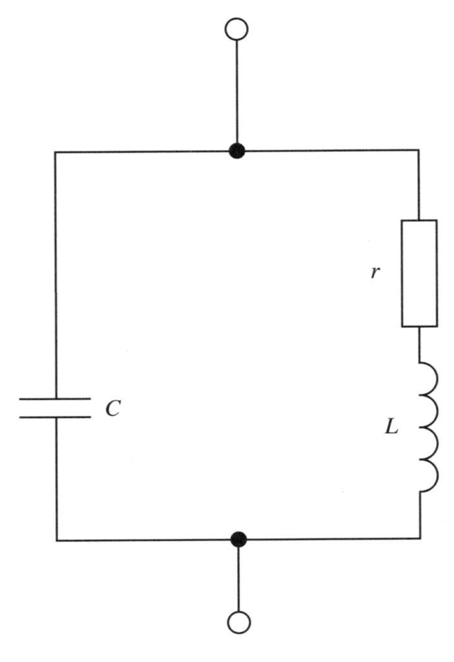

Fig. 4.2

At resonance (f_o) the impedance (Z) is a maximum and the supply current a minimum. The voltage across the circuit is large. As the resonant frequency is given by

$$f_o = \frac{1}{2\pi\sqrt{LC}} \tag{4.1}$$

the tuning may be altered by selecting appropriate values of C and L.

Note that C and L may be fixed for a constant frequency or else C can be made variable or L varied by means of a ferrite slug introduced into the coil. The sharpness of the tuning determines the Q factor of the coil.

4.3 The Q factor

The Q factor refers to the selectivity of a tuned circuit. At the moment we are dealing with the Q factor of the coil, but, as we will see later, various factors alter this value in a tuned amplifier so that the Q factor of the stage at resonance (Q_s) is quite different from the Q of the coil (Q). In fact it generally has a smaller value than that of the coil. Note that the higher the value of Q or Q_s the greater the selectivity of a circuit and the sharper the response.

The Q factor of a coil is given by

$$Q = \frac{2\pi f_o L}{r} = \frac{1}{2\pi C r} \tag{4.2}$$

It can be seen that Q depends on the d.c. resistance of the coil – the smaller the value of r, the better the selectivity. The effects of low and high values are shown in Fig. 4.3. This diagram also shows the significance of what is called the L/C ratio. If this is high then good selectivity is obtained. The lower the coil resistance and the higher the L/C ratio, the higher the dynamic impedance (Z_D) of the tuned circuit.

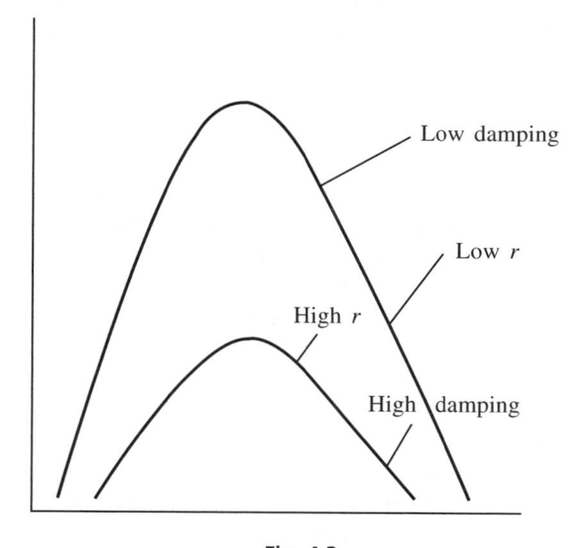

Fig. 4.3

Note also that by placing a resistance in parallel with the original tuned circuit, the value of Z_D and Q can be decreased if desired. This is known as **damping** and is a technique sometimes used to increase the bandwidth. Damping also occurs when a tuned amplifier's output is loaded (say) by a second stage, and this has to be allowed for when designing a stage for a particular Q factor.

4.4 Dynamic impedance

At resonance the actual impedance of the tuned circuit is purely resistive and is known as the dynamic impedance (Z_D). It is generally given by

$$Z_D = \frac{L}{Cr} \tag{4.3}$$

Another expression which will give the same result is:

$$R_{P1} = \frac{(2\pi f_o)^2}{r} L^2 \tag{4.4}$$

where $R_{P1} = Z_D$. This is an important parameter as the ultimate gain and Q of the stage will be affected by the loading of the dynamic impedance. R_{P1} is sometimes referred to as the equivalent parallel load resistance of r.

4.5 Gain and bandwidth

Figure 4.3 shows what is referred to as the bandwidth. In most amplifiers the bandwidth is determined from the 3 dB points on the response. These points set the limit of the useful bandwidth of the amplifier and also indicate that the power has been reduced by half. Not surprisingly, these points are sometimes referred to as the half-power points. The three parameters, resonant frequency, 3 dB bandwidth and stage Q factor, are related as follows:

$$B_{3dB} = \frac{f_o}{Q_s}$$

The gain of the amplifier is always an important consideration as in many cases the incoming signal may be small. There are two gains to be considered: one at resonance and the other off resonance. In most cases we are concerned with the gain at resonance as this is the maximum gain possible at the centre frequency, but as we may be dealing with a wide-band amplifier it is necessary to know the gain at frequencies either side of resonance.

At resonance the maximum voltage gain is given by

$$A_{vo} = -g_m R_{P2} \tag{4.6}$$

where R_{P2} is the total equivalent resistance of the tuned amplifier stage and g_m is the mutual conductance of the active device in milliamps per volt. Note that the negative sign indicates a 180° phase shift between input and output. Hence for a constant value of g_m changes in load impedance with frequency produce proportional changes in amplifier gain at resonance.

Frequently, measurements are required off resonance, particularly at antenna tuning units in transmitter systems sometimes is taken at the 3dB points. The gain at points away from resonance may be given by the following.

$$A_V = \frac{A_{Vo}}{\sqrt{\left(1 + 4Q^2\left(\frac{\omega - \omega_o}{\omega_o}\right)^2\right)}} \tag{4.7}$$

where ω_o is the resonant frequency, ω is the frequency off resonance and A_{Vo} is the gain at resonance. A derivation of this expression is given in Appendix B.

Example 4.1

A tuned collector amplifier uses a transistor with $g_m = 50$ mA/V and a very high value of collector-to-emitter resistance. The coil of the tuned circuit has an inductance of 150 μH and $Q = Q_s = 20$. It is tuned to resonate at 796 kHz. Calculate:

(a) the gain at resonance;

(b) the gain at frequencies of $f_o \pm 10$ kHz;

(c) the effective bandwidth.

Solution

(a)

$$Z_D = R_{P1} = \frac{L}{Cr} = \frac{\omega_o L}{\omega_o Cr}$$

∴

$$Z_D = \frac{Q}{\omega_o C} = Q\omega_o L \tag{4.8}$$

hence, taking $Z_D = Q\omega_o L$ gives,

$$Z_D = \frac{20 \times 2\pi \times 796 \times 150}{10^6} = 15\ \text{k}\Omega$$

and

$$A_{Vo} = g_m Z_D = 750$$

(b)

$$A_V = \frac{A_{Vo}}{\sqrt{(1 + 4Q^2 D^2)}}$$

If

$$D = \left(\frac{\omega - \omega_o}{\omega_o}\right)$$

then we have

$$D = \pm\frac{10}{796}$$

hence

$$A_V = \frac{750}{\sqrt{(1 + 4 \times 20^2 \times (10/796)^2)}} = 670$$

(c)

$$BW = \frac{f_o}{Q_s} = \frac{796}{20} = 39.8 \text{ kHz}$$

This gives us an effective bandwidth of 796 ± 19.9 kHz.

4.6 Effect of loading

In most practical amplifier circuits the value of output resistance compared with the dynamic impedance Z_D is not so high that it can be regarded as negligible. (In Example 4.1 it was ignored as it was assumed to be high.) Also this type of amplifier is probably the first of a number of stages in a receiver. It is for this reason that its tuned circuit must be loaded by the input impedance of the next stage, thus modifying the overall frequency response.

Assuming the input impedance over the operating range of the subsequent stage to be purely resistive and represented by R_L, Z_D will be loaded as shown in Fig. 4.4. Hence the effective load at resonance on the collector is

$$Z' = \frac{R_L \times Z_D}{R_L + Z_D} > Z_D$$

as

$$Z_D = Q\omega_o L = Q/\omega_o C.$$

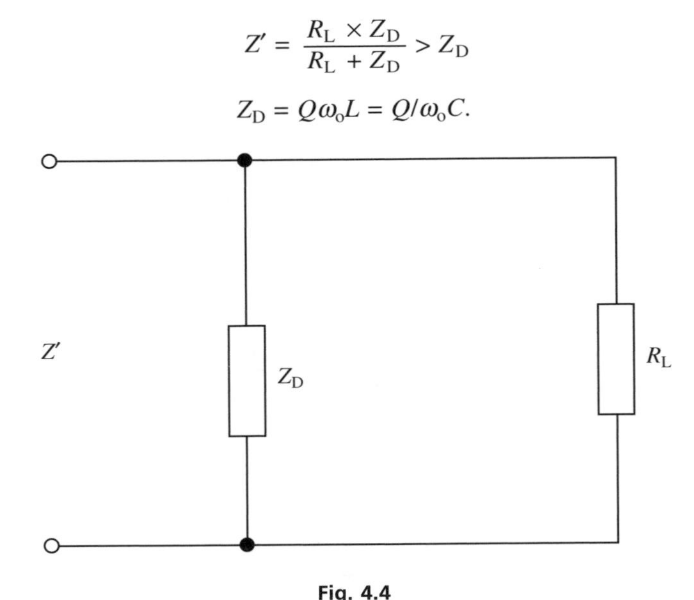

Fig. 4.4

It should be noted in both expressions that the resonant frequency is hardly affected by loading. Therefore it can be deduced that the loaded Q factor of the circuit and the effective load presented to the transistor are proportional. Hence, the loading of the tuned amplifier by another stage not only reduces the gain at resonance but also affects the bandwidth. There is therefore a loss of sensitivity and a reduction in selectivity.

Two further points should be noted. First, the loaded Q factor must be used in calculations involving cascaded stages. This is shown in Example 4.2. Second, in practice more than one stage of amplification is used.

It has been mentioned that the bandwidth BW of a signal stage is also altered when cascading stages and this new bandwidth BW' may be calculated for n identical cascaded stages by using the following simple expression:

$$BW' = BW\sqrt{(2^{1/n} - 1)} \quad (4.9)$$

Example 4.2

Calculate the mutual conductance for the transistor required in a single tuned RF amplifier operating at 465 kHz if its voltage gain at resonance is to be 78.6 dB, neglecting any loading effect. The tuned circuit consists of a 200 pF capacitor in parallel with an inductor of $Q = 100$. What are the new values of maximum gain in decibels and bandwidth when the amplifier supplies another stage whose input resistance (R) is 40 kΩ?

Solution

$$Z_D = R_{P2} = \frac{Q}{\omega_o C} = \frac{100}{2\pi \times 465 \times 10^3 \times 200 \times 10^{-12}} = 171\ \text{k}\Omega$$

$A_{Vo} = 78.6$ dB has to be changed to a decibel value:

$$78.6 = 20\log\left(\frac{V_{out}}{V_{in}}\right)$$

$$\therefore \quad \frac{V_{in}}{V_{out}} = 8511$$

Also

$$g_m R_{P2} = 8511$$

$$\therefore \quad g_m = 49.7\ \text{mA/V}$$

when loaded with 40 kΩ input resistance (R_i),

$$R'_{P2} = \frac{R_{P2} R_i}{R_{P2} + R_i}\ 32.4\ \text{k}\Omega$$

So the gain at resonance is

$$A_{Vo} = g_m R'_{P2} = 49.7 \times 32.4 = 1613$$

Since the loaded Q factor, denoted Q', is given by

$$\frac{Q'}{Q} = \frac{R'_{P2}}{R_{P2}}$$

we have

$$Q' = \frac{100 \times 32.4}{171} = 18.95$$

Similarly,

$$BW = \frac{f_o}{Q'} = \frac{465}{18.95} = 24.5\ \text{kHz}$$

giving an operating frequency of 456 ± 12.25 kHz.

Another example at this stage will help to consolidate the ideas behind designing a single-stage fixed tuned amplifier. This time an FET is used, but the principles are exactly the same. The schematic diagram is shown in Fig. 4.5

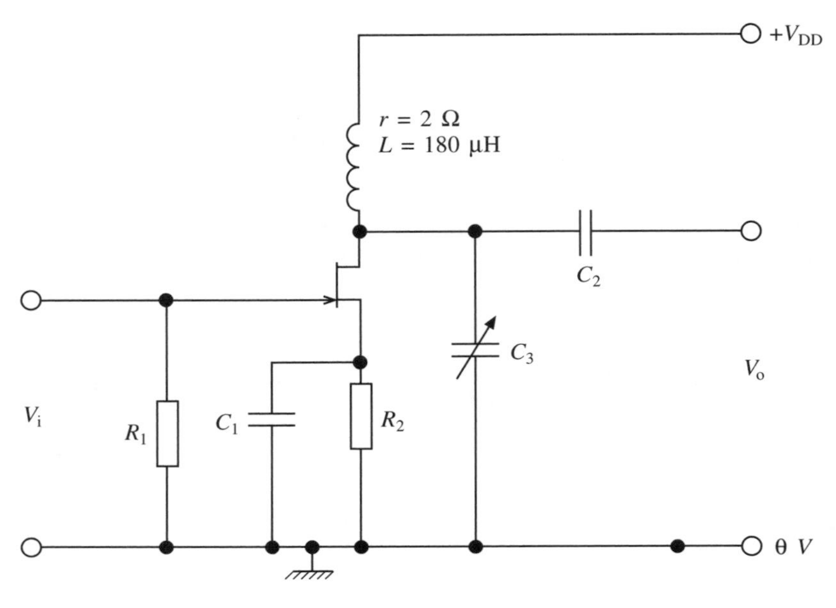

Fig. 4.5

Example 4.3
The circuit in Fig. 4.5 consists of an FET having a mutual conductance $g_m = 3$ mA/V and the drain-to-source resistance $r_{ds} = 120$ kΩ. If the d.c. resistance of the coil is 2.0 Ω, calculate:

(i) the value the tuning capacitor should be set at for a resonant frequency of 850 kHz;

(ii) the maximum voltage gain assuming negligible load impedance;

(iii) the bandwidth of the stage;

(iv) the bandwidth and Q factor if a further two identical stages are connected in cascade.

Solution

(i)

$$C = \frac{1}{4\pi^2 Lf_2} = \frac{10^6}{4\pi^2 \times 180 \times (850 \times 10^3)^2} = 195 \text{ pF}$$

(ii)

$$\frac{L}{Cr} = \frac{180 \times 10^{-6}}{195 \times 10^{-12} \times 2}$$

$\therefore$ $$R_{P1} = 461.54 \text{ k}\Omega$$

Hence the total parallel resistance is given by

$$R_{P2} = \frac{120 \times 461.54}{120 + 461.54} = \frac{55384.8}{581.54} = 95.24\ \text{k}\Omega$$

so

$$A_{Vo} = 95.24 \times 3 = 285.72$$

(iii)

$$Q_s = \frac{95.24 \times 10^3 \times 10^6}{2\pi \times 850 \times 10^3 \times 180} = 99$$

So

$$B = \frac{f_o}{Q_s} = \frac{850 \times 10^3}{99} = 8.59\ \text{kHz}$$

(iv)

$$BW' = BW\ \sqrt{(2^{1/n} - 1)} = 8.59\ \sqrt{(2^{1/3} - 1)} = 4.379\ \text{kHz}$$

So

$$Q_s' = \frac{f_o}{BW'} = \frac{850}{4.379} = 194$$

4.7 Effect of tapping the tuning coil

In practice, tuning coils are tapped in order to match the output impedance of the tuned circuit with the input impedance of the following stage. This requires a different approach which is explained in Fig. 4.6. Here L_1 and L_2 are part of a continuous winding which is tapped as shown:

$$L = L_1 \pm M + L_2 \pm M$$

$$L = L_1 + L_2 \pm 2M$$

R_L in parallel with $L_2 \pm M$ can be resolved in its equivalent series circuit as

$$R_s = \frac{\omega^2 (L_2 + M)^2}{R_L}$$

This is shown in Fig. 4.6(b). Using equation (4.4) will now give

$$R_{P3} = \frac{\omega^2 L^2}{R_s} = \frac{\omega^2 L^2}{\omega^2 (L^2 \pm M)^2 / R_L}$$

hence

$$R_{P3} = \frac{L^2 \times R_L}{(L^2 \pm M)^2} \qquad (4.10)$$

This allows a large R_{P3} in parallel with the tuned circuit and will allow a large Q factor and hence a narrower bandwidth.

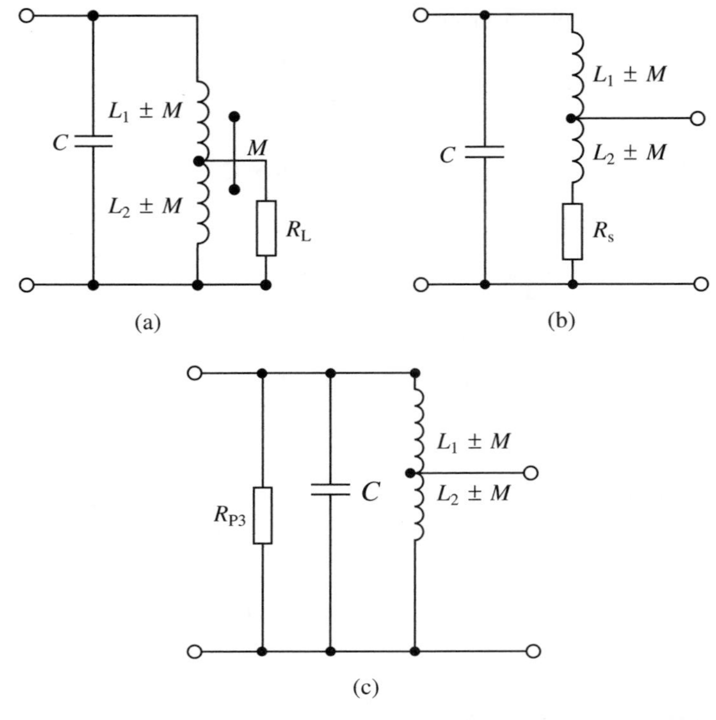

Fig. 4.6

Example 4.4

The circuit shown in Fig. 4.7 incorporates a tapped coil to more closely match the input to the second-stage transistor TR_2. If in this circuit $L = 200\ \mu\text{H}$, $L_1 = L_2$, $R_{in} = 100\ \text{k}\Omega$, $h_{oe} = 10^{-5}$ S (see Section 5.2 for more on h-parameters) and the coil resistance $r = 3\ \Omega$, determine the Q factor and bandwidth of stage one. Compare the result with the Q factor and bandwidth for an untapped coil. Assume that $f_0 = 1$ MHz and the coupling coefficient $k = 0.02$. Also let $L_2' = L_2 - M$.

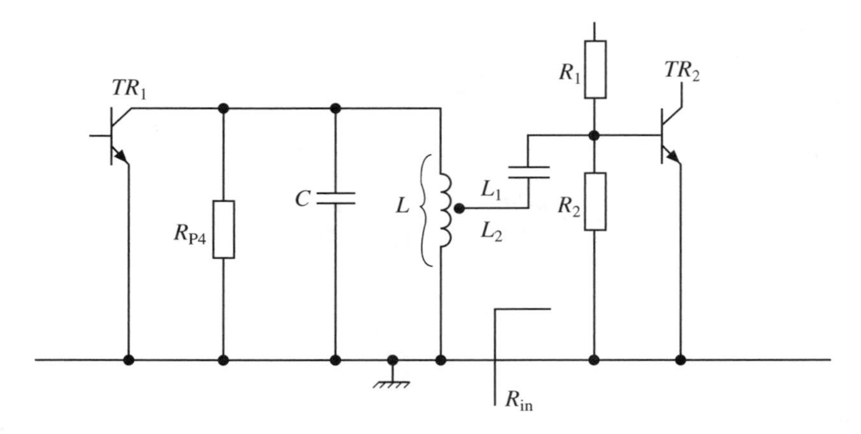

Fig. 4.7

Solution
Convert resistance to its parallel form

$$R_{P1} = \frac{\omega_o^2 L^2}{r} = \frac{4\pi \times 2 \times 4 \times 10^{-8}}{3} \times 10^{12} = 526 \text{ k}\Omega$$

R_{P1} in parallel with $1/h_{oe}$ will give

$$\frac{526 \times 100}{526 + 100} = 84 \text{ k}\Omega = R_{P4}$$

Since $$M = k\sqrt{L_1 L_2} = kL_1 = 0.02 \times 100 = \mu\text{H}$$

then $$R_{P3} = 10^5 \times \frac{200^2}{(100 - 2)^2} = 41.6 \text{ k}\Omega$$

The total resistance R_{P2} is thus R_{P4} in parallel with R_{P3}

$$R_{P2} = \frac{84 \times 41.6}{84 + 41.6} = 27.8 \text{ k}\Omega$$

$$Q_s = \frac{R_{P2}}{\omega_o L} = \frac{27.8 \times 103}{2\pi \times 10^6 \times 200 \times 10^{-6}} = 22.1$$

$$BW = \frac{f_0}{Q_s} = \frac{1000}{22.1} = 45 \text{ kHz}$$

If the untapped coil is now considered, then R_{P4} in parallel with R_{in} gives a different value for R_{P2}:

$$R_{P2} = \frac{84 \times 10}{84 + 10} = 8.94 \text{ k}\Omega$$

$$Q_s = \frac{8.94 \times 103}{400 \times \pi} = 7.1$$

$$BW = \frac{f_0}{Q_s} = \frac{1000}{7.1} = 140 \text{ kHz}$$

This example shows the poor bandwidth and selectivity obtained if tapping is not used. It should also be noted that a 1 : 1 tapping ratio has been used here. A different tapping point would give a different selectivity. This will be tackled in one of the questions at the end of the section.

Example 4.5
A two-stage intermediate frequency amplifier is coupled by means of a tapped transformer. Part of the circuit is shown in Fig. 4.8. If $L_3 = 120$ H, $L_1 = 80$ μH, $L_2 = 40$ μH and the input resistance to the second stage $R_{in} = 4.5$ kΩ, determine the Q factor and bandwidth of stage one. Assume the following parameters are appropriate: $h_{oe} = 10^{-4}$ S, $f_0 = 450$ kHz, $r = 3$, $k = 0.05$ and $L_2' = L_2 - M$.

Solution
The coil resistance is first converted:

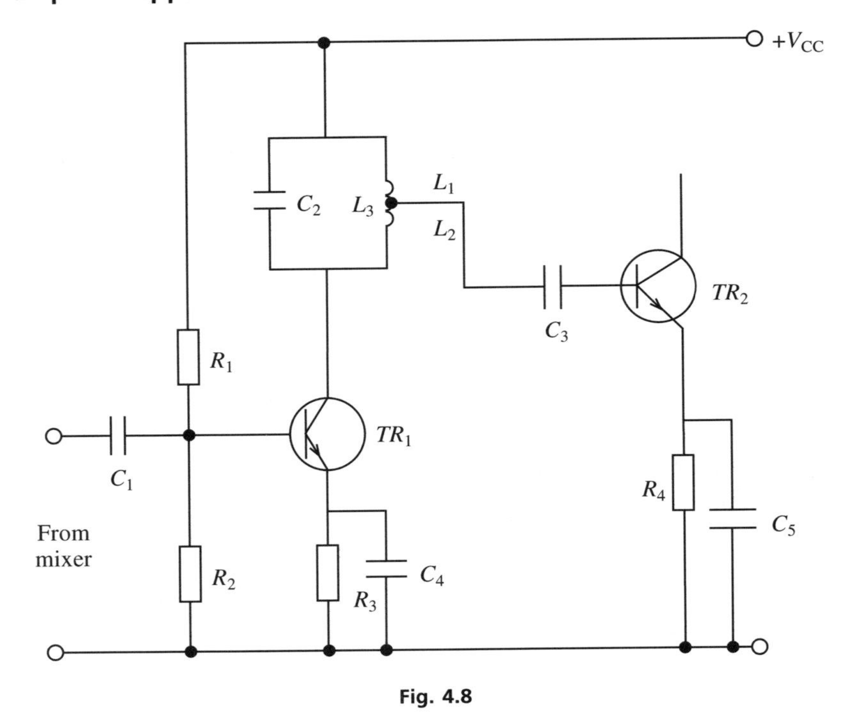

Fig. 4.8

$$R_{P1} = \frac{\omega_0^2 L^2}{r} = \frac{40 \times 450^2 \times 10^6 \times 120^2}{10^{12} \times 3} = 38.9 \text{ k}\Omega$$

R_{P1} in parallel with $1/h_{oe}$ will give R_{P4}

$$R_{P4} = \frac{38.9 \times 10}{38.9 + 10} = \frac{389}{48.9} = 7.95 \text{ k}\Omega$$

Since

$$M = k\sqrt{L_1 L_2} = 0.05\sqrt{40 \times 80} = 2.38 \text{ µH}$$

then

$$R_{P3} = \frac{L^2 R_{in}}{(L_2 - M)^2} = \frac{(120)^2 \times 2.5 \times 10^3}{(40 - 2.83)^2} = 26.1 \text{ k}\Omega$$

The total resistance R_{P2} is thus R_{P4} in parallel with R_{P3}. Hence

$$R_{P2} = \frac{26.1 \times 7.95}{26.1 + 7.95} = 6.1 \text{ k}\Omega$$

So

$$Q_s = \frac{R_{P2}}{\omega_0 L} = \frac{6.1 \times 10^3 \times 10^6}{2\pi \times 450 \times 10^3 \times 120} = 18$$

$$\therefore \qquad BW = \frac{450}{18} = 25 \text{ kHz}$$

4.8 Transformer-coupled amplifier

The tuned amplifiers considered so far have been fixed tuned types with a single collector tuned load. In practice, however, transformer coupling is generally employed in which a radio frequency transformer is used. An example of such an application is shown in Fig. 4.9. The transformer may be connected in different ways depending on the active device used. With a bipolar transistor the tuned primary configuration is frequently used, whereas with an FET the tuned secondary configuration is used. Both these applications are shown below.

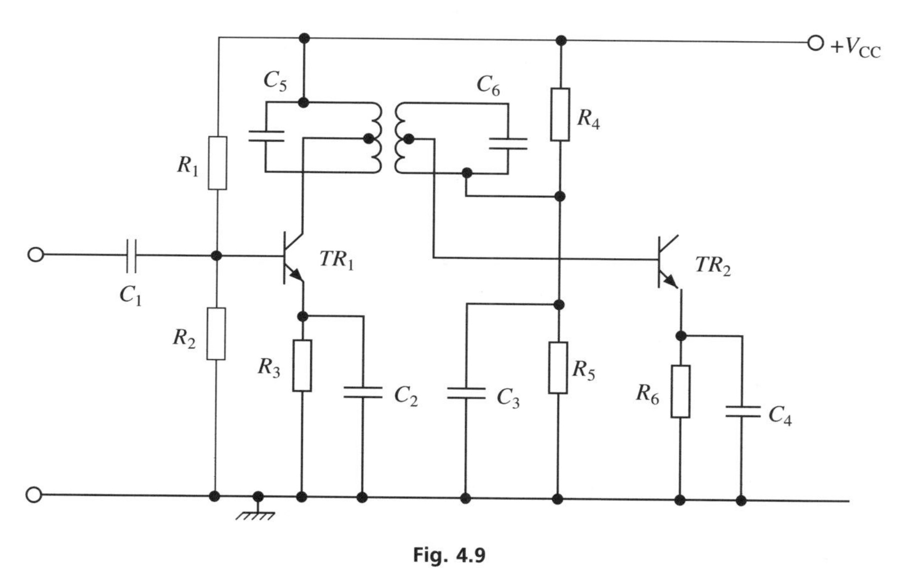

Fig. 4.9

4.9 Tuned primary

This applies where a BJT is used since the output impedance ($1/h_{oe}$) is normally very large. The a.c. circuit is shown in Fig. 4.10 and the equivalent circuit in Fig. 4.11(a). The resonant circuit is connected in the collector as at this point it will act as a high-impedance load. Note that by adjusting C_1 for resonance so that $\omega_o^2 L_1 C_1 = 0$, the circuit impedance becomes pure resistance at the input terminals of the two point network shown in Fig. 4.11(b). The equivalent circuit is shown in Fig. 4.11(c). In transformer-coupled circuits such as this the mutual reactance (ωM) must have an optimum value. Only when this is done can the resistive load in the secondary (R_{22}) be matched to the first stage source. An analysis of this type of tuned amplifier is given in Appendix *C*.

Example 4.6

A primary tuned IF amplifier used in an FM receiver has a coil tapped so that $L_1 - L_2 =$ 120 μH. If the input resistance (R_U) at the primary is 375 kΩ, $R_s = 0.5\ \Omega$, $k = 0.4$ and g_m = 10 mA/V, determine

(a) the gain at resonance for an intermediate frequency of 450 kHz;

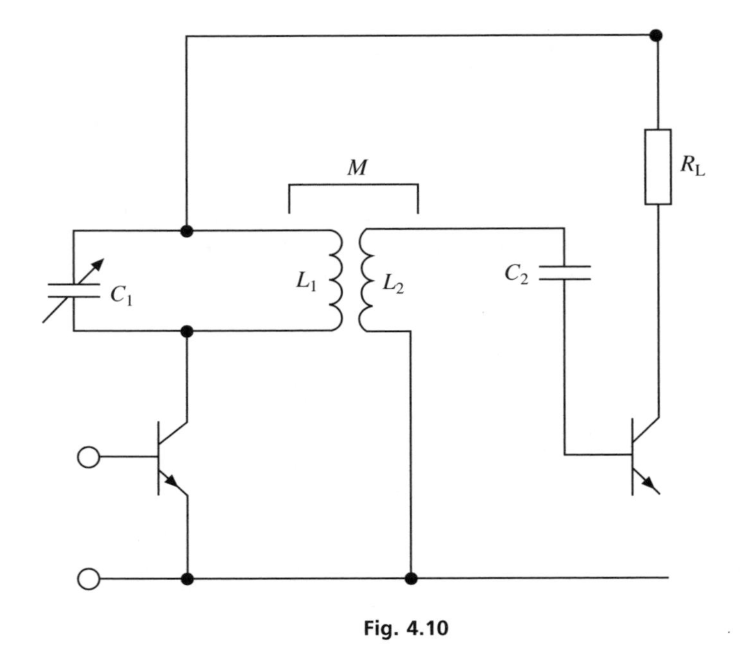

Fig. 4.10

(b) the effective Q factor;

(c) the bandwidth;

(d) the critical coupling for maximum gain;

(e) the Q factor at critical coupling and the corresponding bandwidth.

Solution

(a) Since $M = k\sqrt{L_1 L_2}$, we have $M = 0.4 \times 120 = 48$ μH. Using equation (C.5) in Appendix C,

$$g_m R_{11}(M/L_1) = \frac{10 \times 375 \times 10^3 \times 48 \times 10^6}{10^3 \times 10^6 \times 120} = 1500$$

$$R_{22} = \frac{4\pi^2 \times (450 \times 10^3)^2 \times 48^2}{0.5 \times 10^{12}} = 36.8 \text{ k}\Omega$$

$$\therefore \quad \left(1 + \frac{M^2}{L_1^2}\right)\left(\frac{R_{11}}{R_{22}}\right) = \left(1 + \frac{(48)^2 \times 10^{12}}{10^{12} \times (120)^2}\right)\left(\frac{375 \times 10^3}{36.8 \times 10^3}\right) = 11.8084$$

$$A_{Vo} = \frac{g_m R_{11}(M/L_1)}{(1 + M^2/L_1^2)(R_{11}/R_{22})} = \frac{1500}{11.8084} = 127 = 42.1 \text{ dB}$$

(b)

$$Q_e = \frac{R/\omega_o L_1}{(1 + M^2/L^2)(R_{11}/R_{22})}$$

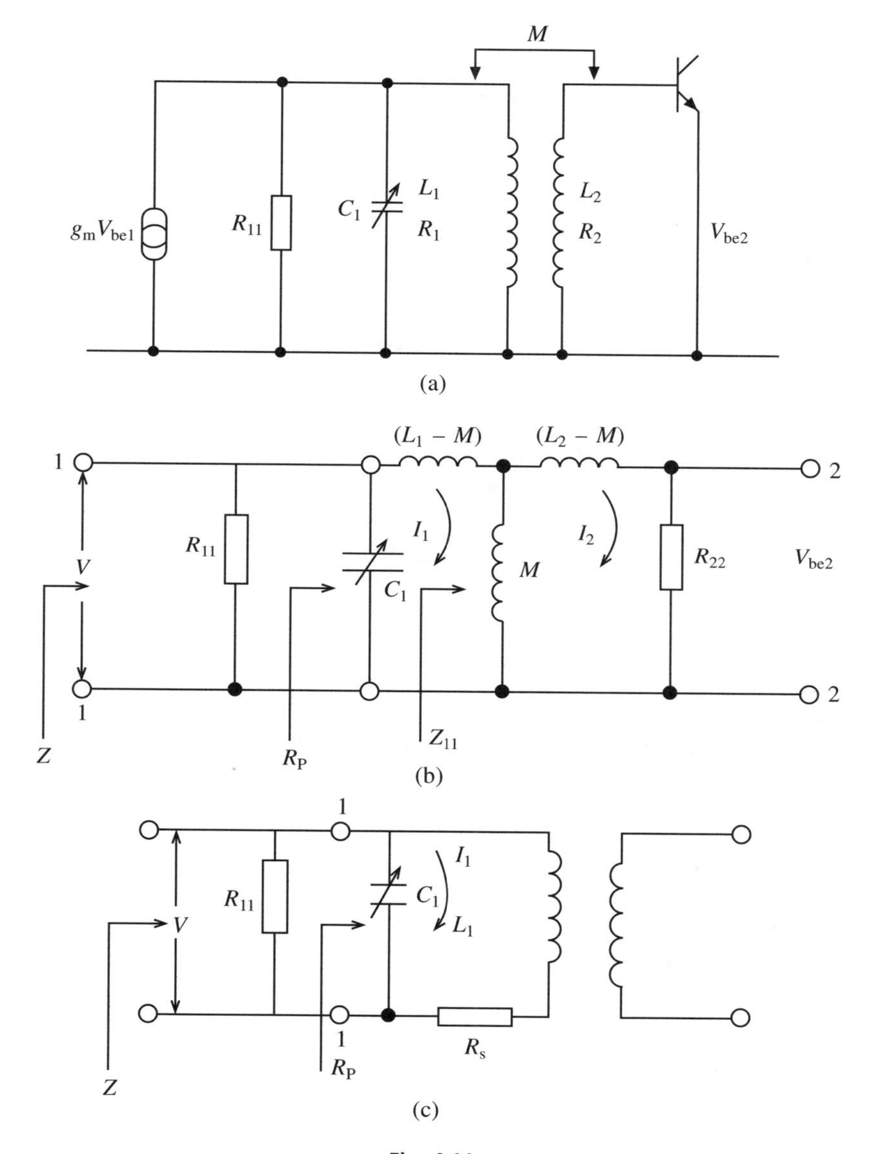

Fig. 4.11

$$R_{11}/\omega_o L_1 = \frac{375 \times 10^3 \times 10^6}{2\pi \times 450 \times 10^3 \times 120} = 1105.2$$

$\therefore$

$$Q_e = \frac{1105.2}{11.8084} = 93.6$$

(c)

$$BW = \frac{f_o}{Q_e} = \frac{450 \times 10^3}{93.6} = 4808 \text{ Hz}$$

(d) Critical coupling occurs when

$$\omega_o M = \omega_o L_1 \sqrt{\frac{R_{22}}{R_{11}}}$$

$$\therefore \qquad M = L_1 \sqrt{\frac{R_{22}}{R_{11}}} = \frac{120}{10^6} \sqrt{\frac{36.8 \times 10^3}{375 \times 10^3}} = 37.6\ \mu\text{H}$$

since $M = k\sqrt{L_1 L_2}$,

$$\therefore \qquad k = \frac{M}{L_1} = \frac{37.6}{120} = 0.3134$$

(e)

$$Q_{e1} = \frac{Q_1}{2} = \frac{R_{11}}{2(1 + M^2/L_1^2)(R_{11}/R_{22})\omega_o L_1}$$

$$= \frac{375 \times 10^3 \times 10^6}{2 \times 11.8084 \times 2\pi \times 450 \times 10^3 \times 120} = 46.8$$

$$\therefore \qquad BW = \frac{450 \times 10^3}{46.8} = 9615.6\ \text{Hz}$$

Example 4.7

A primary tuned RF amplifier is required to have a gain at resonance of 600 and a bandwidth of 3.2 kHz. The following parameters apply: $L_1 = 80\ \mu$H, $L_2 = 60\ \mu$H, $f_o = 285$ kHz, $g_m = 15$ mA/V and $k = 0.32$. Determine:

(a) the value of the secondary load;

(b) the value of input resistance (R_{11});

(c) the optimum value of M which will give maximum gain.

Solution

(a) $M = k\sqrt{L_1/L_2} = 0.32\sqrt{60 \times 80} = 22.2\ \mu$H. Rearranging

$$A_{Vo} = \frac{g_m M R_{22}}{L_1(1 + M^2/L_1^2)}$$

we obtain

$$R_{22} = \frac{A_{Vo} L_1 \left(1 + \dfrac{M^2}{L_1^2}\right)}{g_m M}$$

$$= \frac{600 \times 80 \times 10^{-6}}{0.015 \times 22.2 \times 10^{-6}} \left(1 + \frac{491.52 \times 10^{-12}}{(80 \times 10^{-6})^2}\right) = 155.4\ \text{k}\Omega$$

(b) Since

$$\frac{\omega^2 L_1^2}{R_s} = \frac{\omega^2 L_1^2}{\omega^2 M^2 / R_{22}}$$

we have

$$R_s = \frac{\omega^2 M^2}{R_{22}} = \frac{4\pi^2 \times (285 \times 10^3)^2 \times 491.52 \times 10^{-12}}{155.4 \times 10^3} = 10.14\ \text{M}\Omega$$

Since

$$Q_{el} = \frac{f_o}{BW} = \frac{285 \times 10^3}{3.2 \times 10^3} = 89$$

and

$$Q_{el} = \frac{R_{P2}}{\omega L_1}$$

we have

$$R_{P2} = Q_{e1}\omega L_1 = 89 \times 2\pi \times 285 \times 10^3 \times 80 \times 10^{-6} = 12.76\ \text{k}\Omega$$

Also

$$R_{P1} = \frac{\omega^2 L_1^2}{R_s} = \frac{4\pi^2 \times (285 \times 10^3)^2 \times (80 \times 10^{-6})^2}{10.14 \times 10^{-3}} = 2.02\ \text{M}\Omega$$

Now

$$R_{P2} = \frac{R_{11} R_{P1}}{R_{11} + R_{P1}}$$

and transposing gives

$$R_{11} = \frac{R_{P2} R_{P1}}{R_{P1} - R_{P2}} = \frac{12.76 \times 10^3 \times 2.02 \times 10^6}{2.02 \times 10^6 - 12.76 \times 10^3} = 12.8\ \text{k}\Omega$$

(c) For optimum gain

$$M = L_1 \sqrt{\frac{R_{22}}{R_{11}}} = 80 \times 10^{-6} \sqrt{\frac{155.4 \times 10^3}{12.8 \times 10^3}} = 278.3\ \mu\text{H}$$

4.10 Tuned secondary

This type of coupling (see Fig. 4.12) applies generally to active devices such as FETs since the input impedance is very large. The equivalent circuit is shown in Fig. 4.13. In this case the important parallel resistance is the drain-to-source resistance of the FET. So the resistance R_{11} in this case is r_{ds} in parallel with R_1.

The gain resonance is:

$$A_V = \frac{-g_m \omega_o M Q_2}{1 + \dfrac{\omega_o^2 M^2}{R_{11} R_2}}$$

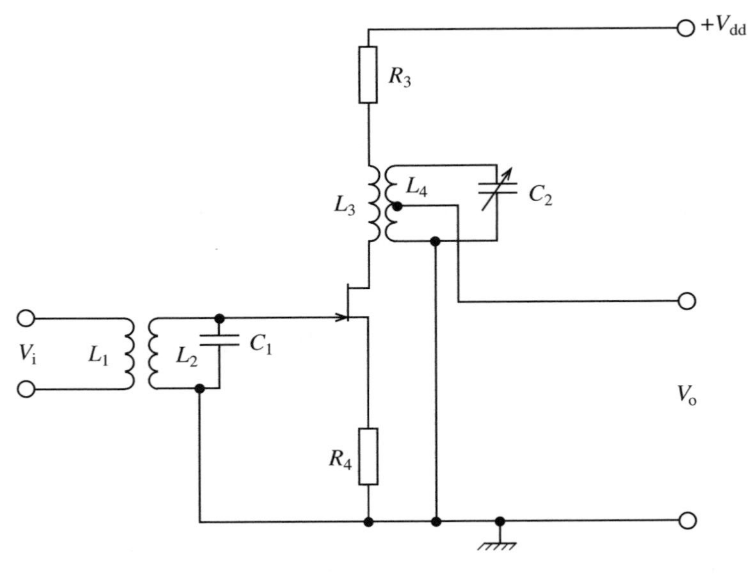

Fig. 4.12

where

$$Q_2 = \frac{1}{\omega_o C_2 R_2}$$

The effective Q factor of the stage is Q_{e2}, so once again there is a similar expression to that for the tuned primary configuration:

$$Q_{e2} = \frac{Q_2}{1 + \dfrac{\omega_o^2 M^2}{R_{11} R_2}}$$

A solution to this circuit, which uses the two-port network and T-configuration is given in Appendix D.

Example 4.8

An RF amplifier has a tuned secondary which has to be tunable over the FM range 87–101 MHz. The following parameters are available: $g_m = 4$ mA/V, $L_1 = 15$ μH, $L_2 = 10$ μH, $R_2 = 2$ Ω and $r_{ds} = 240$ kΩ. Determine:

(a) the equivalent Q of the stage;

(b) the gain at each end of the band;

(c) the value of coupling coefficient which would give maximum gain over this frequency band;

(d) the bandwidth at optimum coupling.

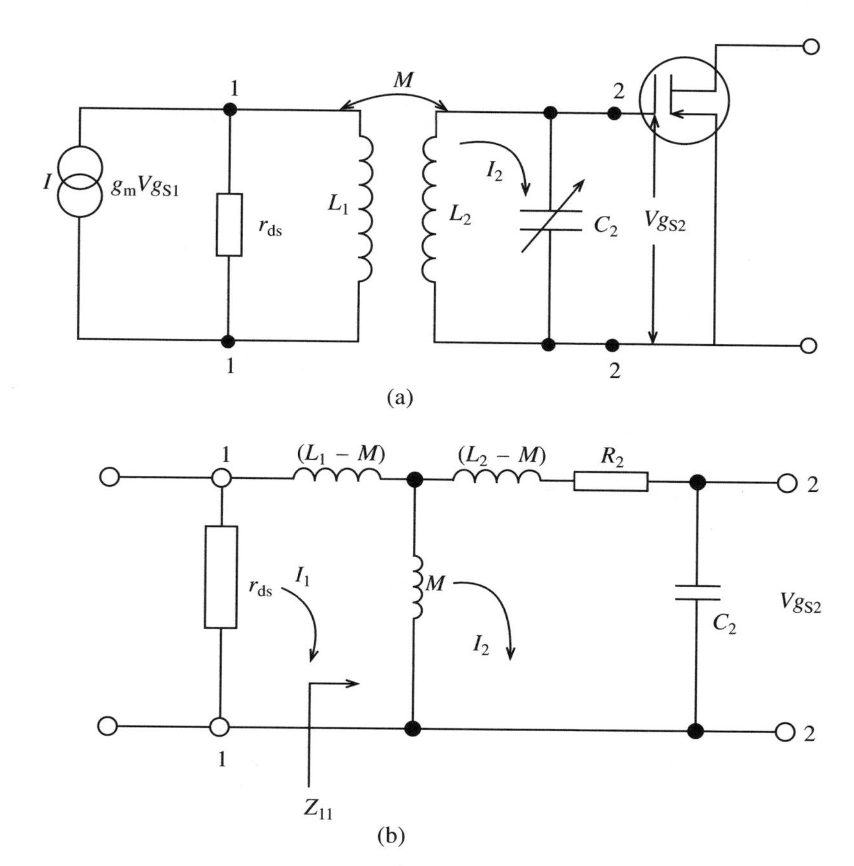

Fig. 4.13

Solution

(a) For the 87 MHz carrier

$$M = k\sqrt{L_1 L_2} = 0.64\sqrt{\frac{10 \times 15}{10^{12}}} = 7.8\ \mu\text{H}$$

$$Q_2 = \frac{\omega_o L_2}{R_2} = \frac{2\pi \times 87 \times 10^6 \times 10}{10^6 \times 2} = 2738.2$$

and

$$Q_{e2} = \frac{Q_2}{1 + \omega_o^2 M^2 / r_{ds} R_2}$$

$$= \frac{2733.2}{1 + (4\pi^2 \times (87 \times 10^6)^2 \times (7.8 \times 10^{-6})^2)/240 \times 10^3 \times 2} = 70$$

(b)

$$A_{Vo} = -g_m \omega_o M_1 \frac{Q_2}{1 + \omega_o^2 M^2 / r_{ds} R_s}$$

$$= \frac{4 \times 2\pi \times 87 \times 10^6 \times 7.8 \times 70}{10^3 \times 10^6} = 1194 = 61.5\ \text{dB}$$

(a) For the 101 MHz carrier

$$Q_2 = \frac{\omega_o L_2}{R_2} = \frac{2\pi \times 101 \times 10^6 \times 10}{10^6 \times 2} = 3173.0$$

As before,

$$Q_{e2} = \frac{3173}{1 + (4\pi^2 \times (101 \times 10^6)^2 \times (7.8 \times 10^{-6})^2 / 240 \times 10^3 \times 2} = 60$$

(b)

$$A_{Vo} = \frac{4 \times 2\pi \times 101 \times 10^6 \times 7.8 \times 60}{10^3 \times 10^6} = 1201 = 61.6 \text{ dB}$$

(c) At a frequency of 87 MHz

$$\omega_o M = \sqrt{r_{ds} R_2}$$

$$M = \sqrt{\frac{r_{ds} R_2}{2\pi f}} = \frac{\sqrt{240 \times 10^3 \times 2}}{2\pi \times 87 \times 10^6} = 1.267\ \mu\text{H}$$

Also

$$k = \frac{M}{\sqrt{L_1 L_2}} = \frac{1.267}{\sqrt{150}} = 0.1$$

At a frequency of 101 MHz it can be shown by the same method that $k = 0.09$. Hence a coupling coefficient of 0.1 would give maximum gain over the band.

(d) Take the middle of the band (94 MHz) to find BW. There $Q_{e2} = 65$.

$$\therefore \qquad BW = \frac{94 \times 10^6}{65} = 1.45 \text{ MHz}$$

Example 4.9

A communications receiver has two identical secondary tuned RF amplifiers connected in cascade. The parameters are: $R_{ds} = 350\ \text{k}\Omega$, $R_2 = 1\ \Omega$, $g_m = 10$ mA/V, $L_1 = 360\ \mu$H and $L_2 = 220\ \mu$H. If the receiver has to be tuned to an emergency channel of 500 kHz, determine the critical coupling factor and the overall bandwidth which will be obtained if critical coupling is applied.

Solution

For critical coupling

$$\omega_o M = \sqrt{r_{ds} R_2}$$

$$\therefore \qquad M = \frac{\sqrt{r_{ds} R_2}}{\omega_o} = \frac{\sqrt{350 \times 10^3 \times 1}}{2\pi \times 500 \times 10^3} = 188.3\ \mu\text{H}$$

Also

$$M = k_c \sqrt{L_1 L_2}$$

$$\therefore \qquad k_c = \frac{M}{\sqrt{L_1 L_2}} = \frac{188.3}{\sqrt{360 \times 220}} = 0.67$$

Now

$$Q_2 = \frac{\omega_o L_2}{R_2} = \frac{2\pi \times 5 \times 10^5 \times 220}{10^6} = 691$$

$$\therefore \qquad Q_{e2} = \frac{Q_2}{2} = 346$$

$$\therefore \qquad BW = \frac{f_o}{Q_{e2}} = \frac{5 \times 10^5}{347} = 1.447 \text{ kHz}$$

The overall bandwidth for the two stages in cascade is

$$BW\sqrt{2^{1/n} - 1} = 1447\sqrt{2^{1/2} - 1} = 931 \text{ Hz}$$

4.11 Double tuning

This type of coupling uses a double tuned transformer and is generally employed in IF applications where a fixed frequency is required. The tuning is therefore set during manufacture and the skirts, i.e. the high- and low-frequency roll-offs, of the frequency response are steep. The bandwidth is narrow and hence the selectivity high. A schematic diagram is shown in Fig. 4.14(a) and the equivalent circuits in Fig. 4.14(b)–(c). The solution to Fig. 4.14(c) can be found in Appendix E. The relevant equations will be used in the following example.

Example 4.10

A double-tuned transformer has $L_1 = L_2 = 150$ μH, $Q_1 = Q_2 = Q = 100$ and $k = 1.2\ k_c$. If the source has $g_m = 200$ mA/V, determine:

(a) the value of C_1 and C_2 at a resonant frequency of 1.4 MHz (where $C = C_1 = C_2$);

(b) the frequency separation of the response peaks;

(c) the voltage gain at a peak;

(d) the practical bandwidth.

Solution

(a)

$$f_0 = \frac{1}{2\pi\sqrt{LC}}$$

$$\therefore \qquad C = \frac{1}{4\pi^2 L f_o^2} = \frac{1}{4\pi^2 \times 150 \times 10^{-6} \times (1.4 \times 10^6)^2} = 0.086 \text{ nF}$$

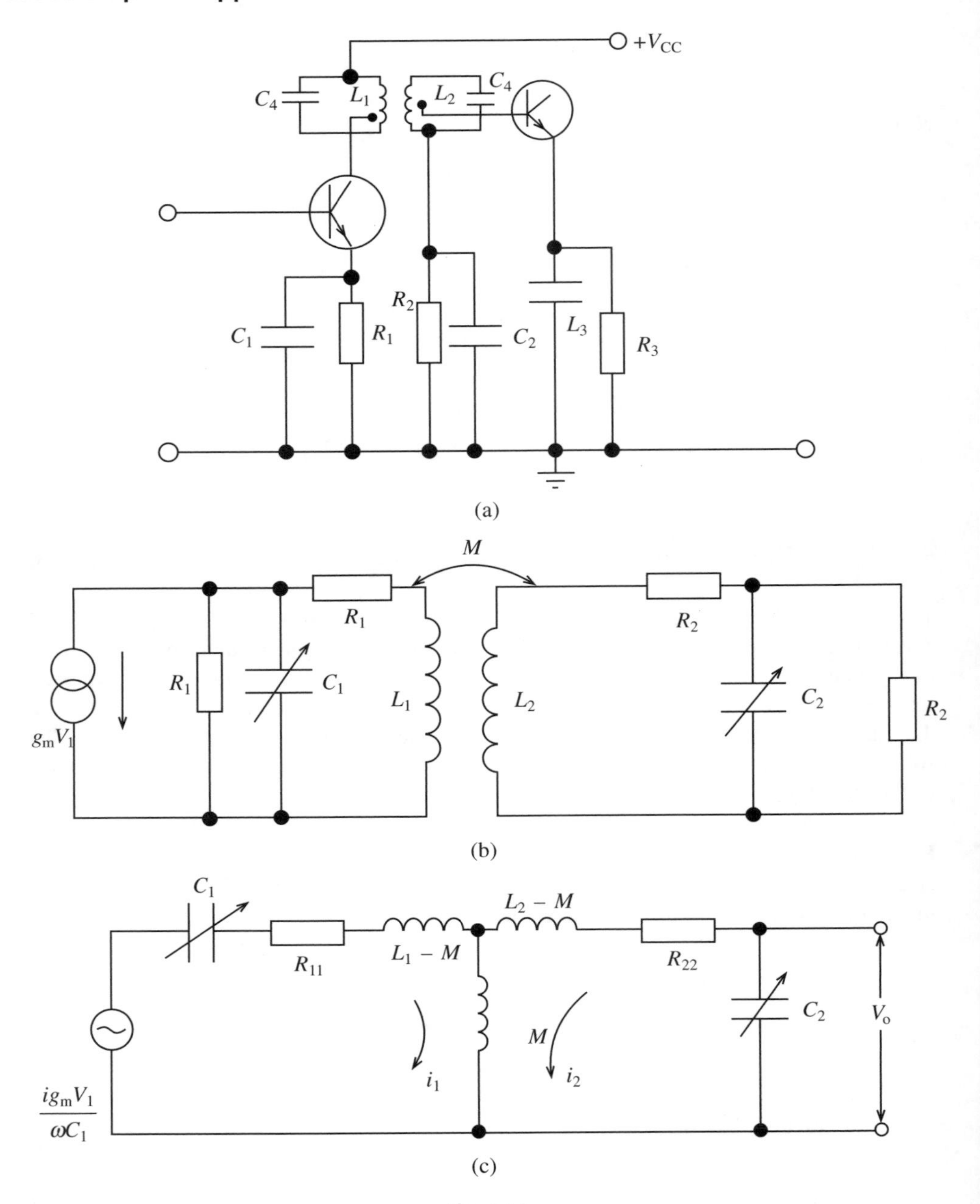

Fig. 4.14

(b)

$$k_c = \frac{1}{Q} = \frac{1}{100} = 0.01$$

$$k = 1.2k_c = 1.5 \times 0.01 = 0.015$$

$$f_2 = f_o\left(1 + \frac{1}{2Q}\sqrt{k^2Q^2 - 1}\right)$$

$$= 1.4 \times 10^6\left(1 + \frac{1}{2 \times 100}\sqrt{(0.015)^2\,(100)^2 - 1}\right) = 1.4078 \text{ MHz}$$

$$f_1 = f_o\left(1 - \frac{1}{2Q}\sqrt{k^2Q^2 - 1}\right) = 1.3922 \text{ MHz}$$

∴ Frequency separation is 15.7 kHz

(c) The voltage gain at a peak is

$$A_p = \frac{g_m\omega_o\sqrt{L_1L_2}kQ}{2}$$

$$= \frac{200 \times 10^{-3} \times 2\pi \times 1.4 \times 10^6 \times 150 \times 10^{-6} \times 0.015 \times 100}{2} = 198$$

(d) The working bandwidth is given by

$$2D_1 = \sqrt{2}(f_2 - f_1) = 1.414(1.4078 - 1.3922) = 22.1 \text{ kHz}$$

Example 4.11

A double-tuned circuit couples an IF amplifier to a second stage with an input resistance of 1 kΩ. If the output resistance of the first stage transistor is 42 kΩ, the resonant frequency 465 kHz and the bandwidth 22 kHz, find the Q factor and the values of L, C and M.

Assume $k = 0.02$ and that the secondary is matched to the input resistance of the next stage by tapping.

Solution

Since $BW = 3.1f_o/Q$ then

$$Q = \frac{465 \times 10^3 \times 3.1}{22 \times 10^3} = 65.5$$

Assuming R_1 is very small, then

$$R_{11} = \frac{\omega_o^2L_1^2}{r_1}$$

Also

$$Q_1 = Q = \frac{\omega_oL_1}{R_{11}}$$

Substituting for R_{11} gives

$$Q_1 = \frac{\omega_oL_1r_1}{\omega_o^2L_1^2}$$

$\therefore$
$$Q = \frac{r_1}{\omega_o L_1}$$

So

$$L_1 = \frac{r_1}{Q_1 \omega_o} = \frac{42 \times 10^3}{65.5 \times 465 \times 2\pi \times 10^3} = 219.4\ \mu\text{H}$$

Also

$$R_{22} = \frac{\omega_o^2 L_2^2}{r_2}$$

Assuming R_2 is very small,

$$Q_2 = Q = \frac{\omega_o L_2}{R_{22}}$$

$\therefore$
$$Q = \frac{\omega_o L_2 r_2}{\omega_o^2 L_2^2} = \frac{r_2}{\omega_o L_2}$$

So

$$L_2 = \frac{r}{Q\omega_o} = \frac{10^3}{65.5 \times 2\pi \times 465 \times 10^3} = 5224\ \text{nH}$$

Substituting, we obtain

$$M = k\sqrt{L_1 L_2} = 0.02\sqrt{219.4 \times 10^{-6} \times 5224 \times 10^{-9}} = 677\ \text{nH}$$

Now

$$f_o = \frac{1}{2\pi\sqrt{L_1 C_1}} = \frac{1}{2\pi\sqrt{L_2 C_2}}$$

hence

$$C_1 = \frac{1}{f_o^2\, 4\pi^2 L_1} = \frac{1}{(465 \times 10^3)^2 \times 4\pi^2 \times 219.4 \times 10^{-6}} = 534\ \text{pF}$$

Similarly,

$$C_2 = \frac{1}{f_o^2\, 4\pi^2 L_2} = \frac{1}{(465 \times 10^3)^2 \times 4\pi^2 \times 5224 \times 10^{-9}} = 22.4\ \text{pF}$$

4.12 Crystal and ceramic tuned amplifiers

Modern communications systems frequently use crystal tuned amplifiers both for stability and high Q factor. Since

$$X_L = 2\pi f L$$

the inductive reactance will be small if a coil with a few turns is used. Also as

$$X_C = \frac{1}{2\pi f C}$$

the capacitive reactance will be small if a parallel tuning capacitor with large capacitance is used. Thus the parallel tuned circuit in a tuned amplifier should have a relatively small L and a large C in order to give a high Q factor for the tuned circuit: the Q of the tuned load is given by Z_l/X, where Z_L is the impedance of any device which loads the parallel tuned circuit and X is the reactance of either the coil or the capacitor in the parallel tuned circuit.

With a given Z_L, the Q factor of the tuned circuit can be made large by making X_L or X_C small. Hence the frequency stability is good when the L/C ratio is low. One way to achieve good frequency stability is to use crystals such as quartz, tourmaline and Rochelle salts, but because of its electric and mechanical properties, quartz is the most widely used.

When a properly cut slice of crystal is bent, electric charges appear on its opposite faces. This is known as the piezo-electric effect. This was discussed in Chapter 3. Furthermore, if a voltage is applied to opposite faces of the crystal plate it bends. Thus applying an alternating voltage across the crystal causes it to vibrate. If the frequency of the applied voltage is near a mechanical resonant frequency of the crystal the vibrations have large amplitude.

The dimensions of the crystal determine the resonant frequency: the thinner the crystal the higher the frequency. Naturally there is a limit to how thin the crystal can be cut, and this sets an upper frequency limit of approximately 10 MHz. The dimensions of the crystal can be equated to the circuit shown in Fig. 4.15(a). This circuit has its analogue in the mechanical world. C is the capacitance of the crystal, L the inertia of the crystal, R the resistance of the crystal material and C_1 the capacitance between the plates of the crystal holder. The high Q factor of the crystal can be seen in Fig. 4.15(b). Between f_1 and f_2 the crystal is highly inductive, with a very steep slope. This rapid change of reactance with frequency is the reason for the high stability.

As can be seen from Fig. 4.15(b), the crystal has two resonant frequencies. The electrical properties of the crystal itself give the series resonant frequency f_1, while the holder capacitance C_1 in parallel with the inductance L of the crystal forms the parallel resonant frequency f_2.

The ceramic filter is similar to the quartz crystal in that the piezo-electric effect is also evident. A three-terminal ceramic disc is shown in Figure 4.16. As can be seen, the ceramic material is mounted between two plates. The main difference is that the crystal is normally earthed. The ceramic filter is cheaper than the crystal filter and requires no series or parallel tuning capacitor for fine-tuning. However, it has a greater loss at the pass band, its selectivity is not as good and it suffers more from temperature instability.

It has been the policy to incorporate ceramic filters in domestic receivers and crystals in communications receivers; however, with improved fabrication, ceramic devices are being used more in communications receivers. A typical example of where ceramic filters may be used is in wide-band IF amplifiers. A general circuit is shown in Fig. 4.17, where an integrated IF amplifier is used. A similar set-up is shown in Fig. 4.18 with discrete tuning circuits. In both cases C_1 and C_2 are decoupling capacitors.

The ceramic filters shape the response of the amplifier and hence give a suitable selectivity and sensitivity.

The quartz crystal, because of its high selectivity, is used in different configurations to give specific responses. In Fig. 4.19 the series resonant frequency has been selected to

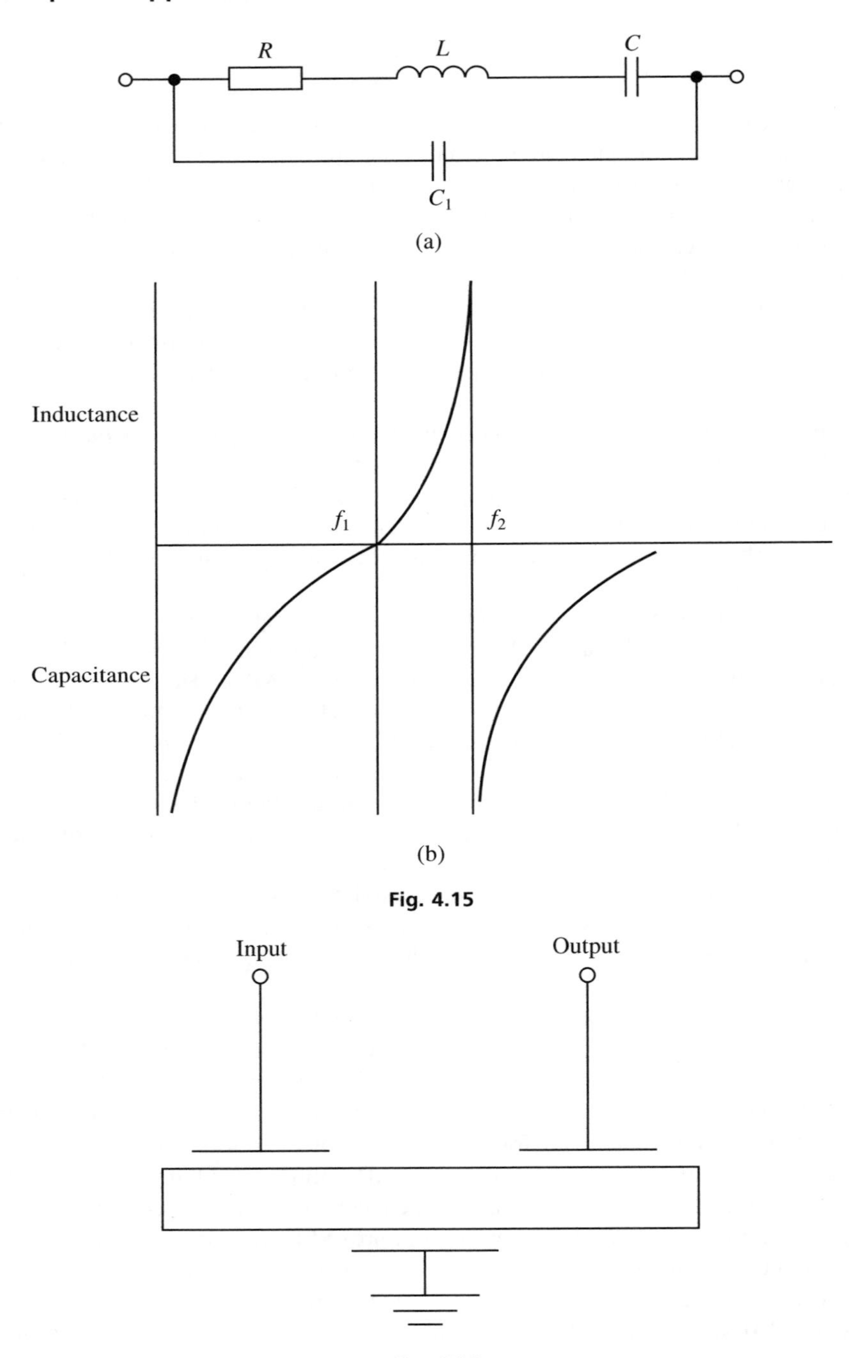

Fig. 4.15

Fig. 4.16

give the lower frequency of the pass-band response, while the parallel frequency of the crystal has been selected as the upper frequency of the pass-band response. This latter frequency is normally adjusted by a variable capacitor as shown. As has been mentioned

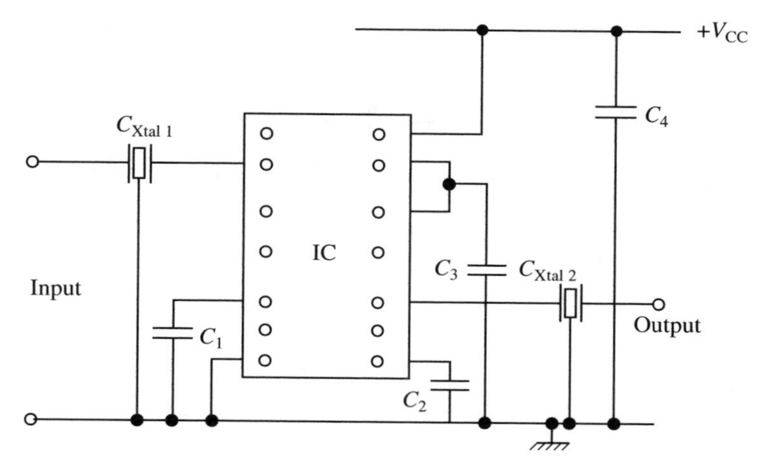

Fig. 4.17

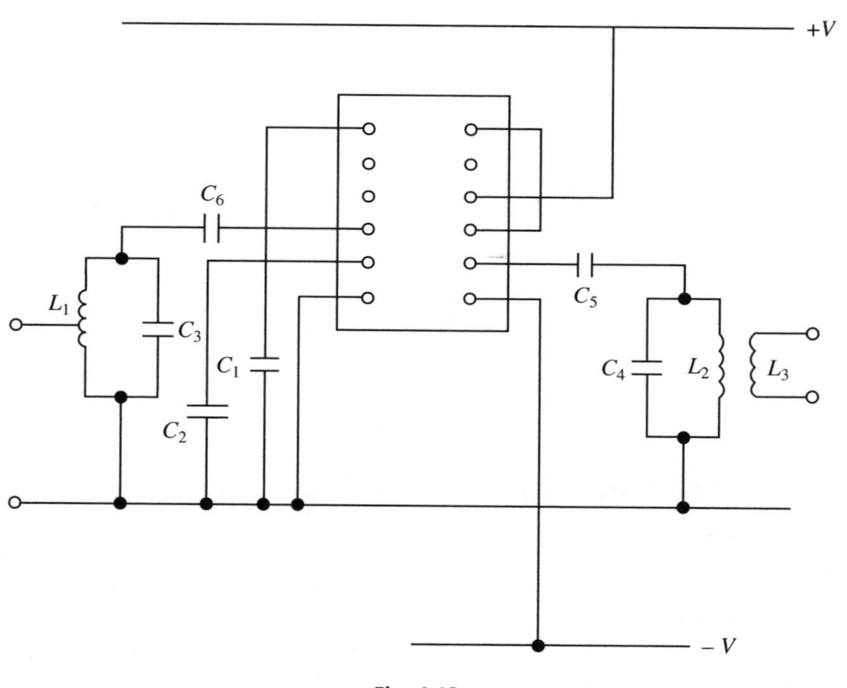

Fig. 4.18

before, it is essential that the impedance of the tuned circuit matches the input impedance of the next stage. The impedance of the crystal also has to be matched by the following stage and for this reason a network such as R_1, C_4 and L_3 is included.

A second configuration which will give a wider pass band is shown in Fig. 4.20.

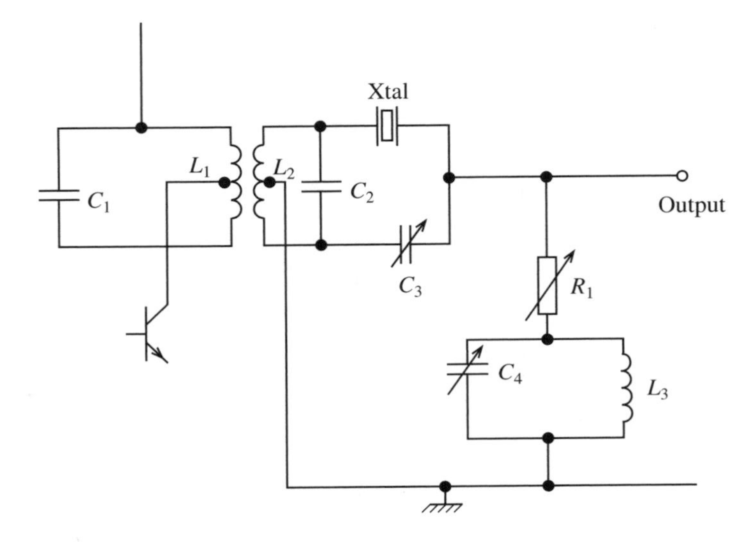

Fig. 4.19

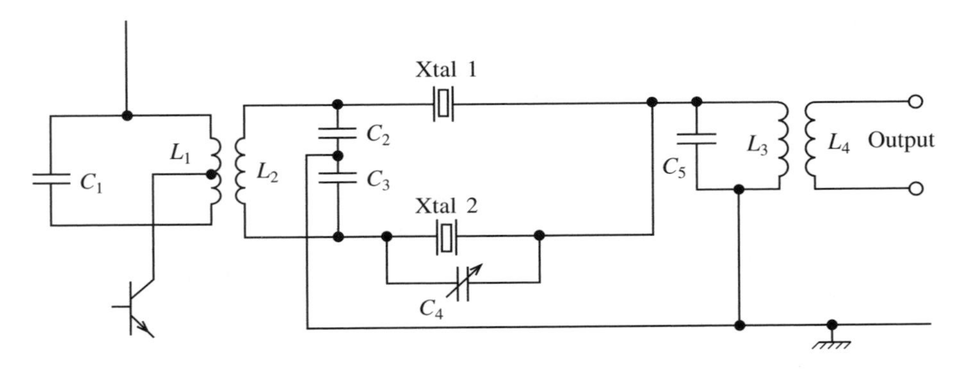

Fig. 4.20

4.13 Integrated tuned amplifiers

The tuned amplifiers studied so far have included a tuned load incorporated into a discrete component circuit. There are many integrated circuits on the market, and most of them include other circuit functions besides that of the IF amplifier. The Motorola MC3372 is such a device and consists of an oscillator, mixer and IF amplifier fabricated on the same chip. All these integrated circuits, however, require external tuning circuits as coils cannot be etched on to substrates. Manufacturers frequently indicate physical dimensions for the tuning coils but occasionally they have to be calculated.

Two integrated circuits will be considered in this text, namely the CA3028A manufactured by Harris and the MC1350 manufactured by Motorola. Both these chips can be used for IF and RF applications and have on board automatic gain facilities. Data sheets for both chips can be found at the end of the chapter.

The CA3028A may be used as a cascade amplifier with the advantages of low input resistance, large bandwidth and low noise; alternatively it may be operated in the differential mode. It is designed to be used in communications and industrial equipment operating at frequencies between d.c. and 120 MHz and may be configured as a mixer or oscillator. Some applications are shown in the data sheets, together with the electrical specifications and performance curves which are frequently used in design.

The MC1350 is generally used as an IF amplifier and has applications in both radio and television. It has a wide bandwidth, ranging from zero up to about 60 MHz, but this is reduced to the required value by the tuned circuits.

A variety of different integrated RF amplifiers are available from several manufacturers, and a few have been mentioned above. Typical parameters for such integrated circuits are:

(i) voltage gain 20 dB, bandwidth 140 MHz, maximum input signal voltage 100 mV;

(ii) voltage gain 26 dB, bandwidth 100 MHz, maximum input signal voltage 50 mV;

(iii) voltage gain 80 dB, maximum input voltage 15 μV, input impedance 15 Ω in parallel with 100 pF.

The range is comprehensive but selection and design are normally determined by these parameters. Some calculations follow which help illustrate the requirements for certain applications.

Example 4.12

An MC1350 integrated circuit is used as an IF amplifier operating at 455 kHz. If the secondary load is 120 Ω, determine

(a) appropriate values for C and L;

(b) the Q factor of the stage;

(c) the differential output impedance;

(d) the differential input impedance;

(e) the value of M for maximum gain.

Solution

(a) Selecting 0.001 μF,

$$L = \frac{1}{(2\pi f_o)^2 C} = \frac{1}{4\pi^2 \times (455 \times 10^3)^2 \times 0.001 \times 10^{-6}} = 122.4\ \mu\text{F}$$

Since the coil is centre-tapped, $L_1 = L_2 = 61.2\ \mu\text{F}$

(b) As the bandwidth is stated as 20 kHz,

$$Q = \frac{f_o}{BW} = \frac{455}{20} = 23$$

(c) From the data sheets the differential output conductance $g_{22} = 4$ and the susceptance $b_{22} = 3$. Hence the admittance is given by

$$\sqrt{3^2 + 4^2} = \sqrt{9 + 16} = 5\ \mu\Omega^{-1}$$

So the magnitude of the impedance is

$$\frac{10^6}{5} = 200 \text{ k}\Omega$$

(d) Also from the data sheets, the single-ended input admittance is given by

$$g_{11} + \text{j}b_{11} = 0.31 + \text{j}0.022$$

hence the admittance magnitude is

$$\sqrt{0.31^2 + 0.022^2} = 0.31 \text{ m}\Omega^{-1}$$

Hence

$$Z = \frac{1000}{0.31} = 3.2 \text{ k}\Omega$$

Since the differential impedance is half this value, $Z = 1.6$ kΩ.
(e) In order to determine M, we require

$$M = L_1 \sqrt{\frac{R_{22}}{R_{11}}}$$

Since $g_{22} = 4$ and $1/g_{22} = 250 \text{ k}\Omega = R_{11}$

$$M = 122.4 \times 10^{-6} \sqrt{\frac{120}{250 \times 10^3}} = 2.68 \text{ }\mu\text{H}.$$

Note that because of the input and output acceptance the frequency will be varied and this may have to be remedied by including trimmer or padder capacitors.

Example 4.13
The Harris CA3028A integrated circuit has to be used in differential mode as a tuned amplifier working in the range 80–90 MHz. The circuit is shown in Fig. 4.21. Select initial values of L and C and assume the d.c. resistance of the coils is 1 Ω. Select input and output transformers such that $L = 0.25$ μH and wind accordingly with SWG. Determine the following:

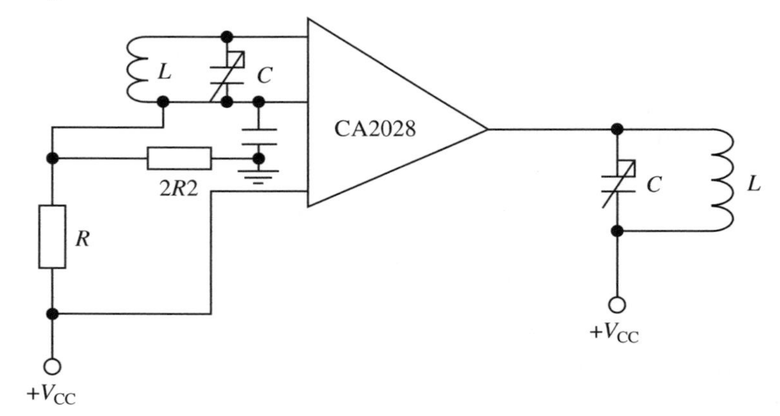

Fig. 4.21

(a) a typical inductance for L;

(b) a tuning ratio for C;

(c) suitable values of bias resistor;

(d) the Q factor for the stage;

(e) the bandwidth.

Solution

(a)

$$C_{80} = \frac{1}{4\pi^2 (f_{80})^2 L} = \frac{1}{4\pi^2 \times (80 \times 10^6)^2 \times 0.25 \times 10^{-6}} = 15.83 \text{ pF}$$

$$C_{90} = \frac{1}{4\pi^2 (f_{90})^2 L} = \frac{1}{4\pi^2 \times (90 \times 10^6)^2 \times 0.25 \times 10^{-6}} = 12.51 \text{ pF}$$

(b) From (a) we observe that $C_{90}/C_{80} = 80^2/90^2 = 64/81$, so the required tuning ratio is 1 : 1.27. Note that the input and output susceptance will necessitate the use of padder or trimmer capacitors.

(c) Select bias resistors of 1 kΩ and 2.2 kΩ.

(d) The Q factor of the stage (Q_s) is

$$Q_s = \frac{R_t}{\omega_o L}$$

where R_t is the total resistance across the tuned circuit. There are two resistive components across the tuned circuit, namely the equivalent parallel resistance of the coil (R_P) and the output conductance (g_{22}) which gives

$$R = \frac{1}{g_{22}} = \frac{10^6}{60} = 16.7 \text{ k}\Omega$$

since g_{22} at 80–90 MHz is approximately 60 µS from the data sheets.

(e) As the series resistance of the coil has been selected as 1 Ω then, letting $C = (C_{80} + C_{90})/2 = 14.17$ pF, the equivalent parallel resistor loading the coil will be

$$R_P = \frac{L}{Cr} = \frac{0.25 \times 10^{-6}}{14.17 \times 10^{-12} \times 1} = 17.643 \text{ k}\Omega$$

So

$$R_t = \frac{17.643 \times 16.7}{17.643 + 16.7} = 8.57 \text{ k}\Omega$$

Hence

$$Q_s = \frac{8.57 \times 10^3}{2\pi \times 0.25 \times 10^{-6} \times 85 \times 10^3} = 64.2$$

where f_0 is the average value. Average bandwidth is then

$$BW = \frac{f_0}{Q} = \frac{85}{64.2} = 1.32 \text{ MHz}$$

4.14 Testing tuned amplifiers

The testing of amplifiers such as RF and IF types requires the typical set-up shown in Fig. 4.22. The matching interface may not be necessary, depending on the output impedance of the RF generator. More modern generators will also have facilities for introducing modulation and marker frequencies which indicate if any frequency drift has occurred in the tuned amplifier under test. Generally the marker frequency is present and the RF generator sweeps through the desired range. Frequently a table similar to those shown in Figs. 4.23 and 4.24 is produced and frequency is plotted against gain.

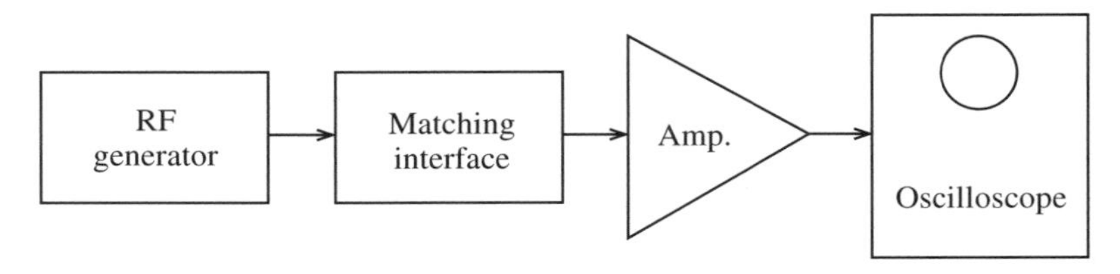

Fig. 4.22

Figure 4.23 shows a fixed tuned amplifier with a centre frequency of 455 kHz. The bandwidth is the 3 dB or half-power bandwidth, and this can be obtained from the frequency response (35 kHz in this case). The Q factor for the unloaded stage can then be calculated.

Remember that a load will lower this Q factor, and the loading effect can be tested by connecting a variable impedance load across the output of the amplifier. This should take account of the practical input capacitance and resistance of the intended load as the capacitance will alter the centre frequency and the resistance the gain and the Q factor.

Frequently a spectrum analyser may be used to determine the output of an RF amplifier at the front end of a receiver where adjacent channel interference is present. This may then be eliminated by filtering or RF traps.

Figure 4.24 was achieved by over-coupling. The centre frequency is 475 kHz and the bandwidth is 890 kHz.

4.15 Further problems

1. The circuit of a tuned amplifier stage is shown in Fig. 4.25. The FET has a mutual conductance of $g_m = 2.0$ mA/V and the drain-to-source resistance is $r_{ds} = 200$ kΩ. With the output terminals open-circuited, calculate:

(i) the value of the tuning capacitor C to give maximum voltage gain at 1 MHz;

(ii) the value of this voltage gain;

(iii) the bandwidth of the stage.

What value of resistor connected across the output terminals would increase the bandwidth to 15 kHz, and what would then be the voltage gain at 1 MHz?
Answer: 101.3 pF, 321.8, 9.766 kHz, 301 kΩ, 209.5

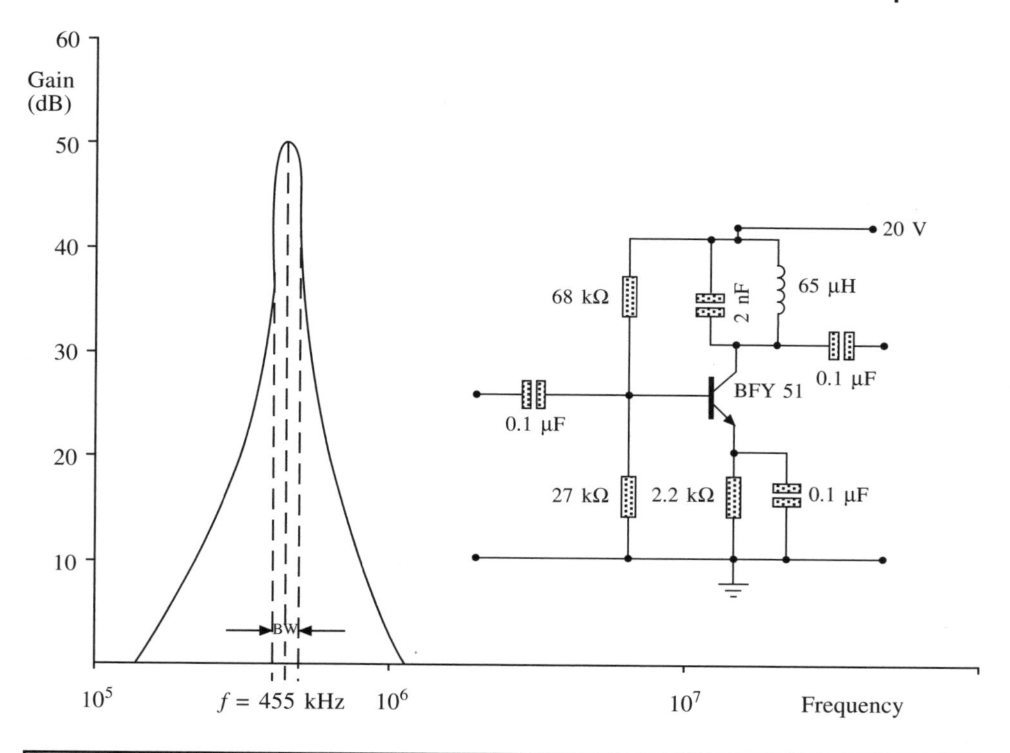

Frequency (kHz)	Input voltage (V_i) mV	Output voltage (V_o) mV	V_o/V_i	20 log V_o/V_i
300	2	14	7	17
325	2	22	11	20.8
335	2	25	12.5	22
350	2	40	20	26
375	2	64	32	30
400	2	110	55	34.8
420	2	225	113	41
440	2	500	250	48
455	2	600	300	49.5
460	2	400	200	46
465	2	240	120	41.6
480	2	200	100	40
500	2	125	62.5	35.9
550	2	70	35	31
580	2	55	27.5	28.7
600	2	50	25	27.9
650	2	24.6	12	22
700	2	20	10	17
750	2	9	4.7	13.5
800	2	6	3	9.8

Fig. 4.23

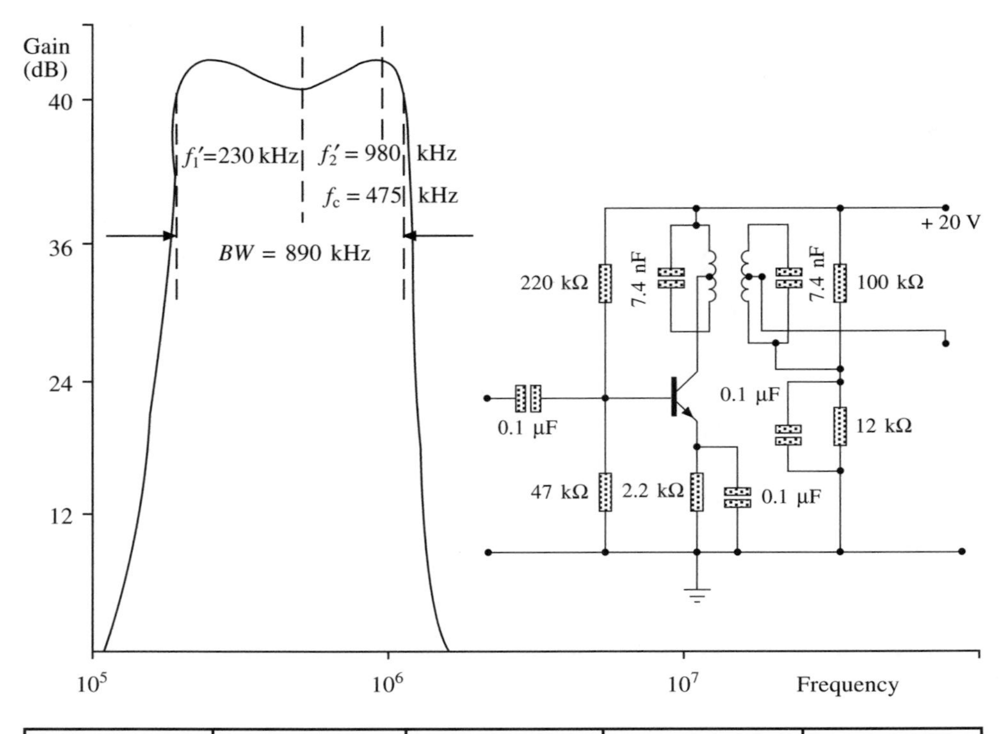

Frequency (kHz)	V_i	V_o	V_o/V_i	dB
150	10	0.0794	7.94	18
155	10	0.112	11.2	21
160	10	0.1584	15.84	24
170	10	0.2238	22.38	27
175	10	0.3162	31.62	30
185	10	0.6309	63.1	36
190	10	1.126	126	42
200	10	2.51	251	48
220	10	3.55	355	51
250	10	2.99	299	49.5
300	10	2.66	266	48.5
450	10	2.51	251	48
600	10	2.72	272	48.7
800	10	3	300	49.7
900	10	3.16	316	50
983	10	3.55	355	51
1090	10	2.51	251	48
1120	10	1.26	126	42
1150	10	0.6309	63.1	36
1200	10	0.2238	22.8	27
1260	10	0.1122	11.2	21
1300	10	0.079	7.9	18
1350	10	0.0562	5.6	15

Fig. 4.24

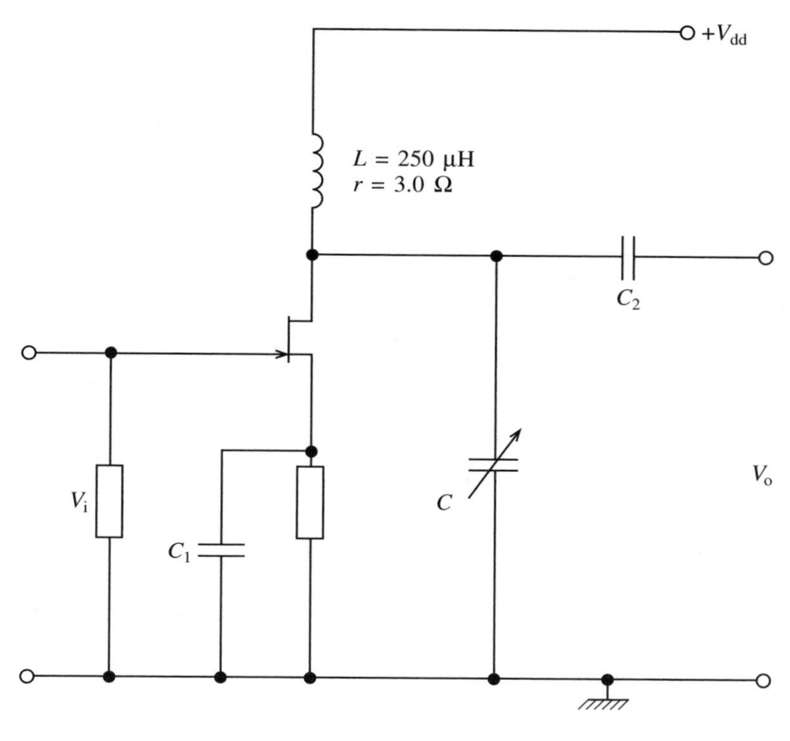

Fig. 4.25

2. The tuned amplifier shown in Fig. 4.26 is required to have a bandwidth sufficient to pass a double sideband amplitude-modulated voltage waveform of carrier frequency

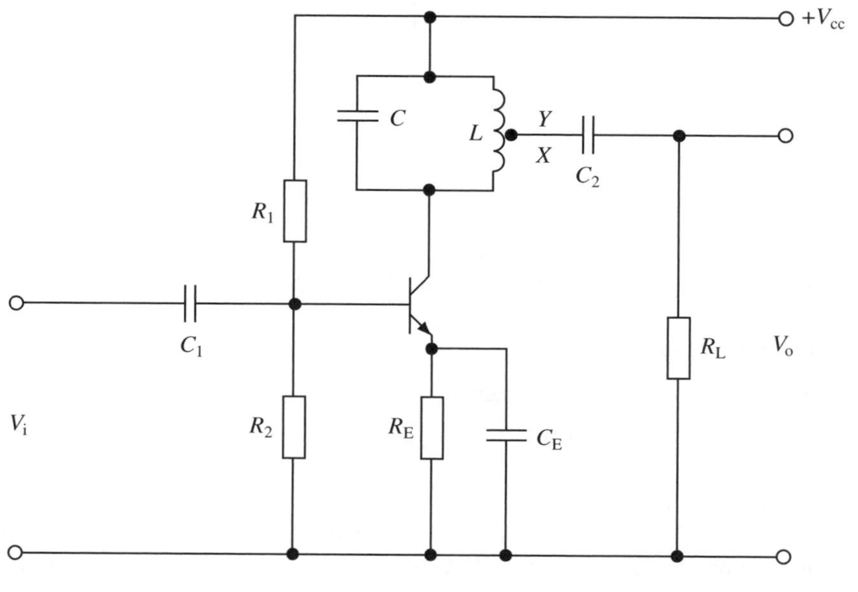

Fig. 4.26

465 kHz and of highest modulating frequency 5 kHz. The inductor has a value of 300 μH and a Q factor at this carrier frequency of 90. The transistor has an h_{ie} of 5.36 kΩ and an h_{oe} of 8 μS. R_L = 5.36 kΩ, and the reactances of C_1, C_2 and C_E are negligible at the carrier frequency. Note also that R_1 and R_2 are very much greater than h_{ie}. Given that the inductor has a total of 70 turns and that the overall gain must be at least 250 at the resonant frequency, determine the number of turns X needed to give the minimum necessary bandwidth and the minimum value of transistor h_{fe} required.
Answer: 230, 60

3. Figure 4.27 shows a single-stage tuned amplifier. The transistor parameters are h_{ie} = 1.5 kΩ and h_{oe} = 20 μS. The coil L has an unloaded Q factor of 45 at the tuned frequency of 460 kHz. Determine the half-power bandwidth of the stage.
Answer: 16 kHz

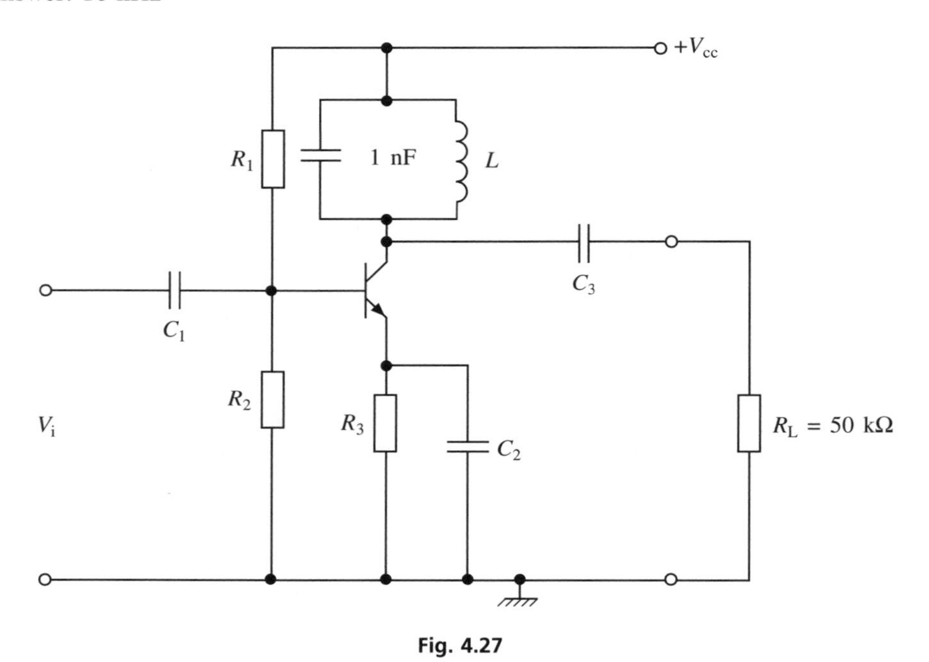

Fig. 4.27

4. The tuned IF amplifier circuit shown in Fig. 4.28 has an IF of 465 kHz. For both transistors h_{ie} = 3.75 kΩ, h_{fe} = 90 and h_{oe} = 15 μS. Calculate:

(i) the value of capacitor C;

(ii) the turns ratio of the transformer;

(iii) the voltage gain at 465 kHz.

Answer: 334.7 pF, 11 : 1, 69

5. The circuit of a tuned amplifier is shown in Fig. 4.29. The transistor has the following parameters: h_{fe} = 200, h_{ie} = 4 kΩ, h_{oe} = 50 μS. Assuming that the reactances of C_1, C_2 and C_3 may be neglected, calculate:

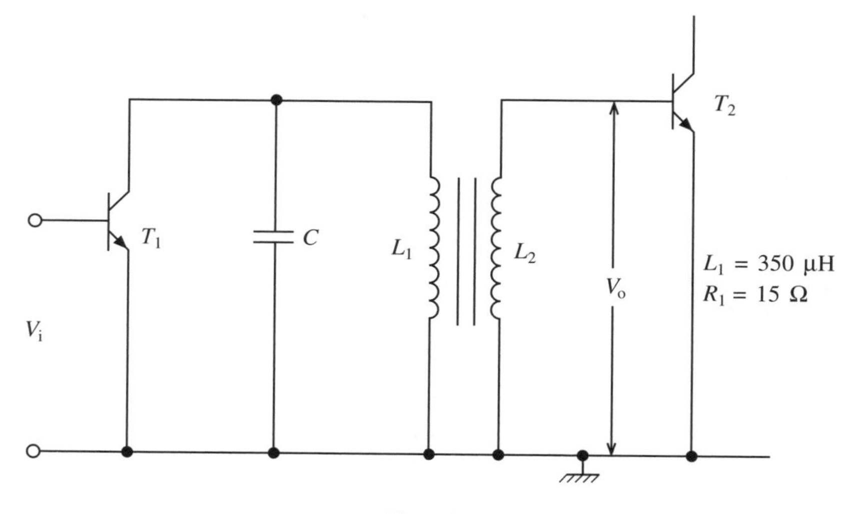

Fig. 4.28

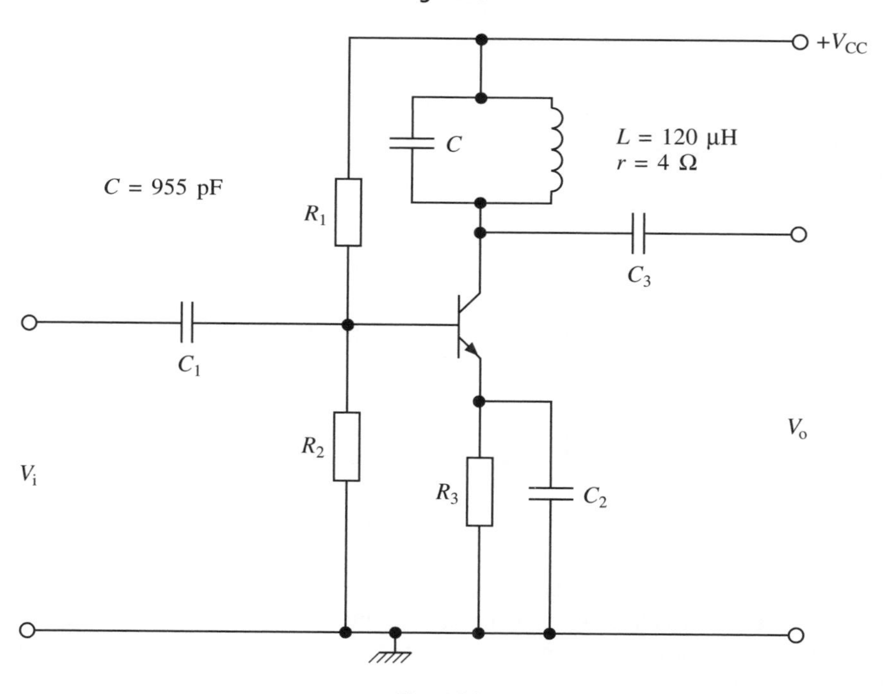

Fig. 4.29

(i) the resonant frequency and the Q factor of the tuned circuit;

(ii) the Q factor and bandwidth of the stage;

(iii) the amplifier voltage gain at the resonant frequency.

Answer: 470 kHz, 88.6, 34.4, 13.8 kHz, 610

6. For the amplifier stage shown in Fig. 4.30, $L = 2.5$ mH, $r = 10\ \Omega$ and the transistor parameters are $h_{fe} = 80$, $h_{ie} = 5\ \text{k}\Omega$, $h_{oe} = 25\ \mu\text{S}$. Calculate:

(i) The value of C to give maximum voltage gain at 110 kHz;

(ii) the value of this voltage gain;

(iii) the bandwidth of the stage;

(iv) the value of the resistance to be connected across the output terminals to give a bandwidth of 10 kHz.

Answer: 837 pF, 564, 5.3 kHz, 39 kΩ

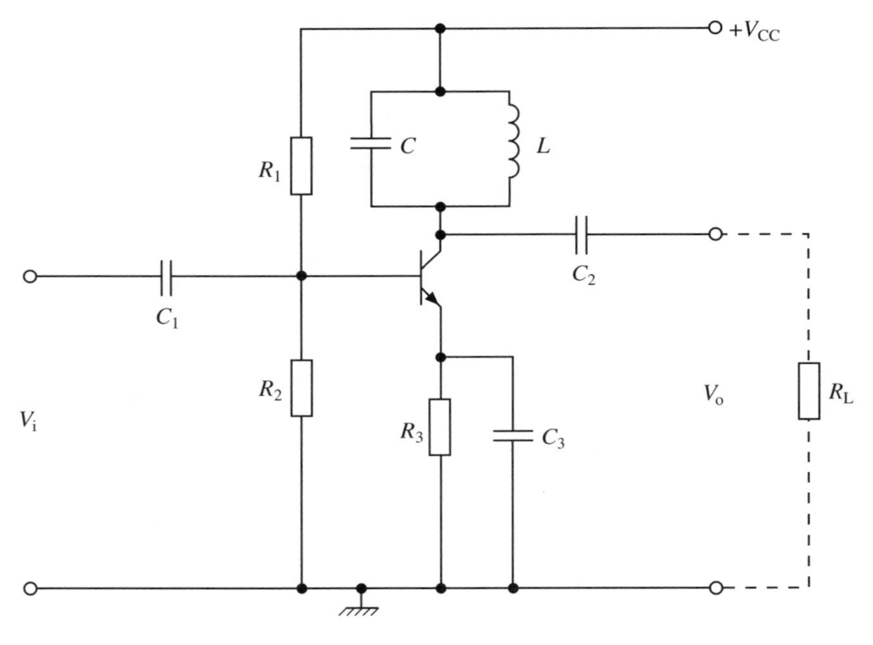

Fig. 4.30

8. A tuned primary amplifier has to be designed to accept a channel with a carrier of 920 kHz and an input voltage of 120 mV. It is coupled to a second stage having an input resistance of 265 kΩ. If the circuit parameters are $L_1 = 420\ \mu\text{H}$, $R_{P2} = 253.6\ \text{k}\Omega$ and $g_m = 2$ mA/V determine:

(a) the value of C required to tune the primary;

(b) the effective Q factor;

(c) the output voltage at the secondary.

Answer: 70.3 pF, 104, 12.1 V

9. A tuned secondary amplifier forms part of a tuner in a television receiver. It has to tune over the range 81–104 MHz. If the resistance of the secondary is 3.2 Ω, the input resistance (r_{ds}) is 120 kΩ and the Q factor of the secondary is 220, determine:

(i) the average value of M for critical coupling;

(ii) the values of $L_1 = L_2$ and C for the frequency range;

(iii) the gain at critical coupling.

Answer: 1.025 μH, 1.2 μH, 1.93–2.75 pF, 73 dB

10. A double-tuned transformer has to be used in an IF amplifier having a centre frequency (f_c) of 475 kHz. If the bandwidth is 100 kHz between peaks, $A_p = 18.6$, $G = \sqrt{2}$, $I_m = 40$ mA/V, determine

(i) the bandwidth when the gain drops to A_c;

(ii) the 3 dB bandwidth if $Q_1 = Q_2 = Q = 60$;

(iii) the coupling coefficient;

(iv) the values of $L_1 = L_2$ and $C_1 = C_2$.

Answer: 141 kHz, 25 kHz, 0.04, 114 μH, 38.9 nF

11. A tuned secondary RF amplifier has an input resistance (r_{ds}) of 350 kΩ. The d.c. resistance of the coil is 2.2 Ω and the Q factor of the secondary tuned circuit is 70. If the resonant frequency is 650 kHz and $g_m = 20$ mA/V, determine L_2 and M for optimum bandwidth and gain. Calculate the values of gain and bandwidth for optimum conditions.
Answer: 37.7 μH, 215 μH, 61.6 dB, 18.6 kHz

12. A tuned amplifier has its primary tuned to 820 kHz. If the input resistance is 12.2 kΩ, the load resistance is 1.2 kΩ and the tuning capacitor has a value of 850 pF, determine:

(i) the value of L_1;

(ii) the effective Q factor if $R_s = 9$ Ω

(iii) the value of M required to give the effective Q factor.

Answer: 43 μH, 29, 20 μH

13. The MC1350 has to be used as a video IF amplifier with a centre frequency of 45 MHz. Show how this integrated circuit is configured for such an application, calculate the value of primary inductance for the input transformer and explain how the secondary winding may be constructed and measured. What will be the differential output impedance?
Answer: 0.62 μH, 2.56 kΩ

14. An RF amplifier has to be configured as a cascade amplifier using a CA3028A. The operating frequency has to be 25 MHz and conventional automatic gain control has to be included. Calculate the tuning components and the value of the bandwidth and voltage gain at resonance. Draw the resultant schematic diagram.

Data sheets

Order this document by MC1350/D

MC1350

Monolithic IF Amplifier

IF AMPLIFIER

SEMICONDUCTOR TECHNICAL DATA

The MC1350 is an integrated circuit featuring wide range AGC for use as an IF amplifier in radio and TV over an operating temperature range of 0° to +75°C.

- Power Gain: 50 dB Typ at 45 MHZ
  50 dB Typ at 58 MHZ
- AGC Range: 60 dB Min, DC to 45 MHz
- Nearly Constant Input & Output Admittance over the Entire AGC Range
- Y_{21} Constant (–3.0 dB) to 90 MHz
- Low Reverse Transfer Admittance: < < 1.0 µmho Typ
- 12 V Operation, Single–Polarity Power Supply

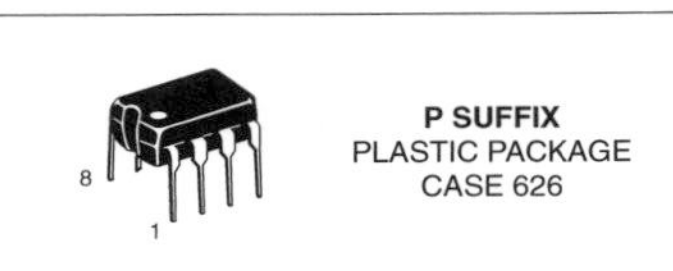

P SUFFIX
PLASTIC PACKAGE
CASE 626

D SUFFIX
PLASTIC PACKAGE
CASE 751
(SO–8)

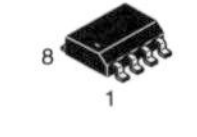

MAXIMUM RATINGS (T_A = +25°C, unless otherwise noted.)

Rating	Symbol	Value	Unit
Power Supply Voltage	V^+	+18	Vdc
Output Supply Voltage	V_1, V_8	+18	Vdc
AGC Supply Voltage	V_{AGC}	V^+	Vdc
Differential Input Voltage	V_{in}	5.0	Vdc
Power Dissipation (Package Limitation) Plastic Package Derate above 25°C	P_D	 625 5.0	 mW mW/°C
Operating Temperature Range	T_A	0 to +75	°C

ORDERING INFORMATION

Device	Operating Temperature Range	Package
MC1350P	T_A = 0° to +75°C	Plastic DIP
MC1350D		SO–8

Figure 1. Typical MC1350 Video IF Amplifier and MC1330 Low–Level Video Detector Circuit

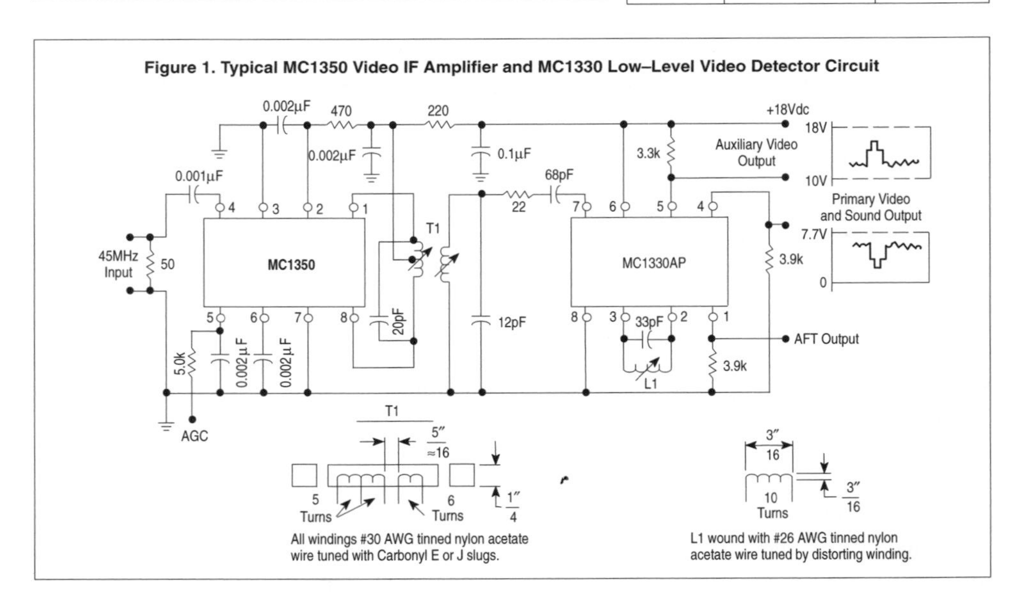

 Rev 3

MC1350

ELECTRICAL CHARACTERISTICS (V^+ = +12 Vdc, T_A = +25°C, unless otherwise noted.)

Characteristics	Symbol	Min	Typ	Max	Unit
AGC Range, 45 MHz (5.0 V to 7.0 V) (Figure 1)		60	68	–	dB
Power Gain (Pin 5 grounded via a 5.1 kΩ resistor) f = 58 MHz, BW = 4.5 MHz See Figure 6(a) f = 45 MHz, BW = 4.5 MHz See Figure 6(a), (b) f = 10.7 MHz, BW = 350 kHz See Figure 7 f = 455 kHz, BW = 20 kHz	A_p	 – 46 – –	 48 50 58 62	 – – – –	dB
Maximum Differential Voltage Swing 0 dB AGC –30 dB AGC	V_O	 – –	 20 8.0	 – –	V_{pp}
Output Stage Current (Pins 1 and 8)	$I_1 + I_8$	–	5.6	–	mA
Total Supply Current (Pins 1, 2 and 8)	I_S	–	14	17	mAdc
Power Dissipation	P_D	–	168	204	mW

DESIGN PARAMETERS, Typical Values (V^+ = +12 Vdc, T_A = +25°C, unless otherwise noted.)

Parameter	Symbol	Frequency				Unit
		455 kHz	10.7 MHz	45 MHz	58 MHz	
Single–Ended Input Admittance	g_{11} b_{11}	0.31 0.022	0.36 0.50	0.39 2.30	0.5 2.75	mmho
Input Admittance Variations with AGC (0 dB to 60 dB)	Δg_{11} Δb_{11}	– –	– –	60 0	– –	μmho
Differential Output Admittance	g_{22} b_{22}	4.0 3.0	4.4 110	30 390	60 510	μmho
Output Admittance Variations with AGC (0 dB to 60 dB)	Δg_{22} Δb_{22}	– –	– –	4.0 90	– –	μmho
Reverse Transfer Admittance (Magnitude)	$\lvert y_{12} \rvert$	< < 1.0	< < 1.0	< < 1.0	< < 1.0	μmho
Forward Transfer Admittance Magnitude Angle (0 dB AGC) Angle (–30 dB AGC)	 $\lvert y_{21} \rvert$ $< y_{21}$ $< y_{21}$	 160 –5.0 –3.0	 160 –20 –18	 200 –80 –69	 180 –105 –90	 mmho Degrees Degrees
Single–Ended Input Capacitance	C_{in}	7.2	7.2	7.4	7.6	pF
Differential Output Capacitance	C_O	1.2	1.2	1.3	1.6	pF

Figure 2. Typical Gain Reduction

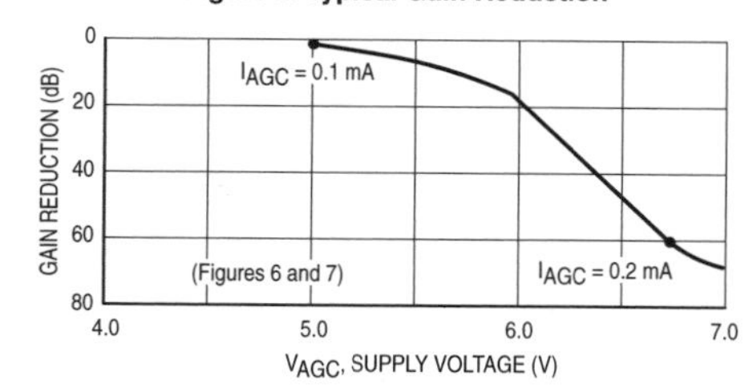

Figure 3. Noise Figure versus Gain Reduction

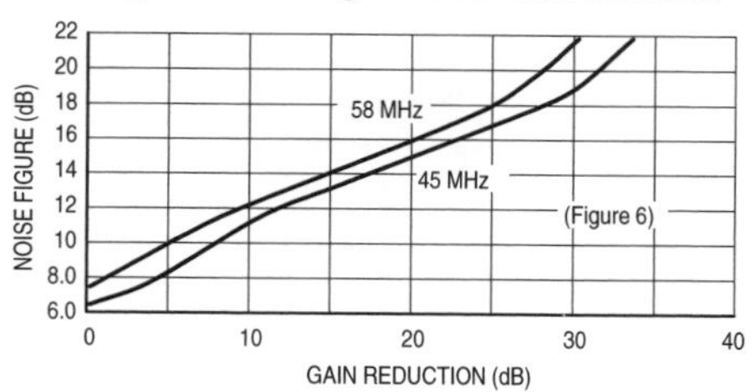

MC1350

GENERAL OPERATING INFORMATION

The input amplifiers (Q1 and Q2) operate at constant emitter currents so that input impedance remains independent of AGC action. Input signals may be applied single–ended or differentially (for ac) with identical results. Terminals 4 and 6 may be driven from a transformer, but a dc path from either terminal to ground is not permitted.

Figure 4. Circuit Schematic

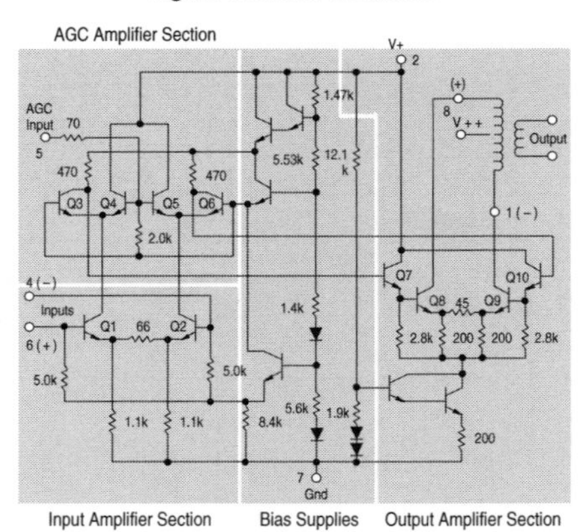

AGC action occurs as a result of an increasing voltage on the base of Q4 and Q5 causing these transistors to conduct more heavily thereby shunting signal current from the interstage amplifiers Q3 and Q6. The output amplifiers are supplied from an active current source to maintain constant quiescent bias thereby holding output admittance nearly constant. Collector voltage for the output amplifier must be supplied through a center–tapped tuning coil to Pins 1 and 8. The 12 V supply (V+) at Pin 2 may be used for this purpose, but output admittance remains more nearly constant if a separate 15 V supply (V+ +) is used, because the base voltage on the output amplifier varies with AGC bias.

Figure 5. Frequency Response Curve (45 MHz and 58 MHz)

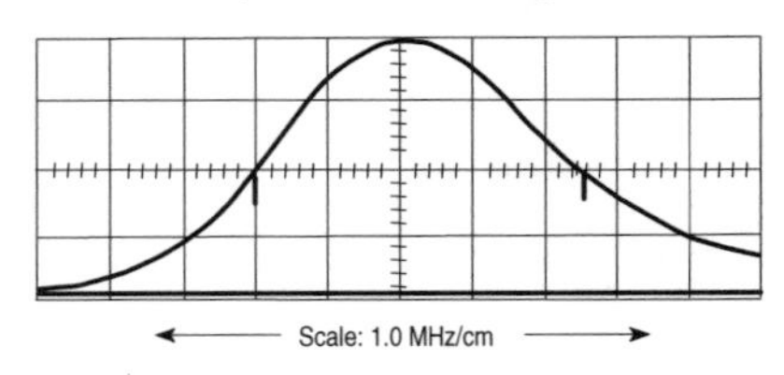

Figure 6. Power Gain, AGC and Noise Figure Test Circuits

(a) 45 MHz and 58 MHz

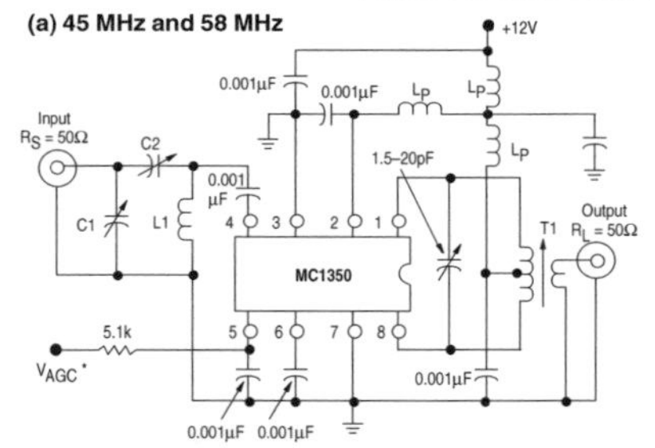

(b) Alternate 45 MHz

*Connect to ground for maximum power gain test.
All power supply chokes (Lp), are self–resonant at input frequency. Lp ≥ 20 kΩ.
See Figure 5 for Frequency Response Curve.

L1 @ 45 MHz = 7 1/4 Turns on a 1/4″ coil form
@ 58 MHz = 6 Turns on a 1/4″ coil form
T1 Primary Winding = 18 Turns on a 1/4″ coil form, center–tapped, #25 AWG
Secondary Winding = 2 Turns centered over Primary Winding @ 45 MHz
= 1 Turn @ 58 MHz
Slug = Carbonyl E or J

L1	Ferrite Core 14 Turns 28 S.W.G.
C1	5–25 pF
C2	5–25 pF
C3	5–25 pF

	45 MHz		58 MHz	
L1	0.4 µH	Q ≥ 100	0.3 µH	Q ≥ 100
T1	1.3 µH to 3.4 µH	Q ≥ 100 @ 2.0 µH	1.2 µH to 3.8 µH	Q ≥ 100 @ 2.0 µH
C1	50 pF to160 pF		8.0 pF to 60 pF	
C2	8.0 pF to 60 pF		3.0 pF to 35 pF	

MC1350

Figure 7. Power Gain and AGC Test Circuit (455 kHz and 10.7 MHz)

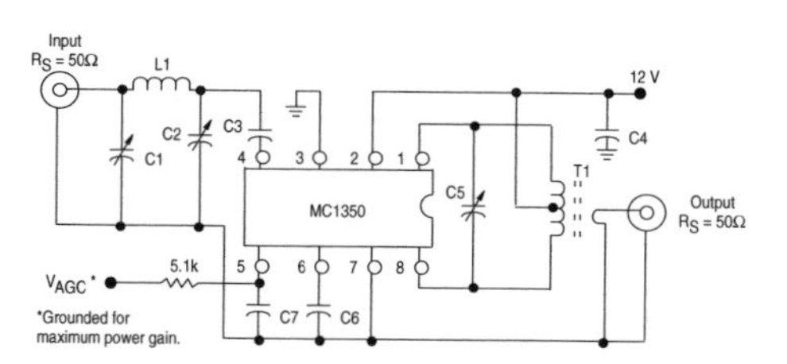

Component	Frequency	
	455 kHz	10.7 MHz
C1	–	80–450 pF
C2	–	5.0–80 pF
C3	0.05 μF	0.001 μF
C4	0.05 μF	0.05 μF
C5	0.001 μF	36 pF
C8	0.05 μF	0.05 μF
C7	0.05 μF	0.05 μF
L1	–	4.6 μH
T1	Note 1	Note 2

NOTES: 1. Primary: 120 μH (center–tapped)
Q_u = 140 at 455 kHz
Primary: Secondary turns ratio ≈ 13
2. Primary: 6.0 μH
Primary winding = 24 turns #36 AWG
(close–wound on 1/4″ dia. form)
Core = Carbonyl E or J
Secondary winding = 1–1/2 turns #36 AWG, 1/4″ dia.
(wound over center–tap)

Figure 8. Single–Ended Input Admittance

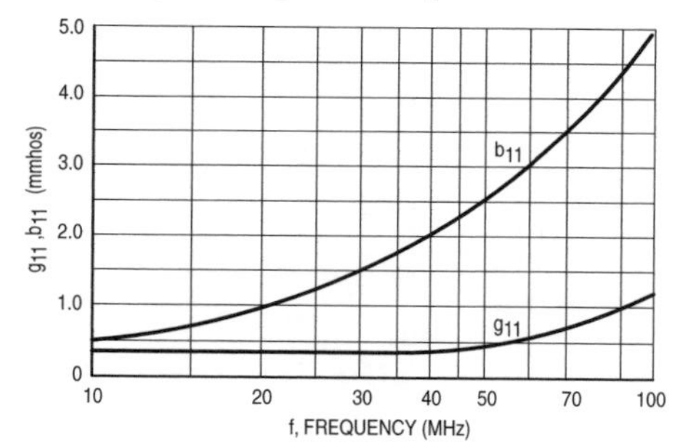

Figure 9. Forward Transfer Admittance

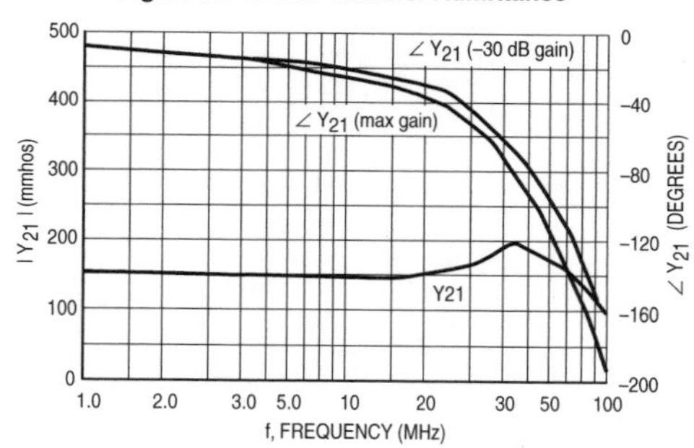

Figure 10. Differential Output Admittance

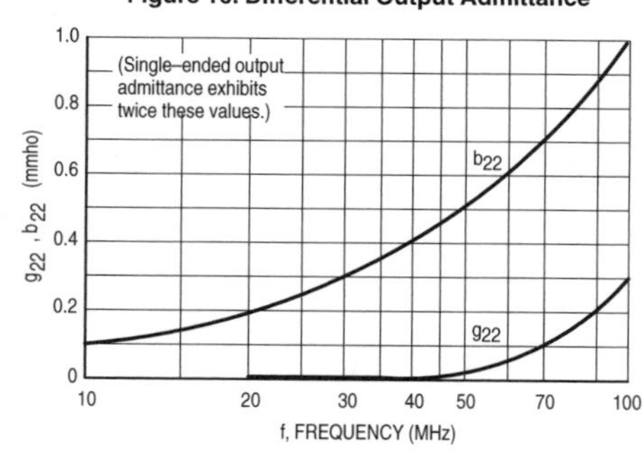

Figure 11. Differential Output Voltage

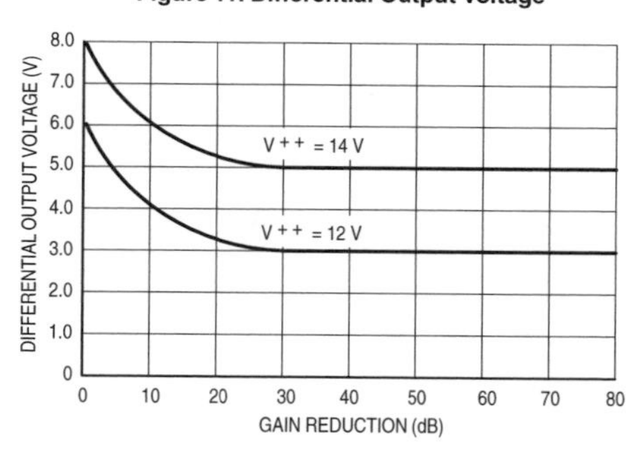

MC1350

OUTLINE DIMENSIONS

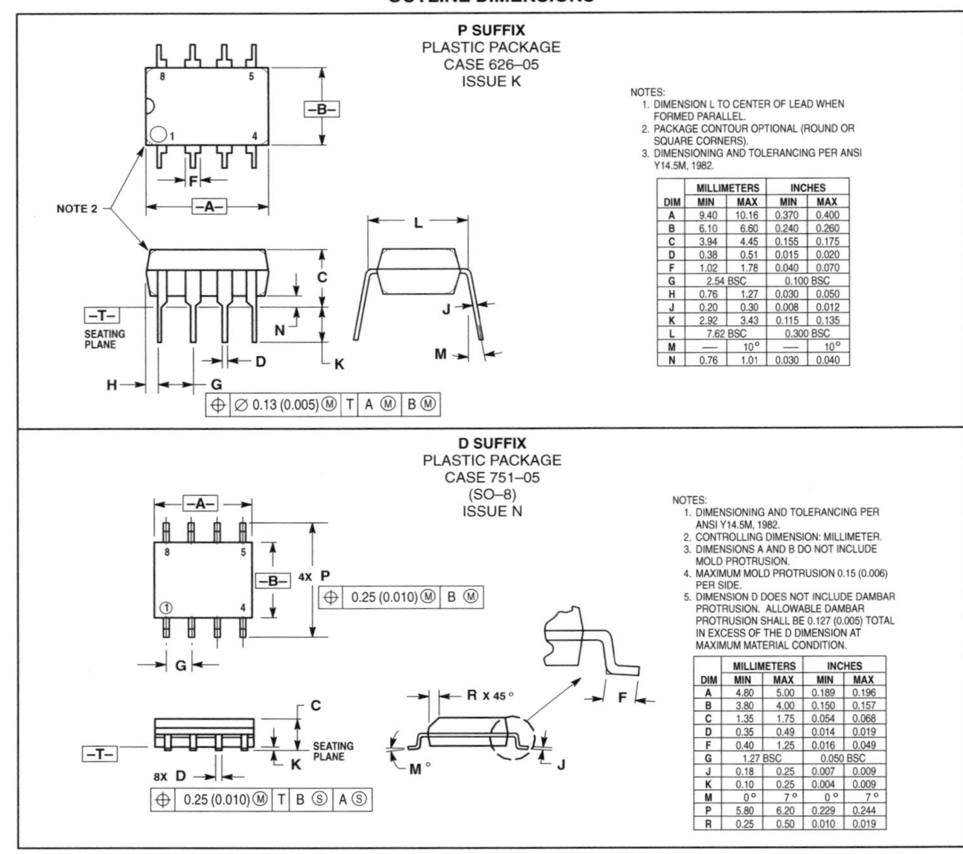

P SUFFIX
PLASTIC PACKAGE
CASE 626–05
ISSUE K

NOTES:
1. DIMENSION L TO CENTER OF LEAD WHEN FORMED PARALLEL.
2. PACKAGE CONTOUR OPTIONAL (ROUND OR SQUARE CORNERS).
3. DIMENSIONING AND TOLERANCING PER ANSI Y14.5M, 1982.

	MILLIMETERS		INCHES	
DIM	MIN	MAX	MIN	MAX
A	9.40	10.16	0.370	0.400
B	6.10	6.60	0.240	0.260
C	3.94	4.45	0.155	0.175
D	0.38	0.51	0.015	0.020
F	1.02	1.78	0.040	0.070
G	2.54 BSC		0.100 BSC	
H	0.76	1.27	0.030	0.050
J	0.20	0.30	0.008	0.012
K	2.92	3.43	0.115	0.135
L	7.62 BSC		0.300 BSC	
M	—	10°	—	10°
N	0.76	1.01	0.030	0.040

D SUFFIX
PLASTIC PACKAGE
CASE 751–05
(SO–8)
ISSUE N

NOTES:
1. DIMENSIONING AND TOLERANCING PER ANSI Y14.5M, 1982.
2. CONTROLLING DIMENSION: MILLIMETER.
3. DIMENSIONS A AND B DO NOT INCLUDE MOLD PROTRUSION.
4. MAXIMUM MOLD PROTRUSION 0.15 (0.006) PER SIDE.
5. DIMENSION D DOES NOT INCLUDE DAMBAR PROTRUSION. ALLOWABLE DAMBAR PROTRUSION SHALL BE 0.127 (0.005) TOTAL IN EXCESS OF THE D DIMENSION AT MAXIMUM MATERIAL CONDITION.

	MILLIMETERS		INCHES	
DIM	MIN	MAX	MIN	MAX
A	4.80	5.00	0.189	0.196
B	3.80	4.00	0.150	0.157
C	1.35	1.75	0.054	0.068
D	0.35	0.49	0.014	0.019
F	0.40	1.25	0.016	0.049
G	1.27 BSC		0.050 BSC	
J	0.18	0.25	0.007	0.009
K	0.10	0.25	0.004	0.009
M	0°	7°	0°	7°
P	5.80	6.20	0.229	0.244
R	0.25	0.50	0.010	0.019

MC1350

How to reach us:
USA / EUROPE / Locations Not Listed: Motorola Literature Distribution; P.O. Box 20912; Phoenix, Arizona 85036. 1–800–441–2447 or 602–303–5454

MFAX: RMFAX0@email.sps.mot.com – TOUCHTONE 602–244–6609
INTERNET: http://Design–NET.com

JAPAN: Nippon Motorola Ltd.; Tatsumi–SPD–JLDC, 6F Seibu–Butsuryu–Center, 3–14–2 Tatsumi Koto–Ku, Tokyo 135, Japan. 03–81–3521–8315

ASIA/PACIFIC: Motorola Semiconductors H.K. Ltd.; 8B Tai Ping Industrial Park, 51 Ting Kok Road, Tai Po, N.T., Hong Kong. 852–26629298

◊

MC1350/D

CA3028A

January 1999 File Number 382.5

Differential/Cascode Amplifier for Commercial and Industrial Equipment from DC to 120MHz

The CA3028A is a differential/cascode amplifier designed for use in communications and industrial equipment operating at frequencies from DC to 120MHz.

Part Number Information

PART NUMBER (BRAND)	TEMP. RANGE (°C)	PACKAGE	PKG. NO.
CA3028A	-55 to 125	8 Pin Metal Can	T8.C
CA3028AE	-55 to 125	8 Ld PDIP	E8.3
CA3028AM96 (3028A)	-55 to 125	8 Ld SOIC Tape and Reel	M8.15

Pinouts

CA3028A
(PDIP, SOIC)
TOP VIEW

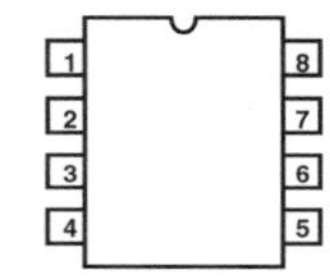

CA3028A
(METAL CAN)
TOP VIEW

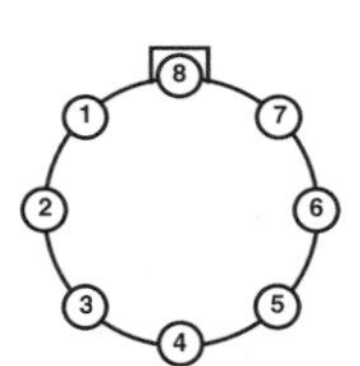

Features

- Controlled for Input Offset Voltage, Input Offset Current and Input Bias Current
- Balanced Differential Amplifier Configuration with Controlled Constant Current Source
- Single-Ended and Dual-Ended Operation

Applications

- RF and IF Amplifiers (Differential or Cascode)
- DC, Audio and Sense Amplifiers
- Converter in the Commercial FM Band
- Oscillator
- Mixer
- Limiter
- Related Literature
  - Application Note AN5337 "Application of the CA3028 Integrated Circuit Amplifier in the HF and VHF Ranges." This note covers characteristics of different operating modes, noise performance, mixer, limiter, and amplifier design considerations

Schematic Diagram

(Terminal Numbers Apply to All Packages)

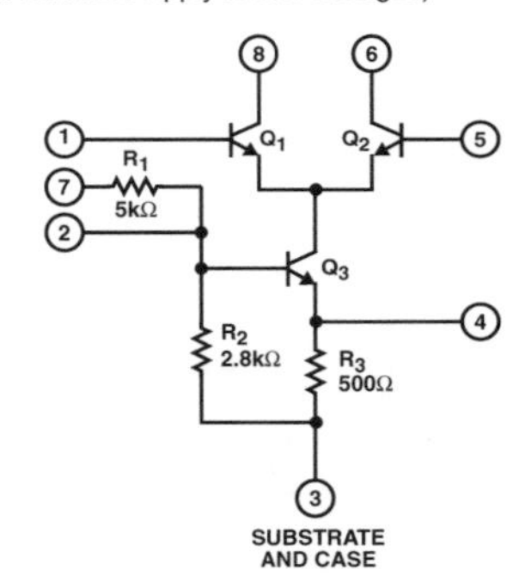

CAUTION: These devices are sensitive to electrostatic discharge; follow proper IC Handling Procedures.

CA3028A

Operating Conditions

Temperature Range . . . . -55°C to 125°C

Thermal Information

Thermal Resistance (Typical, Note 1)	θ_{JA} (°C/W)	θ_{JC} (°C/W)
Metal Can Package	225	140
PDIP Package	155	N/A
SOIC Package	185	N/A

Maximum Junction Temperature (Metal Can Package). . . . . 175°C
Maximum Junction Temperature (Plastic Package) . . . . 150°C
Maximum Storage Temperature Range . . . . -65°C to 150°C
Maximum Lead Temperature (Soldering 10s) . . . . 300°C
(SOIC - Lead Tips Only)

CAUTION: Stresses above those listed in "Absolute Maximum Ratings" may cause permanent damage to the device. This is a stress only rating and operation of the device at these or any other conditions above those indicated in the operational sections of this specification is not implied.

NOTE:

1. θ_{JA} is measured with the component mounted on an evaluation PC board in free air.

Absolute Maximum Voltage Ratings T_A = 25°C

The following chart gives the range of voltages which can be applied to the terminals listed horizontally with respect to the terminals listed vertically. For example, the voltage range of the horizontal Terminal 4 with respect to Terminal 2 is -1V to +5V.

TERM NO.	1	2	3	4	5	6	7	8
1		0 to -15	0 to -15	0 to -15	+5 to -5	Note 3	Note 3	+20 to 0
2			+5 to -11	+5 to -1	+15 to 0	Note 3	+15 to 0	Note 3
3 (Note 2)				+10 to 0	+15 to 0	+24 to 0	+15 to 0	+24 to 0
4					+15 to 0	Note 3	Note 3	Note 3
5						+20 to 0	Note 3	Note 3
6							Note 3	Note 3
7								Note 3
8								

NOTES:

2. Terminal No. 3 is connected to the substrate and case.
3. Voltages are not normally applied between these terminals. Voltages appearing between these terminals will be safe, if the specified voltage limits between all other terminals are not exceeded.

Absolute Maximum Current Ratings

TERM NO.	I_{IN} mA	I_{OUT} mA
1	0.6	0.1
2	4	0.1
3	0.1	23
4	20	0.1
5	0.6	0.1
6	20	0.1
7	4	0.1
8	20	0.1

Electrical Specifications T_A = 25°C

PARAMETER	SYMBOL	TEST CONDITIONS	MIN	TYP	MAX	UNIT
DC CHARACTERISTICS						
Input Bias Current (Figures 1, 10)	I_I	V_{CC} = 6V, V_{EE} = -6V	-	16.6	70	μA
		V_{CC} = 12V, V_{EE} = -12V	-	36	106	μA
Quiescent Operating Current (Figures 1,11, 12)	I_6, I_8	V_{CC} = 6V, V_{EE} = -6V	0.8	1.25	2.0	mA
		V_{CC} = 12V, V_{EE} = -12V	2.0	3.3	5.0	mA
AGC Bias Current (Into Constant Current Source Terminal 7) (Figures 2, 13)	I_7	V_{CC} = 12V, V_{AGC} = 9V	-	1.28	-	mA
		V_{CC} = 12V, V_{AGC} = 12V	-	1.65	-	mA
Input Current (Terminal 7)	I_7	V_{CC} = 6V, V_{EE} = -6V	0.5	0.85	1.0	mA
		V_{CC} = 12V, V_{EE} = -12V	1.0	1.65	2.1	mA
Power Dissipation (Figures 1, 14)	P_T	V_{CC} = 6V, V_{EE} = -6V	24	36	54	mW
		V_{CC} = 12V, V_{EE} = -12V	120	175	260	mW

CA3028A

Electrical Specifications $T_A = 25^oC$ **(Continued)**

PARAMETER	SYMBOL	TEST CONDITIONS		MIN	TYP	MAX	UNIT
DYNAMIC CHARACTERISTICS							
Power Gain (Figures 3, 4, 5, 15, 17, 19)	G_P	f = 100MHz V_{CC} = 9V	Cascode	16	20	-	dB
			Diff. Amp.	14	17	-	dB
		f = 10.7MHz V_{CC} = 9V	Cascode	35	39	-	dB
			Diff. Amp.	28	32	-	dB
Noise Figure (Figures 3, 4, 5, 16, 18, 19)	NF	f = 100MHz, V_{CC} = 9V	Cascode	-	7.2	9.0	dB
			Diff. Amp.	-	6.7	9.0	dB
Input Admittance (Figures 20, 21)	Y_{11}	f = 10.7MHz, V_{CC} = 9V	Cascode	-	0.6 + j1.6	-	mS
			Diff. Amp.	-	0.5 + j0.5	-	mS
Reverse Transfer Admittance (Figures 22, 23)	Y_{12}	f = 10.7MHz, V_{CC} = 9V	Cascode	-	0.0003 - j0	-	mS
			Diff. Amp.	-	0.01 - j0.0002	-	mS
Forward Transfer Admittance (Figures 24, 25)	Y_{21}	f = 10.7MHz, V_{CC} = 9V	Cascode	-	99 - j18	-	mS
			Diff. Amp.	-	-37 + j0.5	-	mS
Output Admittance (Figures 26, 27)	Y_{22}	f = 10.7MHz, V_{CC} = 9V	Cascode	-	0 + j0.08	-	mS
			Diff. Amp.	-	0.04 + j0.23	-	mS
Output Power (Untuned) (Figures 6, 28)	P_O	f = 10.7MHz, V_{CC} = 9V	Diff. Amp., 50Ω Input-Output	-	5.7	-	μW
AGC Range (Maximum Power Gain to Full Cutoff) (Figures 7, 29)	AGC	f = 10.7MHz, V_{CC} = 9V	Diff. Amp.	-	62	-	dB
Voltage Gain (Figures 8, 9, 30, 31)	A	f = 10.7MHz, V_{CC} = 9V, R_L = 1kΩ	Cascode	-	40	-	dB
			Diff. Amp.	-	30	-	dB
Peak-to-Peak Output Current	I_{P-P}	f = 10.7MHz, e_{IN} = 400mV, Diff. Amp.	V_{CC} = 9V	2.0	4.0	7.0	mA
			V_{CC} = 12V	3.5	6.0	10	mA

Test Circuits

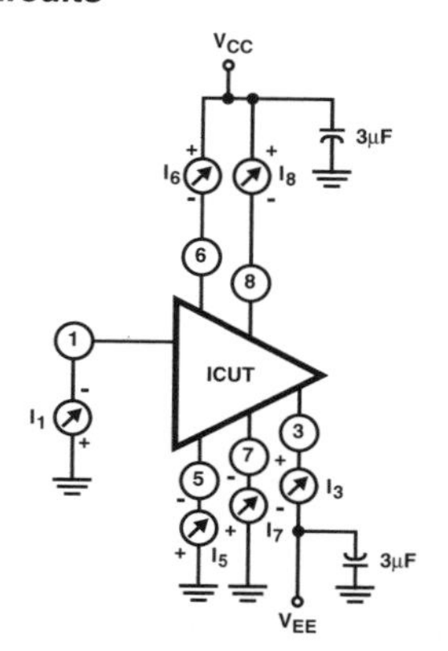

NOTE: Power Dissipation = $I_3 V_{EE} + (I_6 + I_8) V_{CC}$.

FIGURE 1. INPUT OFFSET CURRENT, INPUT BIAS CURRENT, POWER DISSIPATION, AND QUIESCENT OPERATING CURRENT TEST CIRCUIT

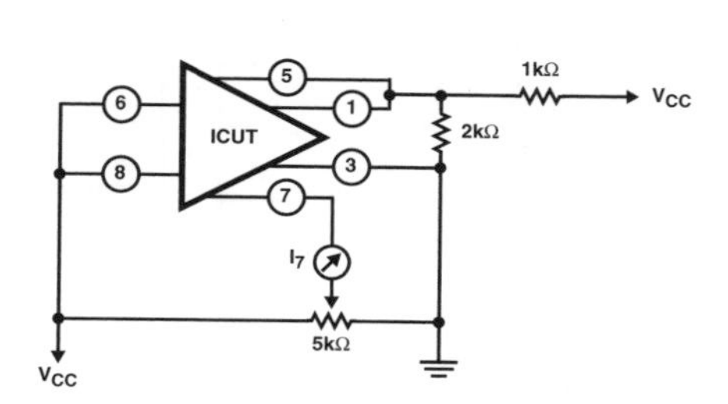

FIGURE 2. AGC BIAS CURRENT TEST CIRCUIT (DIFFERENTIAL AMPLIFIER CONFIGURATION)

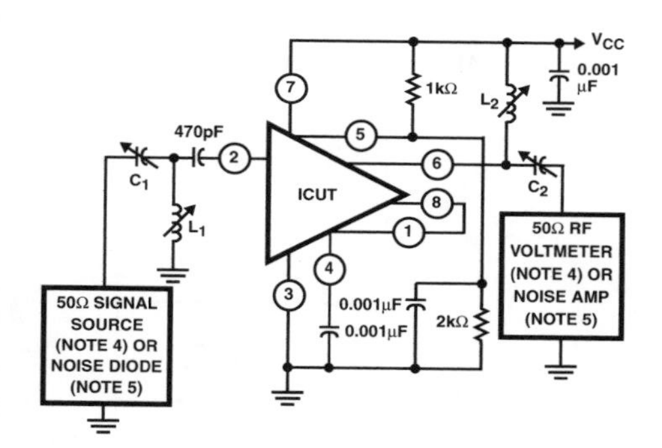

f (MHz)	C_1 (pF)	C_2 (pF)	L_1 (μH)	L_2 (μH)
10.7	20 - 60	20 - 60	3 - 5	3 - 5
100	3 - 30	3 - 30	0.1 - 0.25	0.15 - 0.3

NOTES:

4. For Power Gain Test.
5. For Noise Figure Test.

FIGURE 3. POWER GAIN AND NOISE FIGURE TEST CIRCUIT (CASCODE CONFIGURATION)

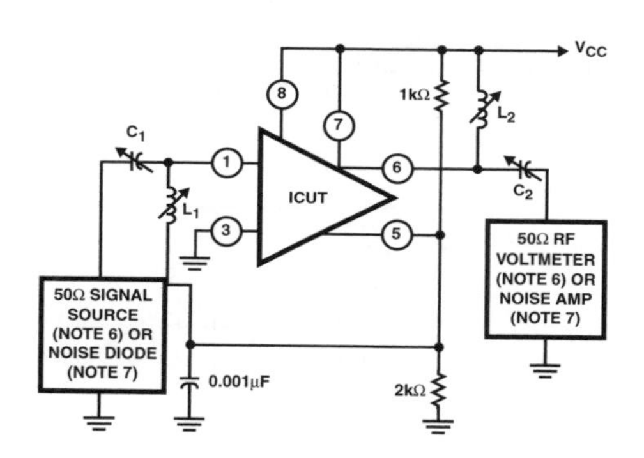

f (MHz)	C_1 (pF)	C_2 (pF)	L_1 (μH)	L_2 (μH)
10.7	30 - 60	20 - 50	3 - 6	3 - 6
100	2 - 15	2 - 15	0.2 - 0.5	0.2 - 0.5

NOTES:

6. For Power Gain Test.
7. For Noise Figure Test.

FIGURE 4. POWER GAIN AND NOISE FIGURE TEST CIRCUIT (DIFFERENTIAL AMPLIFIER CONFIGURATION AND TERMINAL 7 CONNECTED TO V_{CC})

Test Circuits (Continued)

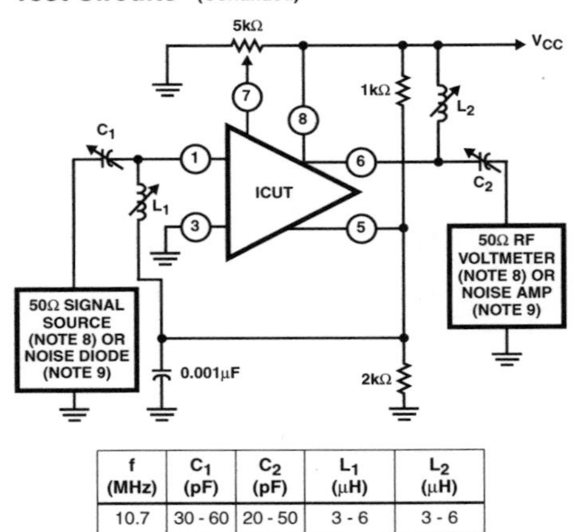

f (MHz)	C_1 (pF)	C_2 (pF)	L_1 (μH)	L_2 (μH)
10.7	30 - 60	20 - 50	3 - 6	3 - 6
100	2 - 15	2 - 15	0.2 - 0.5	0.2 - 0.5

NOTES:

8. For Power Gain Test.
9. For Noise Figure Test.

FIGURE 5. POWER GAIN AND NOISE FIGURE TEST CIRCUIT (DIFFERENTIAL AMPLIFIER CONFIGURATION)

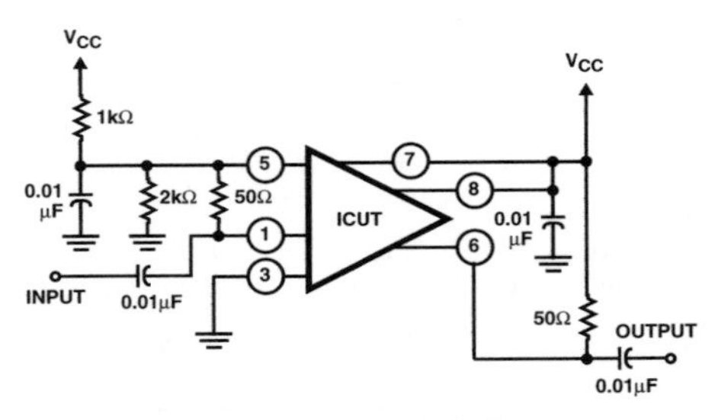

FIGURE 6. OUTPUT POWER TEST CIRCUIT

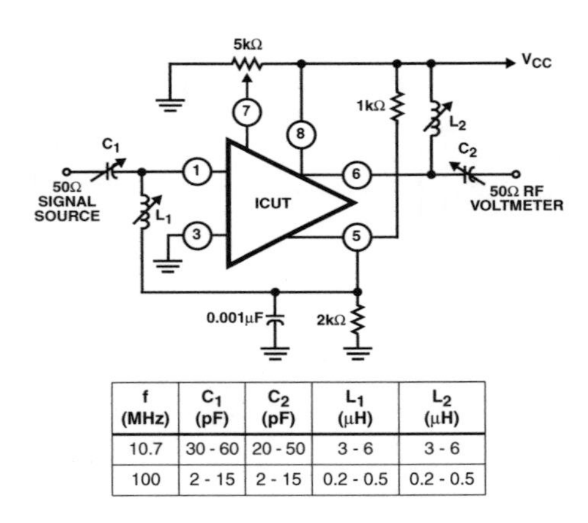

f (MHz)	C_1 (pF)	C_2 (pF)	L_1 (μH)	L_2 (μH)
10.7	30 - 60	20 - 50	3 - 6	3 - 6
100	2 - 15	2 - 15	0.2 - 0.5	0.2 - 0.5

FIGURE 7. AGC RANGE TEST CIRCUIT (DIFFERENTIAL AMPLIFIER)

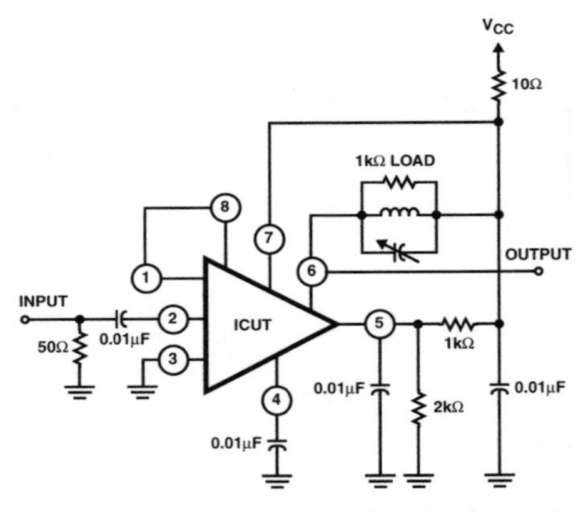

FIGURE 8. TRANSFER CHARACTERISTIC (VOLTAGE GAIN) TEST CIRCUIT (10.7MHz) CASCODE CONFIGURATION

Test Circuits (Continued)

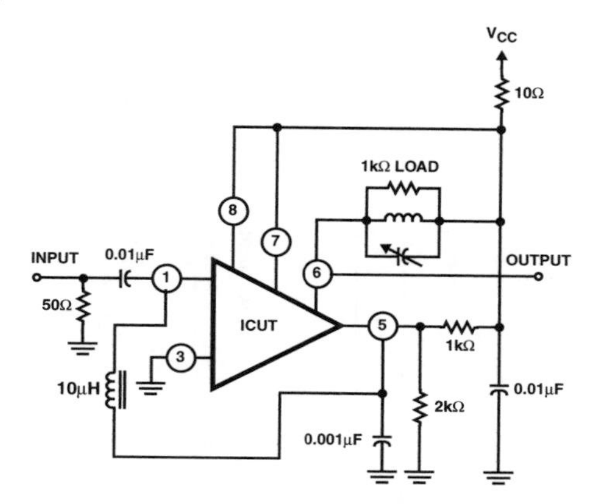

FIGURE 9. TRANSFER CHARACTERISTIC (VOLTAGE GAIN) TEST CIRCUIT (10.7MHz) DIFFERENTIAL AMPLIFIER CONFIGURATION

Typical Performance Curves

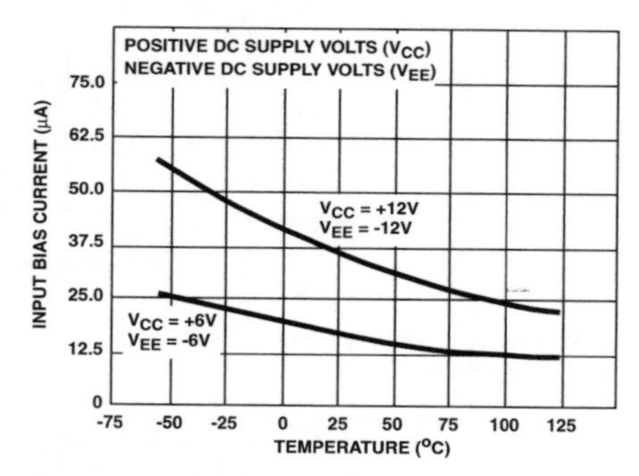

FIGURE 10. INPUT BIAS CURRENT vs TEMPERATURE

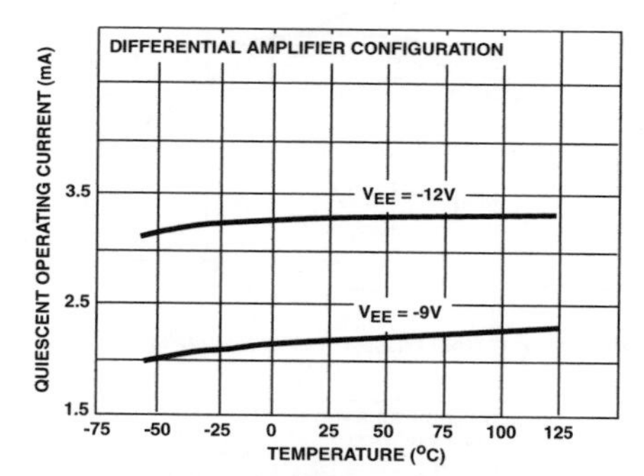

FIGURE 11. QUIESCENT OPERATING CURRENT vs TEMPERATURE

Typical Performance Curves (Continued)

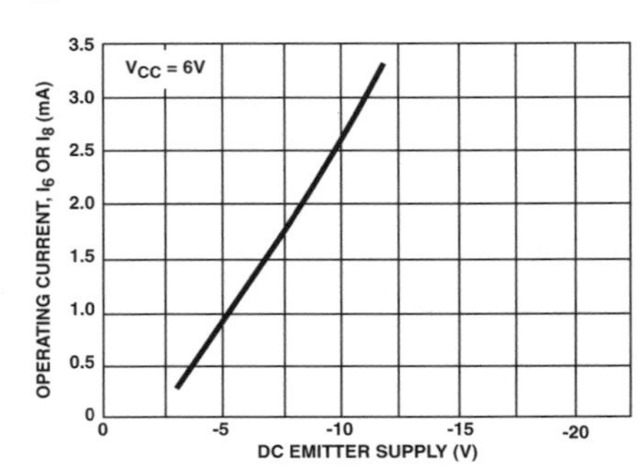

FIGURE 12. OPERATING CURRENT vs V_{EE} VOLTAGE

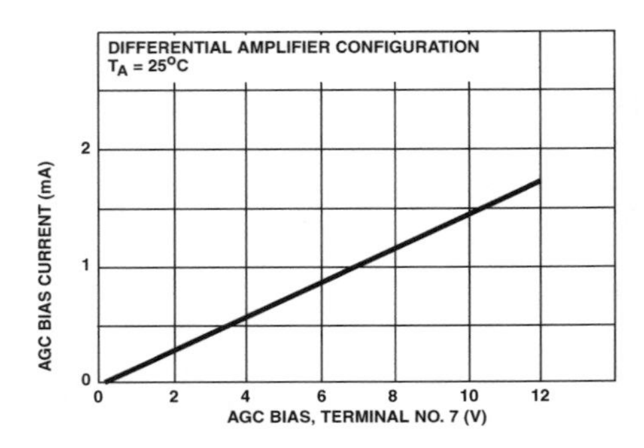

FIGURE 13. AGC BIAS CURRENT vs BIAS VOLTAGE (TERMINAL 7)

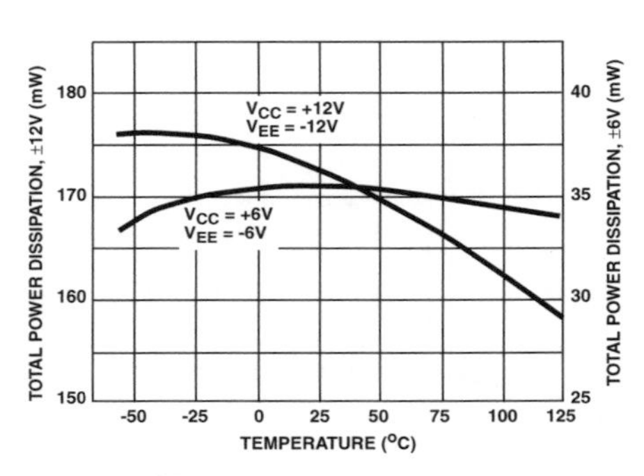

FIGURE 14. POWER DISSIPATION vs TEMPERATURE

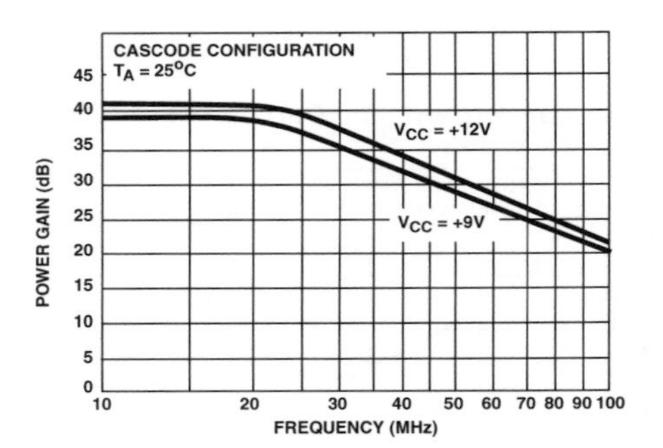

FIGURE 15. POWER GAIN vs FREQUENCY (CASCODE CONFIGURATION)

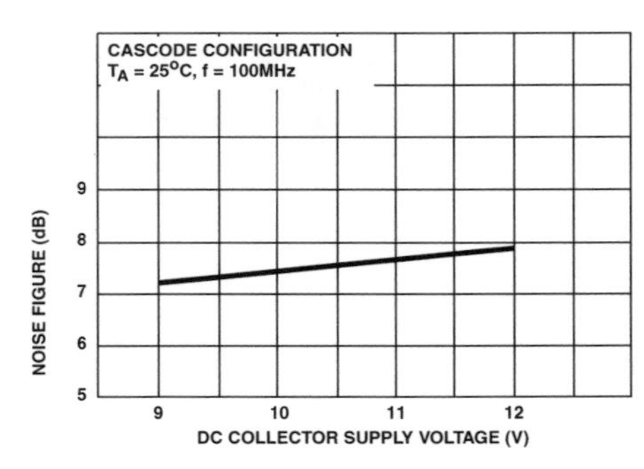

FIGURE 16. 100MHz NOISE FIGURE vs COLLECTOR SUPPLY VOLTAGE (CASCODE CONFIGURATION)

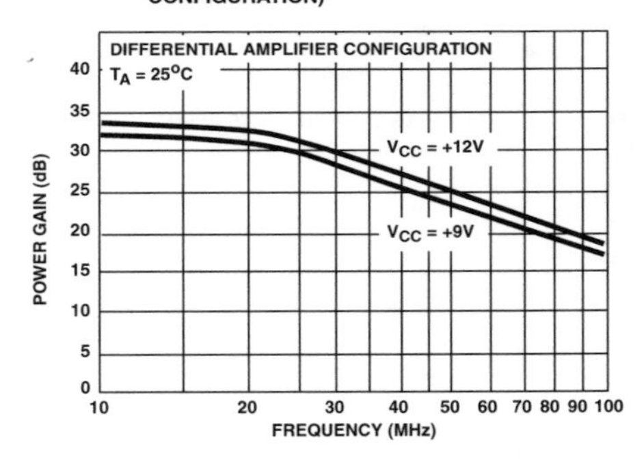

FIGURE 17. POWER GAIN vs FREQUENCY (DIFFERENTIAL AMPLIFIER CONFIGURATION)

Typical Performance Curves (Continued)

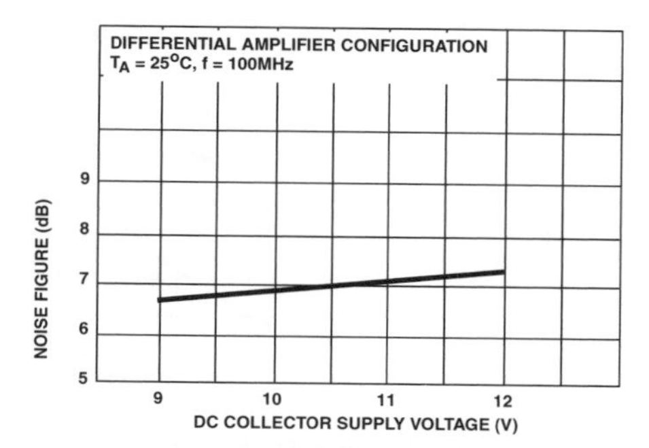

FIGURE 18. 100MHz NOISE FIGURE vs COLLECTOR SUPPLY VOLTAGE (DIFFERENTIAL AMPLIFIER CONFIGURATION)

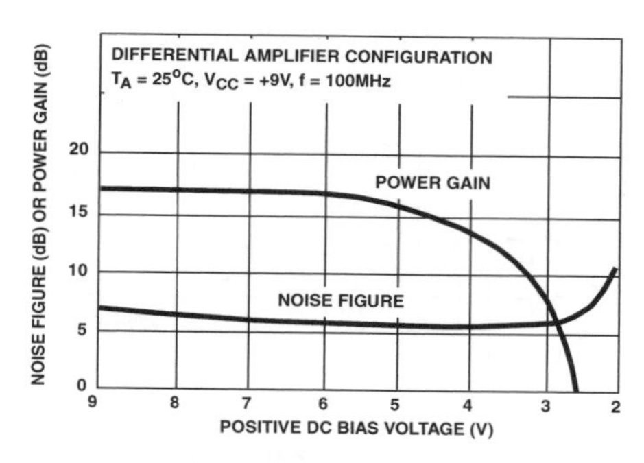

FIGURE 19. 100MHz NOISE FIGURE AND POWER GAIN vs BASE-TO-EMITTER BIAS VOLTAGE (TERMINAL 7)

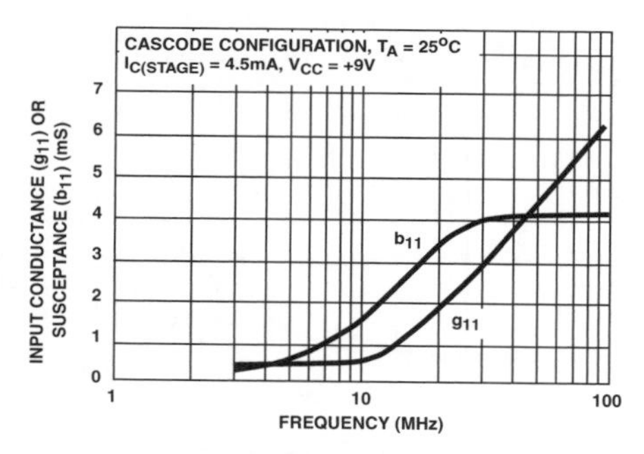

FIGURE 20. INPUT ADMITTANCE (Y_{11}) vs FREQUENCY (CASCODE CONFIGURATION)

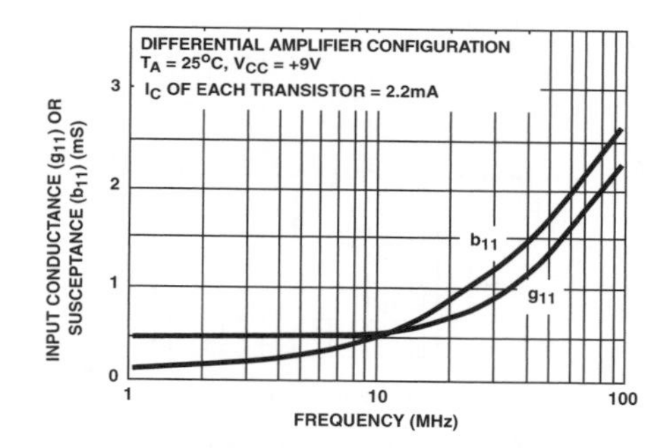

FIGURE 21. INPUT ADMITTANCE (Y_{11}) vs FREQUENCY (DIFFERENTIAL AMPLIFIER CONFIGURATION)

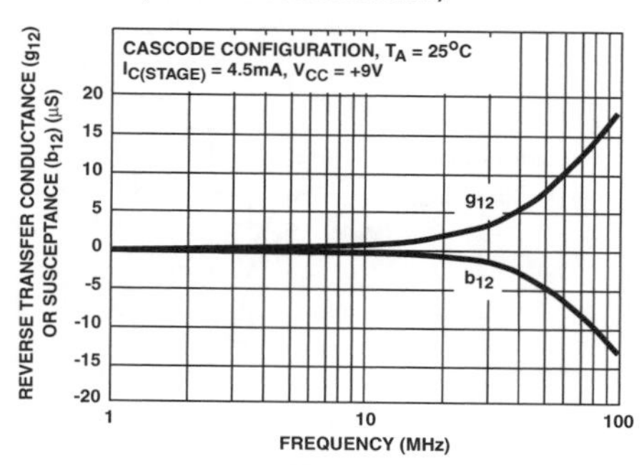

FIGURE 22. REVERSE TRANSADMITTANCE (Y_{12}) vs FREQUENCY (CASCODE CONFIGURATION)

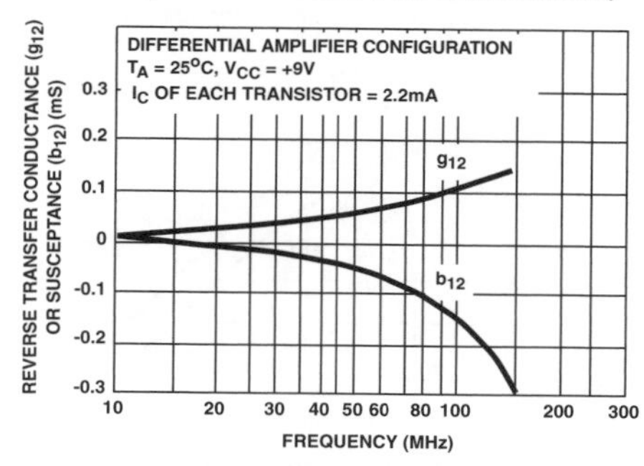

FIGURE 23. REVERSE TRANSADMITTANCE (Y_{12}) vs FREQUENCY (DIFFERENTIAL AMPLIFIER CONFIGURATION)

Typical Performance Curves (Continued)

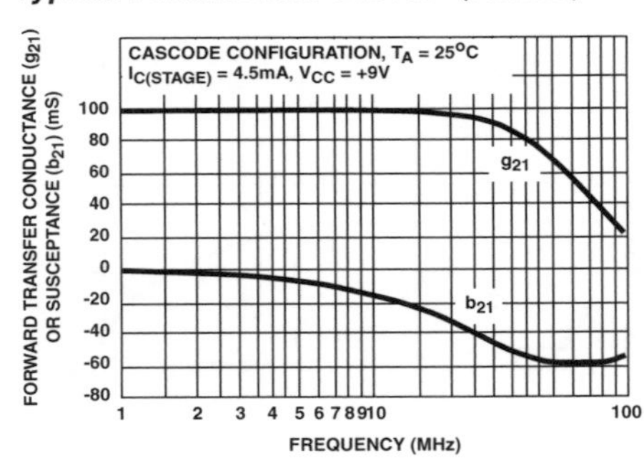

FIGURE 24. FORWARD TRANSADMITTANCE (Y_{21}) vs FREQUENCY (CASCODE CONFIGURATION)

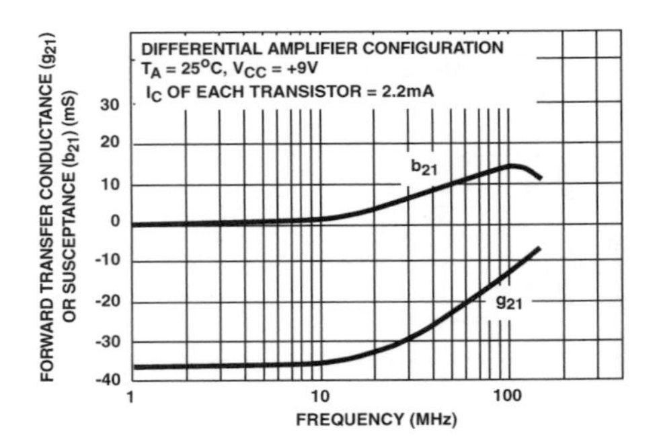

FIGURE 25. FORWARD TRANSADMITTANCE (Y_{21}) vs FREQUENCY (DIFFERENTIAL AMPLIFIER CONFIGURATION)

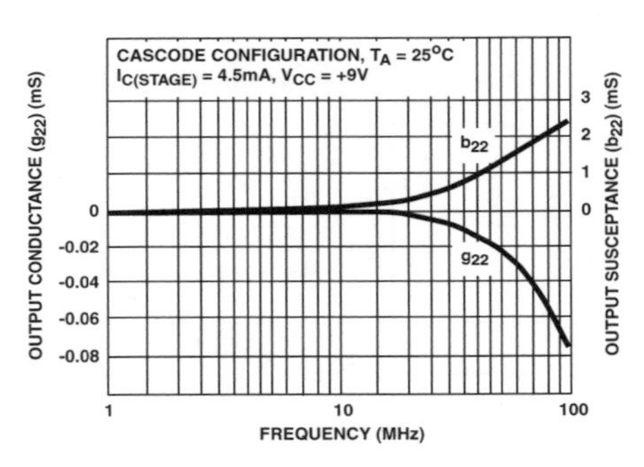

FIGURE 26. OUTPUT ADMITTANCE (Y_{22}) vs FREQUENCY (CASCODE CONFIGURATION)

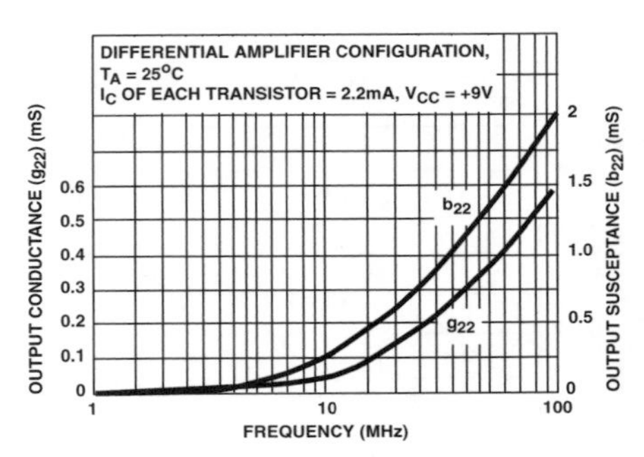

FIGURE 27. OUTPUT ADMITTANCE (Y_{22}) vs FREQUENCY (DIFFERENTIAL AMPLIFIER CONFIGURATION)

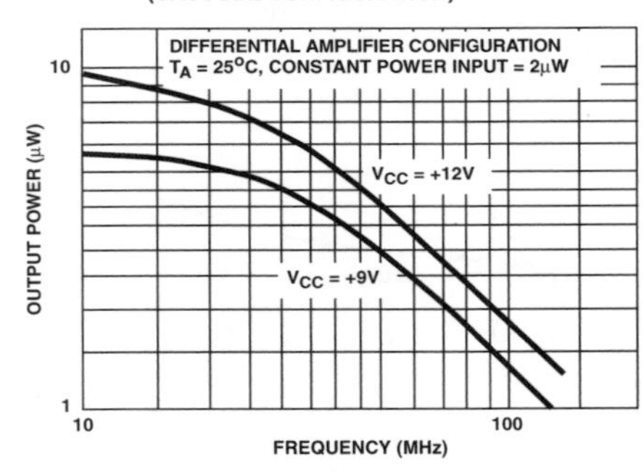

FIGURE 28. OUTPUT POWER vs FREQUENCY - 50Ω INPUT AND 50Ω OUTPUT (DIFFERENTIAL AMPLIFIER CONFIGURATION)

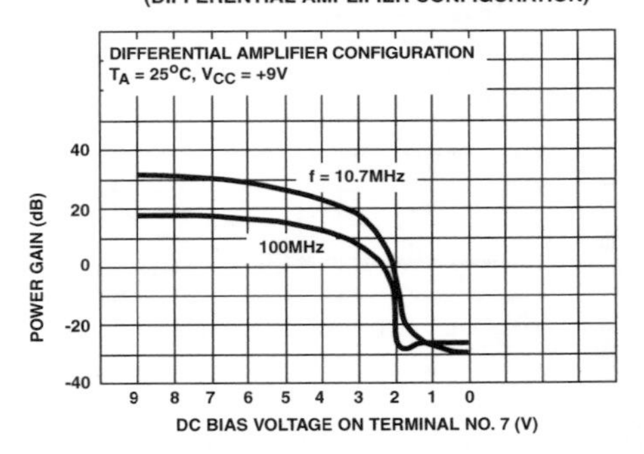

FIGURE 29. AGC CHARACTERISTICS

Typical Performance Curves (Continued)

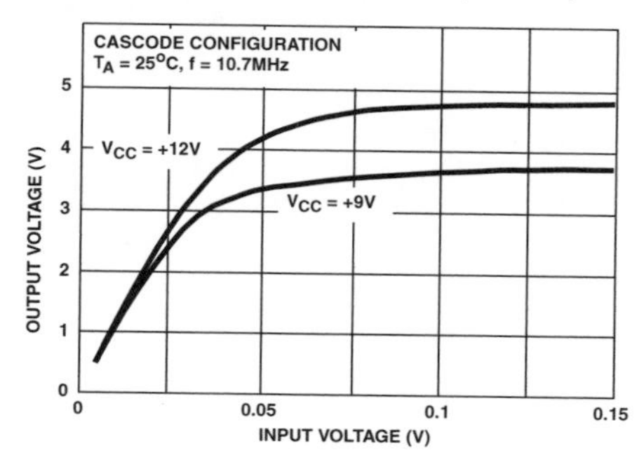

FIGURE 30. TRANSFER CHARACTERISTICS (CASCODE CONFIGURATION)

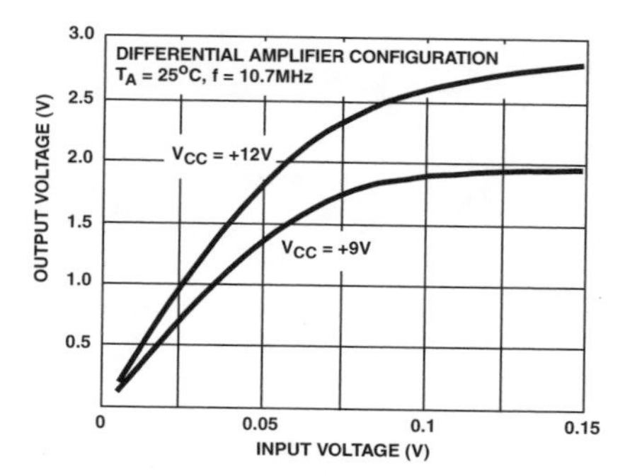

FIGURE 31. TRANSFER CHARACTERISTICS (DIFFERENTIAL AMPLIFIER CONFIGURATION)

Glossary of Terms

AGC Bias Current - The current drawn by the device from the AGC voltage source, at maximum AGC voltage.

AGC Range - The total change in voltage gain (from maximum gain to complete cutoff) which may be achieved by application of the specified range of dc voltage to the AGC input terminal of the device.

Common Mode Rejection Ratio - The ratio of the full differential voltage gain to the common mode voltage gain.

Power Dissipation - The total power drain of the device with no signal applied and no external load current.

Input Bias Current - The average value (one half the sum) of the currents at the two input terminals when the quiescent operating voltages at the two output terminals are equal.

Input Offset Current - The difference in the currents at the two input terminals when the quiescent operating voltages at the two output terminals are equal.

Input Offset Voltage - The difference in the DC voltages which must be applied to the input terminals to obtain equal quiescent operating voltages (zero output offset voltage) at the output terminals.

Noise Figure - The ratio of the total noise power of the device and a resistive signal source to the noise power of the signal source alone, the signal source representing a generator of zero impedance in series with the source resistance.

Power Gain - The ratio of the signal power developed at the output of the device to the signal power applied to the input, expressed in dB.

Quiescent Operating Current - The average (DC) value of the current in either output terminal.

Voltage Gain - The ratio of the change in output voltage at either output terminal with respect to ground, to a change in input voltage at either input terminal with respect to ground, with the other input terminal at AC ground.

5

Power amplifiers

5.1 Introduction

Figure 5.1 shows two applications where a power amplifier may be used. In Fig. 5.1(a) two RF power amplifiers are used to feed power into the antenna of a transmitter. The booster amplifier raises the power level to the value required in order to drive the final RF amplifier which is responsible for feeding large power levels into the antenna via the antenna matching unit. Note that matching stages are required in order to match the output of one stage to the input impedance of the next stage for maximum power transfer. In the case of the antenna and final RF power amplifier there is an impedance mismatch which is addressed by the antenna matching unit which is normally part of the antenna tuning unit required to tune the antenna for optimum resonance and radiation. Figure 5.1(b) shows an audio power amplifier used in a public address system. Once again the output impedance of the microphone is very different from the preamplifier input impedance, and this is remedied by the matching network.

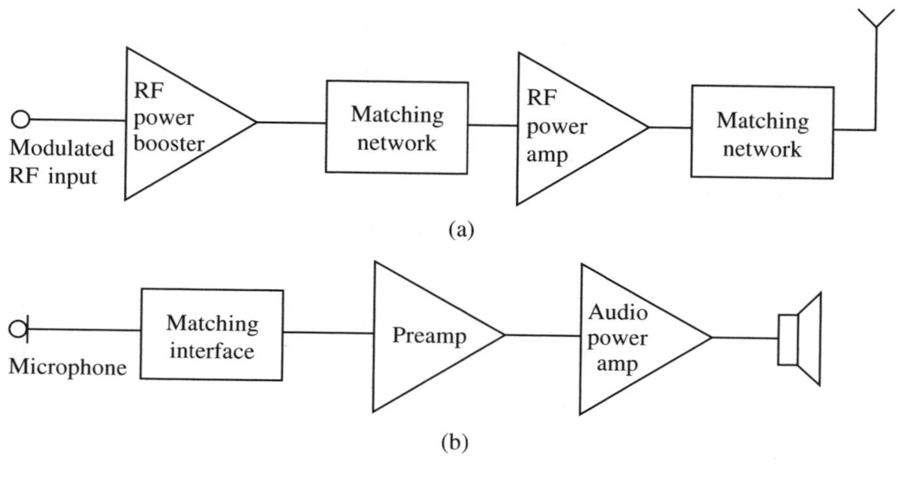

Fig. 5.1

There are many types of amplifiers, and so far only a few have been used in such applications as filters and tuned amplifiers. These are considered as voltage or small signal amplifiers, such as the preamplifier shown in Fig. 5.1(b), but the tuned amplifier can be a large signal amplifier.

A small signal amplifier is normally designed to give a large undistorted output voltage from small input voltages but only small amounts of power are produced in the collector loads. An amplifier may be required to drive a loudspeaker or a motor, and in this application a large signal amplifier generating large swings of current and voltage is required.

Generally a power amplifier has a low internal impedance, enabling large currents to be produced. The voltage amplifier, on the other hand, has a high internal impedance, enabling large voltages to be developed.

A power amplifier must be capable of staying within its rated specifications, distortion must be kept low and the efficiency of the amplifier must be as high as possible. In order to achieve these aims there are many circuits available, some of which are more suitable than others.

While the audio power amplifier will be considered in this chapter, it should be appreciated that other types are available such as the class C tuned amplifier used in radio transmitters and driver power amplifiers for motors. The performance of all power amplifiers, however, depends on certain parameters which are best understood through a knowledge of small signal amplifiers. These will be tackled in the following sections.

5.2 Transistor characteristics and parameters

A small signal amplifier is one in which the amplified output signal remains within the limits of the amplifier to produce an undistorted output. In order to accomplish this many small signal parameters are used, but the most common are the hybrid or h-parameters derived by considering the transistor as a four-terminal network, as shown in Fig. 5.2.

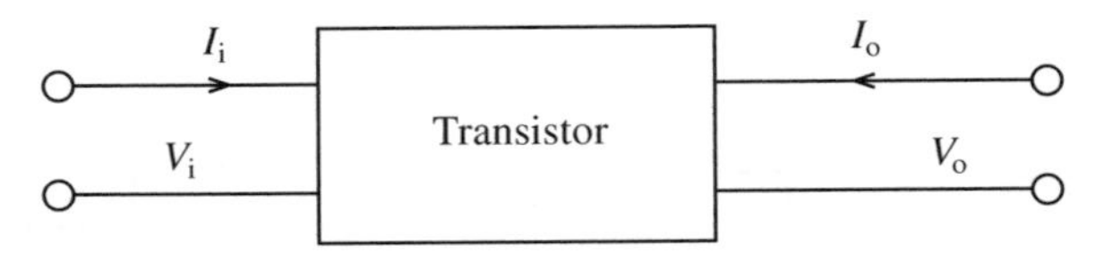

Fig. 5.2

If I_i and V_o are taken to be the independent variables, then the input voltage V_i and the output current I_o can be written as:

$$V_i = I_i h_I + V_o h_R \tag{5.1}$$

$$I_o = I_i h_F + V_o h_o \tag{5.2}$$

With $V_o = 0$ equation (5.1) becomes

$$V_i = I_i h_I$$

therefore

$$h_I = \frac{V_i}{I_i} \text{ (ohms)} \qquad (5.3)$$

h_I is known as the d.c. input resistance with the output terminals short-circuited. Also from equation (5.2),

$$I_o = I_i h_F$$

therefore

$$h_F = \frac{I_o}{I_i} \qquad (5.4)$$

h_F is the forward d.c. current gain with the output terminals short-circuited. With $I_i = 0$ equations (5.1) and (5.2) become

$$h_R = \frac{V_i}{V_o} \qquad (5.5)$$

$$h_O = \frac{I_o}{V_o} \text{ (siemens)} \qquad (5.6)$$

h_R is the reverse d.c. voltage ratio with the input terminals open-circuited; h_O is the d.c. output conductance with the input terminals open-circuited. Note that h_F and h_R in equations (5.4) and (5.5) respectively are dimensionless quantities: they have no units.

When the h-parameters are applied to a bipolar transistor a second suffix is added to indicate the transistor configuration in which the transistor is connected. Thus for the common emitter connection which will be considered here, the h-parameters are h_{IE}, h_{FE}, h_{RE} and h_{OE}. However h-parameters are nearly always associated with circuit performance under small signal conditions where alternating quantities are used. The h-parameters are then given as h_{ie}, h_{fe}, h_{re} and h_{oe}.

Using transistor characteristics

Transistor characteristics are useful in visualizing what happens when operating conditions are set up in a transistor, and indicate the necessary input and output limitations required in order to avoid signal distortion. Figures 5.3 and 5.4 show the input and output characteristics for the common emitter configuration. In Fig. 5.3 two points have been selected as shown, and this gives a change in V_{BE} and I_B which enables h_{IE} to be obtained:

$$h_{IE} = \frac{V_{BE}}{I_B} = \frac{0.725 \times 10^6}{20} = 36.25 \text{ k}\Omega$$

Note that this is for a V_{CE} of 20 V. The input resistance therefore changes for different values of V_{CE}. Note these are d.c. conditions and upper-case characters are used for the subscripts. Lower-case characters will be used under small signal or a.c. conditions, which will be discussed later.

Figure 5.4 shows a set of output characteristics with details of the relevance of each region. The cut-off region occurs when the base–emitter junction is not forward-biased. All of the currents are approximately zero and V_{CE} is approximately equal to the

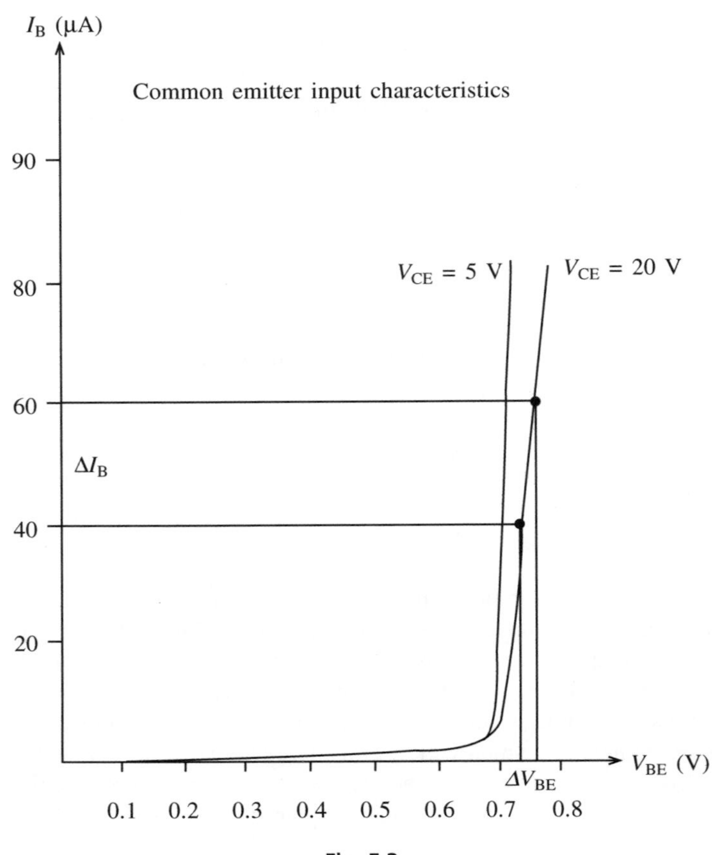

Fig. 5.3

supply voltage V_{CC}. The leakage current (I_{CEO}) is the only current flowing and it is very small.

The saturation condition occurs when the emitter junction is forward-biased and there is enough base current to produce a maximum collector current. Under these conditions the saturation current is calculated by using Ohm's law and using the collector load resistor (R_C) and the supply voltage V_{CC}:

$$I_{C(sat)} = \frac{V_{CC}}{R_C} \tag{5.7}$$

Consider Fig. 5.5, in which a base bias current of 60 μA has been selected. Using this characteristic shows a change of V_{CE} and I_C when solid projection is used. This gives

$$h_{OE} = \frac{\Delta V_{CE}}{\Delta I_C} = \frac{2 \times 10^3}{0.24} = 8.3\ \text{k}\Omega$$

The current gain can be found for a fixed value of V_{CE}. In Fig. 5.5, $V_{CE} = 8.5$ V and the projection gives

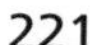

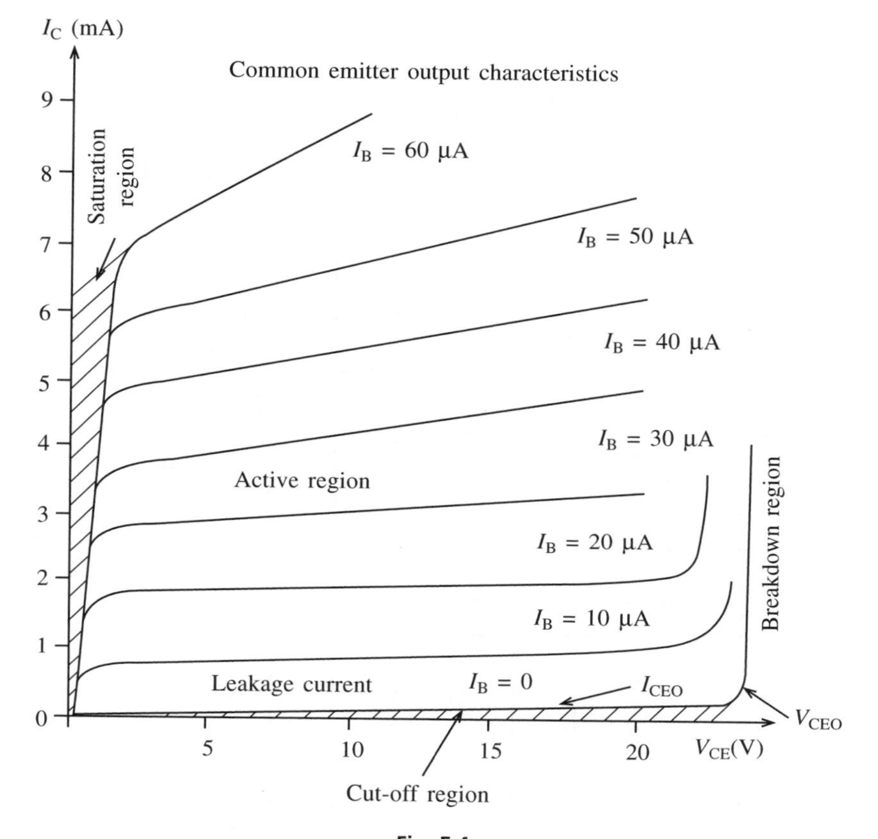

Fig. 5.4

$$h_{FE} = \frac{\Delta I}{\Delta I_B} = \frac{1 \times 10^6}{2 \times 10^4} = 50$$

Occasionally the mutual conductance (g_m) is required, and this is given by

$$g_m = \frac{I_C}{V_{BE}} = \frac{h_{FE}}{h_{IE}} = \frac{50}{36\ 250} = 1.4\ \text{mA/V}$$

5.3 Transistor bias

Up to this point the active device has been considered, namely the transistor. The characteristics under these conditions are used as static curves. However, the ultimate use of the device is in an amplifier circuit and for this purpose the transistor has to be properly set up to accept and amplify the input a.c. signals without distortion. Under these conditions the characteristics are used as dynamic curves.

Figure 5.6 shows three conditions under which an amplifier may function, but of course it is the linear condition which is required as this condition is free of distortion. Mention has already been made of saturation and cut-off with respect to the output

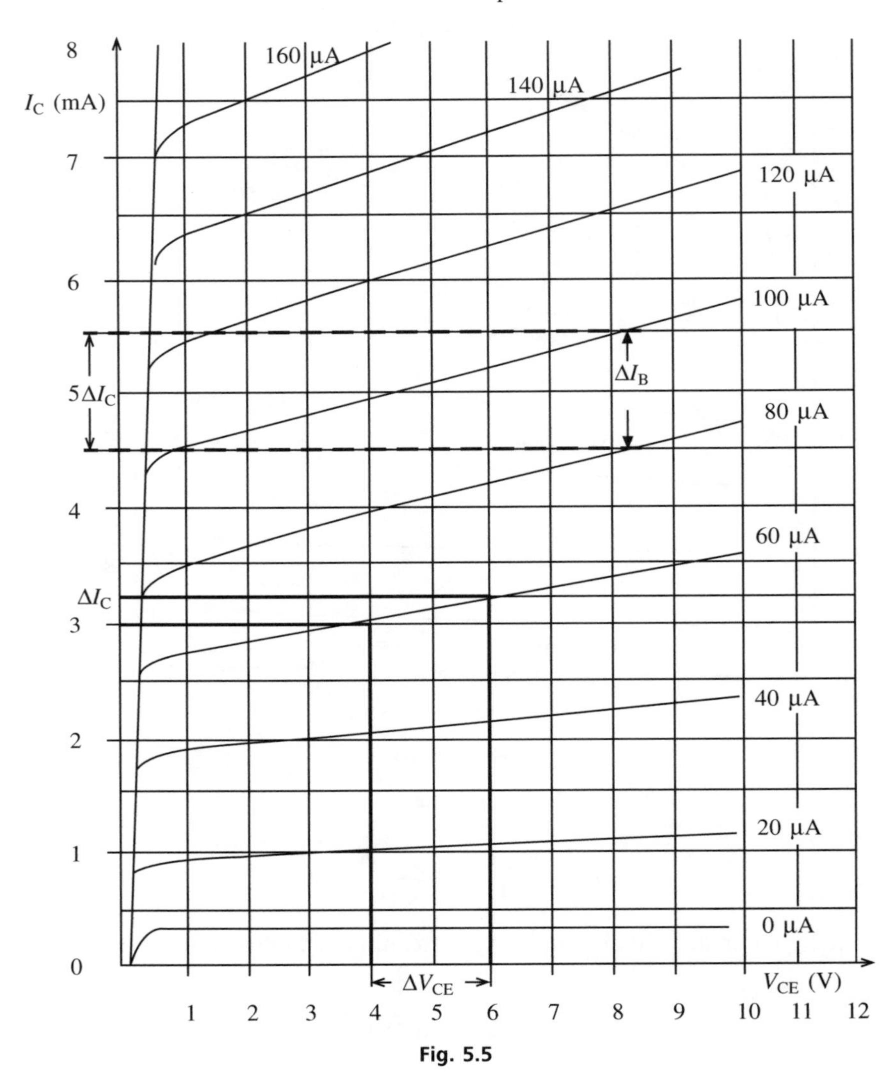

Fig. 5.5

characteristics. We can now see the effects on the output signal if these regions are entered, and this may occur if the input voltage swing is too large. Figure 5.7 shows the effects of saturation and cut-off on an output a.c. waveform using dynamic characteristics. In order to avoid this distortion the transistor has to be properly biased and the use of a load line is helpful in understanding this.

Consider firstly a simple common emitter amplifier, as shown in Fig. 5.8. We obtain

$$I_C = h_{FE}I_B + (I_{CEO}) \tag{5.8}$$

$$I_E = I_B + I_C = I_B + h_{FE}I_B$$

$$= (h_{FE} + 1)I_B \tag{5.9}$$

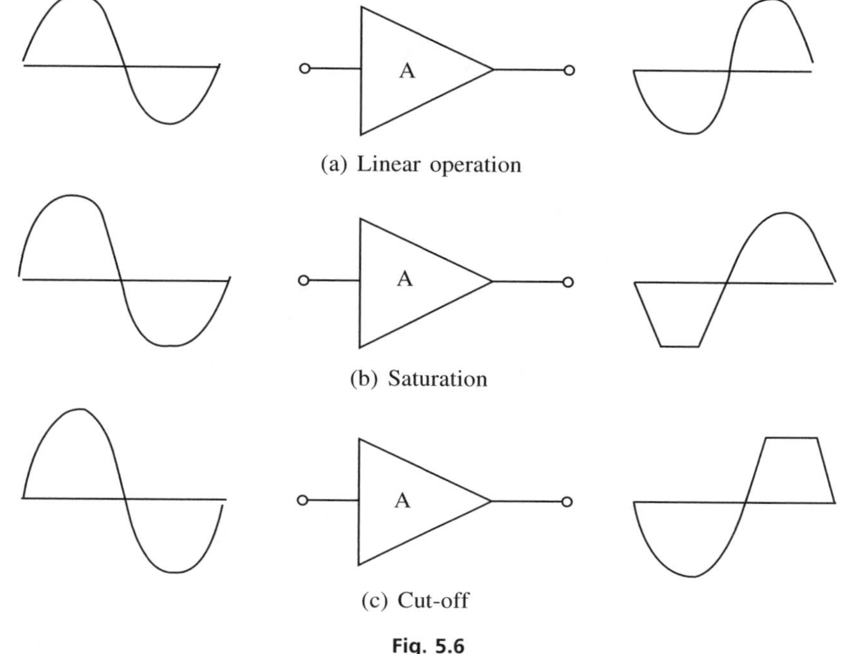

(a) Linear operation

(b) Saturation

(c) Cut-off

Fig. 5.6

$$V_{CC} = I_B R_B + V_{BE}$$

∴
$$I_B = \frac{V_{CC} - V_{BE}}{R_B} \tag{5.10}$$

Also

$$V_{CC} = V_{CE} + I_C R_C \tag{5.11}$$

From (5.11), if $I_C = 0$ then $V_{CE} = V_{CC}$. Also if $V_{CE} = 0$, then

$$I_C = \frac{V_{CC}}{R_C} \tag{5.12}$$

It is now possible to use the above equations to mark the two extremities of the load line on the output characteristics. The d.c. load line can be used to determine the values of current and voltage in the output circuit. An example is shown in Fig. 5.9, where the collector resistor is taken as $R_C = 1.6$ kΩ and the supply voltage $V_{CC} = 10$ V.

The d.c. load line is important in that the Q-point (quiescent point) can be set. A transistor must be d.c.-biased in order to operate as an amplifier and the d.c. operating point, the Q-point, must be set so that signal variations at the input are amplified and accurately reproduced at the output. As the term 'quiescent' implies, the d.c. operating point is determined with no a.c. signal applied. Generally the Q-point is set near the centre of the load line or characteristics, as shown in Fig. 5.9. If, however, the Q-point is moved along the load line between the saturation and cut-off regions then the output will be distorted, as shown in Fig. 5.7. Two extreme conditions, Q_2 and Q_3, are shown in Fig. 5.9. The region along the load line including all points between saturation and cut-off is known as the linear region, and as long as the transistor is operated in this region, the output voltage will be a faithful reproduction of the input voltage.

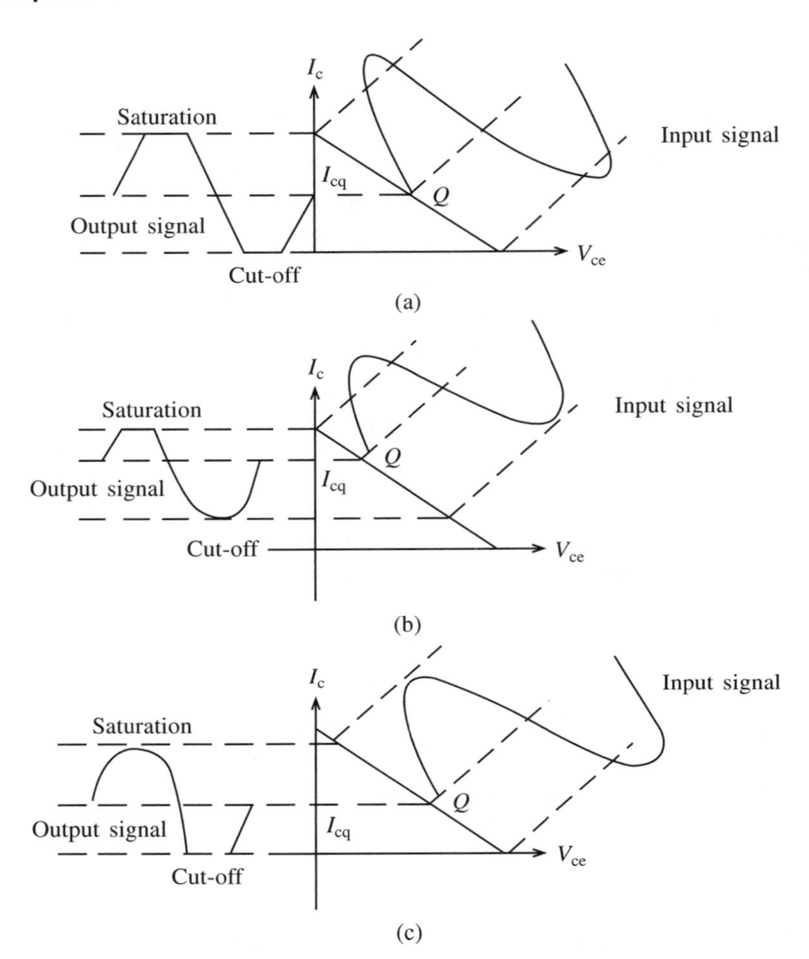
I_c
Saturation
Input signal
I_{cq}
Q
Output signal
V_{ce}
Cut-off
(a)
I_c
Saturation
Input signal
Q
Output signal
I_{cq}
Cut-off
V_{ce}
(b)
I_c
Input signal
Saturation
Q
Output signal
I_{cq}
V_{ce}
Cut-off
(c)

Fig. 5.7

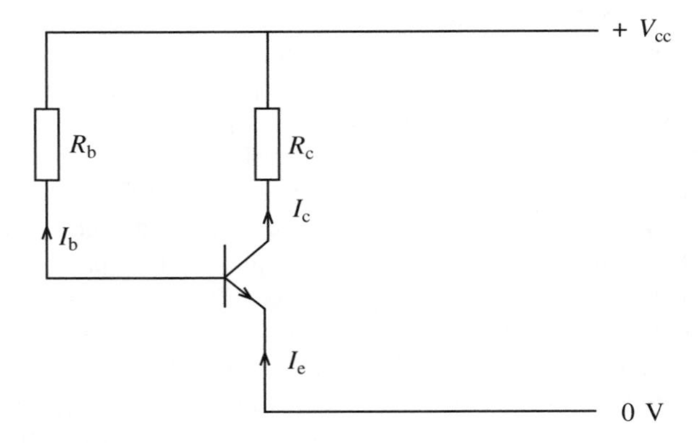
$+ V_{cc}$
R_b
R_c
I_c
I_b
I_e
0 V

Fig. 5.8

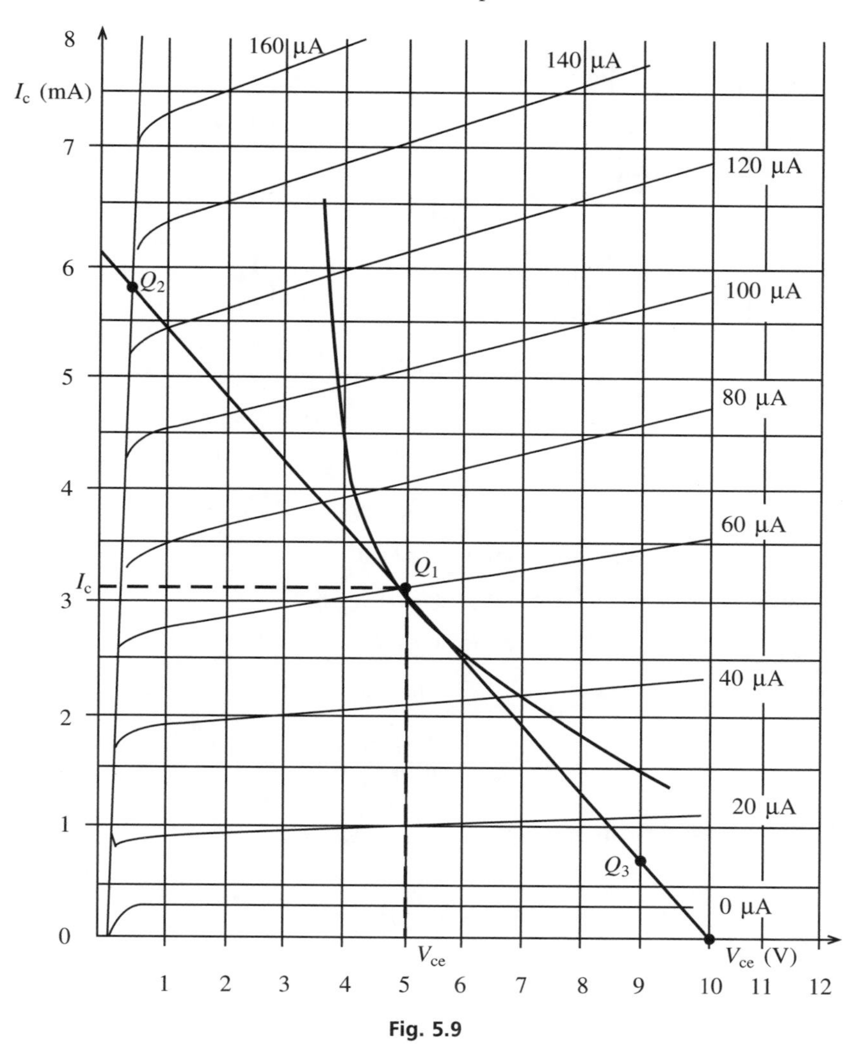

Fig. 5.9

Note that when I_B increases, I_C increases and V_{CE} decreases. Also when I_B decreases, I_C decreases and V_{CE} increases. Hence as the base bias voltage is adjusted the d.c. operating point moves along the load line and the parameters can be selected accordingly.

Voltage divider bias

Figure 5.8 uses what is called base bias, but in practice voltage divider bias is used as it provides better Q-point stability and can be used with a single supply. A typical circuit configuration is shown in Fig. 5.10. An analysis of this circuit must include the input resistance at the base which can be shown to be

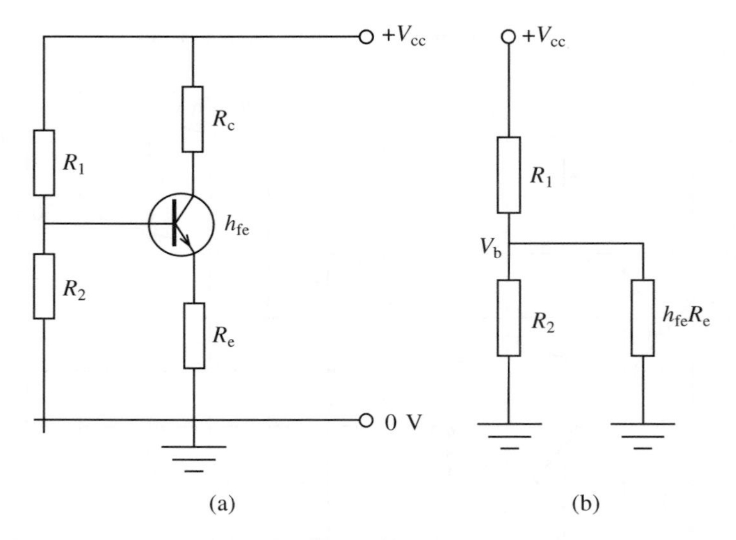

Fig. 5.10

$$R_{in(base)} \cong h_{FE}R_E$$

The total resistance between base and ground is

$$R_2 \parallel h_{FE}R_E$$

A voltage divider is formed now as shown in Fig. 5.10(b), and this gives

$$V_B = \left(\frac{R_2 \parallel h_{FE}R_E}{R_1 + (R_2 \parallel h_{FE}R_E)}\right) V_{CC} \tag{5.13}$$

If $h_{FE}R_E >> R_2$ then this simplifies to

$$V_B = \left(\frac{R_2}{R_1 + R_2}\right) V_{CC} \tag{5.14}$$

Also

$$V_E = V_B - V_{BE}$$

and

$$I_E = \frac{V_E}{R_E}$$

and since

$$I_C \cong I_E$$

then

$$I_C = \frac{V_B - V_{BE}}{R_E} \tag{5.15}$$

$$V_C = V_{CC} - I_CR_C$$

$$V_{CE} = V_C - V_E$$

Also, using Kirchhoff's voltage law,

$$V_{CC} - I_C R_C - I_E R_E - V_{CE} = 0$$

and since $I_C \cong I_E$,

$$V_{CE} \cong V_{CC} - I_C R_C - I_C R_E$$

$$\cong V_{CC} - I_C(R_C + R_E) \quad (5.16)$$

Note that when drawing a d.c. load line the saturation current will no longer be given by

$$I_C = \frac{V_{CC}}{R_C}$$

but instead becomes

$$I_C = \frac{V_{CC}}{R_E + R_C} \quad (5.17)$$

The cut-off point will remain the same when drawing the d.c. load line. This will be shown in the next section, which deals with small signal amplifiers.

5.4 Small signal voltage amplifiers

The biasing of a transistor is purely a d.c. operation, as has been shown in the previous sections. Also it has been seen that the purpose of biasing is to obtain a Q-point about which an input a.c. signal can cause variations in current and voltage. In practice, when dealing with such applications where small signal voltages have to be amplified, the variations about the Q-point are small. Amplifiers used in such applications are called small signal voltage amplifiers.

A typical small signal voltage amplifier is shown in Fig. 5.11. In this circuit the main difference is the inclusion of the capacitors C_1 and C_2, which are included to couple the a.c. signal to the amplifier while blocking the d.c. bias on the transistor from the input and output stages. Hence the capacitor effectively blocks d.c. and appears as a short circuit to a.c. signals, provided the correct value is selected.

Note also that the emitter resistor R_E is included in order to stabilize the bias conditions due to any temperature increases which would cause the collector current to increase. This is called **thermal runaway.** However, the inclusion of this resistor also lowers the a.c. gain of the amplifier. Hence there is a problem in that the emitter resistor is required for stability, but it reduces the a.c. gain. One solution to this is to connect a capacitor (the emitter bypass capacitor) in parallel with R_E as shown. This presents an open circuit to d.c., thus maintaining the effect of R_E for bias purposes, but a short circuit to a.c., thus shorting out R_E for a.c. purposes. This mode of operation is sometimes referred to as a single ended input. When operated in this mode one input is grounded and the signal voltage is applied to the other input. For the common emitter amplifier this causes an inverted signal at the output. Figure 5.12 may help to clarify the operation of this amplifier.

Figure 5.12 shows both the a.c. and the d.c. load lines, together with the expressions necessary to determine the cut-off and saturation points. These have been derived in

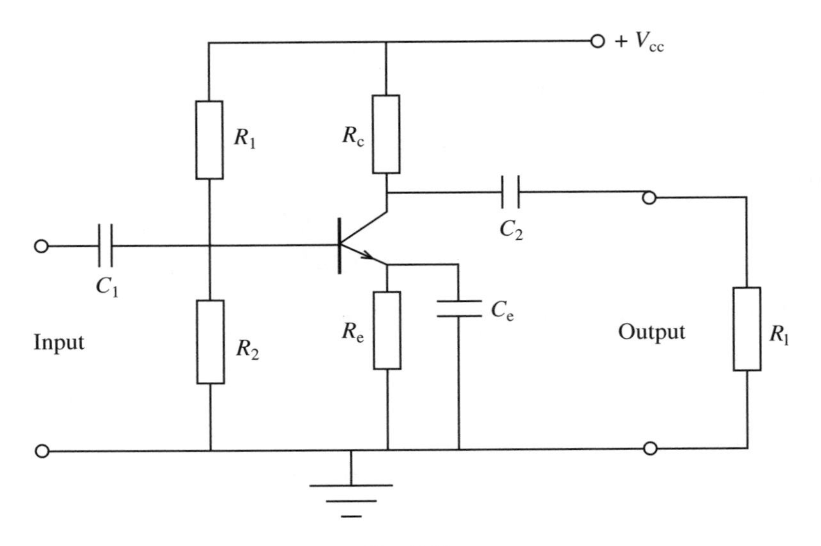

Fig. 5.11

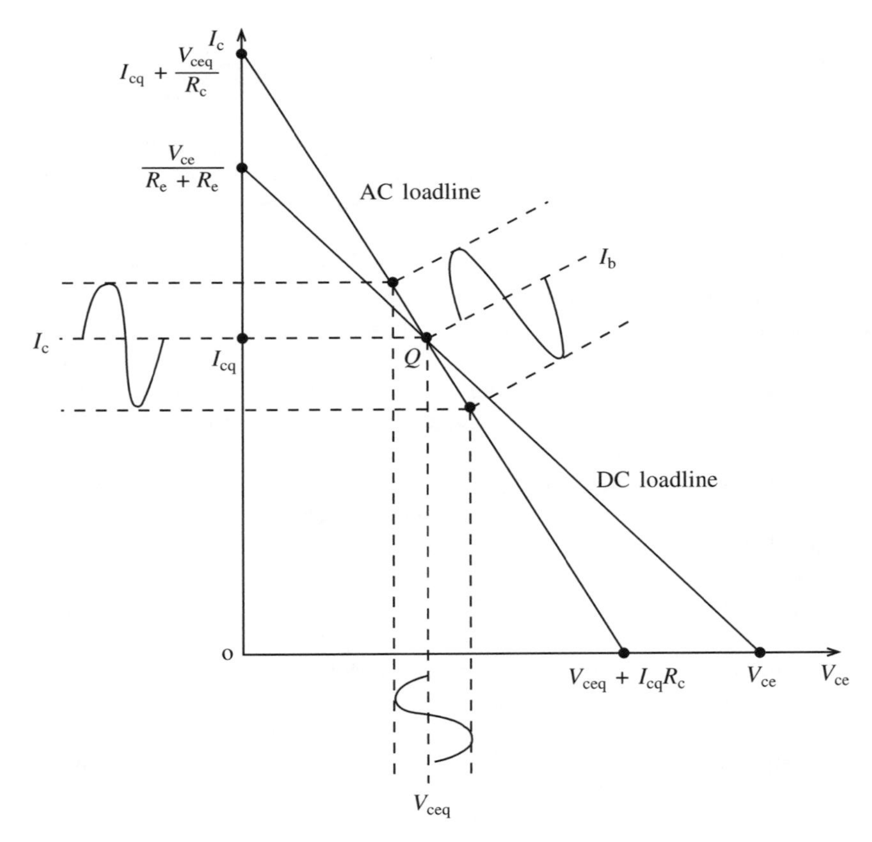

Fig. 5.12

previous sections. Also note that the collector current lies within the linear region of the amplifier; as has already been mentioned, this is necessary for an undistorted output. This is known as class A bias and indicates that the amplifier operates for the full 360° of the input voltage. However, there are two other common types of bias, and these are shown in Fig. 5.13.

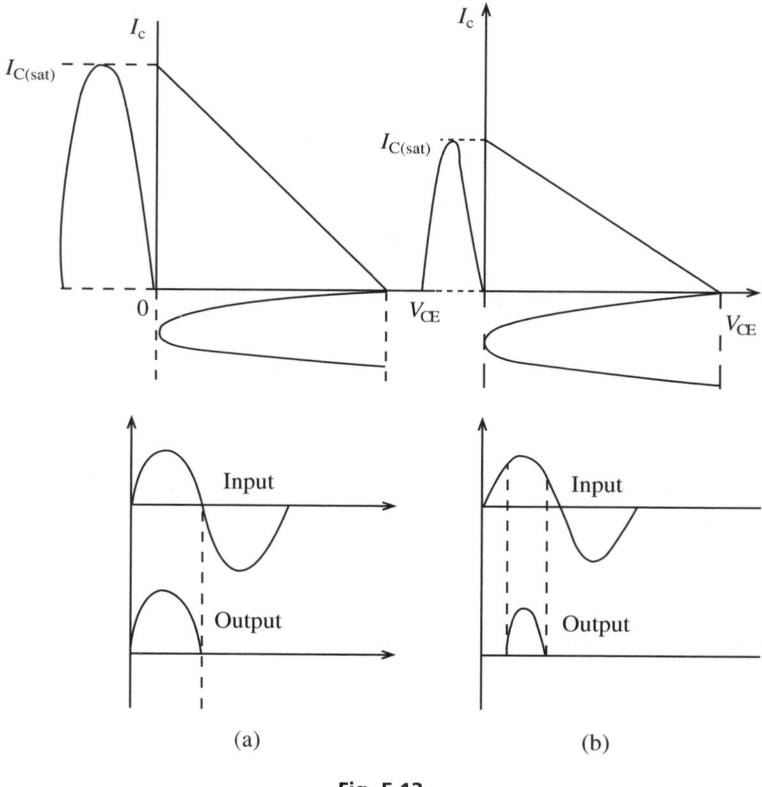

Fig. 5.13

Figure 5.13(a) shows what happens to the output signal when an amplifier is biased at cut-off. It operates in the linear region for 180° of the input cycle and is in cut-off for the other 180°. This is known as class B bias and is used regularly in power amplifiers. A variation of this is the class AB biased power amplifier which is biased to conduct for slightly more than 180°.

Figure 5.13(b) shows the result of biasing so that conduction occurs for less than 180°. Because the output is severely distorted class C amplifiers are limited to such applications as tuned amplifiers and oscillators.

5.5 The use of the decibel

Associated with power amplifiers and voltage amplifiers is the frequency response of the amplifier. This deals with gain, roll-off, bandwidth and other parameters. However, it is customary to use the decibel to determine the gain or attenuation of an amplifier rather than a magnitude gain such as 100 or 500.

The decibel is related to the logarithmic response of the ear to the intensity of sound and is defined as the ratio of one power to another or one voltage to another. Two expressions are generally used:

$$A_{p(dB)} = 10 \log A_p \tag{5.18}$$

where A_p is the power gain P_{out}/P_{in}, and

$$A_{v(dB)} = 20 \log A_v \tag{5.19}$$

where A_v is the voltage gain V_{out}/V_{in}. If the voltage gain A_v is greater than unity the dB gain is positive, and if it is less than unity the dB gain is negative. This is usually called attenuation.

Most amplifiers show a maximum gain over a range of frequencies and a reduced gain outside this range. The maximum range is called the midband range, and the frequency where the gain falls off is called the critical frequency. As described in Chapter 3, in terms of decibels this critical frequency occurs at the 3 dB point and in terms of power we say the power is halved at this point while the output voltage is taken as 70.7% of its midband range.

5.6 Types of power amplifier

Now that the fundamentals of the small signal amplifier have been discussed, we will look at the power amplifier. In many respects this is like the voltage amplifier, except for the large signal swings and larger power dissipation which are not present in the small signal voltage amplifier. There are three popular types of power amplifier frequently used in power applications: the class A (single-ended) amplifier; the class B push-pull (transformer) amplifier; and the class B complementary pair push-pull amplifier. Each of these types will be considered in turn.

Class A (single-ended) amplifier

A single-ended power amplifier uses a single transistor with the usual biasing arrangement as for a voltage amplifier, but there the similarity ends. A typical circuit is shown in Fig. 5.14. Unlike the small signal voltage amplifier this circuit has to drive a current operated load, and this requires large collector current swings to operate the matching transformer and loudspeaker. A voltage amplifier would use a high resistive load but here the load is the low resistance primary of a transformer, the transformer being used to match the high collector resistance of the transistor.

When dealing with power amplifiers it is necessary to know the operating conditions as well as the power rating of the active device. Figure 5.15 shows a set of output characteristics together with the power dissipation curve and a.c. and d.c. load lines. The power curve shows the operational region inside which the device will operate within its rated power value. This area is marked *ABCD*. Often a heat sink has to be attached to the device to achieve the rated value.

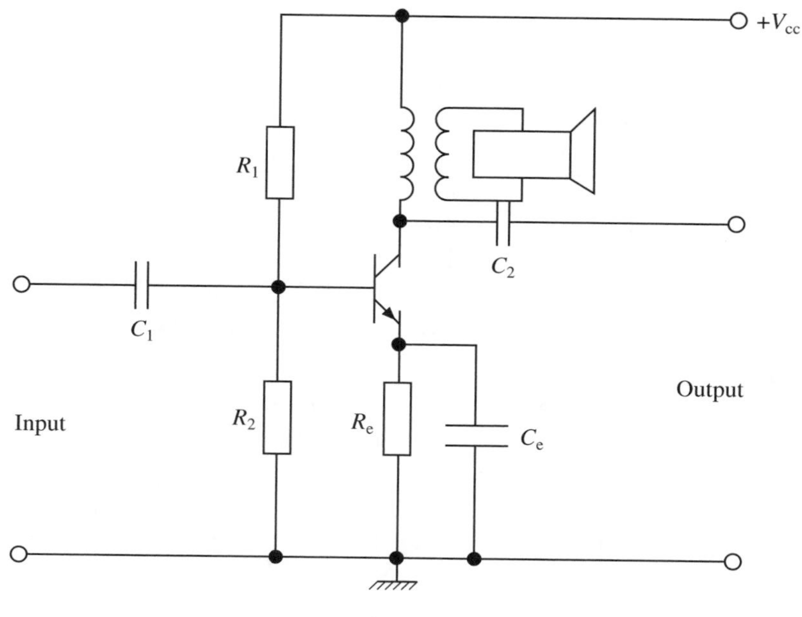

Fig. 5.14

The d.c. load line is determined by knowing the values of the collector resistance (R_C) and emitter resistance (R_E). The slope of the d.c. load line is then given as

$$\text{Slope} = \frac{1}{R_C + R_E} \tag{5.20}$$

The value of R_E is generally between 500 Ω and 1000 Ω. The d.c. load line is drawn for the d.c. operating conditions of the device without the load attached. The a.c. load line takes the load (R_L) into consideration so that

$$R_T = \frac{R_C R_L}{R_C + R_L} \tag{2.21}$$

In summary, the d.c. load line is used to set up the quiescent conditions without an input signal, while the a.c. load line sets up the amplifier for input signal conditions.

While graphical solutions are seldom used in problem-solving, it is instructive to consider the meaning of characteristics under d.c. and a.c. operating conditions. The d.c. power in this case is the product of the supply voltage and the d.c. component of the collector current, i.e.

$$P_{DC} = V_{CC} I_C \tag{5.22}$$

Because of losses due to transformer resistance, emitter resistance and collector dissipation, the output a.c. power delivered to the load is less than the d.c. power. Also since the main power loss is in the collector,

$$P_{DC} = P_{AC} + P_C \tag{5.23}$$

The following points should be observed from Fig. 5.15:

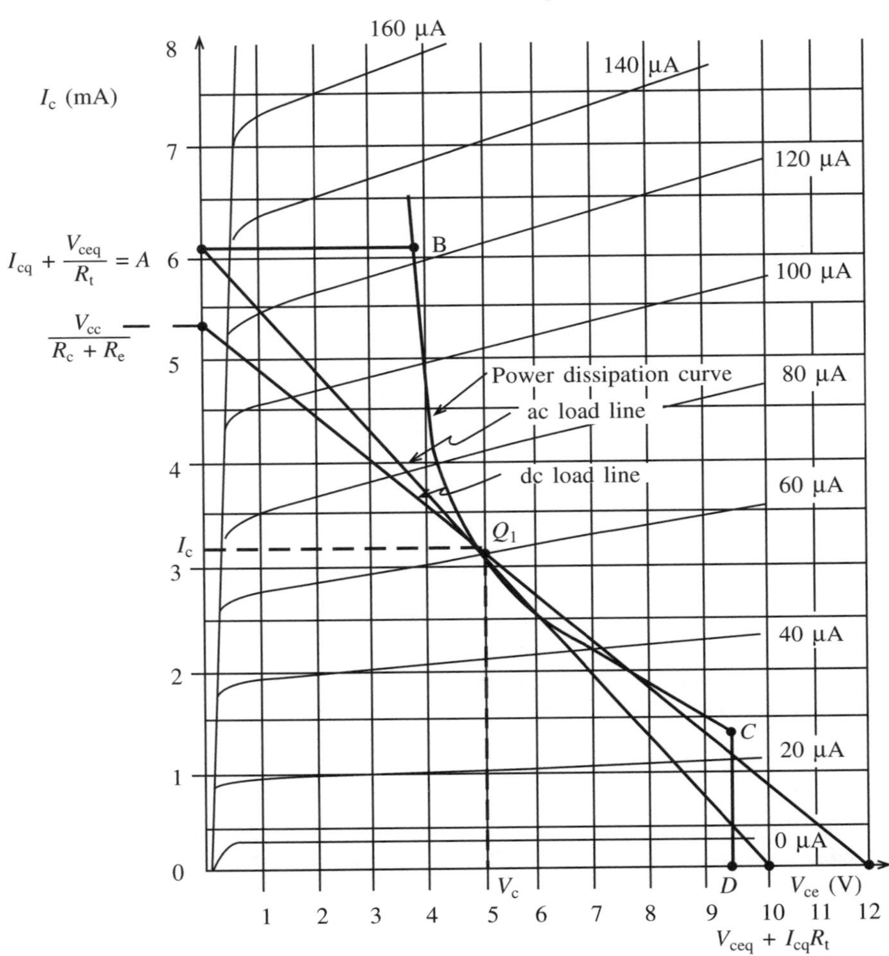

Fig. 5.15

(a) The a.c. load line sets the maximum limits of the current and voltage swings at the output.

(b) The quiescent point sets the d.c. operating conditions for the device and should be chosen to give an undistorted output. If this is not chosen properly, flattening of the waveform could occur.

(c) The intersection of the d.c. and a.c. load lines determines the Q-point, and this is normally achieved by the bias resistors.

(d) When no signal is applied, the quiescent voltage equals the supply voltage.

(e) When no signal is applied, the quiescent current is half the maximum collector current.

(f) The power dissipation curve sets the limits to how much power may be dissipated in the collector of the transistor.

(g) The corresponding base current can be determined for each collector current.

(h) The turns ratio of the transformer (n) transforms the actual load of loudspeakers (R_2) into the optimum load (R_1) of the transistor, i.e.

$$n = \sqrt{\frac{R_1}{R_2}}$$

(i) Maximum collector dissipation occurs when no signal is applied.

Figure 5.16 shows the effect of an input signal. The resultant swings of the collector current and voltage are given as peak-to-peak.

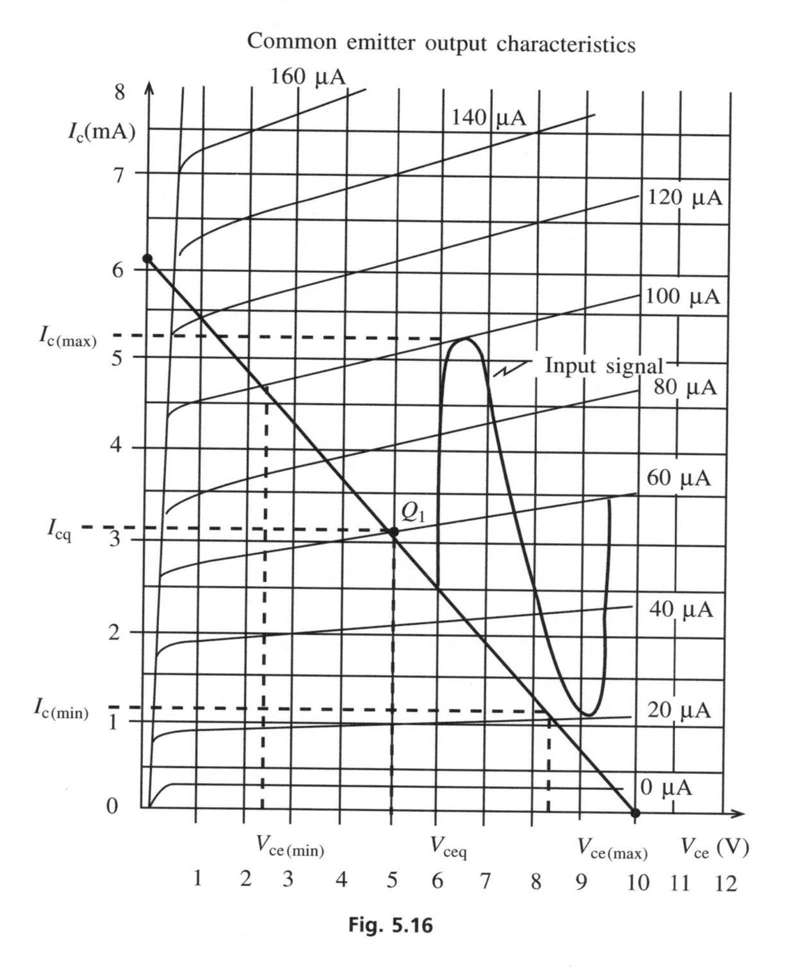

Fig. 5.16

In the above characteristics the current and voltage swing between cut-off and saturation. The efficiency of the stage will be 50% under these conditions but the distortion will be severe so in practice the actual swings are reduced. This results in an efficiency of about 25%. It is for this reason that this type of circuit is not often used.

Practical analysis of class A single-ended parameters

The application of an a.c. signal to a power amplifier causes large current (I_c) and voltage (V_{ce}) swings. The collector current swings between $I_{c(max)}$ and $I_{c(min)}$, while the collector-emitter voltage swings between $V_{ce(max)}$ and $V_{ce(min)}$. The average current and the average voltage are given by

$$\frac{V_{ce\,(max)} - V_{ce\,(min)}}{2} \text{ and } \frac{I_{c(max)} - I_{c(min)}}{2} \tag{5.24}$$

The output power in the load is generally given in r.m.s. values, and if the matching transformer is assumed to be 100% efficient then the output power ($P_{ac(rms)}$) is given by

$$P_{ac(rms)} = \frac{I_{c(max)} - I_{c(min)}}{2\sqrt{2}} \times \frac{V_{c(max)} - V_{c(min)}}{2\sqrt{2}}$$

$$= \frac{(I_{c(max)} - I_{c(min)})\,(V_{c(max)} - V_{c(min)})}{8} \tag{5.25}$$

An alternative expression given in terms of the primary resistance (R_1), the secondary load resistance (R_2) and the turns ratio is $P_{ac(rms)} = I^2R_1$; hence

$$P_{ac(rms)} = \left[\frac{(I_{c(max)} - I_{c(min)})}{8}\right] n^2 R_2 \tag{5.26}$$

The collector efficiency is given by

$$\eta = \frac{P_{dc}}{P_{ac(rms)}} \times 100 \tag{5.27}$$

Class B push-pull (transformer) amplifier

Only class A amplification gives a distortionless output. By using push-pull circuits, however, class B amplification can be obtained with no output distortion. Furthermore, higher efficiencies are achieved. An added bonus is the elimination of even harmonic distortion which drives the collector current beyond its maximum permitted swing. The result of this is that an output power almost double that of a single transistor is produced.

A typical circuit is shown in Fig. 5.17. The driver circuit splits the phase of the input signal into two signals which are in antiphase to one another. This is done by taking the driver transformer secondary winding centre tap to the earth or common line. When no signal is present the only current flowing in the transistors is the quiescent current. This quiescent current does not produce any d.c. magnetization of the output transformer core because the magnetic forces effectively cancel. The output transformer can therefore be made smaller.

When a signal is applied the antiphase base currents will produce antiphase collector currents. These collector currents effectively add in the output transformer and the a.c. power is fed via the secondary to R_L. The signal currents do not pass through the power

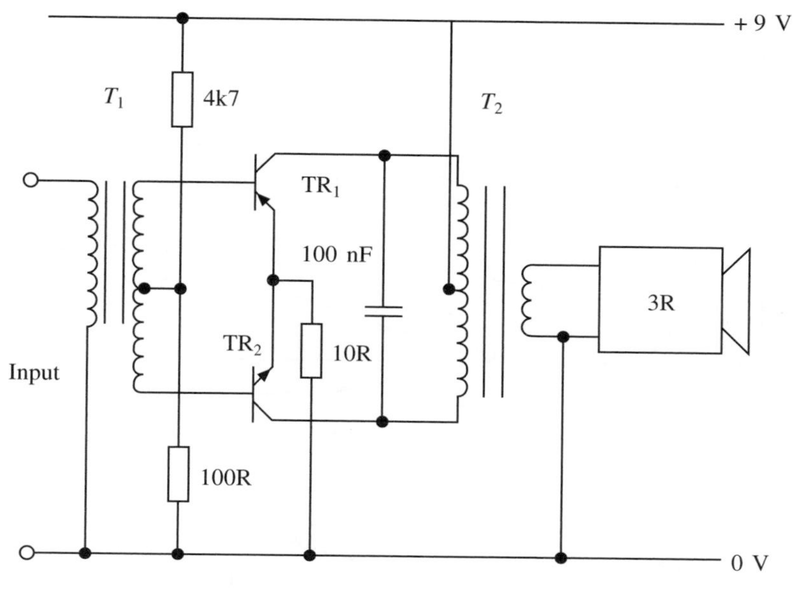

Fig. 5.17

supply or bias components. If the supply is derived from a.c. mains there is no tendency for mains hum to be introduced in this stage.

The operation of this type of amplifier is more clearly seen in Fig. 5.18, which shows the two sets of output characteristics for both transistors combined to form the composite characteristic. The point V_{cc} represents the operating point of both transistors and hence the two a.c. load lines coincide for transistors T_1 and T_2. At point 1 on the input signal for T_1 the collector current is maximum and this is represented by D on the common characteristics. At point 2 the transistor is at cut-off point A and no collector current flows in either transistor. At point 3 the collector current of T_2 is maximum and the collector-emitter voltage is zero. This is point E. If all the variations of collector current and collector-emitter voltage were plotted the composite curve shown in Fig 5.18 would be produced.

Crossover distortion

Push-pull power amplifiers are subject to distortion as shown in Fig. 5.19(a). The mutual characteristics are shown in Fig. 5.19(b), in which the non-linear input characteristics cause distortion as one transistor turns on and another turns off. This effect may be reduced by introducing class AB biasing. Such biasing causes both the output transistors to conduct a small collector current when no signal is fed to the amplifier. Normally 0.6 V is provided by means of two resistors or two diodes. However, diodes are generally preferable because of temperature changes. This means that the potential difference across them will vary in the same way as the base–emitter voltage of the output transistors. One particular configuration is shown in Fig. 5.20, which shows the application of two diodes in the biasing circuit, preventing crossover distortion.

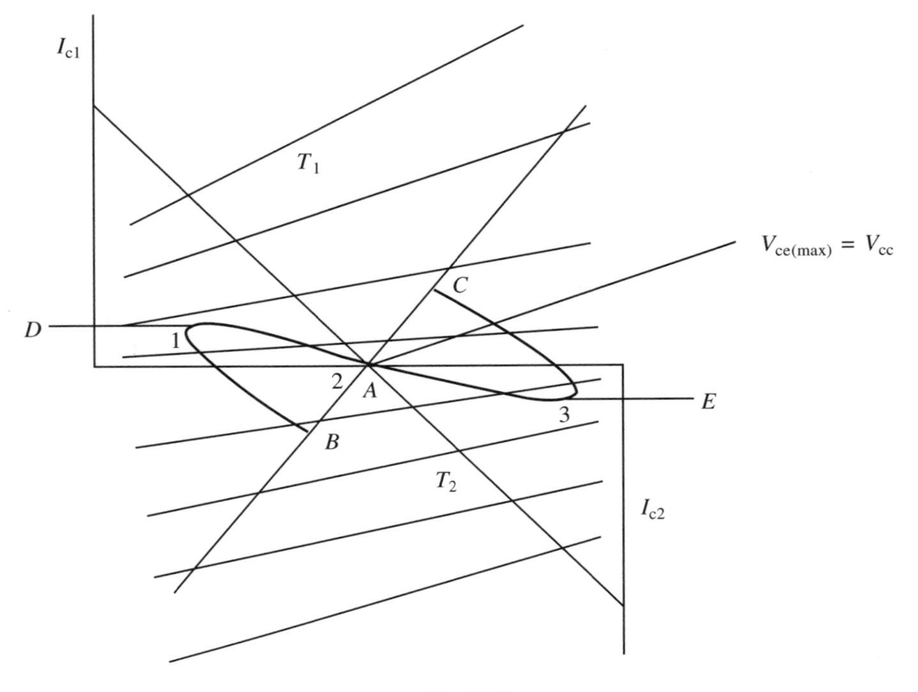

Fig. 5.18

Class B complementary pair push-pull

This type of power amplifier performs in exactly the same way as the transformer type except that two different transistors are now used, i.e. npn and pnp, which complement each other to give a faithful reproduction of the input voltage. A typical practical circuit is shown in Fig. 5.20. Once again each transistor only handles 180° of the input signal, and when there is no input signal both the transistors are switched off. Hence there is no current drawn by the system at the output which once again increases the efficiency of the system. As before, if the input swings positive T_1 will switch on and current will flow through R_l. When the input swings negative T_1 is off while T_2 is switched on, thus causing a current once again through R_l. T_1 thus pushes the current through R_l, while T_2 pulls the current through R_l.

Several practical points should be noted about this circuit:

(a) Figure. 5.18 showed that the supply voltage indicated at the crossover point of the characteristic was V_{cc}. This is not the case with the complementary circuit, where the supply voltage is equal to half the maximum collector-emitter voltage of the transistors used. Otherwise the characteristics are identical for this configuration. In Figure 5.20 the two 10 Ω resistors set the bias conditions so that the junction is at half the supply voltage ($V_{cc}/2$).

(b) Diodes are used rather than bias resistors because they are manufactured from silicon and hence have similar temperature characteristics to the transistors. Also if

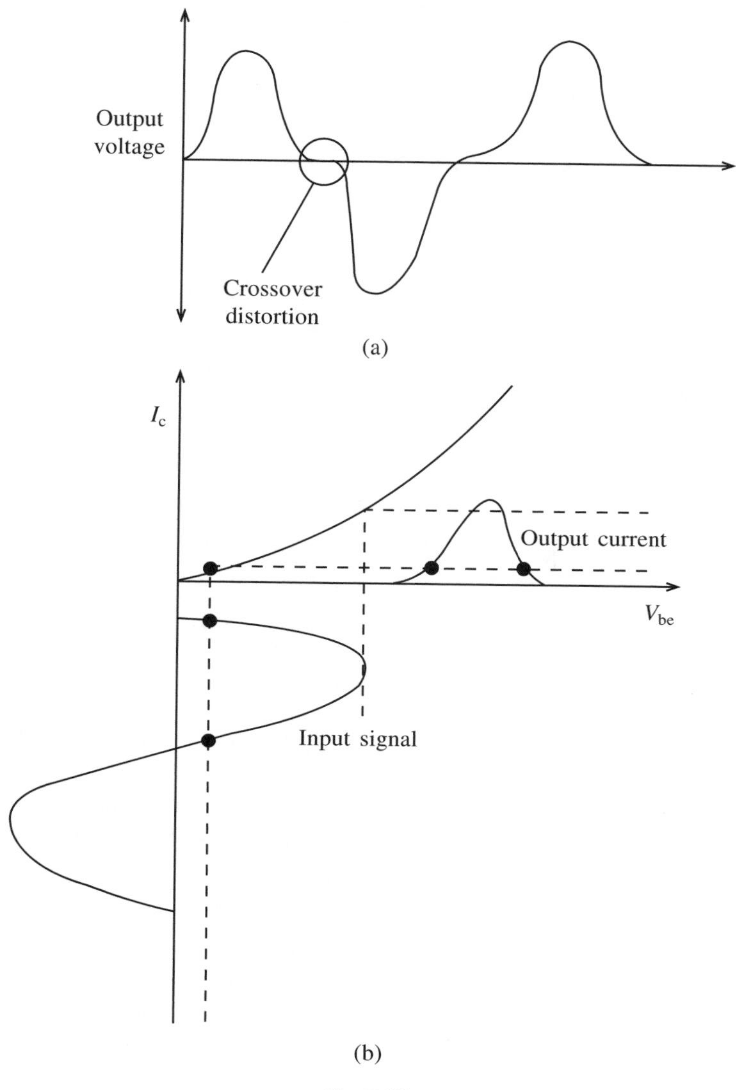

Fig. 5.19

they are mounted on the same heat sink they will undergo similar temperature and voltage changes.

(c) Complementary pair transistors are available in matched pairs in order to ensure they have identical characteristics.

(d) The use of the diodes provides class AB biasing.

Practical analysis of class B push-pull parameters

Both the push-pull transformer transforms and complementary pair types follow the same

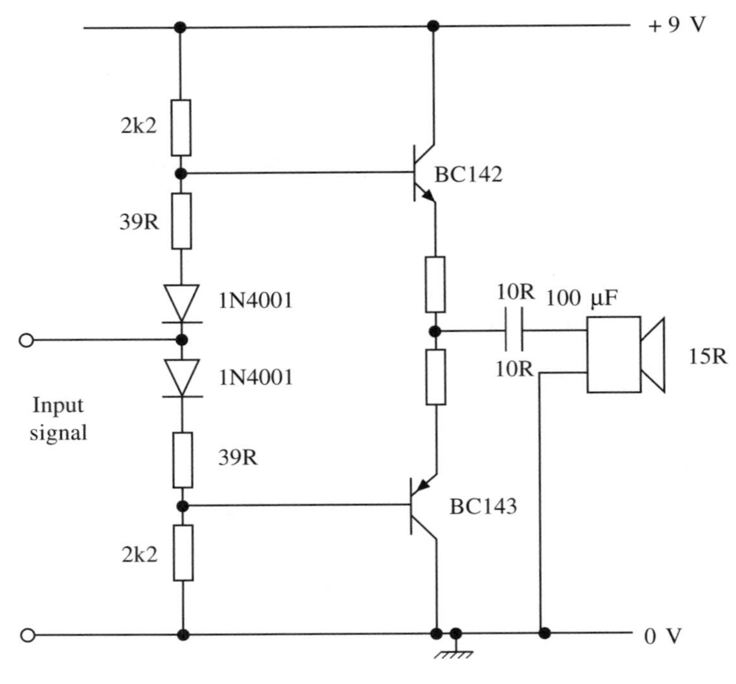

Fig. 5.20

analysis, except that the supply voltage is V_{cc} in the transformer type and $V_{cc}/2$ for the complementary type. The parameters will be derived using the complementary symmetry circuit similar to the type used in Fig. 5.20. In this circuit a single power supply is used and the capacitor C (100 μF in this case) may be calculated by using the expression

$$f = \frac{1}{2\pi C R_l} \tag{5.28}$$

where R_l is the load resistance and f is the lower frequency of the frequency response. Note that if a dual power supply is used the coupling capacitor is not necessary.

The collector current in each transistor takes the form of a series of rectified pulses. A Fourier analysis of this waveform will give an instantaneous value of

$$i_c = \frac{I_{c(max)}}{\pi} + \frac{I_{c(max)}}{2\pi} \sin \omega t - \frac{2 I_{c(max)}}{3\pi} \cos 2\omega t \tag{5.29}$$

The first term in this expansion represents a d.c. component and the second term is the fundamental frequency component. Harmonics can be ignored as these are mostly reduced to a small percentage or eliminated entirely.

The fundamental r.m.s. a.c. power produced by each transistor to a load is

$$P_{ac(rms)} = \left(\frac{I_{c(max)}}{2\sqrt{2}} \right)^2 R_l \tag{5.30}$$

Note that R_l is the load into which the transistor feeds. Therefore total power delivered by both transistors is

$$P_{ac(rms)} = \frac{I^2_{c(max)} R_l}{4} \tag{5.31}$$

Also the d.c. power taken by each transistor is

$$\frac{I_{c(max)} V_{cc}}{2} \tag{5.32}$$

Hence total power is

$$P_{dc} = \frac{I_{c(max)} V_{cc}}{\pi} \tag{5.33}$$

So the efficiency is given by

$$\eta = \frac{P_{ac}}{P_{dc}} \times 100 = \frac{I^2_{c(max)} R_l}{4 I_{c(max)} V_{cc}} = \frac{I_{c(max)} R_l}{4 V_{cc}} \tag{5.34}$$

The same solution can be achieved by using

$$P_{ac(rms)} = \left[\frac{I_{c(max)} - I_{c(min)}}{2\sqrt{2}} \frac{(V_{cc}/2 - V_{ce(min)})}{\sqrt{2}} \right] \times 2$$

As $I_{c(min)}$ and $I_{ce(min)}$ are small then

$$P_{ac(rms)} = \frac{I_{c(max)} \times V_{cc} \times 2}{8} = \frac{I_{c(max)} \times V_{cc}}{4} \tag{5.35}$$

One final consideration when selecting an integrated circuit or transistor for an application is its power rating. The total collector dissipation is given as

$$P_c = P_{dc} - P_{ac}$$

$$= \frac{I_{c(max)} V_{cc}}{\pi} - \frac{I_{c(max)} V_{ce(max)}}{2}$$

Since

$$I_{c(max)} = \frac{V_{ce(max)}}{R_L}$$

$$P_d = \frac{V_{cc} V_{ce(max)}}{\pi R_l} - \frac{V^2_{ce(max)}}{2 R_l} \tag{5.36}$$

Differentiate (5.36) to determine the maximum dissipation:

$$\frac{dP_d}{dV_{ce(max)}} = \frac{V_{cc}}{\pi R_L} - \frac{2 V_{ce(max)}}{2 R_l}$$

Therefore

$$V_{ce(max)} = \frac{V_{cc}}{\pi} \tag{5.37}$$

Substituting (5.37) into (5.36) gives

$$P_{d(max)} = \frac{V_{cc}^2}{\pi^2 R_l} - \frac{V_{cc}^2}{2\pi^2 R_l} = \frac{V_{cc}^2}{2\pi^2 R_l}$$

Since the maximum power occurs when $V_{ce(max)} = V_{cc}/2$ for complementary symmetry, and this occurs at $P_{ac(max)} = V_{cc}^2/8R_l$, then

$$P_{c(max)} = \frac{4P_{ac(max)}}{\pi^2} \simeq 0.4P_{ac(max)} \tag{5.38}$$

This is the total power dissipation for both transistors used in a push-pull arrangement.

Example 5.1
The supply voltage of a complementary amplifier is 30 V and the output feeds into a load of 12 Ω. Determine:

(a) the maximum collector current if an efficiency of 60% is required;

(b) the maximum rated values of the transistors used.

Solution
(a)

$$\eta = \frac{I_{c(max)} R_l}{4V_{cc}}$$

$$\therefore \quad I_{c(max)} = \frac{4V_{cc}\,N}{R_l} = \frac{4 \times 30 \times 0.6}{12\pi} = 1.91 \text{ A}$$

(b)

$$P_{c(max)} = \frac{4P_{ac(max)}}{\pi^2}$$

$$P_{ac(max)} = \frac{V_{cc}^2}{8R_l} = \frac{900}{96}$$

$$\therefore \quad P_{c(max)} = \frac{4 \times 900}{96\pi^2} = 3.80 \text{ W}$$

Hence each transistor rated at 1.90 W.

Example 5.2
A class B push-pull transformer power amplifier has an output configuration as shown in Fig. 5.21. The output transformer is centre tapped such that the turns ratio is $N_1 : N_2 : N_3$ = 5 : 5 : 2. Determine:

(a) the maximum collector current produced in the primary by both transistors if the a.c. power delivered to the primary is 12.5 W;

(b) the value of supply voltage required under these conditions;

(c) the efficiency of the amplifier.

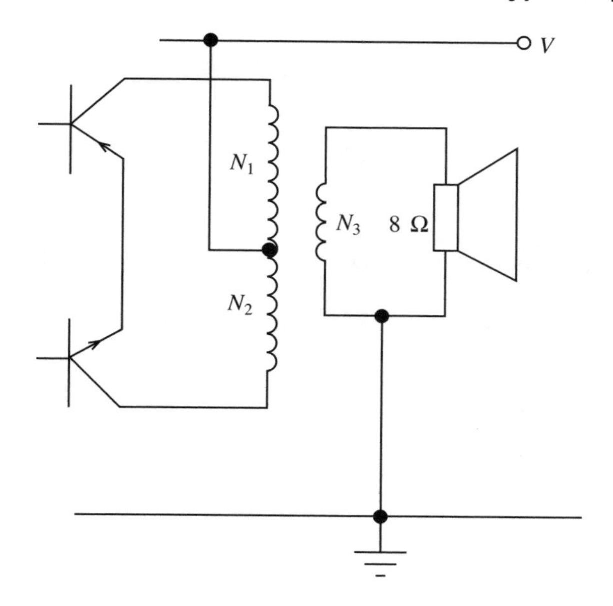

Fig. 5.21

Solution

(a) Since the turns ratio is 5 : 5 : 2, the load into which the transistors feed in the secondary is

$$\frac{N_1}{N_3} = \frac{N_2}{N_3} = \frac{5}{2} = \sqrt{\frac{R_1}{8}}$$

$$\therefore \qquad R_1 = \frac{25}{4} \times 5 = 50$$

Since, for transformer push-pull,

$$P_{\text{ac(rms)}} = \frac{I^2_{\text{c(max)}} R_1}{4}$$

$$I_c = \sqrt{\frac{4 \times 12.5}{50}} = 1 \text{ A}$$

(b)

$$P_{\text{ac(rms)}} = \frac{I_{\text{(max)}} V_{cc}}{2}$$

$$\therefore \qquad V_{cc} = \frac{2P_{\text{ac(rms)}}}{I_{\text{(max)}}} = \frac{2 \times 12.5}{1} = 25 \text{ V}$$

(c)

$$\eta = \frac{P_{\text{ac(rms)}}}{P_{dc}} \times 100$$

$$P_{dc} = \frac{2I_{c(max)}\,V_{cc}}{\pi} = \frac{2 \times 1 \times 25}{\pi} = 15.9\ \text{W}$$

$$\therefore \qquad \eta = 12.5 \times \frac{\pi}{50} \times 100 = 78.5\%$$

Example 5.3

A complementary push-pull amplifier has to deliver 2.5 W into a load of 8 Ω. Determine:

(a) the maximum power dissipation required for each transistor;

(b) the supply voltage required;

(c) the peak load current.

Solution

(a)

$$\frac{4P_{ac(max)}}{\pi^2} = \frac{4 \times 2.5}{\pi^2} = 1.01\ \text{W}$$

(b) Since

$$P_{c(max)} = \frac{V_{cc}^2}{2\pi^2\,R_l}$$

$$V_{cc} = \sqrt{P_{c(max)}\,2\pi^2 R_l} = \sqrt{\frac{10}{\pi^2} \times 2\pi^2 \times 8} = \sqrt{160} = 12.65\ \text{V}$$

(c)

$$P_{ac(rms)} = \frac{V_l^2}{R_l}$$

$$\therefore \qquad V_l = \sqrt{P_{ac(rms)}\,R_l} = \sqrt{2.5 \times 8} = 4.5\ \text{V}$$

$$\therefore \qquad V_{(max)} = \sqrt{2} \times \sqrt{20} = \sqrt{40} = 6.32\ \text{V}$$

∴ Peak load current is

$$I_{(max)} = \frac{V_{(max)}}{R_l} = \frac{6.32}{8} = 0.79\ \text{A}$$

Example 5.4

The complementary push-pull audio amplifier shown in Fig. 5.22 has to provide 5 W into a load of 4 Ω. The 3 dB bandwidth of the amplifier has to be flat between 80 Hz and 18 kHz. The data for the two transistors are as follows: $I_{C(max)} = 4$ A, $P_{(max)} = 36$ W, $h_{FE} = 40$ and $V_{CE(max)} = 45$ V. Determine:

(a) the value of C_1;

(b) the value of V_{CE};

(c) the value of base current;

(d) the values of R_5 and R_6

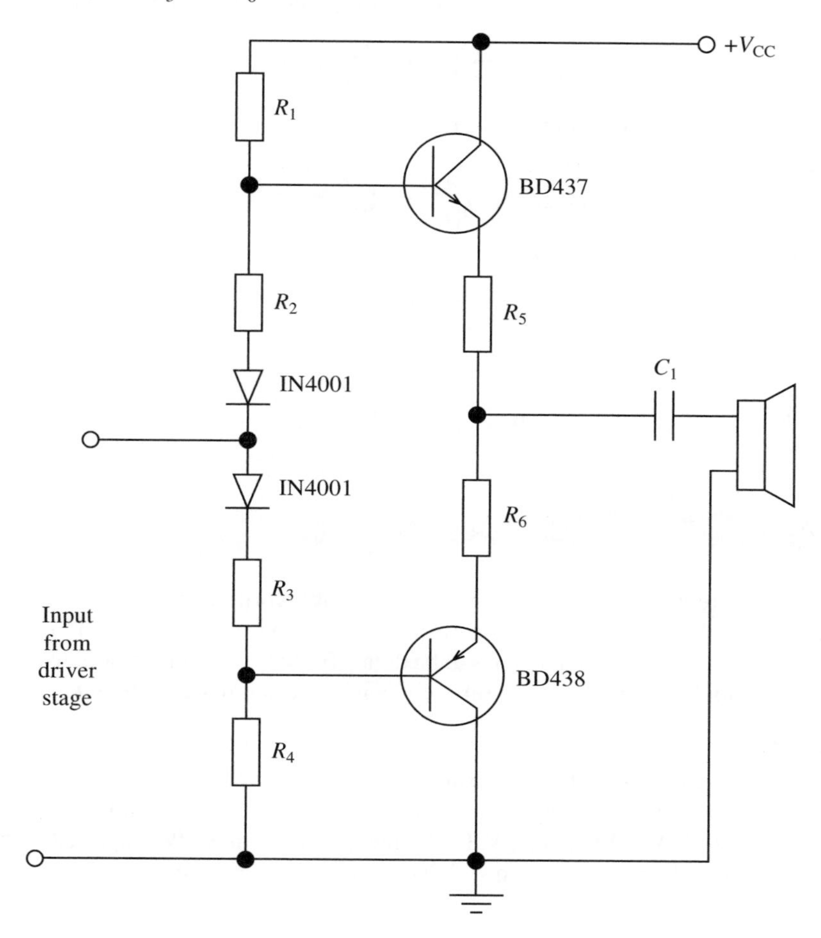

Fig. 5.22

Solution

(a)

$$f = \frac{1}{2\pi R_1 C_1}$$

$$\therefore \qquad C_1 = \frac{1}{2\pi f R_1} = \frac{1}{2\pi \times 80 \times 4} = 497.4\ \mu\text{F}$$

Select a 500 μF capacitor.

(b) The specification is for 5 W into a 4 Ω laod.

$$\therefore \qquad V_L = \sqrt{P_{AC(rms)} \times R_L} = \sqrt{5 \times 4} = 4.47\ \text{V}$$

$$\therefore \qquad V_{max} = \sqrt{2} \times \sqrt{20} = 6.32\ \text{V}$$

Also $I_{(max)} = 6.32/4 = 1.58$ A for each transistor. According to Fig. 5.18, the quiescent voltage is half the supply voltage. Also from the data sheets (not included here) at saturation $V_{CE} = 3.2$ V. If 0.5 V is taken as the drop across R_5 and R_6, then

$$V_{CC} = 6.32 + 3.7 + \frac{V_{CC}}{2} = 20 \text{ V}$$

(c) The maximum allowable base current will be

$$I_{B(max)} = \frac{I_{C(max)}}{h_{FE}} = \frac{1.58}{40} = 39 \text{ mA}$$

(d) The d.c. current through each transistor is

$$I_{DC} = \frac{1.58}{P} = 0.5 \text{ A}$$

Therefore

$$R_5 = R_6 = \frac{0.5}{0.5} = 1 \ \Omega$$

5.7 Calculating power and efficiency

Any machine is required to convert energy into a usable form, and the power amplifier is no exception. In this case the d.c. power from the supply has to be converted to a.c. power to drive an external device, and this has to be done efficiently. Several examples are given below which should clarify the use of different configurations under different applications.

Example 5.5
An amplifier has to work with a 35 V supply and has to be capable of drawing a current of 2.5 A from the supply. A single npn transistor is biased so that under the quiescent conditions $I_C = 1.75$ A and $V_{CE} = 30$ V. Under these conditions 35 W is applied to the load when the output voltage is 18 V r.m.s. Select a suitable transistor.

Solution
From the information given we obtain

$$P_{DC} = 2.5 \times 35 = 87.5 \text{ W}$$

$$P_{AC} = 35 \text{ W}$$

$$P_{CQ} = P_{DC} - P_{AC} = 87.5 - 35 = 52.5 \text{ W}$$

$$I_{rms} = \frac{P_{AC}}{V_{rms}} = \frac{35}{18} = 1.95 \text{ A}$$

Any transistor with ratings higher than these requirements will be suitable – for example, a TIP41 with $P_{tot} = P_{C(max)} = 65$ W, $I_{CQ} = 6$ A and $V_{CEQ} = 40$ V.

Example 5.6
A single-ended transistor power amplifier takes a mean collector current of 1 A from a 12 V supply and delivers an a.c. power of 2.4 W to a transformer-coupled load. Calculate:

(a) the collector efficiency;

(b) the collector dissipation if all other losses are ignored;

(c) the turns ratio of the transformer if the output resistance of the transistor is 1.3 kΩ and the load resistance is 12 Ω.

Solution

(a)

$$P_{dc} = 12 \times 1 = 12 \text{ W}$$

(b)

$$\eta = \frac{2.4 \times 100}{12} = \frac{P_{ac}}{P_{dc}} \times 100 = 20\%$$

(c)

$$n = \sqrt{\frac{R_1}{R_2}} = \sqrt{\frac{1200}{12}} = 10 : 1$$

Example 5.7

A transformer push-pull amplifier has a supply voltage of 20 V. It supplies a load having a resistance of 10 Ω. The total number of turns in the primary winding is 100 and the secondary winding has 50 turns.

(a) Find the maximum power that can be delivered to the load.

(b) Find the power dissipated in each transistor when maximum power is delivered to the load.

(c) Find the power delivered to the load and the power dissipated in each transistor when transistor power dissipation is maximum.

Solution

(a) The turns ratio between each half of the primary and secondary is

$$\frac{N_1}{N_2} = 50 : 50$$

Hence the peak values of primary and secondary voltages are equal, as are the peak values of primary and secondary currents.

$$V_{peak} = V_{CC} = 20 \text{ V}$$

$$I_{peak} = \frac{V_{CC}}{R_L} \frac{20}{10} = 2 \text{ A}$$

Hence

$$P_{AC} = \frac{V_{peak} \times I_{peak}}{2} = \frac{20 \times 2}{2} = 20 \text{ W}$$

(b)

$$P_D = P_{DC} - P_{AC} = \frac{2I_{peak} \times I_{peak} \times V_{CC}}{\pi} - \frac{V_{peak} \times I_{peak}}{2}$$

$$= \frac{2 \times 2 \times 20}{\pi} - 20 = 5.46 \text{ W}$$

Each transistor will therefore take 2.73 W of power.

(c) Transistor power dissipation is a maximum when

$$V_{peak} = 0.636V_{CC} = 0.636 \times 20 = 12.72 \text{ V}$$

Then

$$I_{peak} = \frac{12.72}{10} = 1.272 \text{ A}$$

and

$$P_{ac} = \frac{V_{peak} \times I_{peak}}{2} = \frac{12.72 \times 1.272}{2} = 8.09 \text{ W}$$

$$P_d = \frac{2 \times I_{peak} \times V_{CC}}{\pi} - \frac{V_{peak} \times I_{peak}}{2}$$

$$= \frac{2(1.272 \times 20)}{\pi} - 8.09 = 8.09 \text{ W}$$

Hence the maximum power in each transistor is 4.05 W.

Example 5.8

In a class A transformer power amplifier, the collector current alternates between 5 mA and 115 mA and its quiescent value is 60 mA. A load of 12 Ω is connected to the secondary, and when this is referred to the primary winding it becomes 325 Ω. If the supply voltage is 25 V estimate:

(a) the transformer turns ratio;

(b) the average power output;

(c) the minimum power rating of the transistor;

(d) the efficiency of the stage.

Solution

(a) The turns ratio is given by

$$\frac{N_1}{N_2} = \frac{325}{12} = 27 : 1$$

(b)

$$P_{ac} = I_c^2 R_{eff} = \frac{(115 - 5)^2}{(2 \times 2 \times 10^3)^2} \times 325 = 0.246 \text{ W}$$

(c)

$$P_{dc} = \frac{25 \times 60}{10^3} = 1.5 \text{ W}$$

$$P_c = P_{dc} - P_{ac} = 1.5 - 0.492 = 1.008 \text{ W}$$

(d)

$$\eta = \frac{P_{ac}}{P_{dc}} \times 100 = \frac{0.492}{1.5} \times 100 = 32.8\%$$

Example 5.9

A complementary pair class B push-pull amplifier has a supply voltage of 45 V and the transistors are biased so that they are sinusoidally driven to provide a current which is 80% of the maximum value. Calculate:

(a) the output power supplied to a speaker having a resistance of 15 Ω;

(b) the collector efficiency;

(c) the power rating of the transistors.

Solution

(a) The output power is given by

$$\frac{V_{CC} \times I_{C(max)}}{4} = \frac{V_{CC}^2}{8R_L}$$

Hence

$$I_{C(max)} = \frac{V_{CC}}{2R_L} = \frac{45}{30} = 1.5 \text{ A}$$

Biasing gives 80% swing, so

$$I_{C(max)} = 0.8 \times 1.5 = 1.2 \text{ A}$$

Hence output power is

$$\frac{V_{CC} \times I_{C(max)}}{4} = \frac{45 \times 1.2}{4} = 13.5 \text{ W}$$

(b)

$$P_{dc} = \frac{I_{C(max)} V_{CC}}{\pi} = \frac{1.2 \times 45}{\pi} = 17.2 \text{ W}$$

$$\eta = \frac{13.5}{17.2} \times 100 = 78.5\%$$

(c) The maximum output power is given by

$$P_{(max)} = \frac{V_{CC}^2}{8R_L} = \frac{(45)^2}{8 \times 15} = 16.9 \text{ W}$$

The maximum collector power dissipation is given approximately by $0.4P_{max}$:

$$P_{C(max)} = 0.4 \times 16.9 = 6.8 \text{ W}$$

Therefore power rating of each transistor is 3.4 W.

Example 5.10

An amplifier has to work with a 38 V supply and draw a current of 2 A from the d.c. supply. A single npn transistor is biased so that under quiescent conditions $I_C = 1.64$ A, $V_{CE} = 25$ V. This delivers 35 W into its load when the output voltage is 20 V r.m.s. Select a suitable transistor.

Solution

The steps involved in this type of practical problem are given below

$$P_{dc} = I_C \times V_{CC} = 2 \times 38 = 76 \text{ W.}$$

$$P_{ac} = 35 \text{ W}$$

The quiescent power is given by

$$P_{CQ} = I_{CQ} \times V_{CQ} = P_{dc} - P_{ac} = 41 \text{ W}$$

So

$$I_{C(rms)} = \frac{P_{ac}}{V_{o(rms)}} = \frac{35}{20} = 1.75 \text{ A}$$

Hence any transistor with ratings higher than these requirements will be suitable. A TIP41 is a suitable choice as it has the following characteristics:

$$P_{tot} = P_{C(max)} = 65 \text{ W}$$

$$I_{CQ} = 6 \text{ A}$$

$$V_{CEQ} = 40 \text{ V}$$

$$P_{CQ} = 1.64 \times 25 = 41 \text{ W}$$

5.8 Integrated circuit power amplifiers

Modern chip technology has promoted the fabrication of almost every discrete circuit on to a silicon substrate, and power amplifiers are no exception. All the circuits so far discussed can be etched on to a chip, except inductors and to a certain extent pnp transistors. Where there is a high current and power demand, discrete components still have the edge, and it is possibly this reason which restricts the availability and choice of integrated circuit power amplifiers. There are many power amplifiers on the market but it is a matter of deciding the requirements and parameters, then selecting the most suitable chip.

The integrated circuits discussed in this section are the LM380 (National Semiconductor), TBA820 (SGS–Thomson), TDA2030 (SGS–Thomson) and TDA2006 (SGS–Thomson). Details of characteristics and applications of the TDA 2006 will be found at the end of this chapter; those of the others can be found in appropriate manufacturers' data sheets.

LM380

This is a 14-pin dual in-line package that requires very few external components to make a 2.5 W power amplifier feeding into an 8 Ω load. Typical parameters for this integrated circuit are:

Fixed loop gain	50 (34 dB)
Supply voltage range	8–22 V
Quiescent current	7 mA
Input sensitivity	150 mV r.m.s.
Input resistance	150 kΩ
Load resistance	8 Ω
Bandwidth	100 kHz
Maximum current	1.3 A

The pinout diagram is shown in Fig. 5.23.

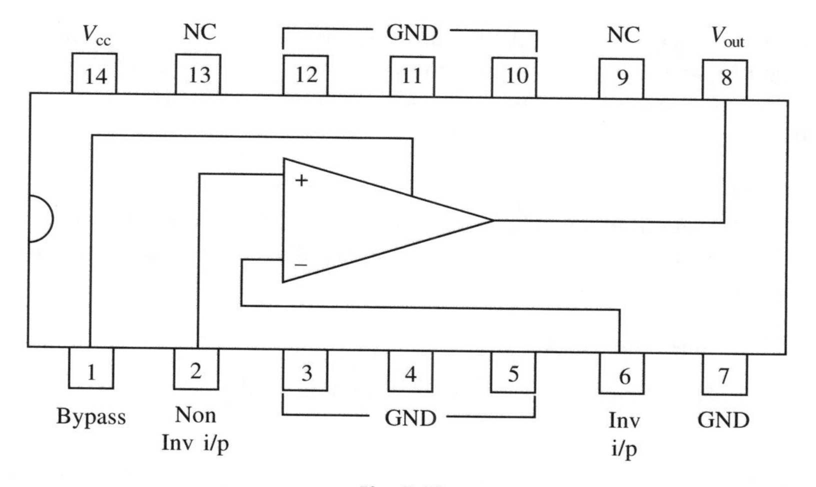

Fig. 5.23

The following points should be noted concerning the connection of this chip:

1. The output signal is automatically centred at half the supply voltage.
2. Pins 3, 4, 5, 10, 11 and 12 should be connected to ground and a heat sink connected directly to the pins for cooling.
3. The amplifier is suitable for low-power audio applications, but when used with a battery the current drain can be excessive.
4. In most applications it is advisable to add a 2.7 Ω resistor and a 0.1 μF polyester

capacitor in series from pin 8 to ground. A 4.7 μF capacitor connected from pin 1 to ground is also recommended.

5. The power-handling capabilities can be increased if a bridge system is used. It is then possible to feed into a 5 W load.

TBA 820M

This is an 8-pin amplifier capable of delivering 2 W into an 8 Ω load, 1.6 W into a 4 Ω load or 1.2 W into an 8 Ω load. A bootstrap circuit is included in order to increase the gain of the amplifier, and high ripple rejection from the power supply is a major feature. This is achieved by including a 47 μF capacitor when required.

Frequency compensation is provided by pin 1. The value of C_B (from the data sheets) is selected with R_f, the gain setting resistor, by using the C_B/R_f graph. Selection of these components will also affect the frequency response, as can be seen from the electrical characteristics.

TDA2006

This chip is identical to the TDA2030 and can be used with single or double power supplies and in single-ended or bridge configurations. The data sheets are set out in a similar manner to those for the TDA2030, and the following example indicates a particular application.

Example 5.11
Design an audio amplifier suitable for feeding into an 8 Ω load and delivering 8 W. The bandwidth has to be 20 kHz and a double power supply is used. The gain has to be set at 30 dB. Determine:

(a) suitable values for components;

(b) the expected efficiency if a supply of ± 12 V is used.

Solution
(a) Suitable values can be achieved from the data sheet applications graph. As the bandwidth has to be 20 kHz and a bandwidth gain of 30 dB is required, the value of C_8 is 250 pF. Also

$$C_8 = \frac{1}{2\pi BWR_1}$$

$$\therefore \qquad R_1 = \frac{1}{2\pi BWC_8} = \frac{10^{12}}{2\pi \times 2 \times 10^4 \times 250} = 31.9\ \text{k}\Omega$$

This will give the larger gain of 30 dB.

(b) From the data sheets and for a load of 8 Ω and $V = \pm$ 12 V, the power dissipation (P_c) is approximately 4.6 W.

$$\therefore \qquad P_{dc} = P_c + P_{ac} = 4.8 + 8 = 12.8 \text{ W}$$

$$\therefore \qquad \eta = \frac{8}{12.8} \times 100 = 62.5\%$$

Note that decreasing the supply voltage would increase the efficiency while decreasing P_{ac} and P_c. Sensitivity will also decrease.

5.9 Radio frequency power amplifiers

These amplifiers normally function under class C bias conditions and are required as conventional amplifiers to boost the final power from the transmitter to the aerial tuning unit and antenna. They are also used as frequency amplifier stages in situations where the carrier frequency is insufficiently high. Two applications where this type of amplifier is used are given in Figs 5.24 and 5.25.

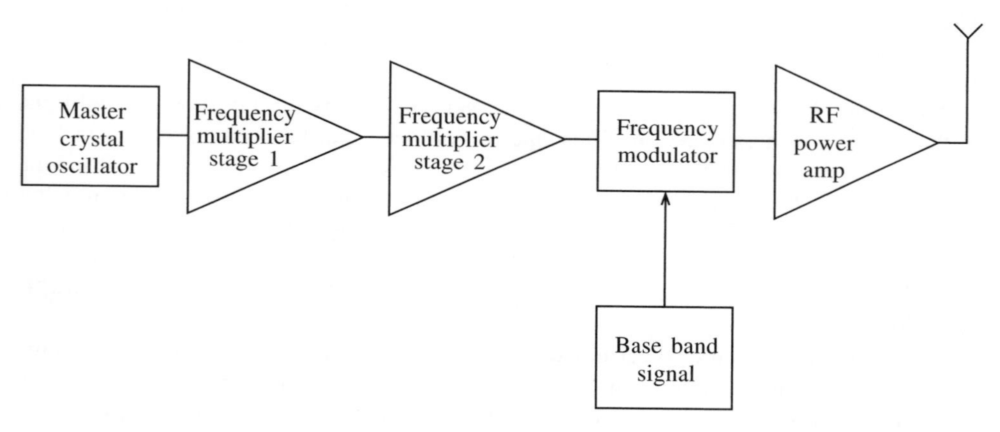

Fig. 5.24

Figure 5.24 shows a simplified block diagram of an FM receiver with frequency multiplication and radio frequency power boosting. The frequency multiplication is necessary

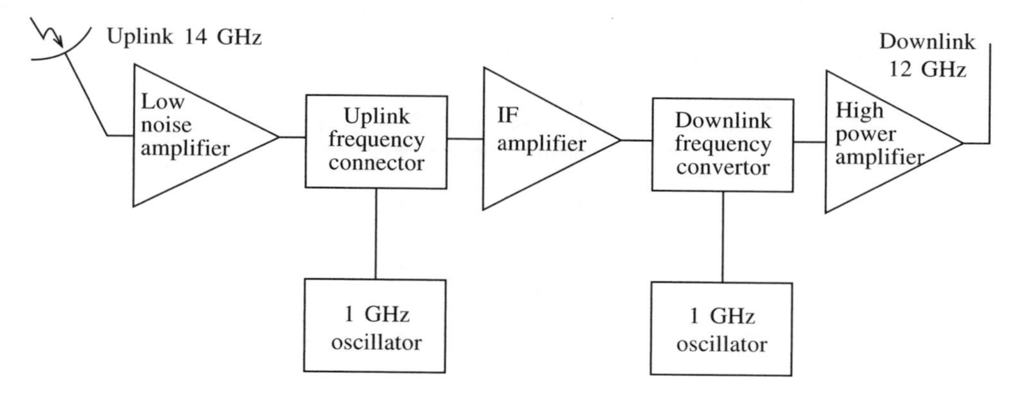

Fig. 5.25

because there is a limit to how thin the crystal in the oscillator can be cut: the thinner the crystal, the higher the frequency. Hence several stages of frequency multiplication are required operating with class C classic bias and having their loads tuned to some harmonic of the crystal oscillator.

Figure 5.25 shows a satellite transponder used for an uplink/downlink of 14/12 GHz. A high-power amplifier is required to accommodate the spectral power used over several channels. The amplifier used here is a microwave type and is highly specialized.

Radio frequency power amplifiers are numerous, and the topic is beyond the scope of this text. However, it is informative to look at one particular versatile chip, the MAX2430 manufactured by Maxim Integrated Products. The data sheets for this chip can be obtained from the manufacturer.

5.10 Power amplifier measurements

There are a number of test procedures which can be carried out on a power amplifier, but perhaps the most important are the measurement of frequency response, output power efficiency and distortion.

Measuring the frequency response for a power amplifier is similar to other amplifiers in that a signal generator and oscilloscope or voltmeter can be used, as shown in Fig. 5.26. A two-channel oscilloscope is generally preferred and, as shown, all instrument grounds should be commoned. Integrated circuits such as the LM380, a response for which is shown in Fig. 5.27, may have several unconnected pins. These should be grounded to avoid signal coupling. The signal generator is set for an undistorted output and swept through a range of frequencies. Figure 5.27 shows the half-power points, the bandwidth and mid-band gain.

The output power can be determined for a given load by using a similar set-up to Fig. 5.26. In this measurement an audio oscillator is set at a test frequency, usually 1 kHz or lower. The input voltage is increased until some distortion is seen in the output waveform. The voltage is then measured by means of a voltmeter and the r.m.s. power calculated. Generally this will be done for a fixed resistance such as the 8 Ω speaker in Fig. 5.27. The output power can also be measured by means of an audio power meter, but these can be expensive and are not necessary in most applications. In order to measure the efficiency the d.c. input power has to be calculated, and this can be done by measuring the d.c. current from the supply when it is delivering its maximum output power.

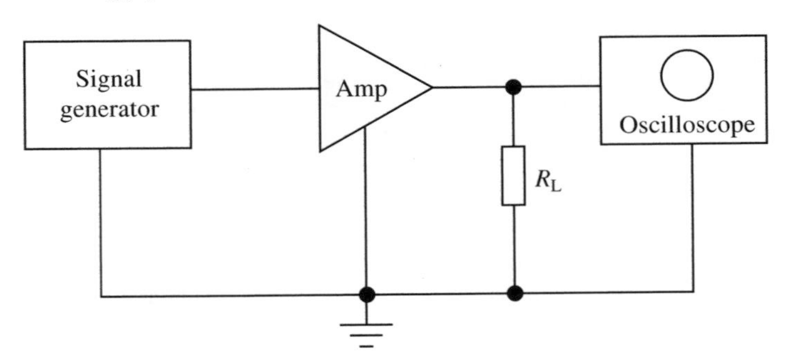

Fig. 5.26

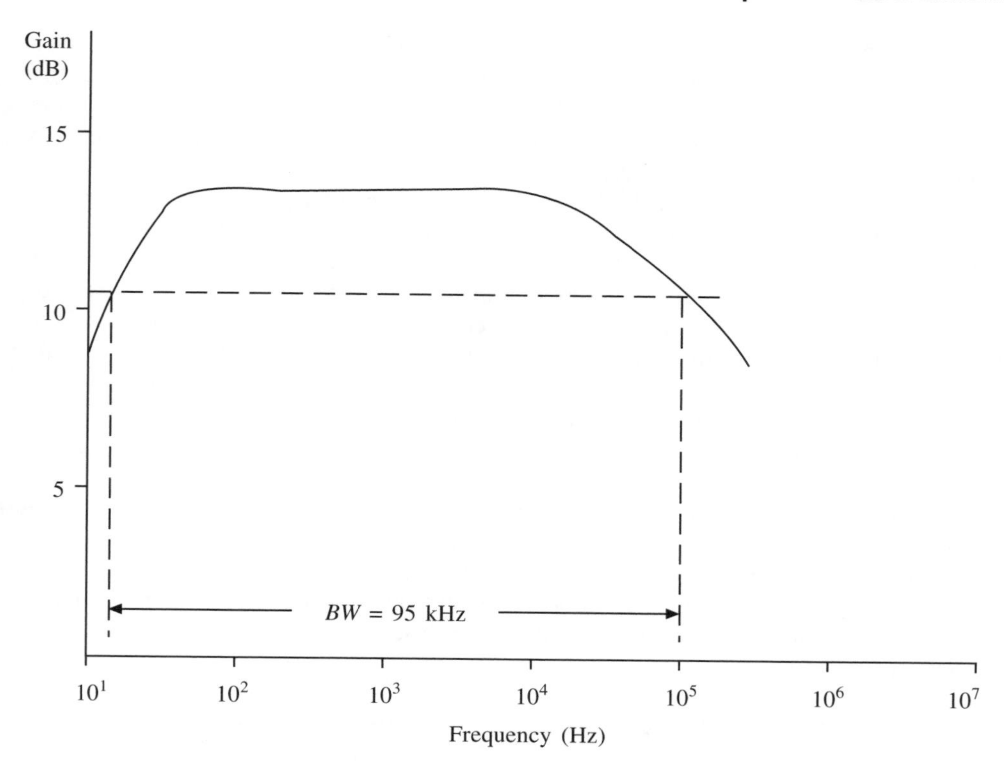

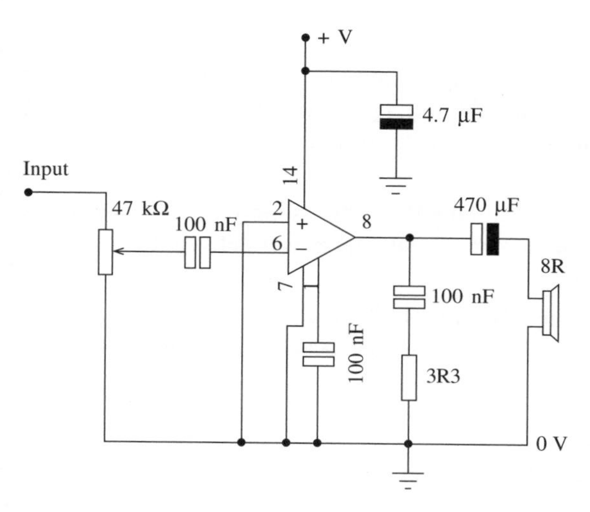

f(Hz)	V_i(V)	V_o(V)	V_i/V_o	dB
0.3		1	3.3	10.5
20	0.3	1.1	3.6	11.3
40	0.3	1.38	4.6	13.2
50	0.3	1.4	4.7	13.3
60	0.3	1.4	4.7	13.4
80	0.3	1.4	4.7	13.4
100	0.3	1.4	4.7	13.4
200	0.3	1.4	4.7	13.4
300	0.3	1.4	4.7	13.4
400	0.3	1.4	4.7	13.4
500	0.3	1.4	4.7	13.4
1000	0.3	1.4	4.7	13.4
3000	0.3	1.4	4.7	13.4
10000	0.3	1.4	4.7	13.4
15000	0.27	1.2	4.6	13.1
20000	0.27	1.19	4.4	13
30000	0.27	1.18	4.2	12.9
50000	0.05	0.2	3.54	11
100000	0.05	0.15	3	9.54

Fig. 5.27

The final test considered here is for harmonic and intermodulation distortion, which is generally expressed as a percentage or so many decibels down on the fundamental (f_1). BJTs and FETs all show some non-linearity in their characteristics, and this causes unwanted frequency components to appear at the output. Generally they are harmonically related to the fundamental ($2f_1$, $3f_1$).

Large signal amplifiers, such as power amplifiers, can have their distorted outputs reduced by negative feedback, which will alter the input and output resistance of the circuit as well as the gain and bandwidth, but a compromise is generally built into the design.

Harmonic distortion can be measured by means of a specturm analyser or distortion meter. A new innovation is the Tektronix power measurement oscilloscope which can be used for power measurement applications, but which also has a bar graph facility for measuring harmonics. Another convenient addition to the oscilloscope is the spectrum analyser adaptor by Thurlby Thander. This instrument operates with any standard oscilloscope having an X-Y mode and features a wide dynamic range and variable centre frequency and scan width.

Intermodulation distortion occurs when new frequencies are introduced at the output which contain sum and difference frequencies from input signals such as $2f_1 + 2f_2$ or $f_3 + 2f_2$. A measurement of this type of distortion can be achieved by applying two frequencies to the input of an amplifier and then observing any additional components at the output using a spectrum analyser or intermodulation analyser.

5.11 Further problems

1. An amplifier has a dc. input power of 4 W and an efficiency of 28%. Calculate its collector dissipation.
   Answer: 2.88 W
2. When a signal is applied to a single-ended amplifier its collector current varies from 28 mA to 225 mA and its collector voltage varies from 1.8 V to 14.8 V. Calculate the a.c. power output and the collector efficiency.
   Answer: 325 mW, 31.13%
3. A complementary symmetry pair audio amplifier has a supply of 30 V and feeds into a load of 16 Ω. Calculate the maximum possible power delivered to the load.
   Answer: 7.03 W
4. Under operating conditions an amplifier draws 800 mA from its 10 V d.c. supply. If 6 W of audio output power is delivered to a loudspeaker, calculate:
   (a) the d.c. power;
   (b) the collector power dissipation;
   (c) the efficiency of the amplifier.
   Assume a single-ended type of audio amplifier.
   Answer: 8 W, 2 W, 75%
5. A class C transistor tuned power amplifier operates from a 30 V collector supply. If the collector dissipation of the transistor is 1.2 W and the mean collector current is 0.1 A, determine:
   (a) the a.c. output power;
   (b) the collector efficiency of the amplifier.
   Answer: 1.8 W, 60%

6. A transducer class C amplifier tuned to 500 kHz uses a coil of inductance 3.5 μH. The effective Q of the circuit is 25. A mean current of 0.12 A is taken from the 50 V collector supply and the collector voltage has a peak value of 47 V. Calculate:
(a) the efficiency of the amplifier;
(b) the collector dissipation;
(c) the peak amplitude of the 500 kHz component of collector current pulses.
Answer: 66.7%, 2 W, 1711 mA

7. In a class A transformer-coupled amplifier, the collector current alternates between 3 mA and 100 mA and its quiescent value is 58 mA. The load is 12 Ω and, when referred to the primary winding, 315 Ω. If the supply voltage is 20 V, estimate:
(a) the transformer turns ratio;
(b) the a.c. power output;
(c) the minimum power rating of the transistor;
(d) the d.c. power supplied to the circuit.
Answer: 5, 0.465 W, 0.695 W, 1.6 W

8. A complementary push-pull amplifier has a supply voltage of 20 V and delivers power to a 10 Ω load. Determine.
(a) the power delivered to the load under maximum signal conditions;
(b) the efficiency under maximum signal conditions;
(c) the value of the capacitor which couples the signal to the load if the amplifier has to be used at signal frequencies down to 20 Hz.
Answer: 5 W, 78.5% 795 μF

9. A single-ended power amplifier has a supply voltage V_{CC} = 20 V and a collector resistance of R_C = 1 Ω. Determine:
(a) the maximum power dissipation rating that the transistor should have;
(b) the value of R_C which should be used to ensure safe operation if an increase of temperature reduces the maximum rating found in (a) by half.
Answer: 0.1 W, 2 kW

10. A push-pull amplifier has a supply voltage V_{CC} = 20 V and supplies a load of 10 Ω. The primary/secondary turns ratio is 100 : 50. Determine:
(a) the maximum power that can be delivered to the load;
(b) the power dissipated in each transistor when maximum power is delivered to the load;
(c) the power delivered to the load and the power dissipated in each transistor when the transistor power dissipated is maximum.
Answer: 20 W, 2.73 W, 4.05 W)

11. Design an audio power amplifier using a TDA2030 chip which has to deliver 14 W/ 8 Ω. The gain has to be 30 dB and the pass band from 20 Hz to 30 kHz. (See Example 5.12.)

12. Design an audio amplifier using a TBA820M chip which has to deliver a maximum of 1 W into an 8 Ω load when the bandwidth is from 20 Hz to 25 kHz. (See Example 5.11.)

13. A 24 W audio amplifier has to be built to feed into a load of 8 Ω. Select a chip which might satisfy this requirement and suggest what practical considerations should be

applied to the circuit if short-circuit protection and thermal shut down have to be applied. The voltage gain has to be 30 dB and the bandwidth from 20 Hz to 25 kHz.

14. Design an RF power amplifier suitable for a cellular phone operating at 915 MHz. State all the practical considerations when fabricating the printed circuit board.
15. An LM380 has to deliver 2 W to an 8 Ω load. The voltage gain has to be set at 33 dB. Explain the function of each component and what dimensions the heatsink should be.

Data sheets

TBA820M

1.2W AUDIO AMPLIFIER

DESCRIPTION

The TBA820M is a monolithic integrated audio amplifier in a 8 lead dual in-line plastic package. It is intended for use as low frequency class B power amplifier with wide range of supply voltage: 3 to 16V, in portable radios, cassette recorders and players etc. Main features are: minimum working supply voltage of 3V, low quiescent current, low number of external components, good ripple rejection, no cross-over distortion, low power dissipation.

Output power: P_o = 2W at 12V/8Ω, 1.6W at 9V/4Ω and 1.2W at 9V/8Ω.

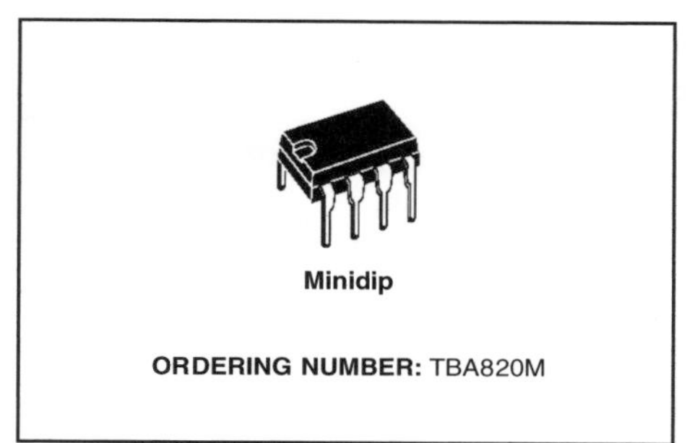

ABSOLUTE MAXIMUM RATINGS

Symbol	Parameter	Value	Unit
V_s	Supply voltage	16	V
I_o	Output peak current	1.5	A
P_{tot}	Power dissipation at T_{amb} = 50°C	1	W
T_{stg}, T_j	Storage and junction temperature	-40 to 150	°C

TEST AND APPLICATION CIRCUITS

Figure 1. Circuit diagram with load connected to the supply voltage

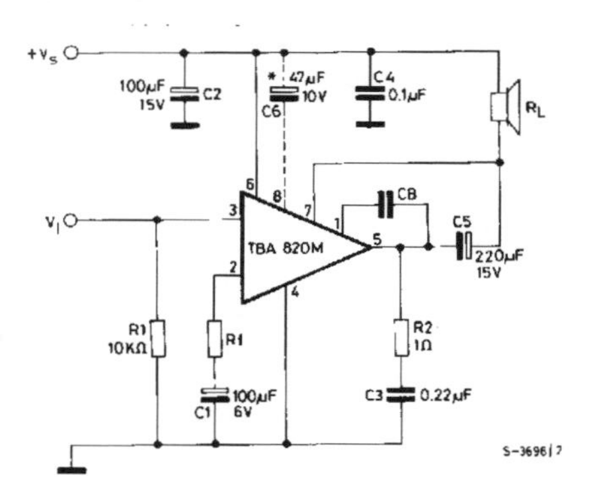

Figure 2. Circuit diagram with load connected to ground

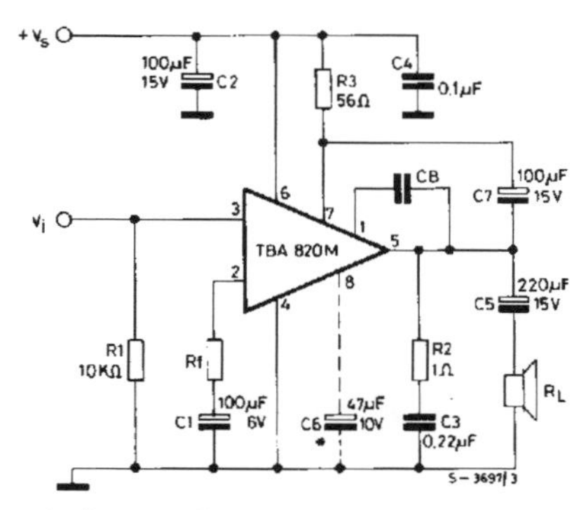

* Capacitor C6 must be used when high ripple rejection is requested.

June 1988

*Now known as STMicroelectronics

TBA820M

PIN CONNECTION (top view)

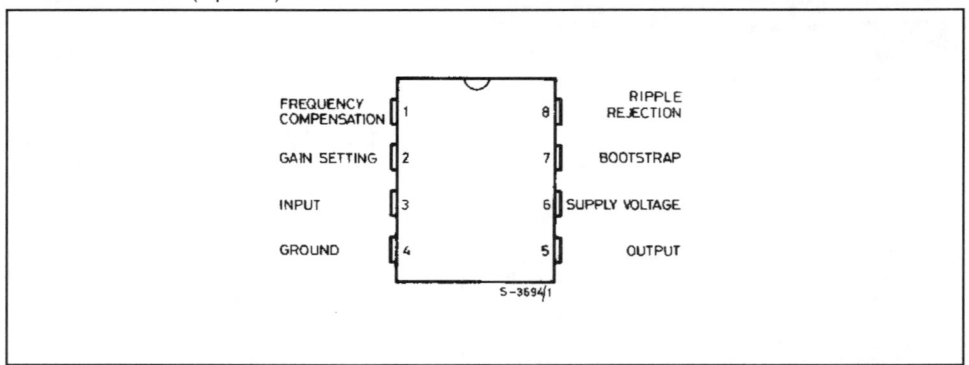

SCHEMATIC DIAGRAM

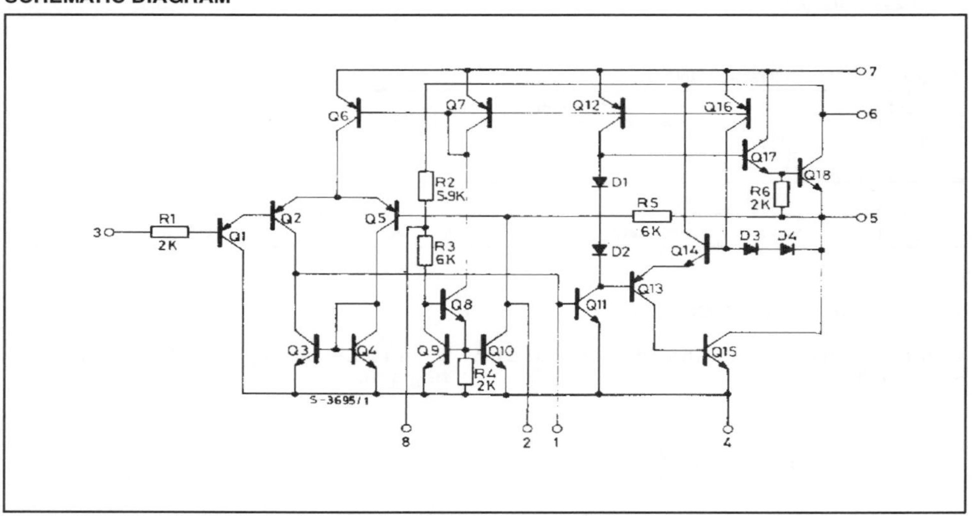

THERMAL DATA

Symbol	Parameter		Value	Unit
$R_{th\text{-}j\text{-}amb}$	Thermal resistance junction-ambient	max	100	°C/W

ELECTRICAL CHARACTERISTICS (Refer to the test circuits Vs = 9V, T_{amb} = 25 °C unless otherwise specified)

Symbol	Parameter	Test conditions		Min.	Typ.	Max.	Unit
V_s	Supply voltage			3		16	V
V_o	Quiescent output voltage (pin 5)			4	4.5	5	V
I_d	Quiescent drain current				4	12	mA
I_b	Bias current (pin 3)				0.1		μA
P_o	Output power	d = 10% R_f = 120Ω V_s = 12V V_s = 9V V_s = 9V V_s = 6V V_s = 3.5V	f = 1 kHz R_L = 8Ω R_L = 4Ω R_L = 8Ω R_L = 4Ω R_L = 4Ω	0.9	2 1.6 1.2 0.75 0.25		W W W W W
Ri	Input resistance (pin 3)	f = 1 kHz			5		MΩ
B	Frequency response (-3 dB)	R_L = 8Ω C_5 = 1000 μF R_f = 120Ω	C_B = 680 pF	25 to 7,000			Hz
			C_B = 220 pF	25 to 20,000			
d	Distortion	P_o = 500 mW R_L = 8Ω f = 1 kHz	R_f = 33Ω		0.8		%
			R_f = 120Ω		0.4		
G_v	Voltage gain (open loop)	f = 1 kHz	R_L = 8Ω		75		dB
G_v	Voltage gain (closed loop)	R_L = 8Ω	R_f = 33Ω		45		dB
		f = 1 kHz	R_f = 120Ω		34		
e_N	Input noise voltage (*)				3		μV
i_N	Input noise current (*)				0.4		nA
$\frac{S+N}{N}$	Signal to noise ratio (*)	P_o = 1.2W R_L = 8Ω G_v = 34 dB	R1 = 10KΩ		80		dB
			R1 = 50 kΩ		70		
SVR	Supply voltage rejection (test circuit of fig. 2)	R_L = 8Ω $f_{(ripple)}$ = 100 Hz C6 = 47 μF R_f = 120Ω			42		dB

(*) B = 22 Hz to 22 KHz

TBA820M

Figure 3. Output power vs. supply voltage

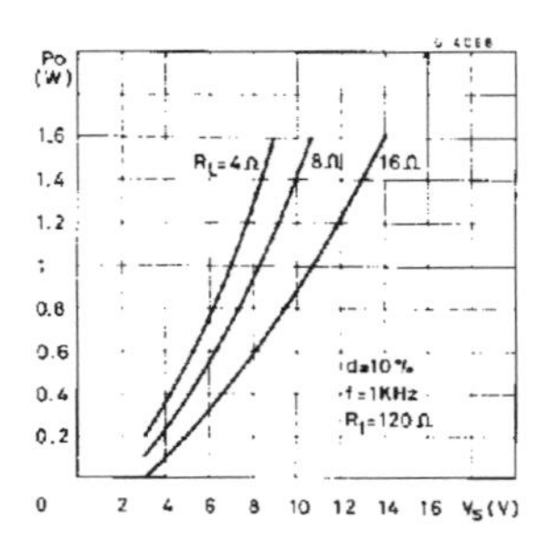

Figure 4. Harmonic distortion vs. output power

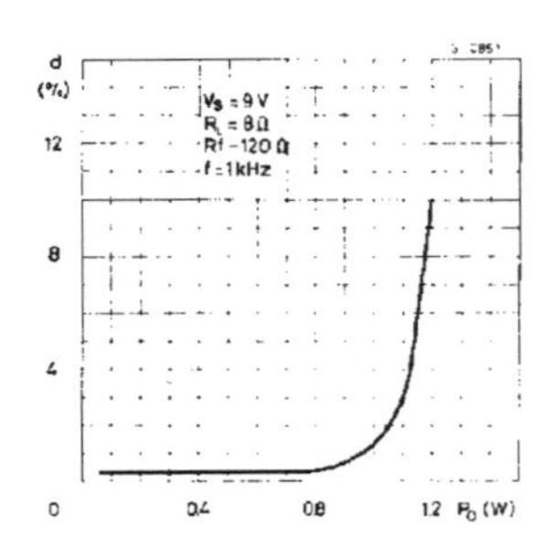

Figure 5. Power dissipation and efficiency vs. output power

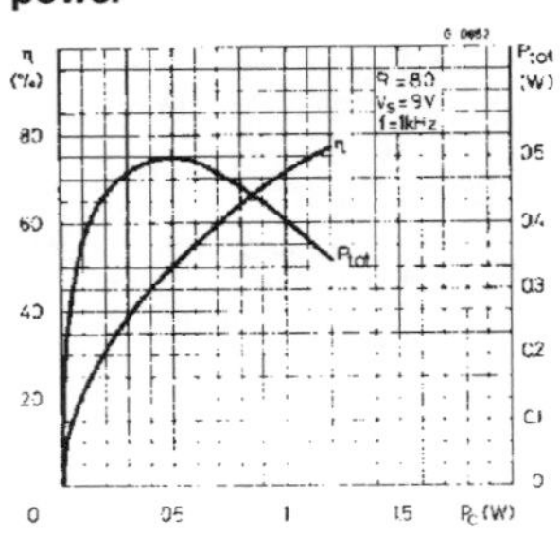

Figure 6. Maximum power dissipation (sine wave operation)

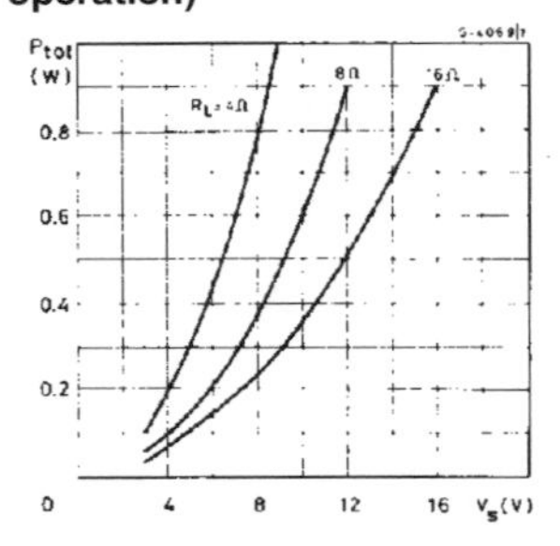

Figure 7. Suggested value of C_B vs. R_f

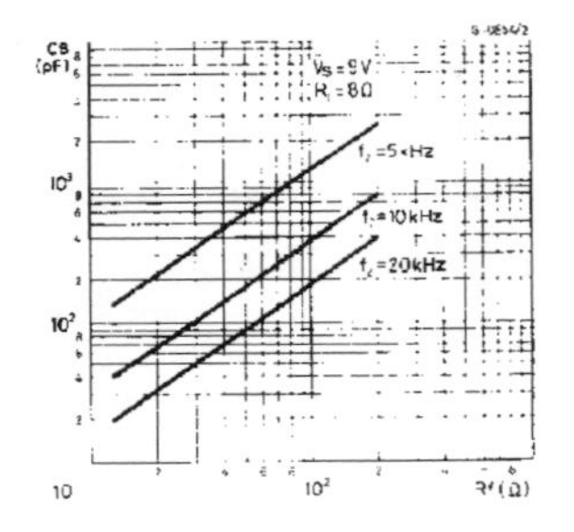

Figure 8. Frequency response

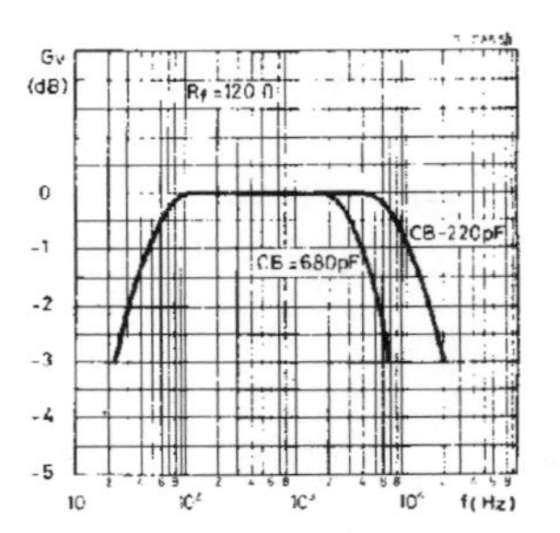

Figure 9. Harmonic distortion vs. frequency

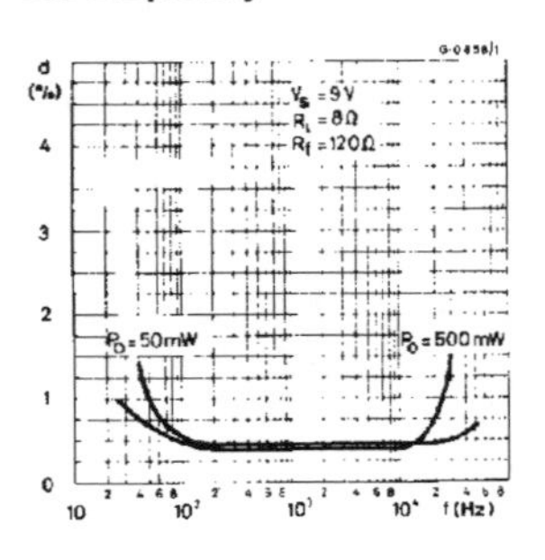

Figure 10. Supply voltage rejection (Fig. 2 circuit)

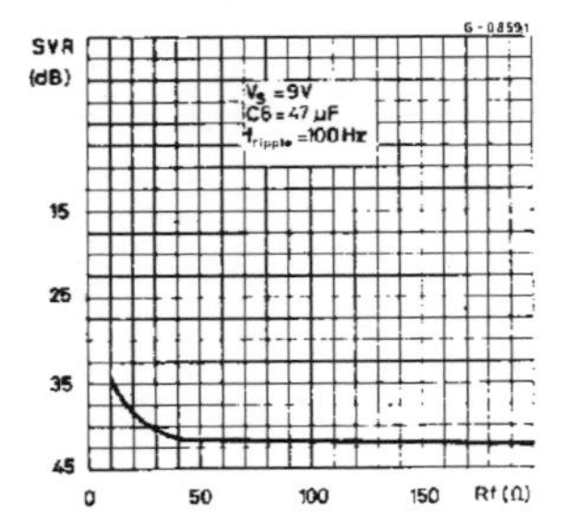

Figure 11. Quiescent current vs. supply voltage

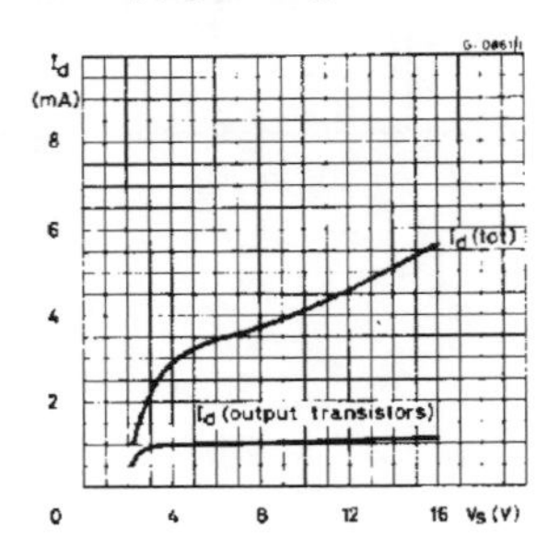

MINIDIP PACKAGE MECHANICAL DATA

DIM.	mm			inch		
	MIN.	TYP.	MAX.	MIN.	TYP.	MAX.
A		3.32			0.131	
a1	0.51			0.020		
B	1.15		1.65	0.045		0.065
b	0.356		0.55	0.014		0.022
b1	0.204		0.304	0.008		0.012
D			10.92			0.430
E	7.95		9.75	0.313		0.384
e		2.54			0.100	
e3		7.62			0.300	
e4		7.62			0.300	
F			6.6			0.260
I			5.08			0.200
L	3.18		3.81	0.125		0.150
Z			1.52			0.060

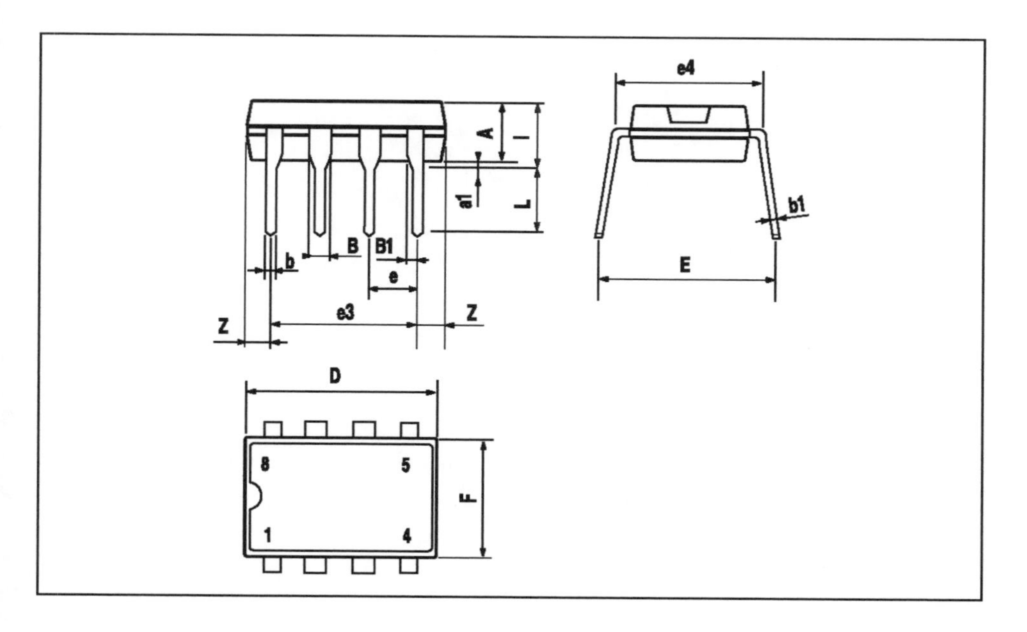

SGS-THOMSON
MICROELECTRONICS

TDA2006

12W AUDIO AMPLIFIER

DESCRIPTION

The TDA2006 is a monolithic integrated circuit in Pentawatt package, intended for use as a low frequency class "AB" amplifier. At ±12V, d = 10 % typically it provides 12W output power on a 4Ω load and 8W on a 8Ω . The TDA2006 provides high output current and has very low harmonic and cross-over distortion. Further the device incorporates an original (and patented) short circuit protection system comprising an arrangement for automatically limiting the dissipated power so as to keep the working point of the output transistors within their safe operating area. A conventional thermal shutdown system is also included. The TDA2006 is pin to pin equivalent to the TDA2030.

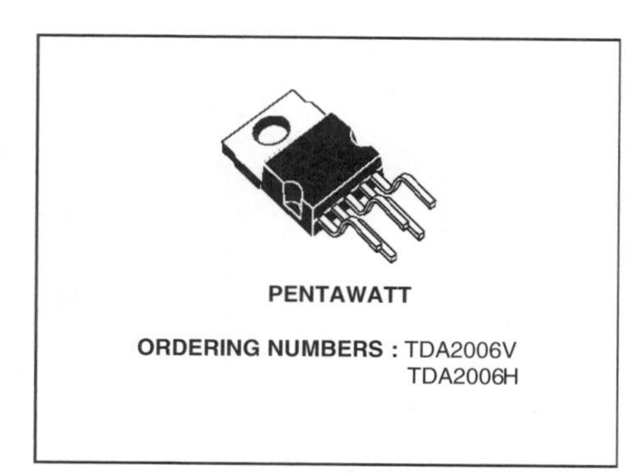

PENTAWATT

ORDERING NUMBERS : TDA2006V
TDA2006H

TYPICAL APPLICATION CIRCUIT

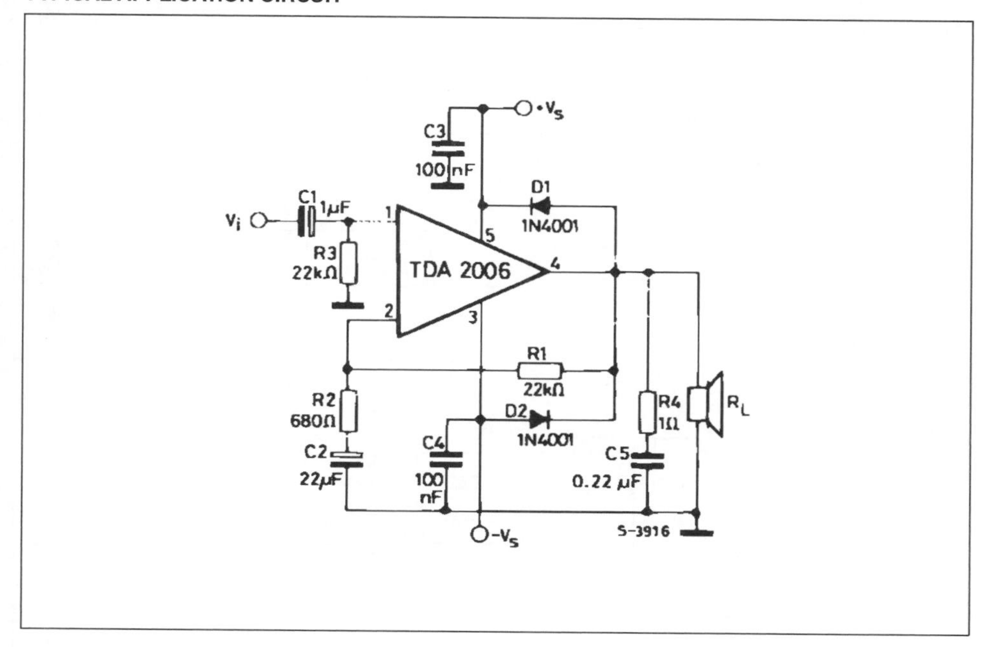

*Now known as STMicroelectronics

TDA2006

SCHEMATIC DIAGRAM

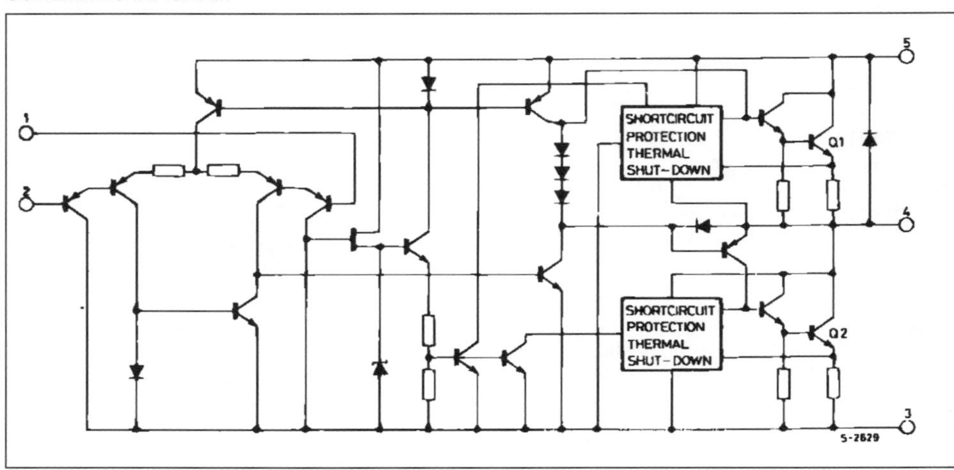

ABSOLUTE MAXIMUM RATINGS

Symbol	Parameter	Value	Unit
V_s	Supply Voltage	± 15	V
V_i	Input Voltage	V_s	
V_i	Differential Input Voltage	± 12	V
I_o	Output Peak Current (internaly limited)	3	A
P_{tot}	Power Dissipation at T_{case} = 90 °C	20	W
T_{stg}, T_j	Storage and Junction Temperature	– 40 to 150	°C

THERMAL DATA

Symbol	Parameter		Value	Unit
$R_{th\ (j\text{-}c)}$	Thermal Resistance Junction-case	Max	3	°C/W

PIN CONNECTION

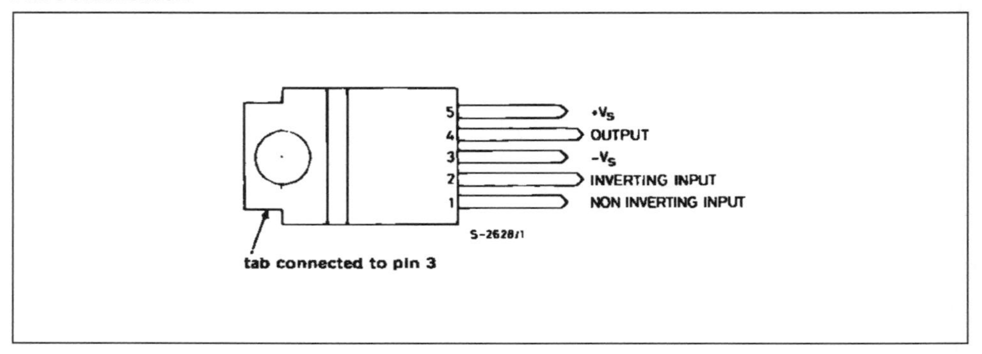

ELECTRICAL CHARACTERISTICS

(refer to the test circuit ; $V_S = \pm$ 12V, $T_{amb} = 25^oC$ unless otherwise specified)

Symbol	Parameter	Test Conditions	Min.	Typ.	Max.	Unit
V_s	Supply Voltage		± 6		± 15	V
I_d	Quiescent Drain Current	$V_s = \pm$ 15V		40	80	mA
I_b	Input Bias Current	$V_s = \pm$ 15V		0.2	3	μA
V_{OS}	Input Offset Voltage	$V_s = \pm$ 15V		± 8		mV
I_{OS}	Input Offset Current	$V_s = \pm$ 15V		± 80		nA
V_{OS}	Output Offset Voltage	$V_s = \pm$ 15V		± 10	± 100	mV
P_o	Output Power	d = 10%, f = 1kHz $R_L = 4\Omega$ $R_L = 8\Omega$	 6	 12 8		W
d	Distortion	P_o = 0.1 to 8W, $R_L = 4\Omega$, f = 1kHz P_o = 0.1 to 4W, $R_L = 8\Omega$, f = 1kHz		0.2 0.1		% %
V_i	Input Sensitivity	P_o = 10W, $R_L = 4\Omega$, f = 1kHz P_o = 6W, $R_L = 8\Omega$, f = 1kHz		200 220		mV mV
B	Frequency Response (– 3dB)	P_o = 8W, $R_L = 4\Omega$	20Hz to 100kHz			
R_i	Input Resistance (pin 1)	f = 1kHz	0.5	5		MΩ
G_v	Voltage Gain (open loop)	f = 1kHz		75		dB
G_v	Voltage Gain (closed loop)	f = 1kHz	29.5	30	30.5	dB
e_N	Input Noise Voltage	B (– 3dB) = 22Hz to 22kHz, $R_L = 4\Omega$		3	10	μV
i_N	Input Noise Current	B (– 3dB) = 22Hz to 22kHz, $R_L = 4\Omega$		80	200	pA
SVR	Supply Voltage Rejection	$R_L = 4\Omega$, $R_g = 22k\Omega$, f_{ripple} = 100Hz (*)	40	50		dB
I_d	Drain Current	P_o = 12W, $R_L = 4\Omega$ P_o = 8W, $R_L = 8\Omega$		850 500		mA mA
T_j	Thermal Shutdown Junction Temperature				145	°C

(*) Referring to Figure 15, single supply.

Figure 1 : Output Power versus Supply Voltage

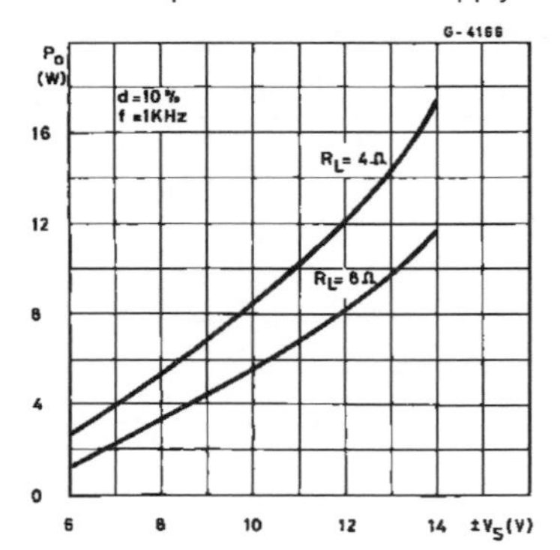

Figure 2 : Distortion versus Output Power

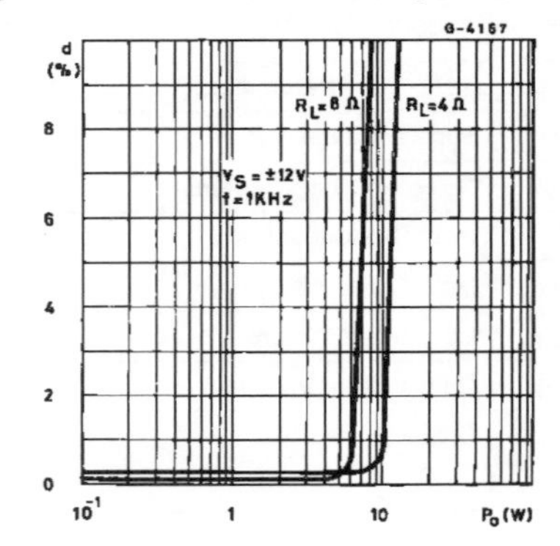

Figure 3 : Distortion versus Frequency

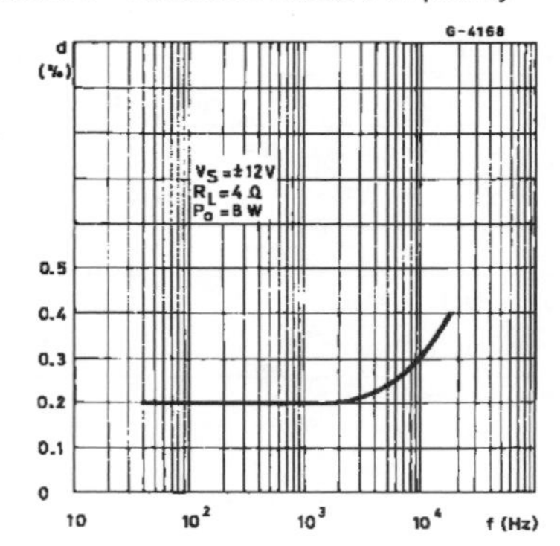

Figure 4 : Distortion versus Frequency

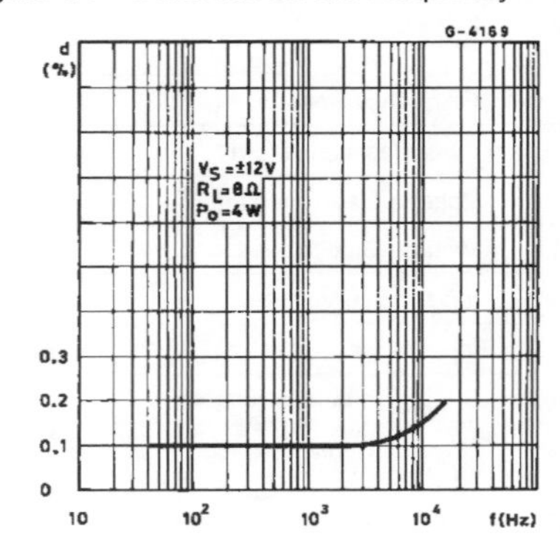

Figure 5 : Sensitivity versus Output Power

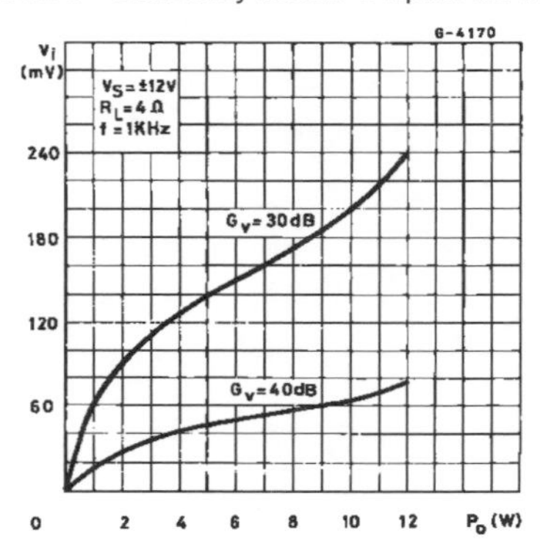

Figure 6 : Sensitivity versus Output Power

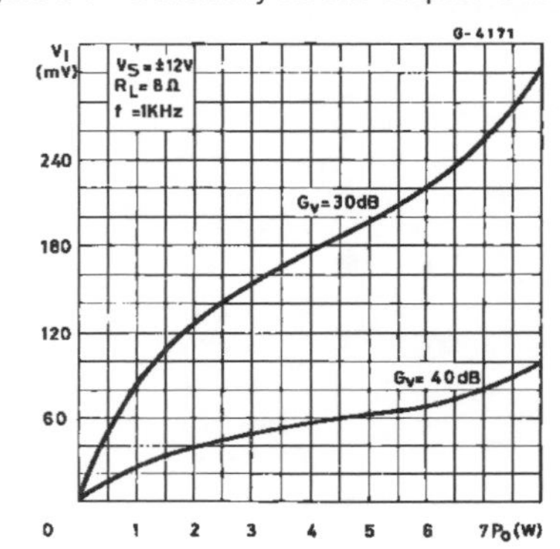

Figure 7 : Frequency Response with different values of the rolloff Capacitor C8 (see Figure 13)

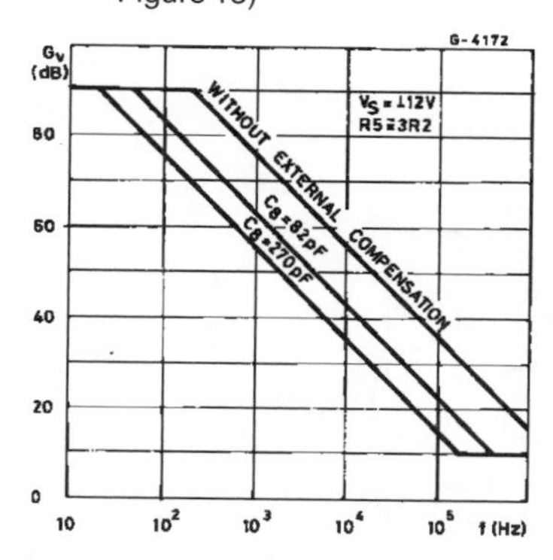

Figure 8 : Value of C8 versus Voltage Gain for different Bandwidths (see Figure 13)

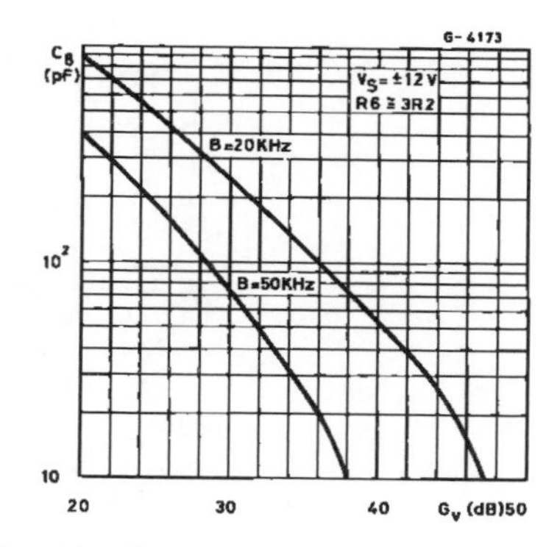

Figure 9 : Quiescent Current versus Supply Voltage

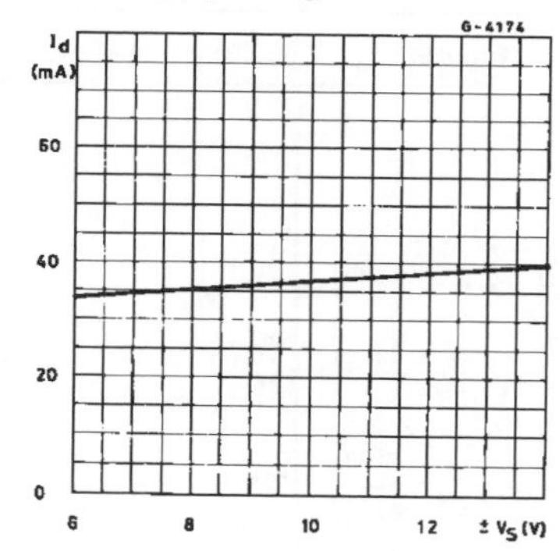

Figure 10 : Supply Voltage Rejection versus Voltage Gain

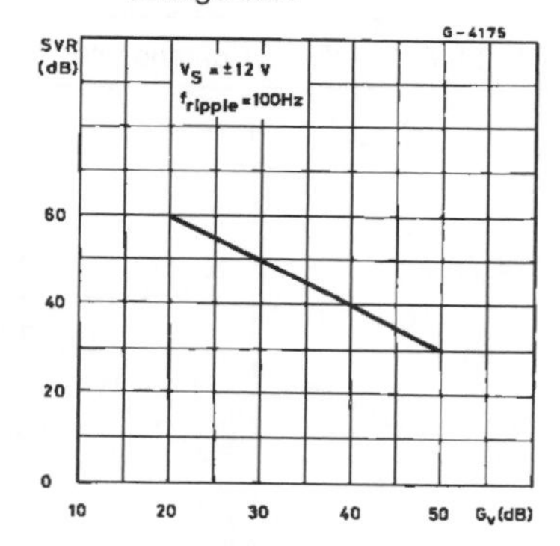

Figure 11 : Power Dissipation and Efficiency versus Output Power

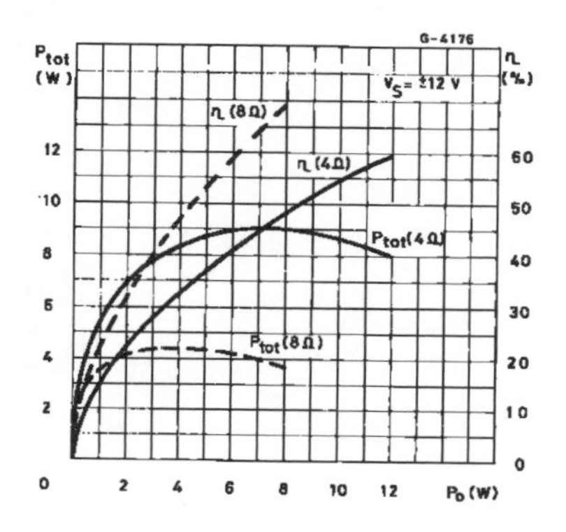

Figure 12 : Maximum Power Dissipation versus Supply Voltage (sine wave operation)

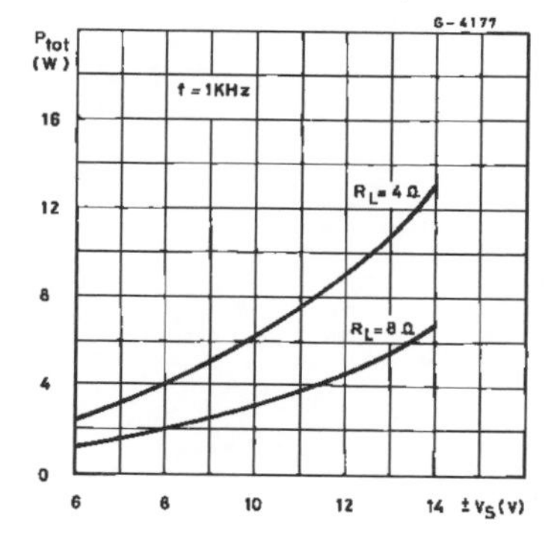

Figure 13 : Application Circuit with Spilt Power Supply

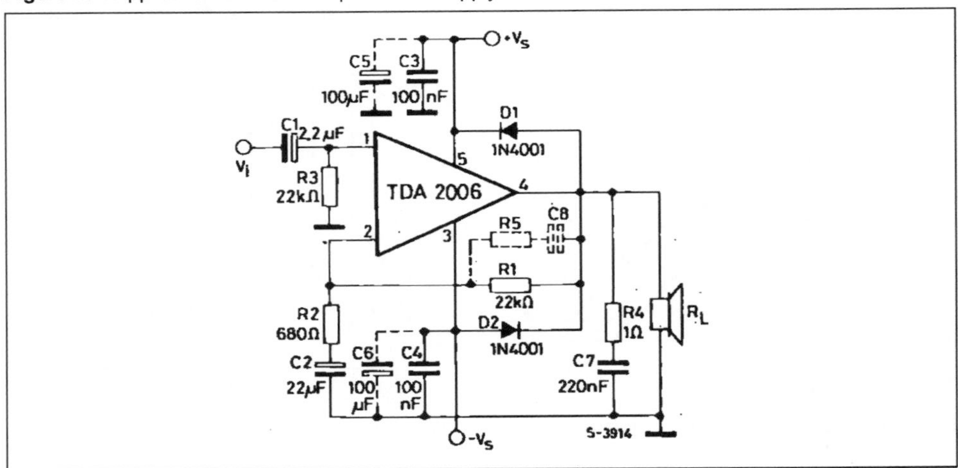

Figure 14 : P.C. Board and Components Layout of the Circuit of Figure 13 (1:1 scale)

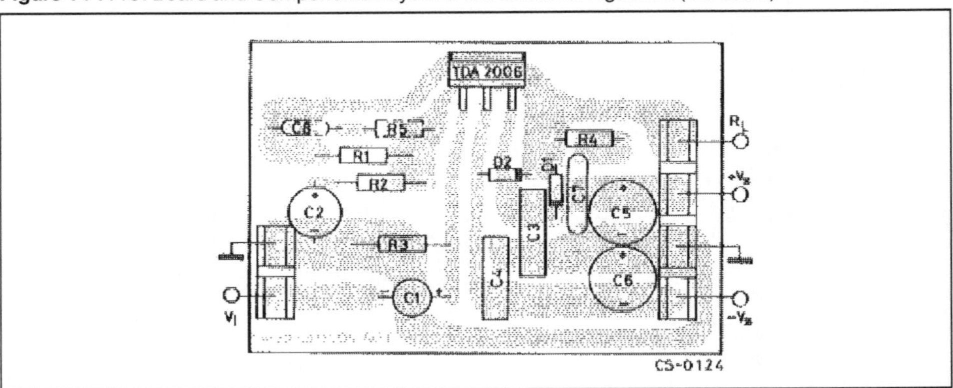

Figure 15 : Application Circuit with Single Power Supply

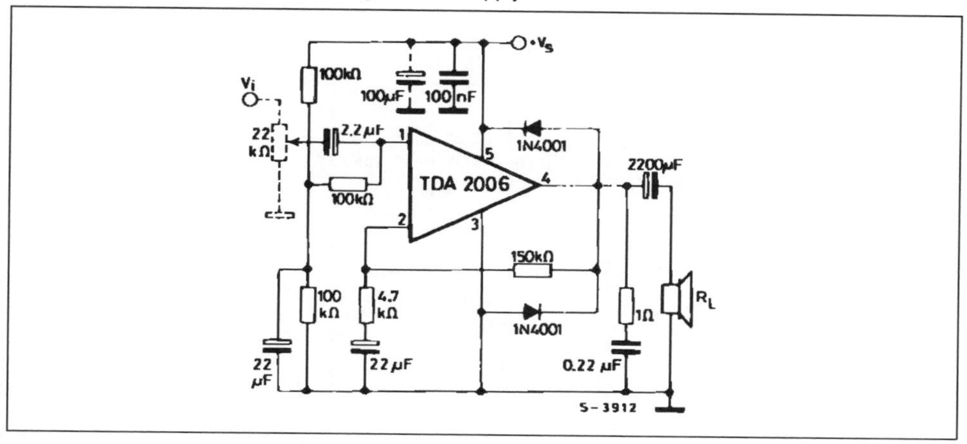

Figure 16 : P.C. Board and Components Layout of the Circuit of Figure 15 (1:1 scale)

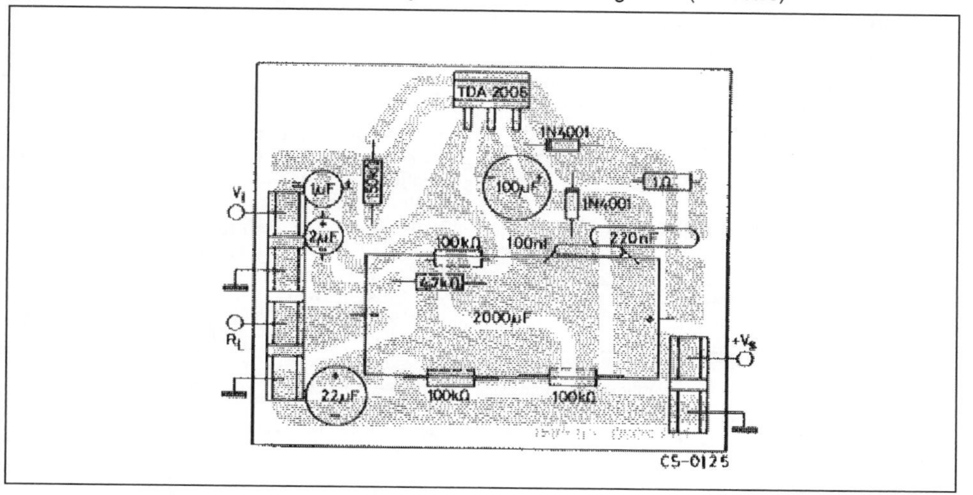

TDA2006

Figure 17 : Bridge Amplifier Configuration with Split Power Supply (P_O = 24W, V_S = ± 12V)

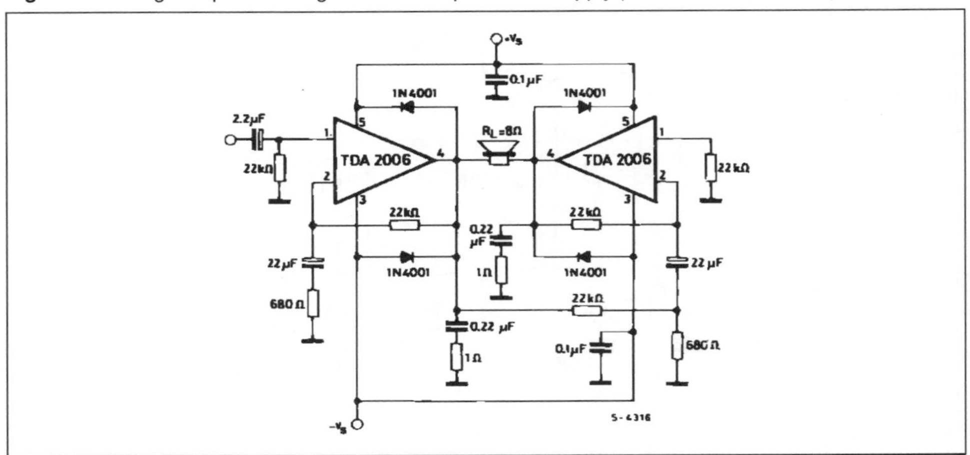

PRACTICAL CONSIDERATIONS

Printed Circuit Board

The layout shown in Figure 14 should be adopted by the designers. If different layout are used, the ground points of input 1 and input 2 must be well decoupled from ground of the output on which a rather high current flows.

Assembly Suggestion

No electrical isolation is needed between the package and the heat-sink with single supply voltage configuration.

Application Suggestion

The recommended values of the components are the ones shown on application circuits of Figure 13. Different values can be used. The table 1 can help the designers.

Table 1

Component	Recommanded Value	Purpose	Larger Than Recommanded Value	Smaller Than Recommanded Value
R_1	22 kΩ	Closed Loop Gain Setting	Increase of Gain	Decrease of Gain (*)
R_2	680 Ω	Closed Loop Gain Setting	Decrease of Gain (*)	Increase of Gain
R_3	22 kΩ	Non Inverting Input Biasing	Increase of Input Impedance	Decrease of Input Impedance
R_4	1 Ω	Frequency Stability	Danger of Oscillation at High Frequencies with Inductive Loads	
R_5	3 R_2	Upper Frequency Cut-off	Poor High Frequencies Attenuation	Danger of Oscillation
C_1	2.2 μF	Input DC Decoupling		Increase of Low Frequencies Cut-off
C_2	22 μF	Inverting Input DC Decoupling		Increase of Low Frequencies Cut-off
C_3C_4	0.1 μF	Supply Voltage by Pass		Danger of Oscillation
C_5C_6	100 μF	Supply Voltage by Pass		Danger of Oscillation
C_7	0.22 μF	Frequency Stability		Danger of Oscillation
C_8	$\frac{1}{2\pi BR_1}$	Upper Frequency Cut-off	Lower Bandwidth	Larger Bandwidth
D_1D_2	1N4001	To Protect the Device Against Output Voltage Spikes.		

(*) Closed loop gain must be higher than 24dB.

SHORT CIRCUIT PROTECTION

The TDA2006 has an original circuit which limits the current of the output transistors. Figure 18 shows that the maximum output current is a function of the collector emitter voltage ; hence the output transistors work within their safe operating area (Figure 19).

This function can therefore be considered as being peak power limiting rather than simple current limiting.

It reduces the possibility that the device gets damaged during an accidental short circuit from AC output to ground.

THERMAL SHUT DOWN

The presence of a thermal limiting circuit offers the following advantages :

1) an overload on the output (even if it is permanent), or an above limit ambient temperature can be easily supported since the T_j cannot be higher than 150°C.
2) the heatsink can have a smaller factor of safety compared with that of a conventional circuit. There is no possibility of device damage due to high junction temperature.

If for any reason, the junction temperature increases up to 150 °C, the thermal shutdown simply reduces the power dissipation and the current consumption.

The maximum allowable power dissipation depends upon the size of the external heatsink (i.e. its thermal resistance) ; Figure 22 shows the dissipable power as a function of ambient temperature for different thermal resistances.

Figure 18 : Maximum Output Current versus Voltage $V_{CE\ (sat)}$ accross each Output Transistor

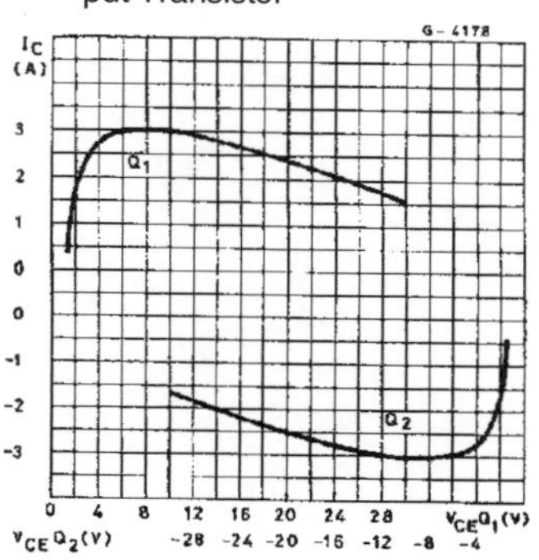

Figure 19 : Safe Operating Area and Collector Characteristics of the Protected Power Transistor

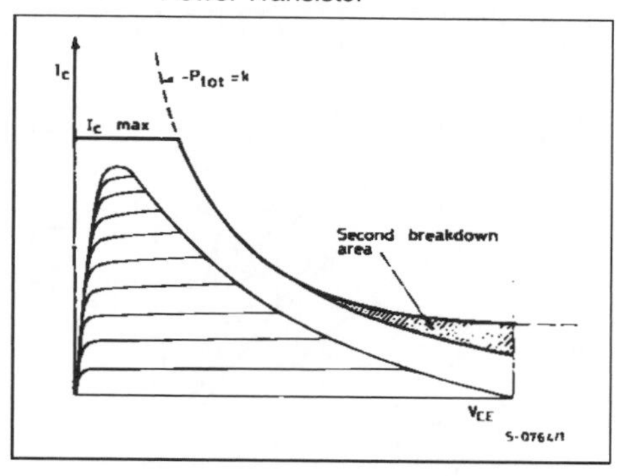

Figure 20 : Output Power and Drain Current versus Case Temlperature (R_L = 4Ω)

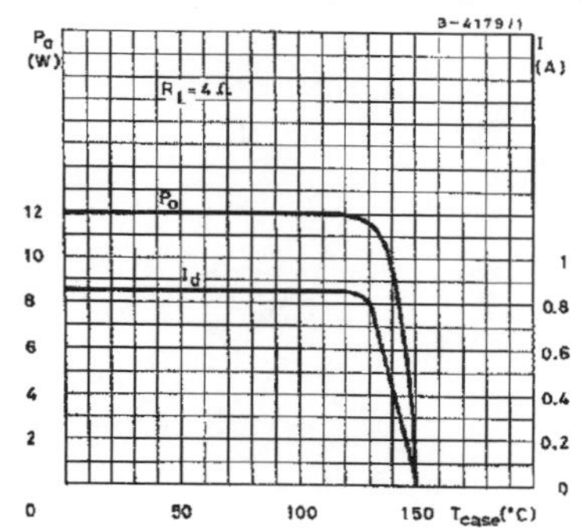

Figure 21 : Output Power and Drain Current versus Case Temlperature (R_L = 8Ω)

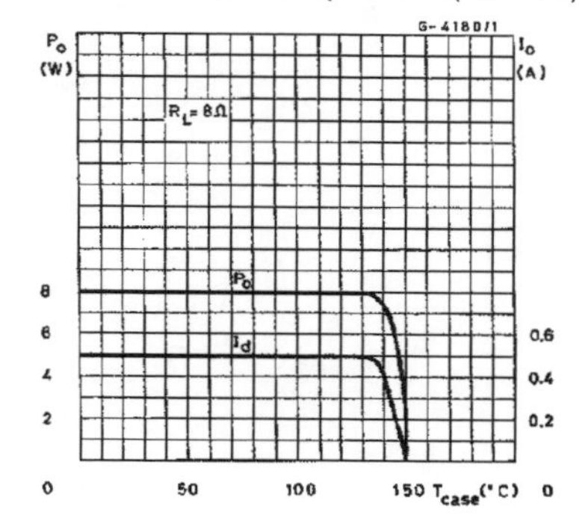

Figure 22 : Maximum Allowable Power Dissipation versus Ambient Temperature

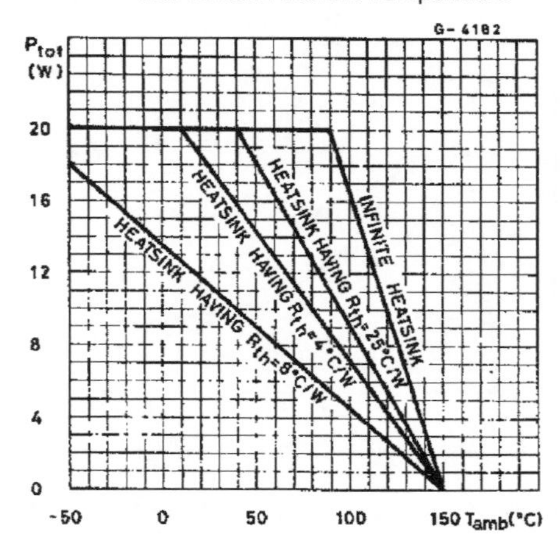

DIMENSION SUGGESTION

The following table shows the length of the heatsink in Figure 23 for several values of P_{tot} and R_{th}.

P_{tot} (W)	12	8	6
Lenght of Heatsink (mm)	60	40	30
R_{th} of Heatsink (°C/W)	4.2	6.2	8.3

Figure 23 : Example of Heatsink

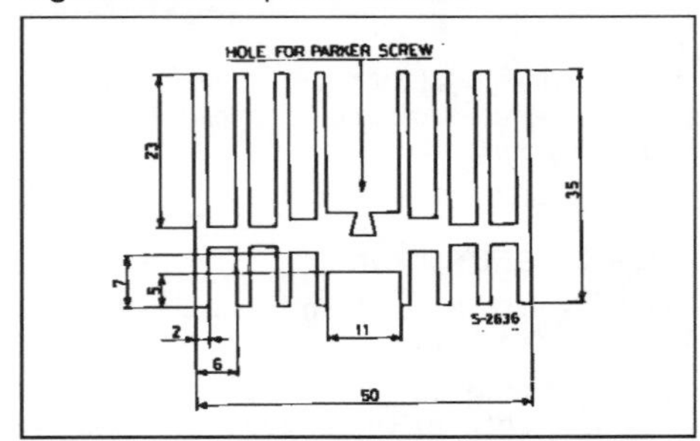

PENTAWATT PACKAGE MECHANICAL DATA

DIM.	mm			inch		
	MIN.	TYP.	MAX.	MIN.	TYP.	MAX.
A			4.8			0.189
C			1.37			0.054
D	2.4		2.8	0.094		0.110
D1	1.2		1.35	0.047		0.053
E	0.35		0.55	0.014		0.022
F	0.8		1.05	0.031		0.041
F1	1		1.4	0.039		0.055
G		3.4		0.126	0.134	0.142
G1		6.8		0.260	0.268	0.276
H2			10.4			0.409
H3	10.05		10.4	0.396		0.409
L		17.85			0.703	
L1		15.75			0.620	
L2		21.4			0.843	
L3		22.5			0.886	
L5	2.6		3	0.102		0.118
L6	15.1		15.8	0.594		0.622
L7	6		6.6	0.236		0.260
M		4.5			0.177	
M1		4			0.157	
Dia	3.65		3.85	0.144		0.152

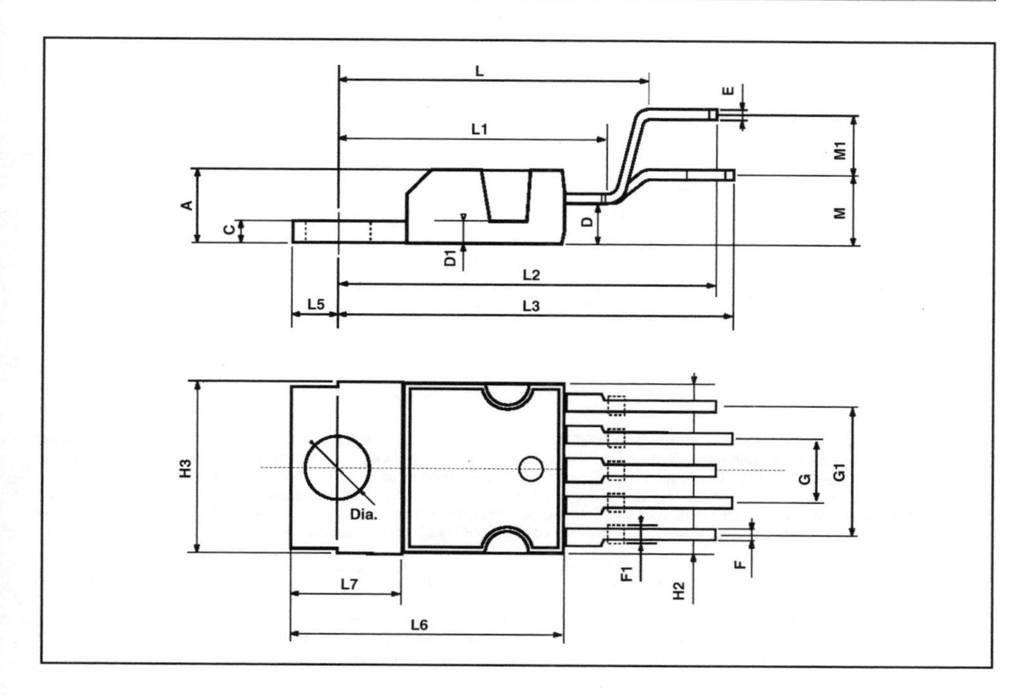

TDA2006

6

Phase-locked loops and synthesizers

6.1 Introduction

Figure 6.1 shows a receiver suitable for detecting signals from fast moving objects where the signal-to-noise ratio is low. The phase detector, amplifier and voltage-controlled oscillator (VCO) form a phase-locked loop (PLL) which is capable of locking on to an incoming carrier and maintaining this lock when the tracking varies. This system is really a feedback loop forming part of a closed-loop control system, and the overall gain is simply the product of all the separate gains for each block in the system.

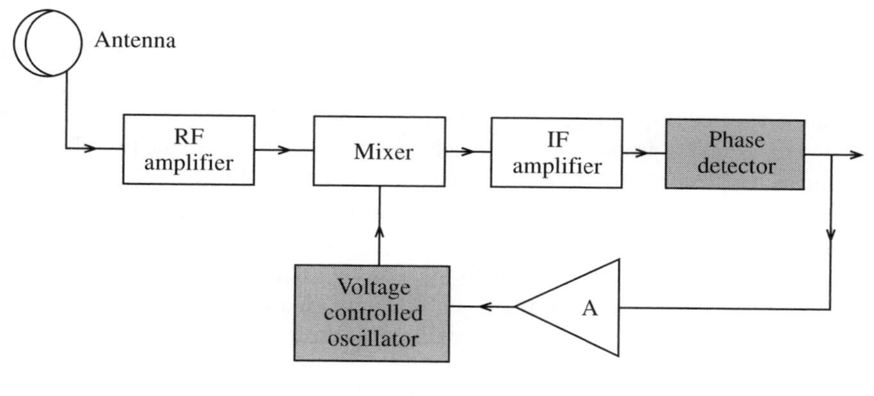

Fig. 6.1

There are many applications for this type of system such as synchronous detection of AM signals, FM modulation, frequency multiplication, automatic frequency control of the television horizontal amplifier and frequency shift keying decoders.

6.2 Operational considerations

A PLL is a form of servo system which consists of a phase detector, low-pass filter and VCO. The phase detector compares the phase of the VCO with the incoming reference signal, giving an output proportional to the difference in phase. This is then filtered to remove unwanted high-frequency components, and the output from the low-pass filter is used to control the frequency of the VCO, locking it to the incoming signal.

It should be noted that phase is the integral of frequency, and there is only a difference in frequency between two signals when the phase is changing between them. When the VCO signal is locked to an incoming reference a steady d.c. error voltage is applied to the VCO, and this is proportional to a constant phase difference between the two signals. Hence there is no actual phase change and the frequency of the VCO must be exactly the same as the reference.

Consider Fig. 6.2. This is a fundamental PLL block diagram which can be assembled discretely or on an integrated chip. It shows what is called a filter tracking circuit.

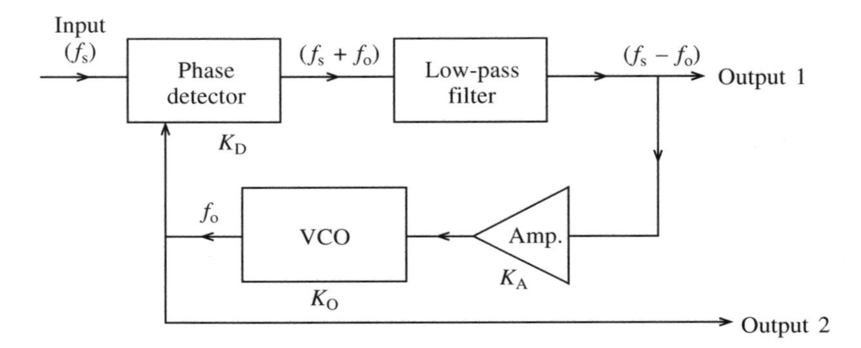

Fig. 6.2

Consider the circuit with no applied signal. There will be no error voltage at the output of the phase detector, and the VCO will run at its natural or free-running frequency (f_o). If an input is applied with a frequency f_s there will be an error voltage generated by the phase detector which is proportional to the phase difference between the two signals. This output consists of sum and difference signals and is filtered through a low-pass filter to leave the difference frequency, which is amplified and applied to the control point on the VCO. The error voltage is such that it reduces the phase, hence the frequency difference between the two signals. Once the loop has locked, the frequency of the oscillator will be exactly the same as that of the reference, but this will be a net phase difference which is necessary to generate the required error voltage to keep the oscillator running at the reference frequency.

When no signal is applied to the PLL it will have no error voltage applied to the oscillator. If a signal is applied and swept towards the oscillator frequency, the phase detector will generate the sum and difference frequencies. If the difference frequency which produces the error voltage falls outside the pass band of the low-pass filter no correction will be applied to the oscillator, but as the reference is swept nearer the oscillator frequency there comes a point where it does fall within the pass band and causes the loop to lock. Therefore, what is called the **capture range** of the loop can be defined as the frequency range over which it can gain acquisition. This range depends largely on the characteristics of the first-order low-pass filter. Once the PLL has achieved capture, it can maintain lock with the input signal over a wider frequency range called the **lock range.**

6.3 Phase-locked loop elements

The four elements of the fundamental block diagram (Fig. 6.2) will now be discussed individually.

Phase detector

The phase detector is really a d.c. coupled mixer with two a.c. input signals. When the signals have the same frequency they are locked and a phase difference results in a d.c. voltage (V_D) appearing at the output of the filter.

The mixer performs a mathematical multiplication. If the two frequencies are given as f_o and f_s then the instantaneous voltages are given by

$$v_s = V_s \sin(\omega_s t + \theta_s)$$

$$v_o = V_o \sin(\omega_o t + \theta_o)$$

where v_s is the instantaneous signal voltage, ω_s is the signal frequency in radians, V_s is the maximum signal voltage, and where v_o is the instantaneous VCO output, ω_o is the oscillator frequency in radians and V_o is the maximum oscillator voltage; thus the output of the detector is given by

$$v = V_o V_s \{\sin(\omega_s t + \theta_s) \sin(\omega_o t + \theta_o)\}$$

Applying the identity

$$\sin A \sin B = \frac{1}{2}\{\cos(A - B) - \cos(A + B)\}$$

we obtain

$$v = \frac{V_s V_o}{2} \cos 2\pi(f_s - f_o + \theta_s - \theta_o)t - \frac{V_s V_o}{2} \cos 2\pi(f_s + f_o + \theta_s + \theta_o)t$$

Hence the output from the detector consists of the sum and difference frequencies. In reality the VCO usually has a square wave output which varies between $+V_o$ and zero. This causes several harmonics to be produced as well as the components mentioned above. However, the low-pass first-order filter response is such that all unwanted frequencies are out with its pass band and only the difference frequency is passed. The output from the filter is therefore given by

$$v = \frac{V_s V_o}{2} \cos\{(2\pi f_s - 2\pi f_o) + (\theta_s - \theta_o)\} \tag{6.1}$$

When $f_s = f_o$ (6.1) becomes

$$v = \frac{V_s V_o}{2} \cos (\theta_s - \theta_o) \tag{6.2}$$

If $(V_s V_o)/2 = K_D$ (the phase detector gain factor) and $\theta_e = \theta_s - \theta_o$ where θ_e is the static phase error, then

$$v = K_D \cos \theta_e \tag{6.3}$$

When $\theta_e = 90°$ the error voltage is zero; hence it is customary to express (6.2) as

$$v = K_D \sin \theta_e \tag{6.4}$$

so that the static phase error is then shown as the variation from a quadrature relationship. If the static phase error is small, as it generally is in practice, then (6.4) becomes

$$v = K_D \theta_e \tag{6.5}$$

Figure 6.3 shows how the d.c. correction voltage due to θ_e is derived between the VCO and input signal. This can be achieved practically by connecting an oscilloscope at the output of the phase detector. Figure 6.3(a) shows the two signals out of phase by 90°. When this is the case a zero d.c. voltage output results. If the phase shift is greater than 90° as in Fig. 6.3(b), a small negative d.c. output is produced. Finally, if the signals are in phase then a small positive d.c. output voltage is produced, as is shown in Fig. 6.3(c). It should be appreciated, however, that small changes of θ_e are considered in practice. From these diagrams the average d.c. value of the unfiltered output is proportional to the phase difference between f_s and f_o. Only the d.c. component is present at the output of the filter.

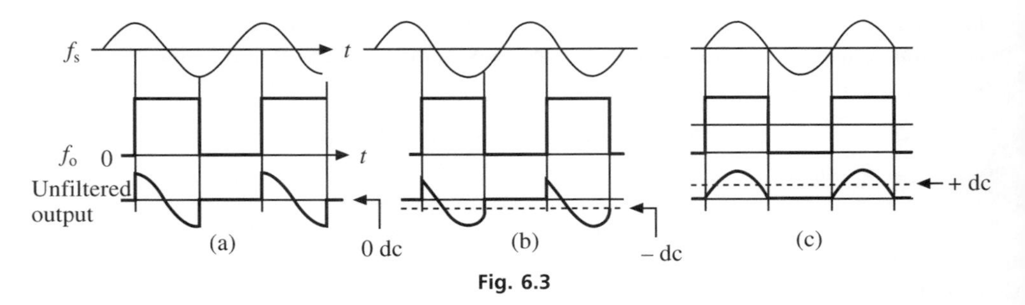

Fig. 6.3

Equation (6.4) showed the phase detector gain factor (K_D). This factor is generally given in volts per radian and is derived from the fact that the phase detector characteristic is a sinusoidal output with period 2π radians. Tracking is limited between ±π/2 otherwise positive feedback would be applied to the loop, with resultant instability. This is shown in Fig. 6.4(a). The slope of the characteristic is given by the derivative of expression (6.4) with $\theta_e = 0$,

$$\frac{d}{d\theta_e} K_D \sin \theta_e \bigg|_{\theta_e = 0} = K_D \cos \text{O} = K_D$$

as indicated in Fig. 6.4(b). This figure shows the response when the VCO output is a sine wave, while Fig. 6.4(c) shows the response when the VCO output is square.

There is one further point which should be understood before leaving the PLL. Once

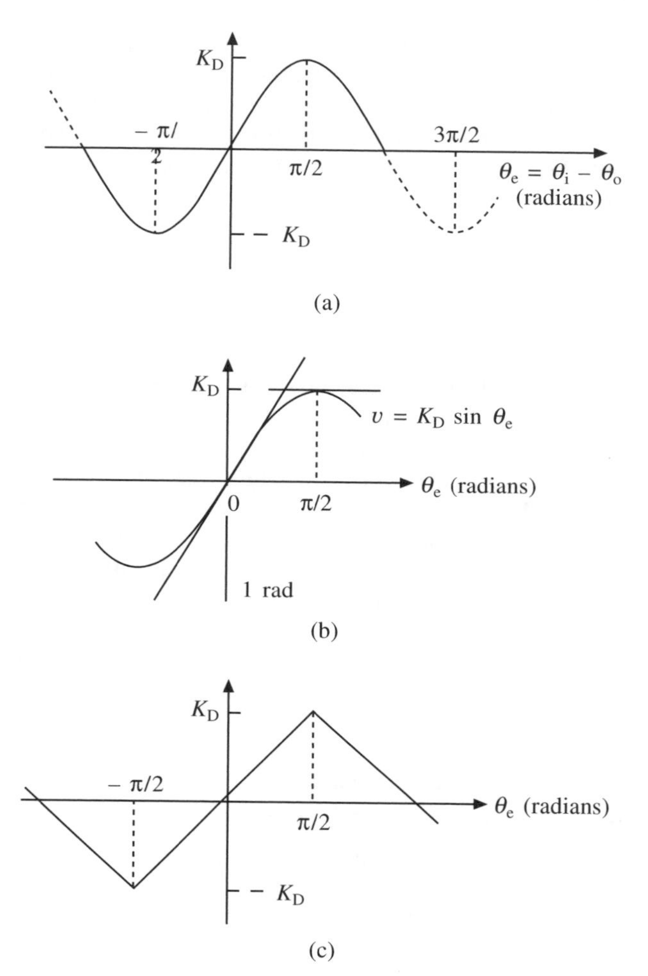

Fig. 6.4

acquisition has been gained the loop will remain locked over the lock range. There are two main factors which affect this. First, the range over which the VCO can swing will be limited, and if the reference goes beyond this then the lock will be lost. Second, the output of the phase detector is sensitive to both phase and amplitude in most cases. Hence if the signal amplitude decreases, the phase difference between the two signals must increase to compensate for this. Therefore at low signal levels the effective lock range is reduced and the phase errors will be larger. Conversely, a large signal will permit use of the full oscillator swing and give smaller phase errors.

Amplifier

The amplifier is a d.c. amplifier which amplifies the phase detector output. It can be a transistor type, or an operational amplifier can be used in its inverting and non-inverting modes. The gain of the amplifier (K_v) can be altered by means of external resistors if a discrete system is used or else by connecting a single resistor externally for chip operation. This will be shown later in a practical example. It is important that the bandwidth of the d.c. amplifier should be very high compared with the loop bandwidth or there will be instability in the loop.

Voltage-controlled oscillator

This element functions virtually as a voltage-to-frequency convertor in which the output frequency is a direct result of the phase error voltage. Generally the VCO incorporates two varactor diodes fabricated into a multivibrator configuration. When a constant d.c. correction voltage is applied the VCO will output its free-running frequency which is determined by an external *RC* network. When this free-running frequency is plotted against the phase error voltage the conversion gain (K_o) of the VCO can be obtained. This conversion gain is defined as the change of free-running frequency to the applied input error voltage. It should be a linear characteristic, and operation is generally centred around the midpoint of the characteristic.

The LM566 is a general-purpose VCO, the data sheets for which are given at the end of the chapter. A typical configuration for this chip is shown in Fig. 6.5. In this configuration the chip operates as a function generator. If this integrated circuit is used the following points should be noted:

(a) R_1 should lie between 2 kΩ and 20 kΩ.

(b) V_s should lie between (+*V*) and 0.75 (+*V*).

(c) The free-running frequency should be less than 1 MHz.

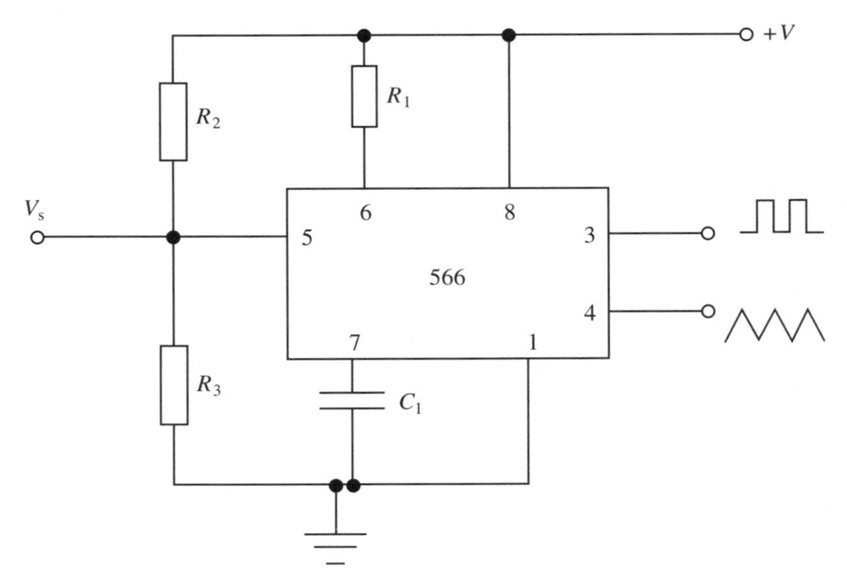

Fig. 6.5

(d) The supply voltage should lie between 10 V and 24 V.

(e) R_2 and R_3 set the modulating voltage at a fixed level by use of a voltage divider network.

(f) The free-running frequency is given by

$$f_o = \frac{2}{R_1 C_1} \frac{(+V) - V_S}{(+V)} \tag{6.6}$$

This chip has been introduced to show how the VCO may be used for a separate function. The conversion gain is given in the data sheets.

Filter

The function of the low-pass filter is to pass the difference frequency while suppressing any noise or interference. However, it also improves the frequency response; this can be done by using either a lag compensation or lead-lag compensation filter.

All the PLL elements are d.c. coupled except for the phase detector input and the VCO output. Also the bandwidth of a phase detector can be very high which means the *RC* output time constant of the phase detector limits the frequency response. This is the filter which passes the difference frequency. Also the VCO frequency response is mainly responsible for limiting the loop bandwidth. For a PLL which has no compensation built into it, the overall uncompensated loop gain is determined by the product of the individual element responses. This is given by

$$K_L = 2\pi K_D K_A K_O \tag{6.7}$$

where K_L is the overall loop gain, K_A is the amplifier gain and K_O is the VCO loop gain. It can also be shown that for an uncompensated PLL the bandwidth is given by

$$K_L = BW \tag{6.8}$$

6.4 Compensation

For many applications the uncompensated loop is too slow, and if there is a sudden change in input frequency the output waveform will be rounded instead of square. The loop response can be improved by introducing some form of loop compensation in the form of a phase delay network.

These compensation networks are generally of three types, referred to as **phase lag, phase lead** or **lead-lag** circuits, and the amount of compensation is best shown in the form of a **Bode plot.** An explanation of these terms will be given initially before their application to compensation is examined.

The Bode plot

It is convenient and generally accepted in practice to plot the logarithmic value of frequency response rather than the raw frequency itself. This is because a greater range of frequencies can then be plotted on suitable graph paper without losing the resolution of the response curve. Log-linear or log-log graph paper may be conveniently used. Figure 6.8 uses log-linear or semilog paper, with the frequency plotted horizontally and the gain (dB) plotted vertically. Log-linear graph paper is also used to plot the phase shift between the input and output responses.

A typical single pole transfer function is given by

$$A_H = \frac{A_M}{1 + j\left(\dfrac{f}{f_H}\right)} \tag{6.9}$$

Taking logs of both sides of (6.9) gives

$$20 \log_{10} | A_H | = 20 \log_{10} | A_M | - 20 \log_{10} \sqrt{1 + \left(\frac{f}{f_H}\right)^2} \tag{6.10}$$

where A_H represents the high frequency gain and A_M represents the midband gain of the circuit.

If (6.10) is plotted on semilog graph paper then a response similar to that shown in Fig. 6.6(a) will be obtained. This figure shows the response which would be obtained if f_H was plotted against A_H (shown as a dotted line), and also the straight-line approximation known as the Bode plot. It consists of an idealized plot of two straight lines which represent the midband gain and the high-frequency gain. Obviously there will be a slight error here, but it gives a good representation of the expected response without a lengthy plotting procedure.

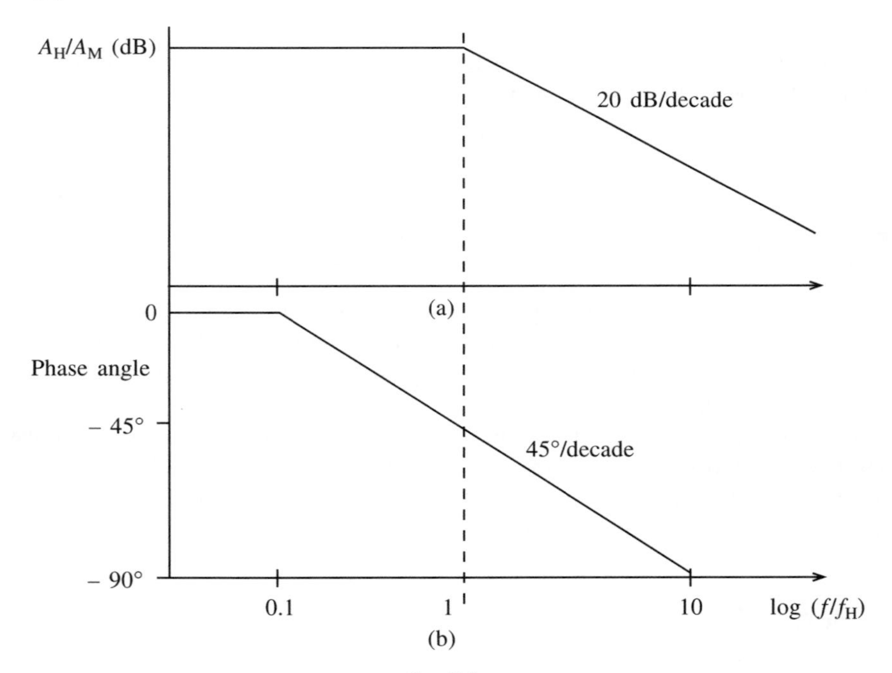

Fig. 6.6

Note that the response rolls off at 20 dB/decade after the cut-off frequency (f). This is characteristic of a single-pole transfer function. A two-pole function would have a roll-off of 40 dB/decade. This type of response has already been discussed in Chapter 3.

A phase/frequency plot is also shown in Fig. 6.6(b). This is also a straight-line approximation, and as can be seen, for frequencies below 0.1 f_H the phase is assumed to be zero. For frequencies above 10 f_H the phase is assumed to be –90°. This plot is assumed to be a straight line between these two points, passing through –45° at $f = f_H$. The slope is therefore 45°/decade. For a two-pole function it would be 90°/decade.

Delay networks

These networks were encountered in our discussion of passive low and high-pass filter networks. The basic types used in compensation networks are shown in Fig. 6.7.

A lead network is an R_1C_1 circuit in which the output voltage across R_1 leads the input voltage, while a lag network is an R_2C_2 network in which the output voltage across C_2 lags the input voltage. The lead-lag network is a combination of these two. At lower frequencies the lead circuit (R_1C_1) dominates due to the high reactance of C_1. As the frequency increases X_{C1} decreases, thus allowing the output voltage to increase. At a particular frequency the response of the lag circuit (R_2C_2) takes over and the decreasing value of X_{C2} causes the output voltage to decrease. Hence the lead-lag network has a particular frequency at which the phase shift through the circuit is zero degrees and the attenuation falls. Below a particular frequency the lead circuit causes the output to lead the input, and above a particular frequency the lag circuit causes the output to lag the input. This principle is used for frequency compensation.

Compensation analysis

If a first-order filter similar to that shown in Fig. 6.7(b) is added to the system then a second-order system is produced because the VCO acts as an integrator.

The transfer function of this filter for a sinusoidal input is given by

$$F(\mathrm{j}\omega) = \frac{V_o(\mathrm{j}\omega_c)}{V_i(\mathrm{j}\omega_c)}$$

$$= \frac{1}{\mathrm{j}\omega_c C_2} \Big/ \left(R_2 + \frac{1}{\mathrm{j}\omega_c C_2} \right)$$

$$= \frac{1}{1 + \mathrm{j}\omega_c C_2 R_2} \tag{6.11}$$

At the 3 dB or corner frequency the magnitude of the filter transfer function is

$$|F(\mathrm{j}\omega_c)| = \frac{1}{\sqrt{2}} = \frac{1}{\sqrt{1 + \omega_c^2 C_2^2 R_2^2}}$$

Hence

$$f_c = \frac{1}{2\pi R_2 C_2} \tag{6.12}$$

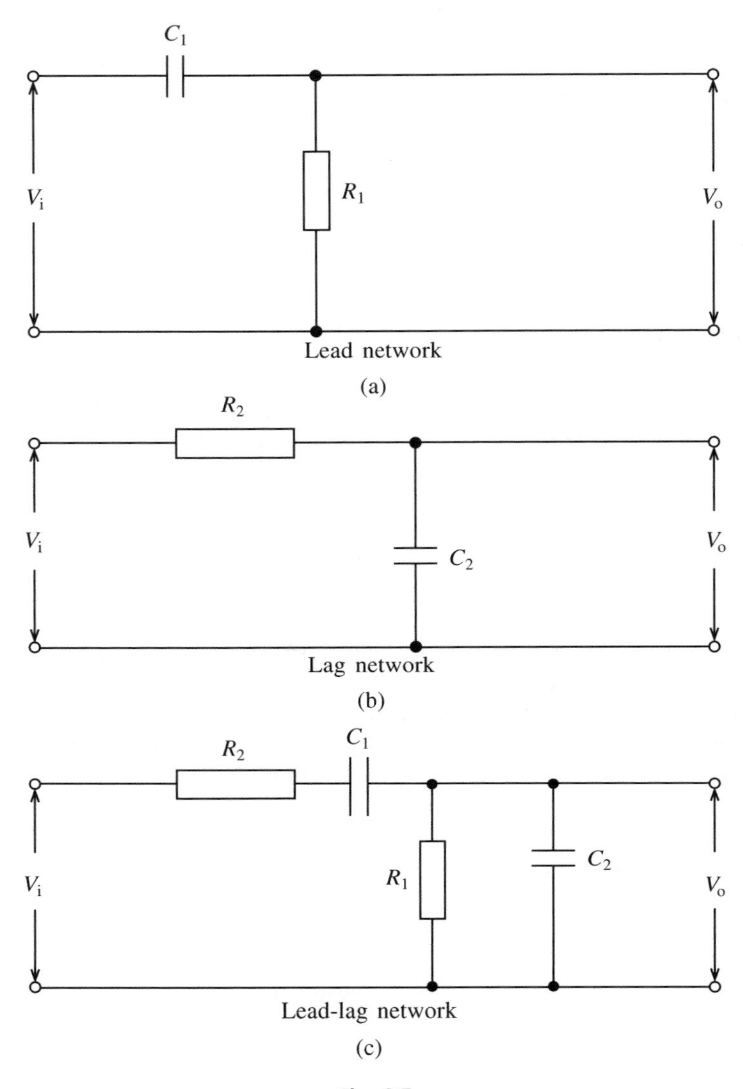

Fig. 6.7

This compensation is generally shown in the form of a Bode plot similar to that of Fig. 6.8. This plot shows how the output voltage remains constant at 0 dB up to the cut-off frequency (f_c) and how thereafter attenuation occurs at 20 dB/decade.

Lead-lag compensation is also used, and a variation on the circuit of Fig. 6.7(c) is shown in Fig. 6.9. The filter transfer function for this filter is given by

$$F(j\omega_c) = \left(R_2 + \frac{1}{j\omega_a C}\right) \bigg/ \left(R_1 + R_2 + \frac{1}{j\omega_b C}\right)$$

$$= \frac{1}{1 + j\omega_a CR_2} \bigg/ \{1 + j\omega_b C(R_1 + R_2)\}$$

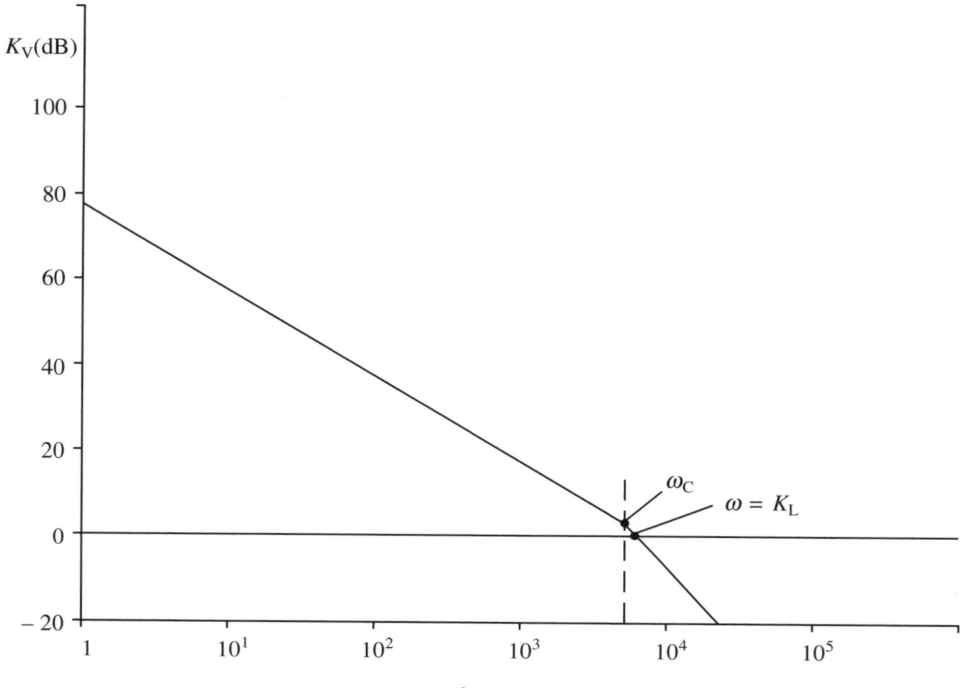

Fig. 6.8

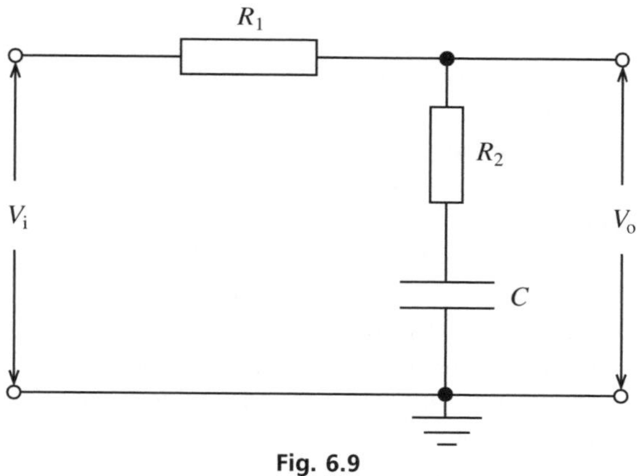

Fig. 6.9

Two cut-off frequencies are indicated here at the 3 dB points:

$$\omega_a = \frac{1}{CR_2} \tag{6.13}$$

and

$$\omega_b = \frac{1}{C(R_1 + R_2)} \tag{6.14}$$

Once again this is shown on the Bode plot of Fig. 6.10. In this case the output voltage is attenuated at the rate of 20 dB/decade until the second cut-off frequency (ω_b). The output voltage then remains constant.

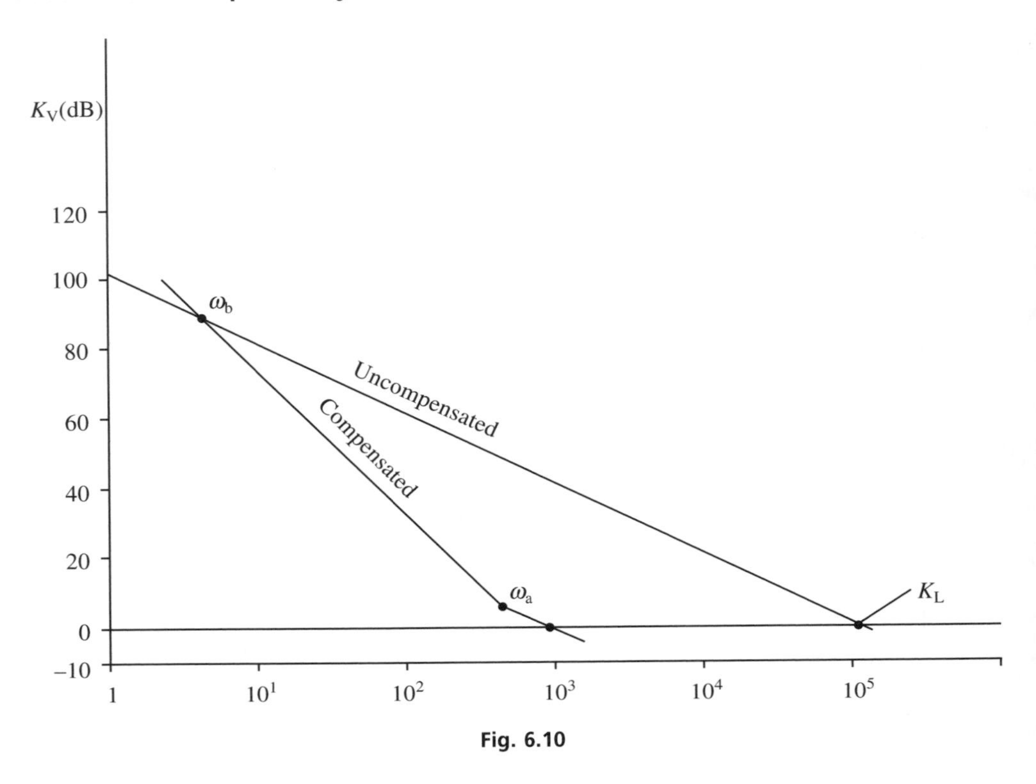

Fig. 6.10

The advantage of lead-lag compensation is that all the parameters (K_L, damping factor and bandwidth) can be varied independently. This is not the case with lag compensation, and it would be a distinct disadvantage in such applications as demodulators where these parameters have to be independently controlled.

Obviously the filter bandwidth can be varied according to the design of the compensating filter. There are advantages in making the pass band small. The narrower the filter pass band, the less likely the loop is to be affected by noise. Also if the *RC* time constant is long, the loop can store the signal frequency if it is temporarily lost. If, however, the signal frequency is lost it is more difficult to regain lock as the capture range will be reduced. Generally a compromise has to be found. The low-pass filter does not alter the lock frequency but it does alter the tracking rate. If tracking is required at high frequencies, the cut-off frequency should be high, but not too high, in order to avoid noise interference.

Example 6.1

A PLL has the following component gains: $K_D = 0.4$ V/rad, $K_A = 5$ and $K_O = 25$ kHz/V. Determine the loop gain.

Solution

$$K_L = 2\pi K_D K_A K_O = 2\pi \times 0.4 \times 5 \times 25 = 314 \times 10^3 \text{ Hz/rad}$$

$$K_L(\text{dB}) = 20 \log 314 \times 10^3 = 109.9 \text{ dB}$$

Note that this answer can also be expressed in kilohertz per radian:

$$K_L = K_D K_A K_O = 0.4 \times 5 \times 25 = 50 \text{ kHz/rad}$$

Example 6.2
A PLL has parameters $K_A = 5$, $K_O = -35$ kHz/V, $K_D = 0.3$ V/rad and $f_o = 220$ kHz. If the input signal (f_s) is 180 kHz and the loop is locked, determine:

(a) the loop gain in kHz/rad;

(b) the VCO output frequency;

(c) the static phase error;

(d) the lock range;

(e) the maximum possible value of v.

Solution
(a)

$$K_L = K_O K_D K_A = -35 \times 0.3 \times 5 = -52.3 \text{ kHz/rad}$$

(b) When the loop is locked, $f_s = f_o = 220$ kHz.
(c) As the loop is locked the input voltage to the VCO must be

$$V_i = \frac{180 - 220}{-35} = \frac{-40}{-35} = 1.143 \text{ V}$$

(This is a d.c. voltage, of course.) Since $K_A = 5$,

$$v = \frac{V_i}{K_A} = \frac{1.143}{5} = 0.229 \text{ V}$$

Also

$$\theta_e = \frac{v}{K_D} \text{ from expression (6.5)}$$

$$= \frac{0.229}{0.3} = 0.763 \text{ rad}$$

(d) It has already been seen from (c) that V_i is required to keep the system in lock, so the phase comparator must produce a voltage of

$$v = \frac{V_i}{K_A} = \frac{\Delta f}{K_O K_A} \tag{6.15}$$

where Δf is the lock range. Also $\theta_e = v/K_D$, hence

$$v = K_D \theta_e \tag{6.16}$$

Substituting (6.13) into (6.14) gives

$$\theta_e K_D = \frac{\Delta f}{K_O K_A}$$

$$\therefore \qquad \theta_e = \frac{\Delta f}{K_L} \tag{6.17}$$

Figure 6.4(b) shows the VCO characteristics for a sinusoidal input. v will reach a maximum at $\theta_e = \pi/2$. Hence the total lock range is $\pm\pi/2$ or π radians. This is the maximum static phase error $\theta_{e(max)}$. Hence from expression (6.17)

$$\Delta f = \theta_e K_L = \pi \times 52.5 = 165 \text{ kHz}$$

(e) v is maximum when $\theta_e = \pi/2$ so

$$v = K_D \theta_e = 0.3 \times \frac{\pi}{2} = 0.47 \text{ V}$$

Example 6.3

A PLL is compensated by means of a lead-lag compensation network. If $\omega_a = 350$ rad and $\omega_b = 5$ rad, determine the values of the filter components. Assume the output resistance (R_o) of the phase comparator is 35 kΩ.

Solution

Select $C = 0.1$ µF. From (6.13),

$$\omega_a = \frac{1}{R_2 C}$$

$$\therefore \qquad R_2 = \frac{1}{\omega_a C} = \frac{10^6}{350 \times 0.1} = 28.57 \text{ k}\Omega$$

From (6.12),

$$\omega_b = \frac{1}{C(R_1 + R_2)}$$

$$\therefore \qquad R_1 = \frac{1 - \omega_b C R_2}{C\omega_b} = \frac{1 - 5 \times 0.1 \times 10^{-6} \times 28.57 \times 10^3}{0.1 \times 10^{-6} \times 5} = 1.97 \text{ M}\Omega$$

Since $R_o = 35$ kΩ the value of the external resistor R_1 should be 1.97 MΩ.

Example 6.4

Figure 6.11 shows the characteristics of a PLL. The low-pass filter consists of an *RC* network having a cut-off frequency of 40 kHz, and the characteristics for each stage are shown in the annotated diagram. Determine:

(a) K_D, K_V, K_O and K_L;

(b) the values of *R* and *C* for the filter;

(c) the value of V_i, v and θ_e under lock conditions;

(d) the lock range.

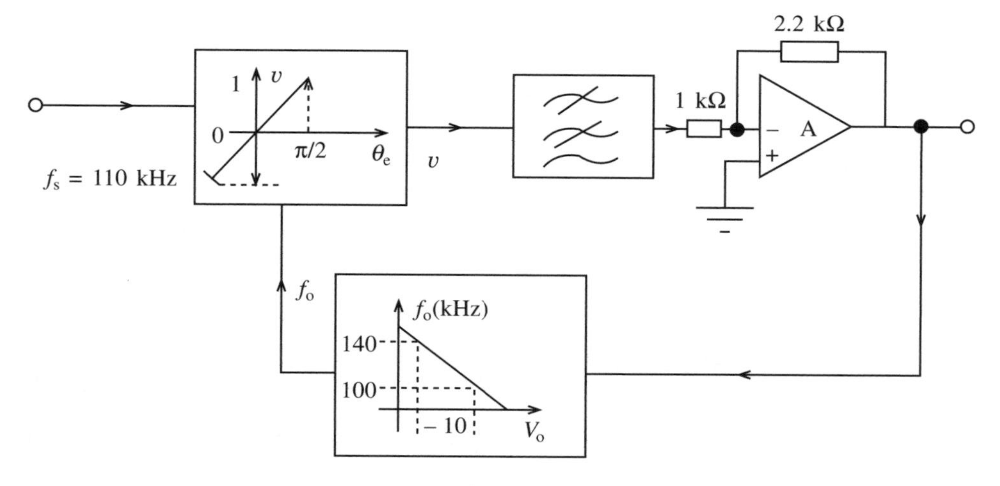

Fig. 6.11

Solution

(a) From the characteristics

$$K_D = 1\ \text{V}\Big/\frac{\pi}{2} = 0.64\ \text{V/rad}$$

$$K_A = \frac{2.2}{1} = 2.2$$

For K_O it is assumed from the characteristics that the output voltage (V_O) is linear between 100 kHz and 140 kHz.

$$K_O = \frac{100 - 140}{(1 - (-1))} = -20\ \text{kHz/V}$$

Hence $K_L = K_A K_D K_O = -2.2 \times 0.64 \times -20 = 28.16$ kHz/rad

(b) This is a first-order *RC* filter, hence

$$f_c = \frac{1}{2\pi RC}$$

Select $C = 0.01\ \mu\text{F}$

$$\therefore \qquad R = \frac{1}{2\pi f_c C} = \frac{10^6}{2\pi \times 40 \times 10^3 \times 0.01} = 398\ \Omega$$

(c)

$$V_i = \frac{f_s - f_o}{K_O}$$

From the characteristics the free-running frequency is 120 kHz. So

$$V_i = \frac{110 - 120}{-20} = 0.5\ \text{V}$$

$$v = \frac{V_i}{K_A} = \frac{0.5}{-2.2} = -0.227\ \text{V}$$

Also

$$\theta_e = \frac{v}{K_D} = \frac{-0.227}{0.64} = -0.355 \text{ rad}$$

(d) As for the previous example the lock range is $\pm \pi/2$ or π rad.

$$\therefore \quad f_L = \pi \times K_L = \pi \times 28.16$$

$$= 88.46 \text{ kHz}$$

6.5 Integrated phase-locked loops

Figure 6.2 showed the various elements of a PLL. However, it is normal practice to use an integrated circuit in most applications. Two commonly used PLL chips are the 565 manufactured by National Semiconductor and the 4046 manufactured by SGS-Thomson Microelectronics. Details of these integrated circuits are given in the data sheets, and these should be consulted when studying the worked examples considered later. Initially the following points should be noted when using the data sheets for the 565.

(a) The loop gain is given as $K_O K_D$, but this is for $K_A = 1$. Typically a value of 1.4 is used, and when this is the case the loop gain is given as $K_O K_D K_A$. From the data sheets

$$K_O K_D = \frac{33.6 f_o}{V_c} \tag{6.18}$$

but

$$K_O K_D K_A = \frac{50 f_o}{V_c} \tag{6.19}$$

(b) Where formulae are stated as $\pm$ values it is easier to keep parameter values in this form. For example, if a lock range (f_L) of $\pm$ 600 Hz is used in a problem in conjunction with a free-running frequency of 1200 Hz, then

$$f_L \pm \frac{8 f_o}{V_C} \tag{6.20}$$

and f_L = 1200 Hz while f_o = 1200 Hz. This accommodates a swing of 600 Hz on either side of the free-running frequency.

(c) The capture range for a simple lag filter is given by

$$f_C = \pm \frac{1}{2\pi} \sqrt{\frac{2\pi f_L}{\tau}} \tag{6.21}$$

where $\tau = 3.6 \times 10^3 C_1$. For a lag-lead filter

$$f_C = \pm \frac{1}{2\pi} \sqrt{\frac{2\pi f_L}{\tau_1 + \tau_2}} \tag{6.22}$$

where $\tau_1 + \tau_2 = (R_1 + R_2)C_1$.

(d) A typical configuration for this chip is given in Fig. 6.12. In this circuit the 680 Ω resistors are included to ensure that both input terminals receive equal bias conditions. R_O should have a value between 2 kΩ and 20 kΩ, and finally the loop gain can be reduced by connecting a resistor between pins 6 and 7.

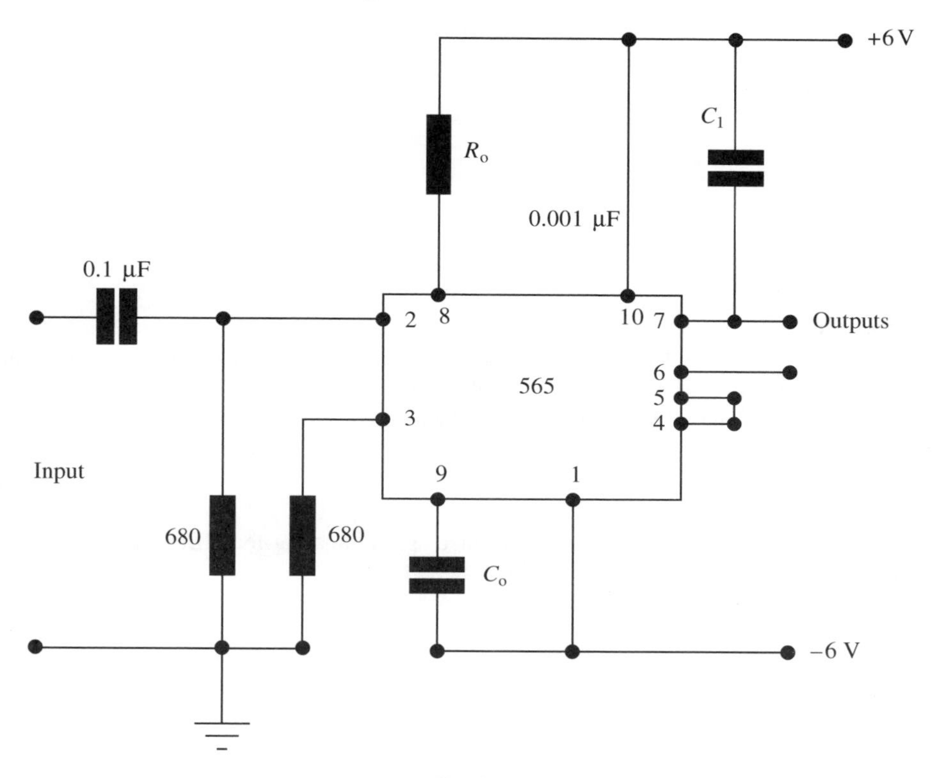

Fig. 6.12

Example 6.5

A 565 PLL is designed for use in a wide closed-loop bandwidth application. If f_o = 1100 Hz, f_L = ±700 Hz and f_C = 400 Hz, calculate R_O, C_O, C_1 and suitable values for the power supply.

Solution

$$f_o = \frac{0.3}{R_O C_O}$$

Selecting C_O = 100 nF,

$$R_O = \frac{0.3}{f_o C_O} = \frac{0.3 \times 10^9}{1100 \times 100} = 2.73 \text{ k}\Omega$$

Also

$$f_C = \pm \frac{1}{2\pi} \sqrt{\frac{2\pi f_L}{\tau}}$$

$$\therefore \qquad \tau = \frac{2\pi f_L}{(2\pi f_C)^2} = \frac{2\pi \times 1400}{(2\pi \times 400)^2} = 1.39 \times 10^{-3} \text{ s}$$

Now $\tau = 3.6 \times 10^3 \times C_1$

$$\therefore \qquad C_1 = \frac{1.39 \times 10^{-3}}{3.6 \times 10^3} = 0.387 \ \mu\text{F}$$

Using equation (6.18) gives

$$f_L = \pm\frac{8f_O}{V_C}$$

$$V_C = \pm\frac{8 \times 1100}{1400} = +6.3 \text{ V}$$

Example 6.6

A 565 PLL has to be used for a narrow-band application in order to reduce noise. The parameters are set at f_o = 1000 Hz, f_L = ±500 Hz and f_C = ±200 Hz. Determine all the required values to satisfy this specification and also the loop gain when the amplifier gain is unity.

Solution

The configuration used in Fig. 6.12 is used once again but a lag-lead filter is used for this application. Since

$$f_o = \frac{0.3}{R_O C_O}$$

selecting C_O = 120 nF gives

$$R_O = \frac{0.3}{f_o C_O} = \frac{0.3}{10^3 \times 120 \times 10^{-9}} = 2.5 \text{ k}\Omega$$

Also

$$f_L = \pm\frac{8 f_o}{V_C}$$

Hence

$$V_C = \pm\frac{8 f_o}{f_L} = \pm\frac{8 \times 10^3}{10^3} = \pm 8 \text{ V}$$

$$f_C = \mp\frac{1}{2\pi}\sqrt{\frac{2\pi f_L}{\tau_1 + \tau_2}}$$

$$\tau = \tau_1 + \tau_2 = \frac{2\pi f_L}{(2\pi f_C)^2} = \frac{2\pi \times 10^3}{(2\pi \times 400)^2} = 9.95 \times 10^{-4} \text{ s}$$

So

$$R_1 C_1 + R_2 C_1 = \tau_1 + \tau_2 = \tau$$

$$\therefore \qquad 9.95 \times 10^{-4} = R_1C_1 + R_2C_1$$

$R_1 = 3.6\ \text{k}\Omega$; selecting $C_1 = 100$ nF gives

$$9.95 = 3.6 \times 10^3 \times 100 \times 10^{-9} + R_2 \times 100 \times 10^{-9}$$

Hence

$$R_2 = \frac{9.95 \times 10^{-4} - 3.6 \times 10^3 \times 100 \times 10^{-9}}{100 \times 10^{-9}} = 63.5\ \text{k}\Omega$$

$C_2 = 0.1\,C_1$; hence

$$C_2 = 10\ \text{nF}$$

For $K_A = 1$,

$$K_O K_D K_A = \frac{33.6 f_o}{V_C} = \frac{33.6 \times 10^3}{16} = 2.1\ \text{kHz/rad}$$

6.6 Phase-locked loop design using the HCC4046B

This chip is a monolithic integrated circuit using a 16-pin package. It is a CMOS micropower PLL device having extra on-board facilities which the 565 does not have. These facilities, together with the circuit description, are explained in the data sheets at the end of the chapter. The functional diagram in the data sheets shows the component connections for the VCO and filter sections. These values are calculated using the same equations as for the 565 and a design information sheet is also provided. The following examples will demonstrate how this PLL is designed for certain specifications.

Example 6.7

An HCC4046B CMOS PLL has to be used with the following parameters: $f_L = \pm 400$ kHz, $f_C = \pm 200$ kHz and $f_o = 500$ kHz. Design a PLL having no offset using comparator 1. Also select a suitable power supply.

Solution

$$f_o = \frac{0.3}{C_1 R_1}$$

(note that the symbols on the data sheet are used). Select $C_1 = 50$ pF. Hence

$$R_1 = \frac{0.3 \times 10^{12}}{5 \times 10^5 \times 50} = 12\ \text{k}\Omega$$

$$f_C = \frac{1}{2\pi}\sqrt{\frac{2\pi f_L}{\tau}}$$

where $\tau = R_3 C_2$.

$$\therefore \qquad \tau = \frac{2\pi f_L}{(2\pi f_C)^2}$$

$$= \frac{2\pi \times 8 \times 10^5}{(2\pi \times 4 \times 10^5)^2} = 796 \text{ nS}$$

Select $C_2 = 10$ nF. Then

$$\tau = R_3 C_2$$

so

$$R_3 = \frac{\tau}{C_2} = \frac{796 \times 10^{-9}}{10 \times 10^{-9}} = 79.6 \ \Omega$$

A suitable value for V_{DD} is 10 V. The schematic diagram is shown in Fig. 6.13.

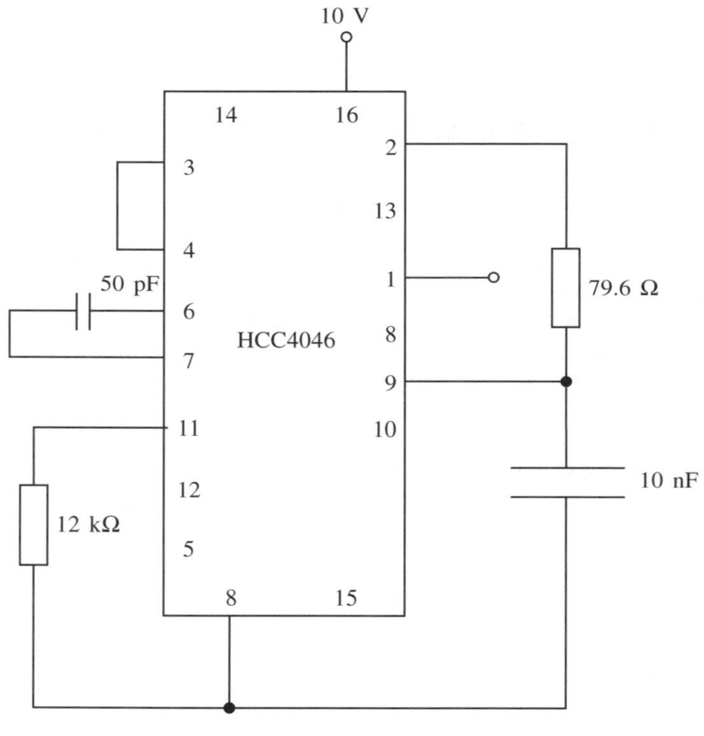

Fig. 6.13

Example 6.8

An HCC4046B chip has to be configured to the following specification: $f_o = 100$ kHz, $f_L = \pm 300$ kHz and $f_C = \pm 300$ kHz. The power supply has to be 15 V and the VCO has to be operated with a variable offset. Calculate all components and draw a schematic diagram which would be suitable for this specification and capable of monitoring the locked condition.

Solution

$$f_o = \frac{0.3}{R_1 C_1}$$

Select $R_1 = 47\ \text{k}\Omega$. Then

$$C_1 = \frac{0.3}{R_1 f_o} = \frac{0.3 \times 10^{12}}{47 \times 10^3 \times 100 \times 10^3} = 63.8\ \text{pF}$$

$$f_C = \pm \frac{1}{2\pi} \sqrt{\frac{2\pi f_L}{\tau}}$$

$$\therefore \qquad \tau = \frac{2\pi f_L}{(2\pi f_C)^2} = \frac{2\pi \times 600 \times 10^3}{(2\pi \times 600 \times 10^3)^2} = 0.2653\ \mu\text{S}$$

Select $R_3 = 10\ \text{k}\Omega$

$$C_2 = \frac{\tau}{R_3} = \frac{0.2653 \times 10^{-6}}{10^4} = 26.53\ \text{pF}$$

As $f_L = f_C$, comparator 2 has to be used in this design. Also, in order to operate a variable-frequency offset R_2 should be variable between 5 kΩ and 1 MΩ, hence a 1 MΩ potentiometer should be connected to pin 12. The final design is shown in Fig. 6.14. Note that some trimming of the components may be required.

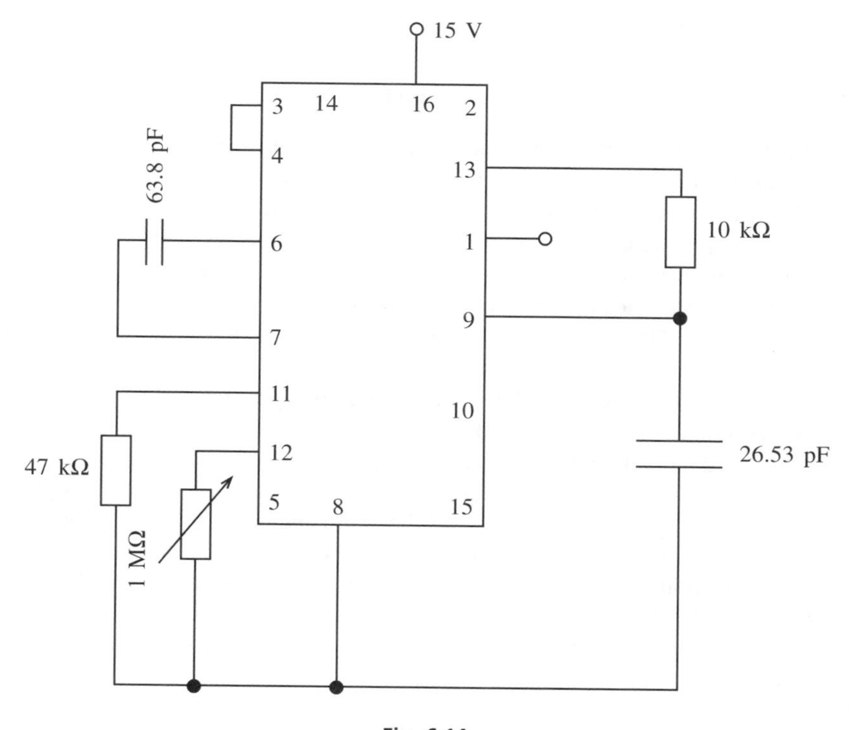

Fig. 6.14

6.7 Frequency synthesis

Frequency synthesis is a method of producing a range of frequencies with crystal-controlled stability. Synthesizers are used in test equipment, communication equipment and television and radio channel selection.

A communications receiver is shown in Fig. 6.15. This receiver uses a principle known as **heterodyning** in which two frequencies are mixed together in a demodulator to produce a single frequency known as the intermediate frequency. Each time the receiver is tuned to a particular channel or carrier, conversion takes place to a low or intermediate frequency. In order to achieve this the local oscillator is usually tuned to a frequency above that of the receiver RF signal. The difference frequency is selected while all other frequencies are suppressed. The difference frequency is then passed on to the IF amplifier.

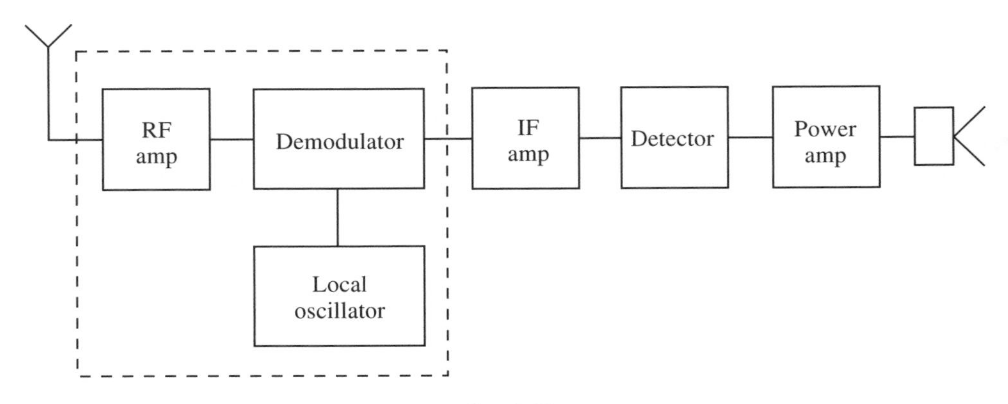

Fig. 6.15

Modern digital receivers use the same principle described above but the use of a synthesizer in the local oscillator stage can provide a system which is highly stable and has very little frequency drift. The synthesizer is variable in discrete steps but uses a crystal for very high stability at each of the discrete frequencies.

Frequency multiplication is generally involved in frequency **synthesis**, and this can be achieved in two ways. The first is to lock the oscillator to some harmonic of the input or reference signal (f_s). This is generally the simplest method, but it has the disadvantage that the range becomes smaller with an increase in harmonics.

The most frequent method is the indirect method in which a PLL and frequency division network is used. This is shown in Fig. 6.16: the general block diagram is shown in Fig. 6.16(a) and a practical system is shown in Fig. 6.16(b). In this case the VCO runs at some harmonic of the input signal (f_s) and comparison takes place at this input signal. The filter has to be carefully designed as the phase detector produces both sum and difference frequencies and when locked the difference frequency is d.c., but the sum is twice the fundamental. If this is not adequately filtered there will be a ripple on the d.c. error voltage which will cause a frequency modulation on the output.

Figure 6.16 compares two frequencies and produces a d.c. error voltage which causes the two frequencies to be locked. This gives

$$f_s = \frac{f_o}{N} \tag{6.23}$$

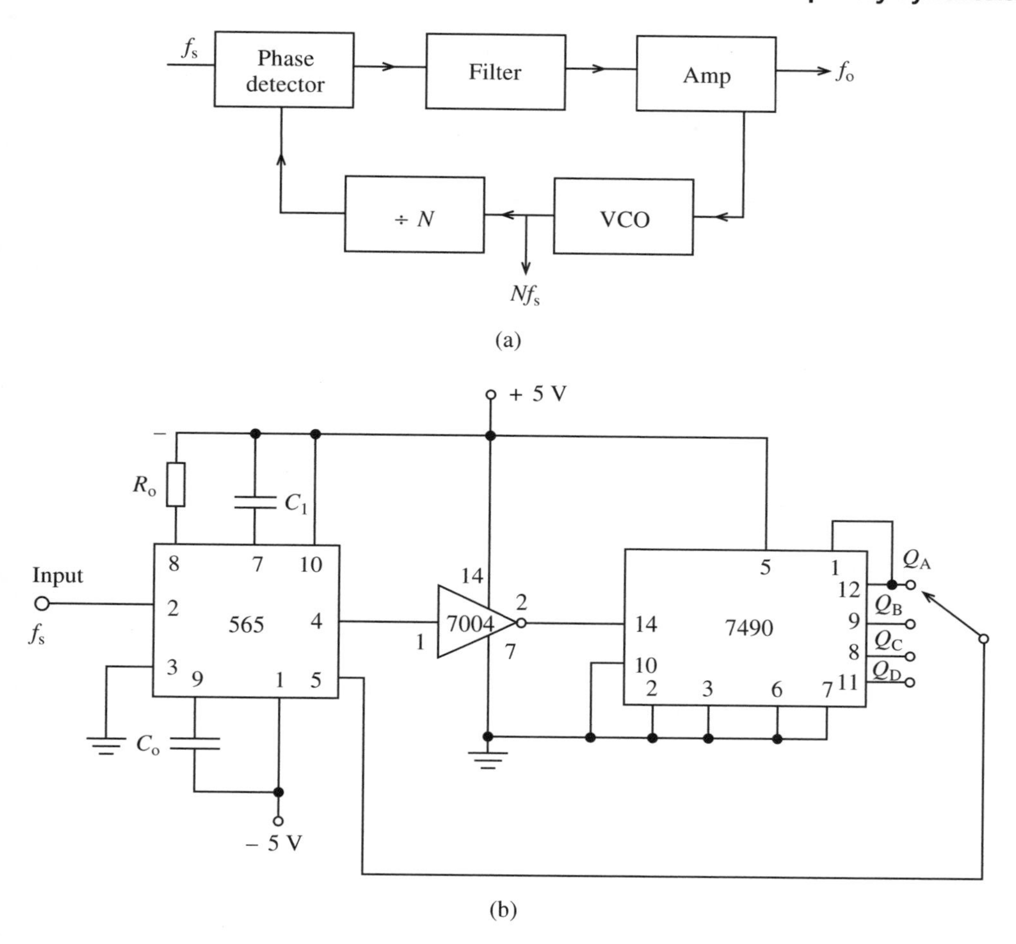

Fig. 6.16

and using a programmable divider will enable N to be varied. This type of divider is different from a presettable or fixed-modulus divider in that its moduli are programmed for the particular input channels which may be required.

The presettable divider is the basic building block which can be used to construct a divider which has any modulus. Practically all presettable dividers are constructed by using four JK flip-flops, and they may be connected in such a way that they function as decade counters or in the count-up or count-down mode. A flip-flop is a bistable device, that is, it has two stable states, namely, a high (1) state and a low (0) state. The output of a flip-flop indicates which bit it is storing.

A practical circuit is shown in Fig. 6.16(b). In this case a decade counter has been used. The output signal from the 565 is passed through an inverter to the decade counter. The output is taken from pin 12 and so the output from the PLL is twice the free-running frequency of the VCO. The VCO can only be varied over a limited range depending on its linear characteristic, and so it may be necessary to change the VCO frequency by varying the timing capacitor C_1.

The circuit shown in Fig. 6.16 can only generate multiples of f_s and if it is necessary to produce frequencies below f_s then a fixed-modulus divider can be used to divide down. This is shown in Fig. 6.17.

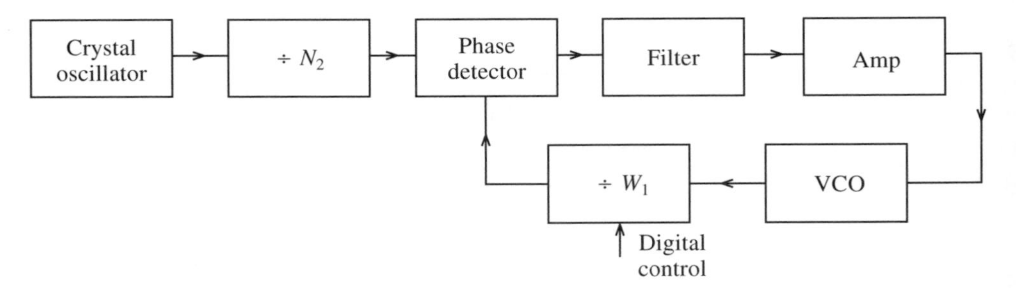

Fig. 6.17

Example 6.9

A radio station using 5 kHz channel separation broadcasts in the range 600–1200 kHz. Design a simple indirect synthesizer which would be suitable for this application if a 5 MHz crystal is used.

Solution

Since the channel separation is 5 kHz, each channel will be a multiple of this frequency. Hence taking f_s = 5 kHz,

$$N_1 = \frac{f_o}{f_s} = \frac{5 \times 10^6}{5 \times 10^3} = 1000$$

N_2 can be specified by assuming that the frequency will change by 5 kHz each time N_1 is incremented or decremented by 1: at 600 Hz

$$N = \frac{f_o}{f_s} = \frac{600}{5} = 120$$

and at 1200 kHz

$$N = \frac{f_o}{f_s} = \frac{1200}{5} = 240$$

A suitable system design might be as shown in Fig. 6.18. Note that the loop gain for a PLL is given by equation (6.7), and this will be changed by the frequency divider network to give

$$K_L = \frac{2\pi K_D K_A K_O}{N} \tag{6.24}$$

The loop gain is therefore reduced so that the other stages must have higher gain. This should be remedied by using linear amplifiers.

Prescaling

Because of the limitation of programmable dividers at much higher frequencies, it is not

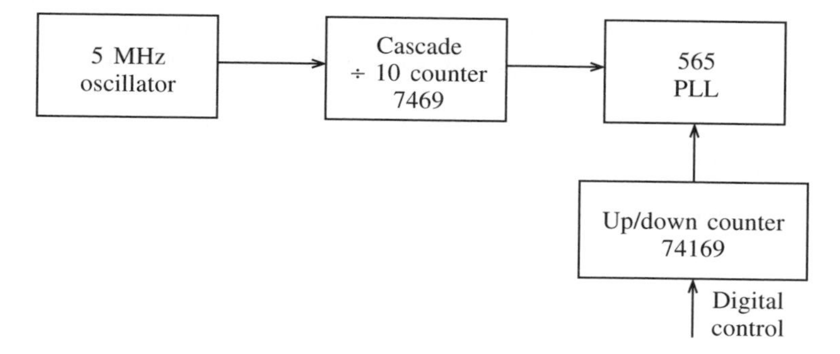

Fig. 6.18

possible to use the simple synthesizer design so far considered. The problem may be partially solved if a fixed-modulus divider is placed before N_1 in Fig. 6.15, but this means that the frequency can only be changed by a minimum amount depending on the value of the new fixed-modulus divider. A method currently used is the inclusion of a two-modulus prescaler similar to the one shown in Fig. 6.19. In this diagram the counter N_1 performs the same function as before, but the prescaler can divide by either N_3 or $(N_3 + 1)$. N_4 switches the prescaler between two different moduli and $N_1 > N_4$.

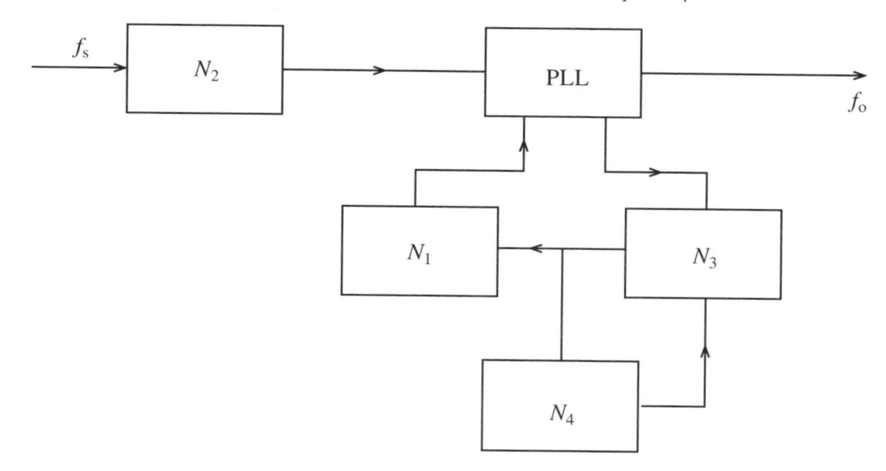

Fig. 6.19

Consider initially the prescaler set to the $(N_3 + 1)$ position and the N_4 counter reset. When the N_4 counter has received the appropriate number of transitions from the VCO the prescaler changes state to N_3. Hence the prescaler will divide by N_3 until the N_4 counter changes state again. Initially the prescaler is set to divide by $(N_3 + 1)$, hence this is the number of transitions required from the VCO to apply one count to the N_4 counter. This counter reaches zero after $N_4(N_3 + 1)$ transitions from pin 4 of the 565. Also the same pulses that go to the N_4 counter are also routed to the N_1 counter. When the N_4 counter reaches zero, the N_1 counter has reached $(N_1 - N_4)$. This value has to be greater than zero and this is why $N_1 > N_4$. The prescaler now switches to its other position and the next N_3 input from the VCO produces one transition from the prescaler output; this is repeated until the N_1 counter reaches zero. This counter begins with a count of $(N_1 + N_4)$ it must receive $(N_1 - N_4)$ transitions from the prescaler. Hence there are $N_3(N_1 - N_4)$ changes from the output of the VCO. Finally, both counters reset (N_1 and N_4) and the prescaler switches back to $(N_3 + 1)$.

The total number of pulses at the output of the VCO for one pulse at the input of the 565 (pin 5) is given by

$$N_4(N_3 + 1) + N_3(N_1 + N_4) = N_4N_3 + N_4 + N_1N_3 - N_4N_3$$
$$= N_4 + N_1N_3 \quad (6.25)$$

Since f_o has been divided by equation (6.23),

$$f_o = (N_4 + N_1N_3)f_s \quad (6.26)$$

Example 6.10
A frequency synthesizer with the design shown in Fig. 6.20 has a prescaler ratio of 15 : 16. Determine the output frequency (f_o).

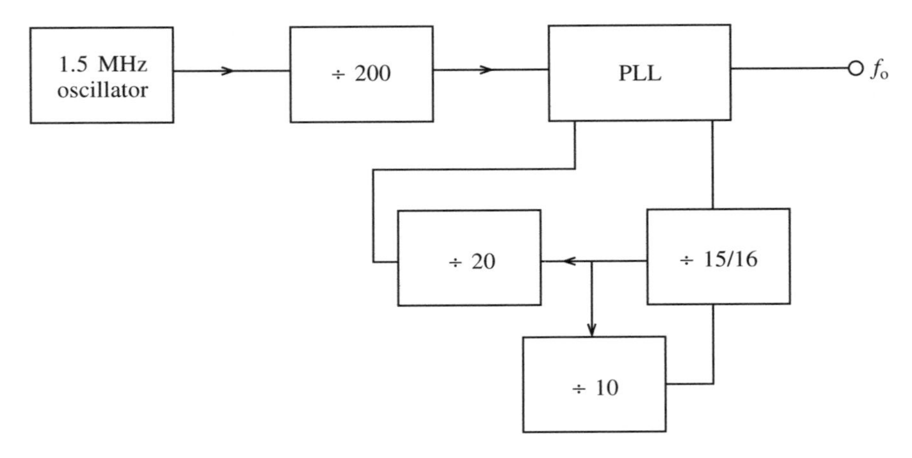

Fig. 6.20

Solution

$$f_o = (N_4 + N_1N_3)f_s$$
$$= (10 + 20 \times 15)\ \frac{1.5 \times 10^6}{200}$$
$$= \frac{310 \times 1.5 \times 10^6}{200}$$
$$= 2.325\ \text{MHz}$$

Example 6.11
A synthesizer uses a two-modulus prescaler with a ratio of 10 : 11. If the input signal to the PLL has to be 5 kHz and the output range has to be 500–900 kHz, determine suitable moduli for the other counters. Assume the incoming signal (f_s) is 400 kHz.

Solution

For a two-modulus system assume N_1 is constant and N_4 is incremented by $(N_4 + 1)$. Initially,

$$\begin{aligned} f_{o1} &= (N_4 + N_1N_3)f_s \\ &= (N_4 + N_1N_3)5 \end{aligned}$$

If N_4 is incremented by 1, then

$$\begin{aligned} f_{o2} &= (N_4 + 1 + N_1N_3)5 \\ f_{o2} - f_{o1} &= (N_4 + 1 + N_1N_3)5 - (N_4 + N_1N_3)5 \\ &= 5 \text{ kHz} \end{aligned}$$

Hence the step size is the same as if no prescaler was used. The resolution is unchanged, and since the incoming signal is 400 kHz, $N_2 = 80$ for $f_s = 5$ kHz. For 500 kHz

$$\begin{aligned} f_o &= (N_4 + N_1N_3)f_s \\ 500 &= (N_4 + N_1 \times 10)5 \\ 100 &= N_4 + 10 \times N_1 \\ N_4 &= 100 - 10 \times N_1 \end{aligned}$$

As $N_1 > N_4$ select $N_1 = 9.5$, hence $N_4 = 5$. For 900 kHz

$$\begin{aligned} 900 &= (N_4 + 10 \times N_1)5 \\ 180 &= (N_4 + 10 \times N_1) \end{aligned}$$

As $N_1 > N_4$ select $N_1 = 19.5$, hence $N_4 = 15$.

6.8 Further problems

1. A PLL has the following element gains: $K_D = 0.3$ V/rad, $K_A = 4$ and $K_O = 30$ kHz/V. Determine the loop gain.
   Answer: 36 kHz/rad

2. A PLL has the following parameters: $K_A = 3$, $K_O = 45$ kHz/V, $K_D = 0.45$ V/rad and $f_o = 365$ kHz. If the input signal is 220 kHz and the loop is locked, determine:
   (a) the loop gain in kHz/rad;
   (b) the VCO output frequency;
   (c) the static phase error;
   (d) the lock range;
   (e) the maximum possible value of v.
   Answer: 60.75 kHz/rad, 365 kHz, 402 rad, 1113.1 kHz, 0.47 V

3. A PLL has to be compensated by using a lead-lag compensation network. If $\omega_a = 260$ rad/s and $\omega_b = 10$ rad/s, determine the value of the filter components. Assume the output resistance (R_o) of the phase comparator is 20 kΩ and select $C = 0.22$ μF.
   Answer: $R_2 = 17.48$ kΩ, $R_1 = 417.1$ kΩ

4. A PLL is compensated by means of a lead-lag compensation network. If $\omega_a = 210$ rad/s and $\omega_b = 6$ rad/s, determine the values of the filter components assuming the phase comparator output resistance is 15 kΩ and the value of C is 47 nF. Also determine the component values if a phase lag compensation network was used with the same value of C.
   Answer: $R_1 = 4.134$ MΩ, $R_2 = 101.32$ kΩ, $R_1 = 101.37$ kΩ

5. Figure 6.21 shows the characteristics of a PLL. The low-pass filter consists of an RC phase lag network having a cut-off frequency of 30 kHz and $C = 0.15$ μF. Determine:
   (a) K_D, K_V, K_O, K_L;
   (b) the values of R and C for the filter;
   (c) the value of V_i, v and θ_e under lock conditions;
   (d) the lock range.
   Answer: 0.64 V/rad, 1.46, –15 kHz/V, 14 kHz/rad, 35 Ω, 0.667 V, –0.457 V, 0.713 rad, 44 kHz

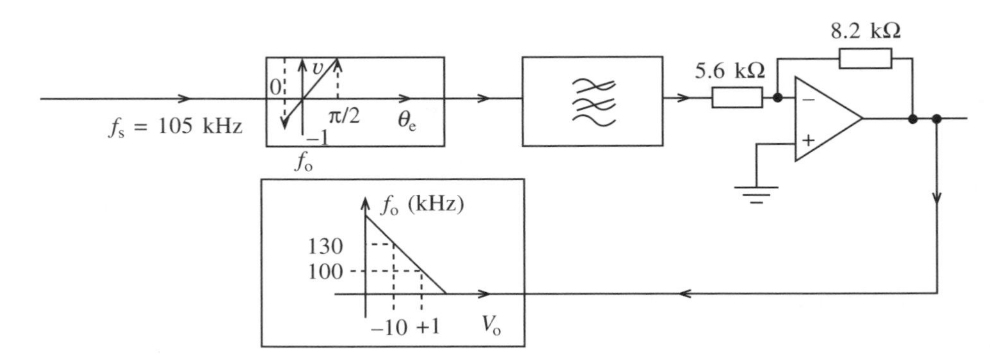

Fig. 6.21

6. Determine typical values for a 565 PLL wide-bandwidth application if $f_o = 1$ kHz, $f_L = \pm 800$ Hz, and $f_C = \pm$ 300 Hz. Select $C_O = 250$ nF.
   Answer: $R_1 = 101.37$ kΩ, $R_o = 1.2$ kΩ, $C_1 = 0.39$ μF, $V_C = \pm 5$ V, $R_2 = 17.6$ kΩ

7. A 565 PLL is used as a narrow-band application in order to correct for jitter in a data communication system. The parameters are $f_o = 1.2$ kHz, $f_L = \pm 600$ Hz and $f_C = \pm 500$ Hz. Calculate the values of the required components and also the loop gain when the amplifier gain is unity. Use a 150 nF capacitor for C_O and 9 nF for C_1.
   Answer: $R_O = 1.67$ kΩ, $V_C = \pm 8$ V, $R_2 = 17.6$ kΩ, 2.52 kHz/rad

8. Design a circuit incorporating an HCC4046B CMOS PLL, to be used with the following parameters: $f_L = \pm 300$ kHz, $f_C = \pm 100$ kHz, and $f_o = 400$ kHz. Select $C_1 = 2$ nF and $C_2 = 4.7$ nF.
   Answer: $R_1 = 375$ Ω, $R_3 = 508.5$ Ω, $V_{DD} = 10$ V

9. Design an indirect synthesizer which could be used in an FM broadcast system operating at 98–110 MHz. The channel separation is 1 MHz and a 20 MHz crystal is used.
   Answer: $N_1 = 20$, $N_2 = 98$

10. Figure 6.22 shows a synthesizer which uses a two-modulus prescaler with a ratio of 8 : 9. Determine the output frequency and explain the operation of this type of synthesizer.
    Answer: 14.9 MHz

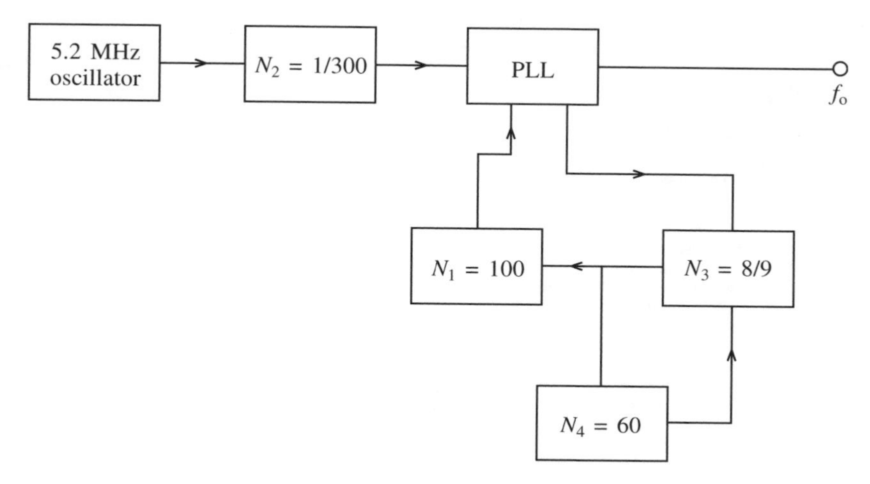

Fig. 6.22

11. A musical synthesizer uses a two-modulus prescaler with a ratio of 10 : 11. If the input signal to the PLL has to be 2.2 kHz and the output range is from 100 Hz to 22 kHz, determine appropriate counters for this application.

12. Explain how a PLL may be tested for free-running frequency, lock range and capture range. Give reasons why a spectrum analyser should not be used in this application.

Data sheets

May 1999

LM565/LM565C Phase Locked Loop

General Description

The LM565 and LM565C are general purpose phase locked loops containing a stable, highly linear voltage controlled oscillator for low distortion FM demodulation, and a double balanced phase detector with good carrier suppression. The VCO frequency is set with an external resistor and capacitor, and a tuning range of 10:1 can be obtained with the same capacitor. The characteristics of the closed loop system — bandwidth, response speed, capture and pull in range — may be adjusted over a wide range with an external resistor and capacitor. The loop may be broken between the VCO and the phase detector for insertion of a digital frequency divider to obtain frequency multiplication.

The LM565H is specified for operation over the −55˚C to +125˚C military temperature range. The LM565CN is specified for operation over the 0˚C to +70˚C temperature range.

Features

- 200 ppm/˚C frequency stability of the VCO
- Power supply range of ±5 to ±12 volts with 100 ppm/% typical
- 0.2% linearity of demodulated output
- Linear triangle wave with in phase zero crossings available
- TTL and DTL compatible phase detector input and square wave output
- Adjustable hold in range from ±1% to > ±60%

Applications

- Data and tape synchronization
- Modems
- FSK demodulation
- FM demodulation
- Frequency synthesizer
- Tone decoding
- Frequency multiplication and division
- SCA demodulators
- Telemetry receivers
- Signal regeneration
- Coherent demodulators

Connection Diagrams

Metal Can Package

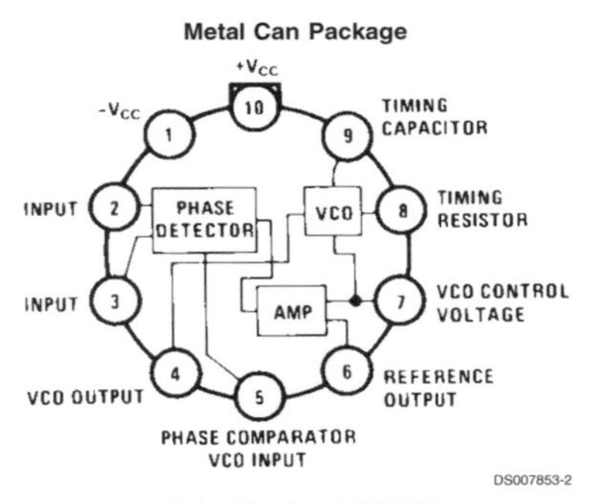

Order Number LM565H
See NS Package Number H10C

Dual-in-Line Package

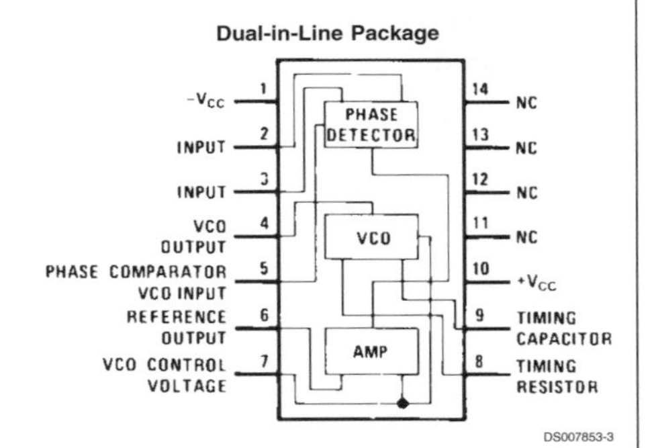

Order Number LM565CN
See NS Package Number N14A

Absolute Maximum Ratings (Note 1)

If Military/Aerospace specified devices are required, please contact the National Semiconductor Sales Office/ Distributors for availability and specifications.

Supply Voltage	±12V
Power Dissipation (Note 2)	1400 mW
Differential Input Voltage	±1V
Operating Temperature Range	
LM565H	−55°C to +125°C
LM565CN	0°C to +70°C
Storage Temperature Range	−65°C to +150°C
Lead Temperature (Soldering, 10 sec.)	260°C

Electrical Characteristics

AC Test Circuit, T_A = 25°C, V_{CC} = ±6V

Parameter	Conditions	LM565			LM565C			Units
		Min	Typ	Max	Min	Typ	Max	
Power Supply Current			8.0	12.5		8.0	12.5	mA
Input Impedance (Pins 2, 3)	$-4V < V_2, V_3 < 0V$	7	10			5		kΩ
VCO Maximum Operating Frequency	C_o = 2.7 pF	300	500		250	500		kHz
VCO Free-Running Frequency	C_o = 1.5 nF R_o = 20 kΩ f_o = 10 kHz	−10	0	+10	−30	0	+30	%
Operating Frequency Temperature Coefficient			−100			−200		ppm/°C
Frequency Drift with Supply Voltage			0.1	1.0		0.2	1.5	%/V
Triangle Wave Output Voltage		2	2.4	3	2	2.4	3	V_{p-p}
Triangle Wave Output Linearity			0.2			0.5		%
Square Wave Output Level		4.7	5.4		4.7	5.4		V_{p-p}
Output Impedance (Pin 4)			5			5		kΩ
Square Wave Duty Cycle		45	50	55	40	50	60	%
Square Wave Rise Time			20			20		ns
Square Wave Fall Time			50			50		ns
Output Current Sink (Pin 4)		0.6	1		0.6	1		mA
VCO Sensitivity	f_o = 10 kHz		6600			6600		Hz/V
Demodulated Output Voltage (Pin 7)	±10% Frequency Deviation	250	300	400	200	300	450	mV_{p-p}
Total Harmonic Distortion	±10% Frequency Deviation		0.2	0.75		0.2	1.5	%
Output Impedance (Pin 7)			3.5			3.5		kΩ
DC Level (Pin 7)		4.25	4.5	4.75	4.0	4.5	5.0	V
Output Offset Voltage $\|V_7 - V_6\|$			30	100		50	200	mV
Temperature Drift of $\|V_7 - V_6\|$			500			500		μV/°C
AM Rejection		30	40			40		dB
Phase Detector Sensitivity K_D			0.68			0.68		V/radian

Note 1: Absolute Maximum Ratings indicate limits beyond which damage to the device may occur. Operating Ratings indicate conditions for which the device is functional, but do not guarantee specific performance limits. Electrical Characteristics state DC and AC electrical specifications under particular test conditions which guarantee specific performance limits. This assumes that the device is within the Operating Ratings. Specifications are not guaranteed for parameters where no limit is given, however, the typical value is a good indication of device performance.

Note 2: The maximum junction temperature of the LM565 and LM565C is +150°C. For operation at elevated temperatures, devices in the TO-5 package must be derated based on a thermal resistance of +150°C/W junction to ambient or +45°C/W junction to case. Thermal resistance of the dual-in-line package is +85°C/W.

Typical Performance Characteristics

Power Supply Current as a Function of Supply Voltage

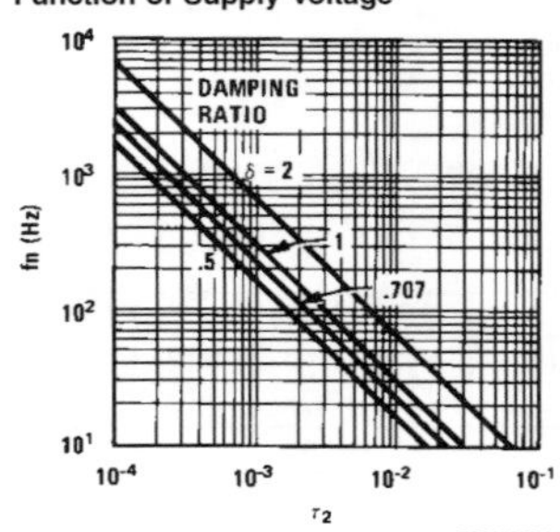

DS007853-14

Lock Range as a Function of Input Voltage

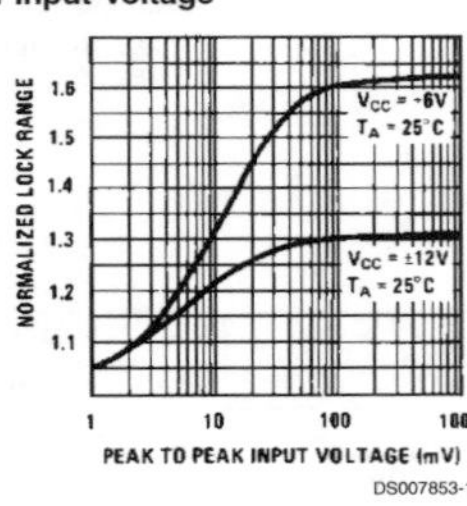

DS007853-15

VCO Frequency

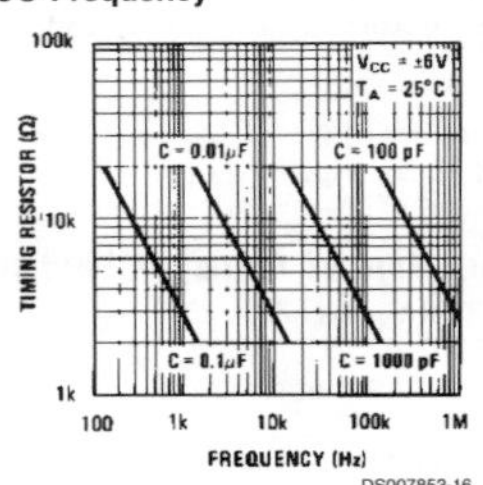

DS007853-16

Oscillator Output Waveforms

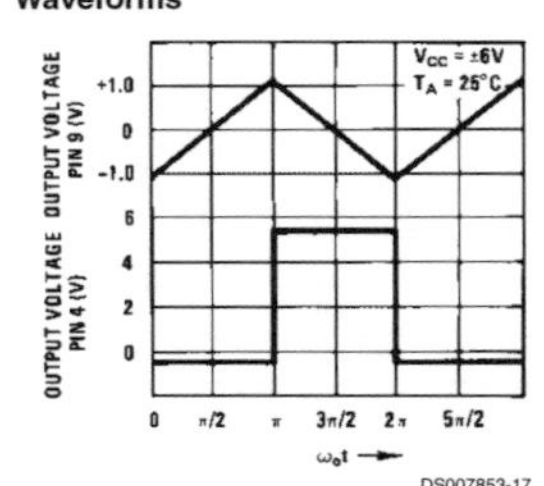

DS007853-17

Phase Shift vs Frequency

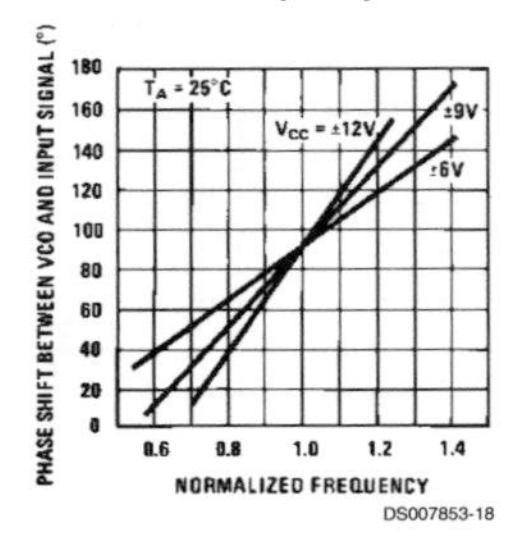

DS007853-18

VCO Frequency as a Function of Temperature

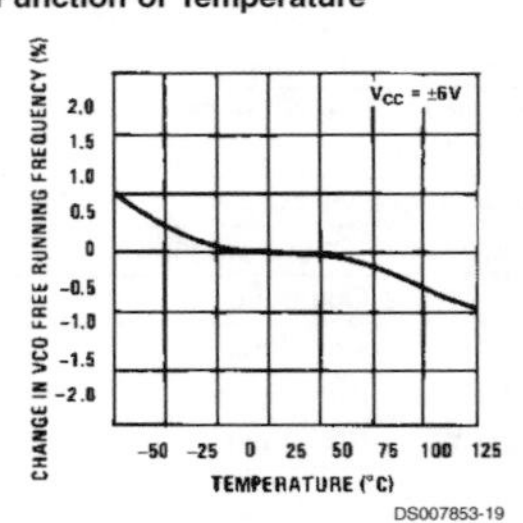

DS007853-19

Loop Gain vs Load Resistance

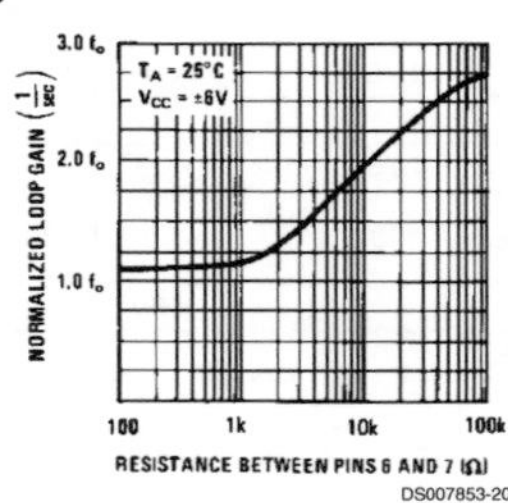

DS007853-20

Hold in Range as a Function of R_{6-7}

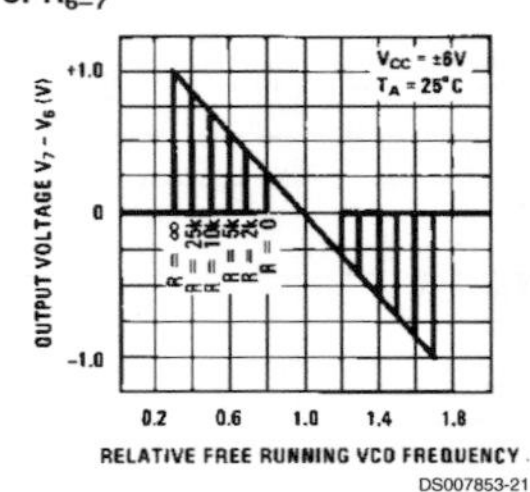

DS007853-21

Schematic Diagram

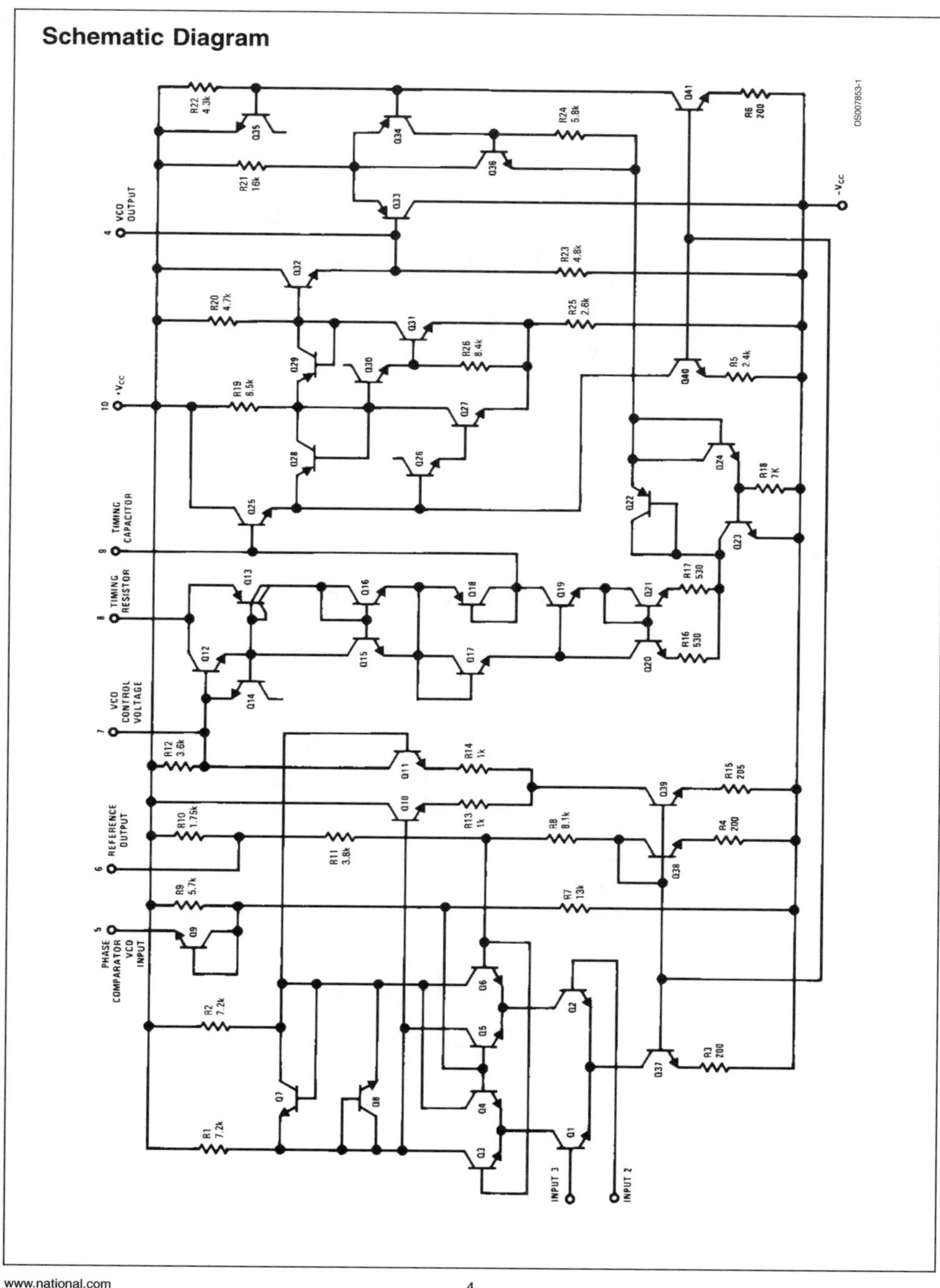

AC Test Circuit

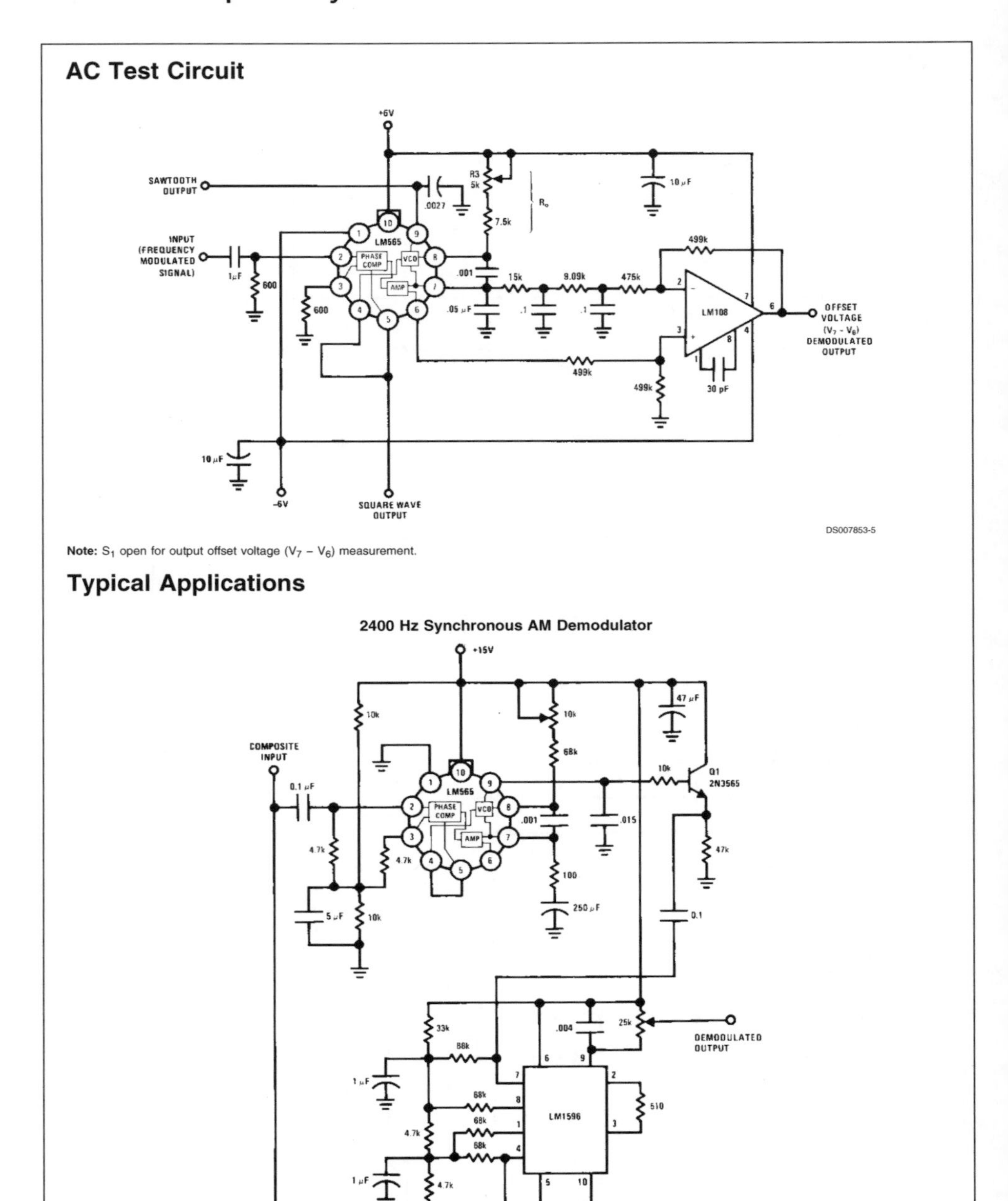

Note: S_1 open for output offset voltage ($V_7 - V_6$) measurement.

Typical Applications

2400 Hz Synchronous AM Demodulator

Typical Applications (Continued)

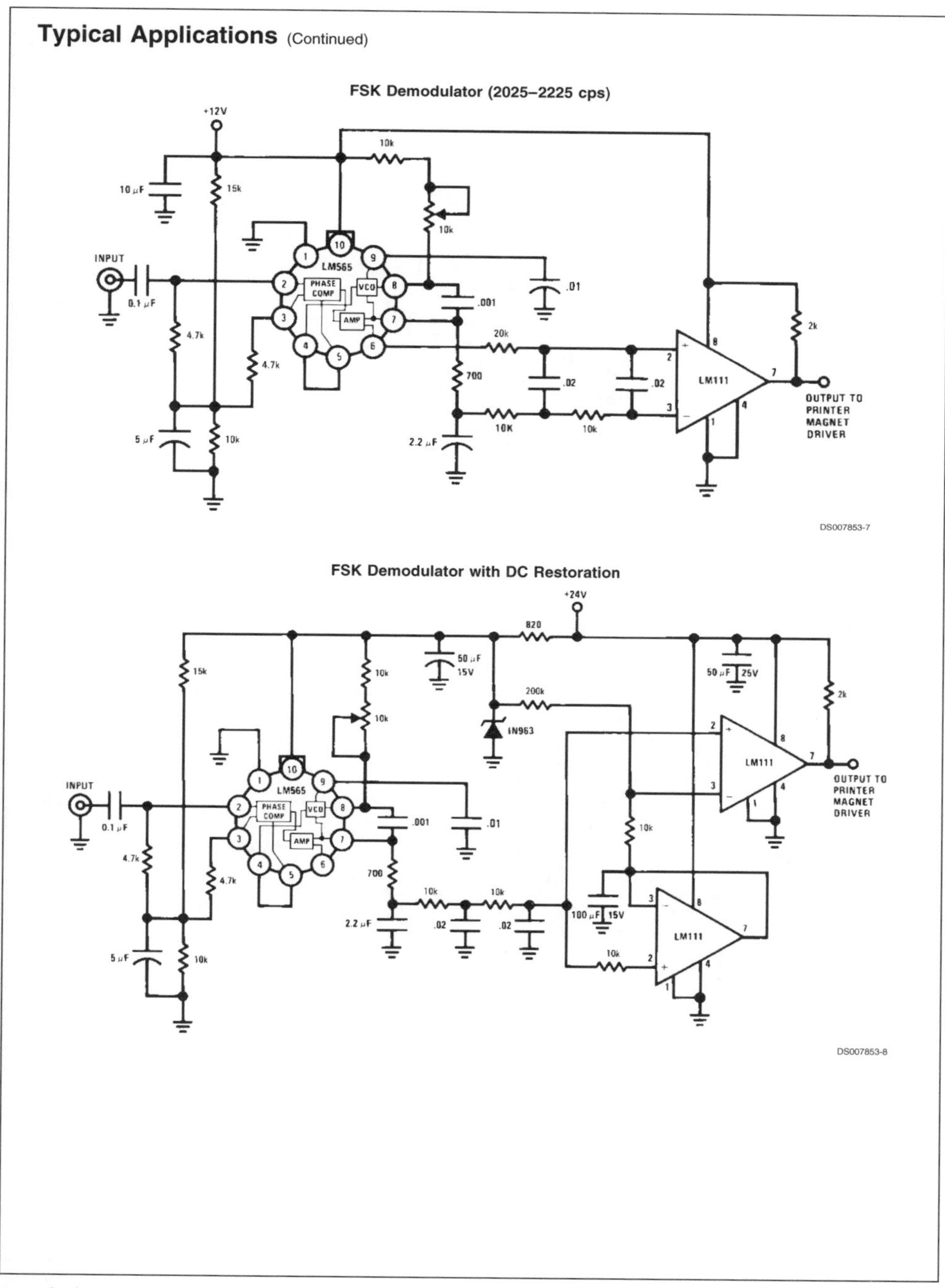

Typical Applications (Continued)

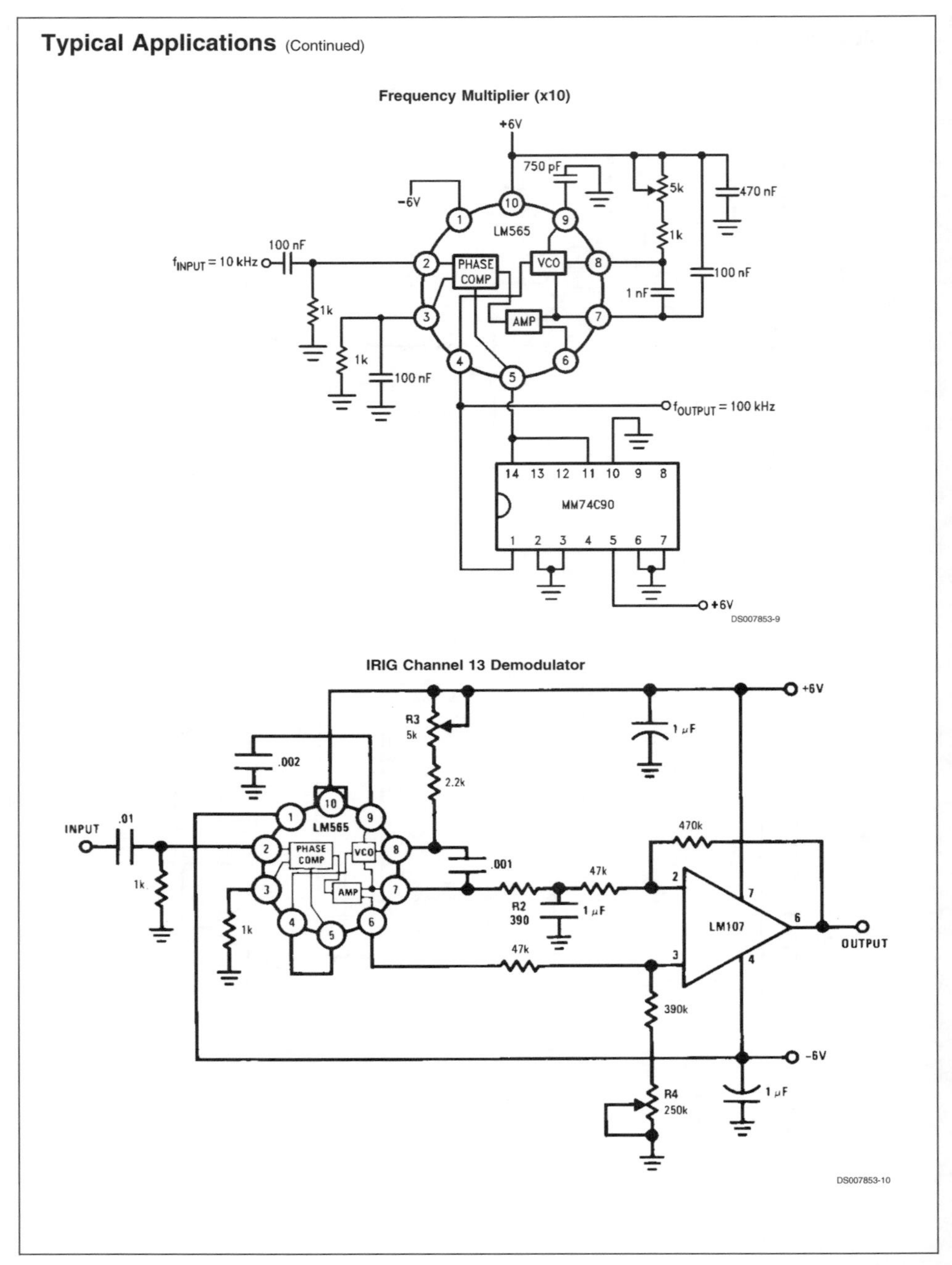

Applications Information

In designing with phase locked loops such as the LM565, the important parameters of interest are:

FREE RUNNING FREQUENCY

$$f_o \cong \frac{0.3}{R_o C_o}$$

LOOP GAIN: relates the amount of phase change between the input signal and the VCO signal for a shift in input signal frequency (assuming the loop remains in lock). In servo theory, this is called the "velocity error coefficient."

$$\text{Loop gain} = K_o K_D \left(\frac{1}{\text{sec}}\right)$$

$$K_o = \text{oscillator sensitivity} \left(\frac{\text{radians/sec}}{\text{volt}}\right)$$

$$K_D = \text{phase detector sensitivity} \left(\frac{\text{volts}}{\text{radian}}\right)$$

The loop gain of the LM565 is dependent on supply voltage, and may be found from:

$$K_o K_D = \frac{33.6\, f_o}{V_C}$$

f_o = VCO frequency in Hz

V_c = total supply voltage to circuit

Loop gain may be reduced by connecting a resistor between pins 6 and 7; this reduces the load impedance on the output amplifier and hence the loop gain.

HOLD IN RANGE: the range of frequencies that the loop will remain in lock after initially being locked.

$$f_H = \pm \frac{8\, f_o}{V_C}$$

f_o= free running frequency of VCO

V_c= total supply voltage to the circuit

THE LOOP FILTER

In almost all applications, it will be desirable to filter the signal at the output of the phase detector (pin 7); this filter may take one of two forms:

Simple Lead Filter

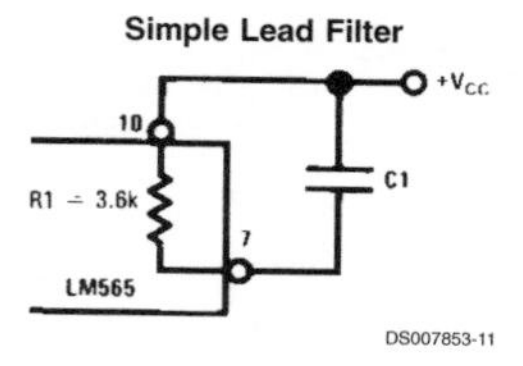

DS007853-11

Lag-Lead Filter

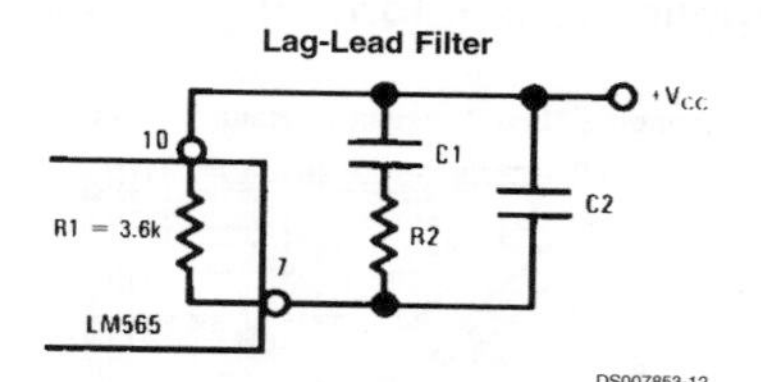

DS007853-12

A simple lag filter may be used for wide closed loop bandwidth applications such as modulation following where the frequency deviation of the carrier is fairly high (greater than 10%), or where wideband modulating signals must be followed.

The natural bandwidth of the closed loop response may be found from:

$$f_n = \frac{1}{2\pi}\sqrt{\frac{K_o K_D}{R_1 C_1}}$$

Associated with this is a damping factor:

$$\delta = \frac{1}{2}\sqrt{\frac{1}{R_1 C_1 K_o K_D}}$$

For narrow band applications where a narrow noise bandwidth is desired, such as applications involving tracking a slowly varying carrier, a lead lag filter should be used. In general, if $1/R_1C_1 < K_o K_D$, the damping factor for the loop becomes quite small resulting in large overshoot and possible instability in the transient response of the loop. In this case, the natural frequency of the loop may be found from

$$f_n = \frac{1}{2\pi}\sqrt{\frac{K_o K_D}{\tau_1 + \tau_2}}$$

$$\tau_1 + \tau_2 = (R_1 + R_2)\,C_1$$

R_2 is selected to produce a desired damping factor δ, usually between 0.5 and 1.0. The damping factor is found from the approximation:

$$\delta \;)\; \pi \tau_2 f_n$$

These two equations are plotted for convenience.

Filter Time Constant vs Natural Frequency

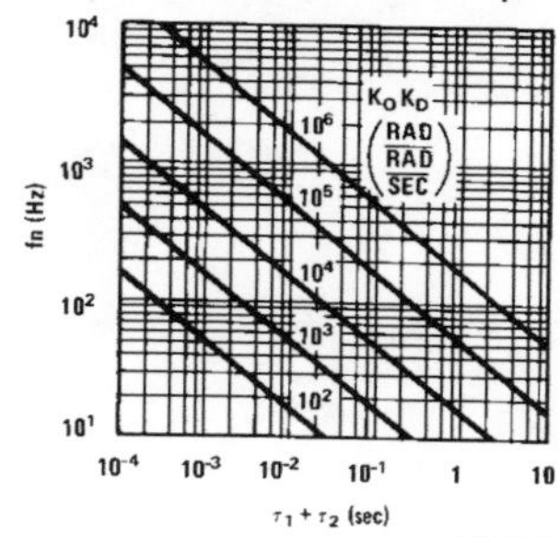

DS007853-13

Applications Information (Continued)

Damping Time Constant vs Natural Frequency

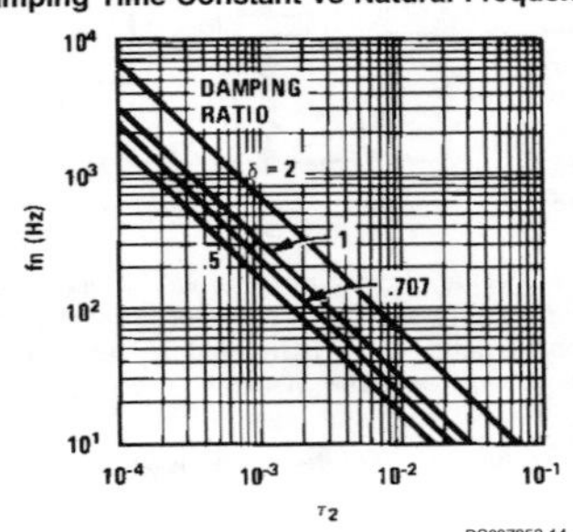

Capacitor C_2 should be much smaller than C_1 since its function is to provide filtering of carrier. In general $C_2 \leq 0.1\ C_1$.

Physical Dimensions inches (millimeters) unless otherwise noted

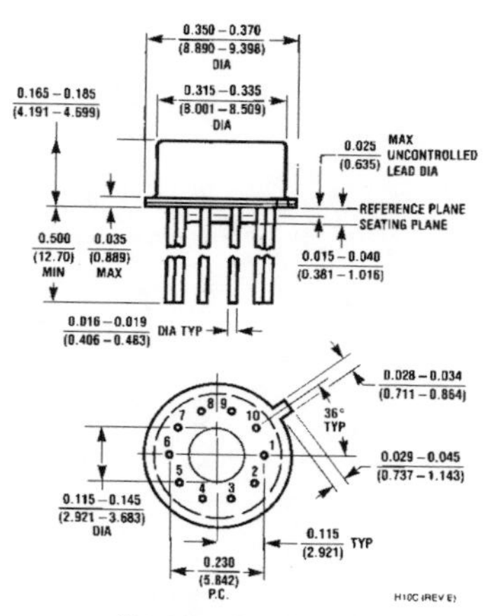

Metal Can Package (H)
Order Number LM565H
NS Package Number H10C

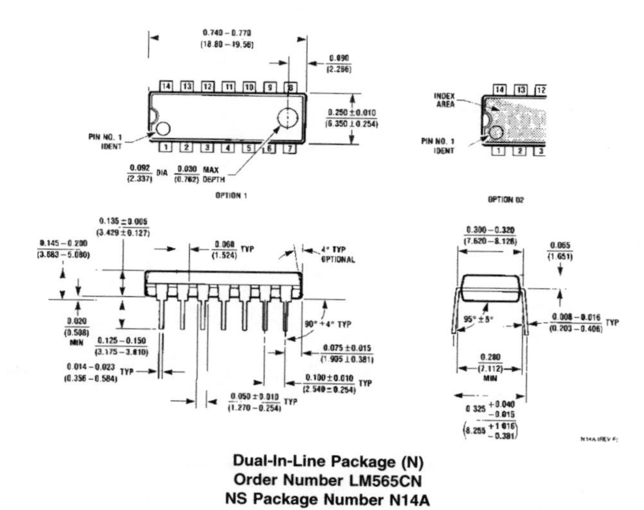

Dual-In-Line Package (N)
Order Number LM565CN
NS Package Number N14A

Notes

National Semiconductor Corporation
Americas
Tel: 1-800-272-9959
Fax: 1-800-737-7018
Email: support@nsc.com

www.national.com

National Semiconductor Europe
Fax: +49 (0) 1 80-530 85 86
Email: europe.support@nsc.com
Deutsch Tel: +49 (0) 1 80-530 85 85
English Tel: +49 (0) 1 80-532 78 32
Français Tel: +49 (0) 1 80-532 93 58
Italiano Tel: +49 (0) 1 80-534 16 80

National Semiconductor Asia Pacific Customer Response Group
Tel: 65-2544466
Fax: 65-2504466
Email: sea.support@nsc.com

National Semiconductor Japan Ltd.
Tel: 81-3-5639-7560
Fax: 81-3-5639-7507

HCC/HCF4046B

MICROPOWER PHASE-LOCKED LOOP

- QUIESCENT CURRENT SPECIFIED TO 20V FOR HCC DEVICE
- VERY LOW POWER CONSUMPTION : 100µW (TYP.) AT VCO f_o = 10kHz, V_{DD} = 5V
- OPERATING FREQUENCY RANGE : UP TO 1.4MHz (TYP.) AT V_{DD} = 10V
- LOW FREQUENCY DRIFT : 0.06%/°C (typ.) AT V_{DD} = 10V
- CHOICE OF TWO PHASE COMPARATORS :
  1) EXCLUSIVE - OR NETWORK
  2) EDGE-CONTROLLED MEMORY NETWORK WITH PHASE-PULSE OUTPUT FOR LOCK INDICATION
- HIGH VCO LINEARITY : 1% (TYP.)
- VCO INHIBIT CONTROL FOR ON-OFF KEYING AND ULTRA-LOW STANDBY POWER CONSUMPTION
- SOURCE-FOLLOWER OUTPUT OF VCO CONTROL INPUT (demod. output)
- ZENER DIODE TO ASSIST SUPPLY REGULATION
- 5V, 10V AND 15V PARAMETRIC RATING
- INPUT CURRENT OF 100nA AT 18V AND 25°C FOR HCC DEVICE
- 100% TESTED FOR QUIESCENT CURRENT
- MEETS ALL REQUIREMENTS OF JEDEC TENTATIVE STANDARD N°. 13A, "STANDARD SPECIFICATIONS FOR DESCRIPTION OF "B" SERIES CMOS DEVICES"

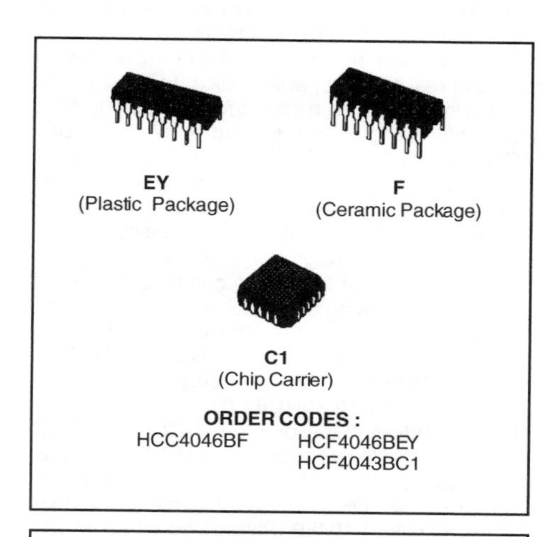

DESCRIPTION

The **HCC4046B** (extended temperature range) and **HCF4046B** (intermediate temperature range) are monolithic integrated circuits, available in 16-lead dual in-line plastic or ceramic package. The **HCC/HCF4046B** COS/MOS Micropower Phase-Locked Loop (PLL) consists of a low-power, linear voltage-controlled oscillator (VCO) and two different phase comparators having a common signal-input amplifier and a common comparator input. A 5.2V zener diode is provided for supply regulation if necessary.

PIN CONNECTIONS

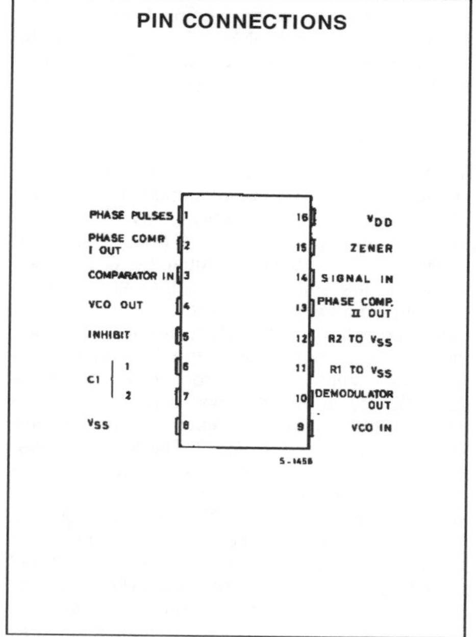

*Now known as STMicroelectronics

VCO Section

The VCO requires one external capacitor C1 and one or two external resistors (R1 or R1 and R2). Resistor R1 and capacitor C1 determine the frequency range of the VCO and resistor R2 enables the VCO to have a frequency offset if required. The high input impedance ($10^{12\Omega}$) of the VCO simplifiers the design of low-pass filters by permitting the designer a wide choice of resistor-to-capacitor ratios. In order not to load the low-pass filter, a source-follower output of the VCO input voltage is provided at terminal 10 (DEMODULATED OUTPUT). If this terminal is used, a load resistor (R_S) of 10 kΩ or more should be connected from this terminal to V_{SS}. If unused this terminal should be left open. The VCO can be connected either directly or through frequency dividers to the comparator input of the phase comparators. A full COS/MOS logic swing is available at the output of the VCO and allows direct coupling to COS/MOS frequency dividers such as the **HCC/HCF4024B**, **HCC/HCF4018B**, **HCC/HCF4020B**, **HCC/HCF4022B**, **HCC/HCF4029B**, and **HBC/HBF4059A**. One or more **HCC/HCF4018B** (Presettable Divide-by-N Counter) or **HCC/HCF4029B** (Presettable Up/Down Counter), or **HBC/HBF4059A** (Programmable Divide-by-"N" Counter), together with the **HCC/HCF4046B** (Phase-Locked Loop) can be used to build a micropower low-frequency synthesizer. A logic 0 on the INHIBIT input "enables" the VCO and the source follower, while a logic 1 "turns off" both to minimize stand-by power consumption.

Phase Comparators

The phase-comparator signal input (terminal 14) can be direct-coupled provided the signal swing is within COS/MOS logic levels [logic "0" $\leq$ 30 % ($V_{DD} - V_{SS}$), logic "1" $\geq$ 70 % ($V_{DD} - V_{SS}$)]. For smaller swings the signal must be capacitively coupled to the self-biasing amplifier at the signal input. Phase comparator I is an exclusive-OR network; it operates analagously to an over-driven balanced mixer. To maximize the lock range, the signal-and comparator-input frequencies must have a 50% duty cycle. With no signal or noise on the signal input, this phase comparator has an average output voltage equal to $V_{DD}/2$. The low-pass filter connected to the output of phase comparator I supplies the averaged voltage to the VCO input, and causes the VCO to oscillate at the center frequency (f_o). The frequency range of input signals on which the PLL will lock if it was initially out of lock is defined as the frequency capture range ($2 f_c$). The frequency range of input signals on which the loop will stay locked if it was initially in lock is defined as the frequency lock range ($2 f_L$). The capture range is $\leq$ the lock range. With phase comparator I the range of frequencies over which the PLL can acquire lock (capture range) is dependent on the low-pass-filter characteristics, and can be made as large as the lock range. Phase-comparator I enables a PLL system to remain in lock in spite of high amounts of noise in the input signal. One characteristic of this type of phase comparator is that it may lock onto input frequencies that are close to harmonics of the VCO center-frequency. A second characteristic is that the phase angle between the signal and the comparator input varies between 0° and 180°, and is 90° at the center frequency. Fig. (a) shows the typical, triangular, phase-to-output response characteristic of phase-comparator I. Typical waveforms for a COS/MOS phase-locked-loop employing phase comparator I in locked condition of f_o is shown in fig. (b). Phase-comparator II is an edge-controlled digital memory network. It consists of four flip-flop stages, control gating, and a three-stage output-circuit comprising p- and n-type drivers having a common output node. When the p-MOS or n-MOS drivers are ON they pull the output up to V_{DD} or down to V_{SS}, respectively. This type of phase comparator acts only on the positive edges of the signal and comparator inputs. The duty cycles of the signal and comparator inputs are not important since positive transitions control the PLL system utilizing this type of comparator. If the signal-input frequency is higher than the comparator-input frequency, the p-type output driver is maintained ON most of the time, and both the n- and p-drivers OFF (3 state) the remainder of the time. If the signal-input frequency is lower than the comparator-input frequency, the n-type output driver is maintained ON most of the time, and both the n- and p-drivers OFF (3 state) the remainder of the time. If the signal and comparator-input frequencies are the same, but the signal input lags the comparator input in phase, the n-type output driver is maintained ON for a time corresponding to the phase difference. If the signal and comparator-input frequencies are the same, but the comparator input lags the signal in phase, the p-type output driver is maintained ON for a time corresponding to the phase difference. Subsequently, the capacitor voltage of the low-pass filter connected to this phase comparator is adjusted until the signal and comparator inputs are equal in both phase and frequency. At this stable point both p- and n-type output drivers remain OFF and thus the phase comparator output becomes an open circuit and holds the voltage on the capacitor of the low-pass filter constant. Moreover the signal at the "phase pulses" output is a high level which can be used for indicating a locked condition. Thus, for phase comparator II, no phase difference exists between signal and comparator input over the full VCO frequency range. Moreover, the power dissipation due to the low-pass filter is reduced when this type of phase comparator is used because both the p- and n-type output drivers are OFF for most of the signal input cycle. It should be noted that the PLL lock range for this type of phase comparator is equal to the capture range, independent of the low-pass filter. With no signal present at the signal input, the VCO is adjusted to its lowest frequency for phase comparator II. Fig. (c) shows typical waveforms for a COS/MOS PLL employing phase comparator II in a locked condition.

Figure a : Phase-Comparator I Characteristics at Low-Pass Filter Output.

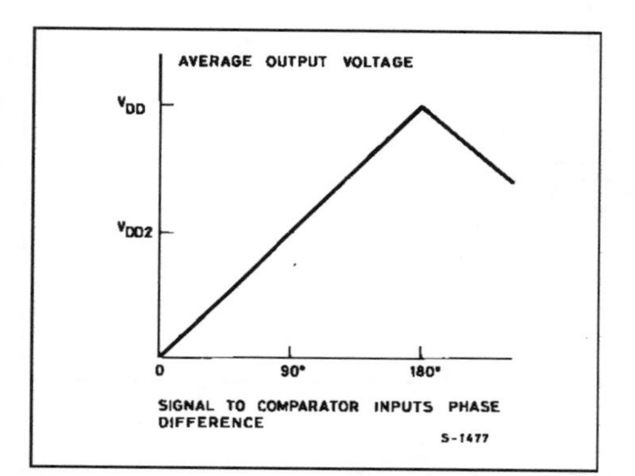

Figure b : Typical Waveforms for COS/MOS Phase Locked-Loop Employing Phase Comparator I in Locked Condition of f_o.

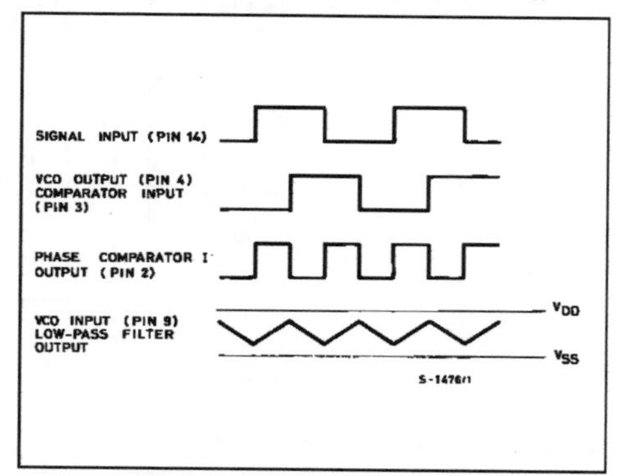

Figure C : Typical Waveforms For COS/MOS Phase-locked Loop Employing Phase Comparator II In Locked Condition.

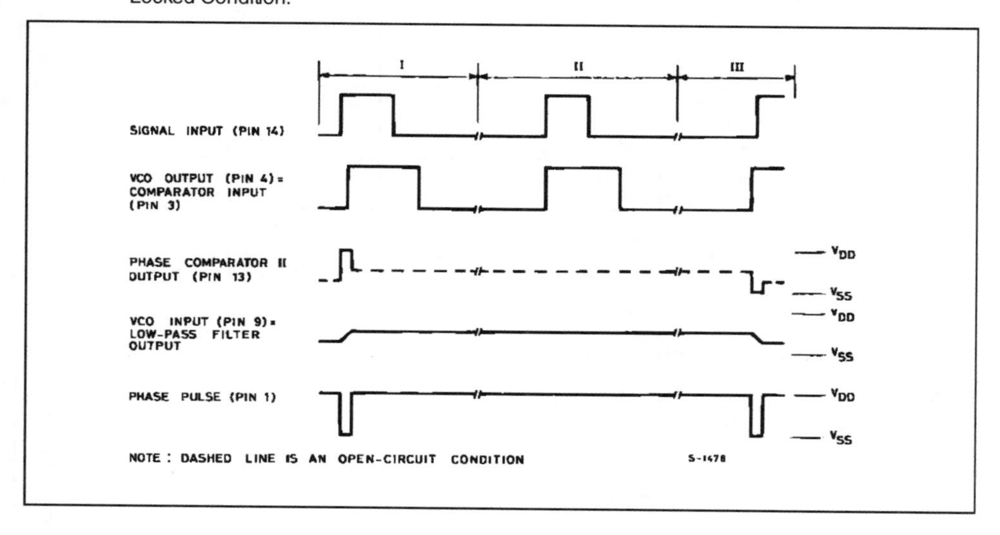

FUNCTIONAL DIAGRAM

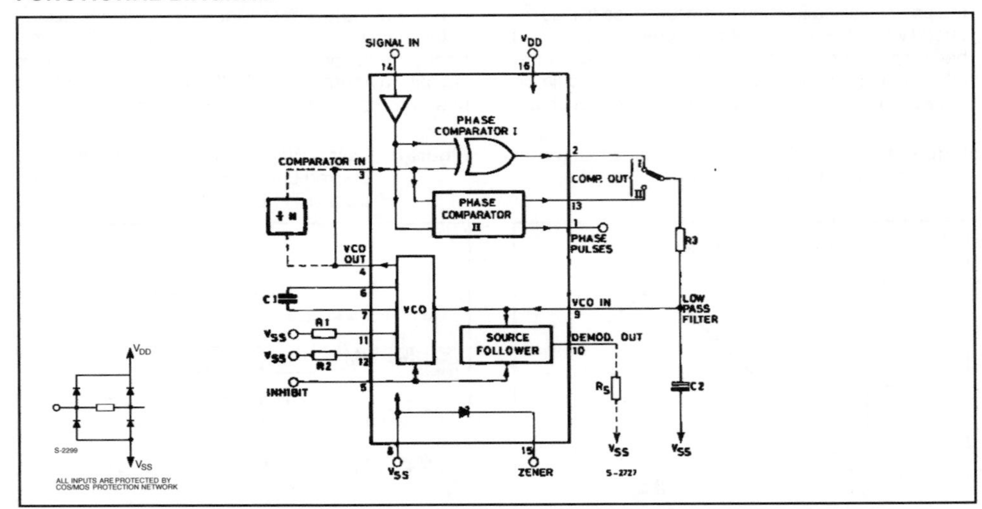

ABSOLUTE MAXIMUM RATINGS

Symbol	Parameter	Value	Unit
V_{DD}*	Supply Voltage : **HCC** Types **HCF** Types	– 0.5 to + 20 – 0.5 to + 18	V V
V_i	Input Voltage	– 0.5 to V_{DD} + 0.5	V
I_I	DC Input Current (any one input)	± 10	mA
P_{tot}	Total Power Dissipation (per package) Dissipation per Output Transistor for T_{op} = Full Package-temperature Range	200 100	mW mW
T_{op}	Operating Temperature : **HCC** Types **HCF** Types	– 55 to + 125 – 40 to + 85	°C °C
T_{stg}	Storage Temperature	– 65 to + 150	°C

Stresses above those listed under "Absolute Maximum Ratings" may cause permanent damage to the device. This is a stress rating only and functional operation of the device at these or any other conditions above those indicated in the operational sections of this specification is not implied. Exposure to absolute maximum rating conditions for external periods may affect device reliability.
* All voltage values are referred to V_{SS} pin voltage.

RECOMMENDED OPERATING CONDITIONS

Symbol	Parameter	Value	Unit
V_{DD}	Supply Voltage : **HCC** Types **HCF** Types	3 to 18 3 to 15	V V
V_I	Input Voltage	0 to V_{DD}	V
T_{op}	Operating Temperature : **HCC** Types **HCF** Types	– 55 to + 125 – 40 to + 85	°C °C

STATIC ELECTRICAL CHARACTERISTICS (over recommended operating conditions)

Symbol	Parameter		Test Conditions				Value							Unit
			V_I (V)	V_O (V)	$\lvert I_O \rvert$ (μA)	V_{DD} (V)	T_{Low}* Min.	T_{Low}* Max.	25°C Min.	25°C Typ.	25°C Max.	T_{High}* Min.	T_{High}* Max.	
VCO SECTION														
V_{OH}	Output High Voltage		0/ 5		< 1	5	4.95		4.95	5		4.95		V
			0/10		< 1	10	9.95		9.95	10		9.95		V
			0/15		< 1	15	14.95		14.95	15		14.95		V
V_{OL}	Output Low Voltage		5/0		< 1	5		0.05			0.05		0.05	V
			10/0		< 1	10		0.05			0.05		0.05	V
			15/0		< 1	15		0.05			0.05		0.05	V
I_{OH}	Output Drive Current	HCC Types	0/ 5	2.5		5	– 2		– 1.6	– 3.2		– 1.15		mA
			0/ 5	4.6		5	– 0.64		– 0.51	– 1		– 0.36		mA
			0/10	9.5		10	– 1.6		– 1.3	– 2.6		– 0.9		mA
			0/15	13.5		15	– 4.2		– 3.4	– 6.8		– 2.4		mA
		HCF Types	0/ 5	2.5		5	– 1.53		– 1.36	– 3.2		– 1.1		mA
			0/ 5	4.6		5	– 0.52		– 0.44	– 1		– 0.36		mA
			0/10	9.5		10	– 1.3		– 1.1	– 2.6		– 0.9		mA
			0/15	13.5		15	– 3.6		– 3.0	– 6.8		– 2.4		mA
I_{OL}	Output Sink Current	HCC Types	0/ 5	0.4		5	0.64		0.51	1		0.36		mA
			0/10	0.5		10	1.6		1.3	2.6		0.9		mA
			0/15	1.5		15	4.2		3.4	6.8		2.4		mA
		HCF Types	0/ 5	0.4		5	0.52		0.44	1		0.36		mA
			0/10	0.5		10	1.3		1.1	2.6		0.9		mA
			0/15	1.5		15	3.6		3.0	6.8		2.4		mA
I_{IH}, I_{IL}	Input Leakage Current	HCC Types	0/18	Any Input		18		± 0.1		$\pm 10^{-5}$	± 0.1		± 1	μA
		HCF Types	0/15	Any Input		15		± 0.3		$\pm 10^{-5}$	± 0.3		± 1	μA
PHASE COMPARATOR SECTION														
I_{DD}	Total Device Current Pin 14 = Open Pin 5 = V_{DD}		0/ 5			5		0.1		0.05	0.1		0.1	mA
			0/10			10		0.5		0.25	0.5		0.5	mA
			0/15			15		1.5		0.75	1.5		1.5	mA
			0/20			20		4		2	4		4	mA
	Pin 14 = V_{SS} or V_{DD} Pin 5 = V_{DD}	HCC Types	0/ 5			5		5		0.04	5		150	μA
			0/10			10		10		0.04	10		300	μA
			0/15			15		20		0.04	20		600	μA
			0/20			20		100		0.08	100		3000	μA
		HCF Types	0/ 5			5		20		0.04	20		150	μA
			0/10			10		40		0.04	40		300	μA
			0/15			15		80		0.04	80		600	μA
I_{OH}	Output Drive Current	HCC Types	0/ 5	2.5		5	– 2		– 1.6	– 3.2		– 1.15		mA
			0/ 5	4.6		5	– 0.64		– 0.51	– 1		– 0.36		mA
			0/10	9.5		10	– 1.6		– 1.3	– 2.6		– 0.9		mA
			0/15	13.5		15	– 4.2		– 3.4	– 6.8		– 2.4		mA
		HCF Types	0/ 5	2.5		5	– 1.53		– 1.36	– 3.2		– 1.1		mA
			0/ 5	4.6		5	– 0.52		– 0.44	– 1		– 0.36		mA
			0/10	9.5		10	– 1.3		– 1.1	– 2.6		– 0.9		mA
			0/15	13.5		15	– 3.6		– 3.0	– 6.8		– 2.4		mA

* T_{Low} = – 55°C for **HCC** device : – 40°C for **HCF** device.
* T_{High} = + 125°C for **HCC** device : + 85°C for **HCF** device.
The Noise Margin for both "1" and "0" level is : 1V min. with V_{DD} = 5V, 2V min. with V_{DD} = 10V, 2.5V min. with V_{DD} = 15V.

STATIC ELECTRICAL CHARACTERISTICS (continued)

Symbol	Parameter		Test Conditions				Value							Unit
			V_I (V)	V_O (V)	$\lvert I_O \rvert$ (µA)	V_{DD} (V)	T_{Low}* Min.	T_{Low}* Max.	25°C Min.	25°C Typ.	25°C Max.	T_{High}* Min.	T_{High}* Max.	
I_{OL}	Output Sink Current	HCC Types	0/ 5	0.4		5	0.64		0.51	1		0.36		mA
			0/10	0.5		10	1.6		1.3	2.6		0.9		
			0/15	1.5		15	4.2		3.4	6.8		2.4		
		HCF Types	0/ 5	0.4		5	0.52		0.44	1		0.36		
			0/10	0.5		10	1.3		1.1	2.6		0.9		
			0/15	1.5		15	3.6		3.0	6.8		2.4		
V_{IH}	Input High Voltage			0.5/4.5	< 1	5	3.5		3.5			3.5		V
				1/9	< 1	10	7		7			7		
				1.5/13.5	< 1	15	11		11			11		
V_{IL}	Input Low Voltage			4.5/0.5	< 1	5		1.5			1.5		1.5	V
				9/1	< 1	10		3			3		3	
				13.5/1.5	< 1	15		4			4		4	
I_{IH}, I_{IL}	Input Leakage Current (except. pin 14)	HCC Types	0/18	Any Input		18		± 0.1		$\pm 10^{-5}$	± 0.1		± 1	µA
		HCF Types	0/15			15		± 0.3		$\pm 10^{-5}$	± 0.3		± 1	
I_{OUT}	3-state Leakage Current	HCC Types	0/18	0/18		18		± 0.4		$\pm 10^{-4}$	± 0.4		± 12	µA
		HCF Types	0/15	0/15		15		± 1.0		$\pm 10^{-4}$	± 1.0		± 7.5	
C_I	Input Capacitance			Any Input						5	7.5			pF

* T_{Low} = – 55°C for **HCC** device : – 40°C for **HCF** device.
* T_{High} = + 125°C for **HCC** device : + 85°C for **HCF** device.
The Noise Margin for both "1" and "0" level is : 1V min. with V_{DD} = 5V, 2V min. with V_{DD} = 10V, 2.5V min. with V_{DD} = 15V.

ELECTRICAL CARACTERISTICS (T_{amb} = 25 °C)

Symbol	Parameter	Test Conditions		Value			Unit
			V_{DD} (V)	Min.	Typ.	Max.	
VCO SECTION							
P_D	Operating Power Dissipation	fo = 10 KHz R1 = 10 MΩ R2 = ∞ $V_{COIN} = \frac{V_{DD}}{2}$	5		70	140	μW
			10		800	1600	
			15		3000	6000	
f_{max}	Maximum Frequency	R1 = 10 KΩ C1 = 50 pF R2 = ∞ V_{COIN} = V_{DD}	5	0.3	0.6		MHz
			10	0.6	1.2		
			15	0.8	1.6		
		R1 = 5 KΩ n C1 = 50 pF R2 = ∞ V_{COIN} = V_{DD}	5	0.5	0.8		
			10	1	1.4		
			15	1.4	2.4		
	Center Frequency (f_o) and Frequency Range f_{max} - f_{min}	Programmable with external components R1, R2 and C1					
	Linearity	V_{COIN}=2.5V ± 0.3 R1=10 KΩ	5		1.7		%
		V_{COIN}=5V ± 1 R1=100 KΩ	10		0.5		
		V_{COIN}=5V ± 2.5 R1=400 KΩ	10		4		
		V_{COIN}=7.5V ± 1.5 R1=100 KΩ	15		0.5		
		V_{COIN}=7.5V ± 5 R1=1 MΩ	15		7		
	Temperature Frequency Stability (no frequency offset) f_{min} = 0		5		±0.12		%/°C
			10		±0.04		
			15		±0.015		
	Temperature Frequency Stability (frequency offset) $f_{min} \neq 0$		5		±0.09		
			10		±0.07		
			15		±0.03		
V_{CO}	Output Duty Cycle		5, 10, 15		50		%
t_{THL} t_{TLH}	VCO Output Transition Time		5		100	200	ns
			10		50	100	
			15		40	80	
	Source Follower Output (demodulated Output): Offset Voltage V_{COIN} - V_{DEM}	R_S > 10 KΩ	5, 10, 15		1.8	2.5	V
	Source Follower Output (demodulated Output): Linearity	R_S=100 KΩ V_{COIN}=$2.5^{\pm 0.3}$V	5		0.3		%
		R_S=300 KΩ V_{COIN}=$5^{\pm 2.5}$V	10		0.7		
		R_S=500 KΩ V_{COIN}=$7.5^{\pm 5}$V	15		0.9		
V_Z	Zener Diode Voltage	I_Z = 50 μA		4.45	5.5	7.5	V
R_Z	Zener Dynamic Resistance	I_Z = 1 mA			40		Ω
PHASE COMPARATOR SECTION							
R14	Pin 14 (signal in) Input Resistance		5	1	2		MΩ
			10	0.2	0.4		
			15	0.1	0.2		
	A.C. Coupled Signal Input Voltage Sensitivity * (peak to paek)	f_{in} = 100 KHz sine wave	5	180	360		mV
			10	330	660		
			15	900	1800		

HCC/HCF4046B

ELECTRICAL CHARACTERISTICS (continued)

Symbol	Parameter	Test Conditions		Value			Unit
			V_{DD} (V)	Min.	Typ.	Max.	
PHASE COMPARATOR SECTION (cont'd)							
T_{PHL}	Propagation Delay Time High to Low Level Pins 14 to 13		5	225	450		ns
			10	100	200		
			15	65	130		
T_{PLH}	Propagation Delay Time Low to High, Level		5		350	700	ns
			10		150	300	
			15		100	200	
T_{PHZ}	Propagation Delay Time 3-state High Level to High Impedance Pins 14 to 13		5		225	450	ns
			10		100	200	
			15		65	130	
T_{PLZ}	Low Level to High Impedance		5		285	570	ns
			10		130	260	
			15		95	190	
t_r, t_f	Input Rise or Fall Time Comparator Pin 3		5			50	µs
			10			1	
			15			0.3	
	Signal Pin 14		5			500	µs
			10			20	
			15			2.5	
T_{THL}, T_{TLH}	Transition Time		5		100	200	ns
			10		50	100	
			15		40	80	

* For sine wave the frequency must be greater than 10KHz for Phase Comparator II.

DESIGN INFORMATION

This information is a guide for approximating the values of external components for the **HCC/HCF 4046B** in a Phase-Locked-Loop system. The selected external components must be within the following ranges :

$5k\Omega \leq R1, R2, R_S \leq 1M\Omega$ $C1 \geq 100pF$ at $V_{DD} \geq 5V$ $C1 \geq 50pF$ at $V_{DD} \geq 10V$

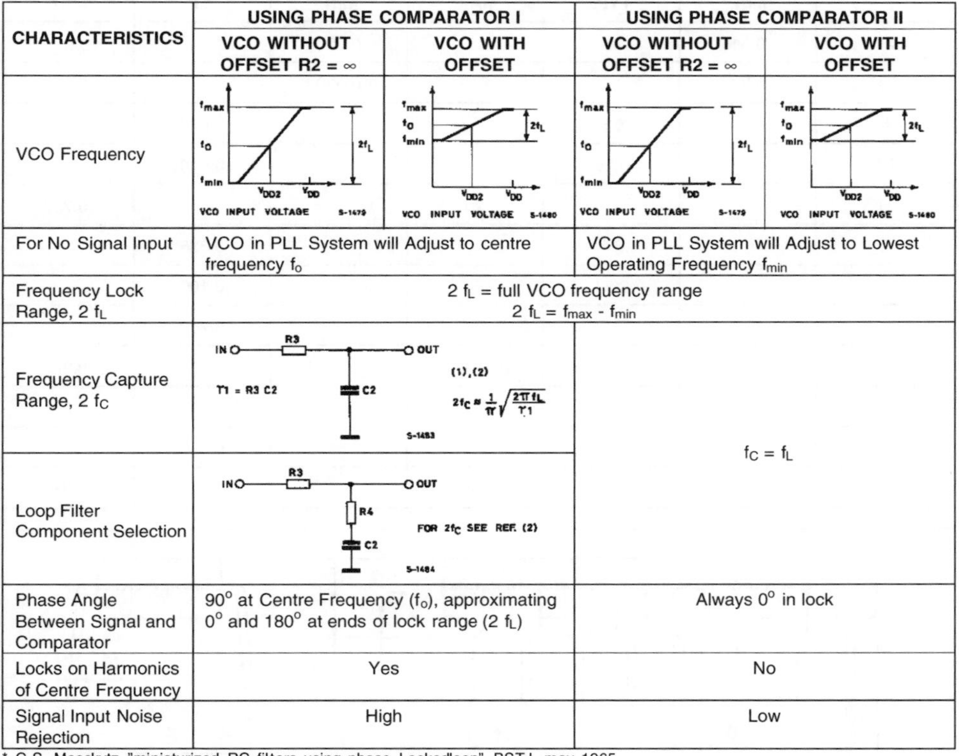

CHARACTERISTICS	USING PHASE COMPARATOR I		USING PHASE COMPARATOR II	
	VCO WITHOUT OFFSET R2 = ∞	VCO WITH OFFSET	VCO WITHOUT OFFSET R2 = ∞	VCO WITH OFFSET
VCO Frequency	f_{max}, f_o, f_{min}, $2f_L$, V_{DD2}, V_{DD}; VCO INPUT VOLTAGE S-1479	f_{max}, f_o, f_{min}, $2f_L$, V_{DD2}, V_{DD}; VCO INPUT VOLTAGE S-1480	f_{max}, f_o, f_{min}, $2f_L$, V_{DD2}, V_{DD}; VCO INPUT VOLTAGE S-1479	f_{max}, f_o, f_{min}, $2f_L$, V_{DD2}, V_{DD}; VCO INPUT VOLTAGE S-1480
For No Signal Input	VCO in PLL System will Adjust to centre frequency f_o		VCO in PLL System will Adjust to Lowest Operating Frequency f_{min}	
Frequency Lock Range, $2 f_L$	$2 f_L$ = full VCO frequency range $2 f_L = f_{max} - f_{min}$			
Frequency Capture Range, $2 f_C$	IN, R3, OUT, C2; $\tau 1 = R3\,C2$; (1),(2) $2f_C \approx \frac{1}{\pi}\sqrt{\frac{2\pi f_L}{\tau 1}}$; S-1483		$f_C = f_L$	
Loop Filter Component Selection	IN, R3, OUT, R4, C2; FOR $2f_C$ SEE REF. (2); S-1484			
Phase Angle Between Signal and Comparator	90° at Centre Frequency (f_o), approximating 0° and 180° at ends of lock range ($2 f_L$)		Always 0° in lock	
Locks on Harmonics of Centre Frequency	Yes		No	
Signal Input Noise Rejection	High		Low	

* G.S. Mosckytz "miniaturized RC filters using phase Lockedloop" BSTJ, may 1965

Plastic DIP16 (0.25) MECHANICAL DATA

DIM.	mm			inch		
	MIN.	TYP.	MAX.	MIN.	TYP.	MAX.
a1	0.51			0.020		
B	0.77		1.65	0.030		0.065
b		0.5			0.020	
b1		0.25			0.010	
D			20			0.787
E		8.5			0.335	
e		2.54			0.100	
e3		17.78			0.700	
F			7.1			0.280
I			5.1			0.201
L		3.3			0.130	
Z			1.27			0.050

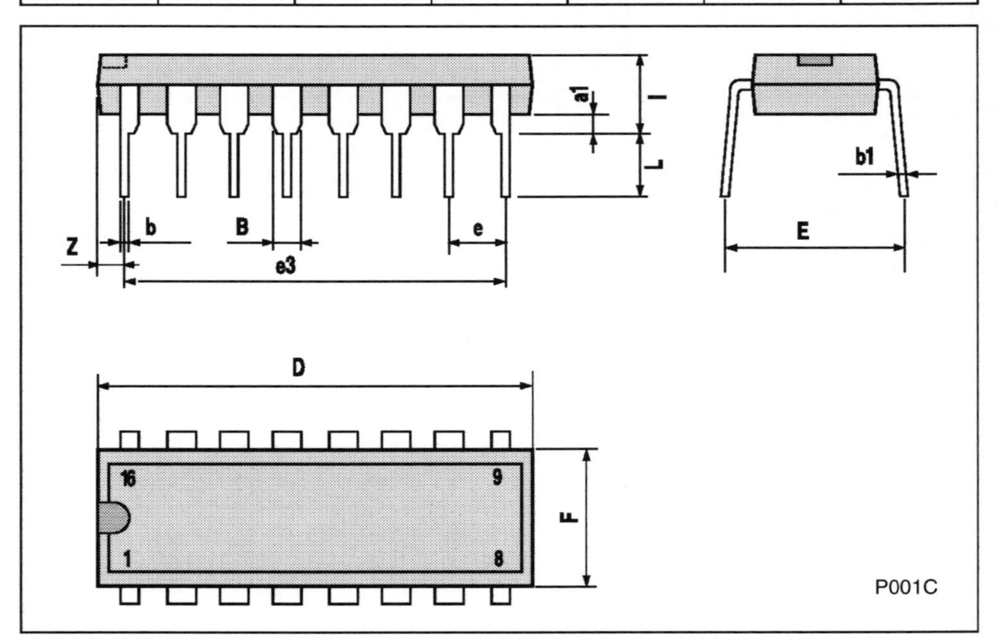

Ceramic DIP16/1 MECHANICAL DATA

DIM.	mm			inch		
	MIN.	TYP.	MAX.	MIN.	TYP.	MAX.
A			20			0.787
B			7			0.276
D		3.3			0.130	
E	0.38			0.015		
e3		17.78			0.700	
F	2.29		2.79	0.090		0.110
G	0.4		0.55	0.016		0.022
H	1.17		1.52	0.046		0.060
L	0.22		0.31	0.009		0.012
M	0.51		1.27	0.020		0.050
N			10.3			0.406
P	7.8		8.05	0.307		0.317
Q			5.08			0.200

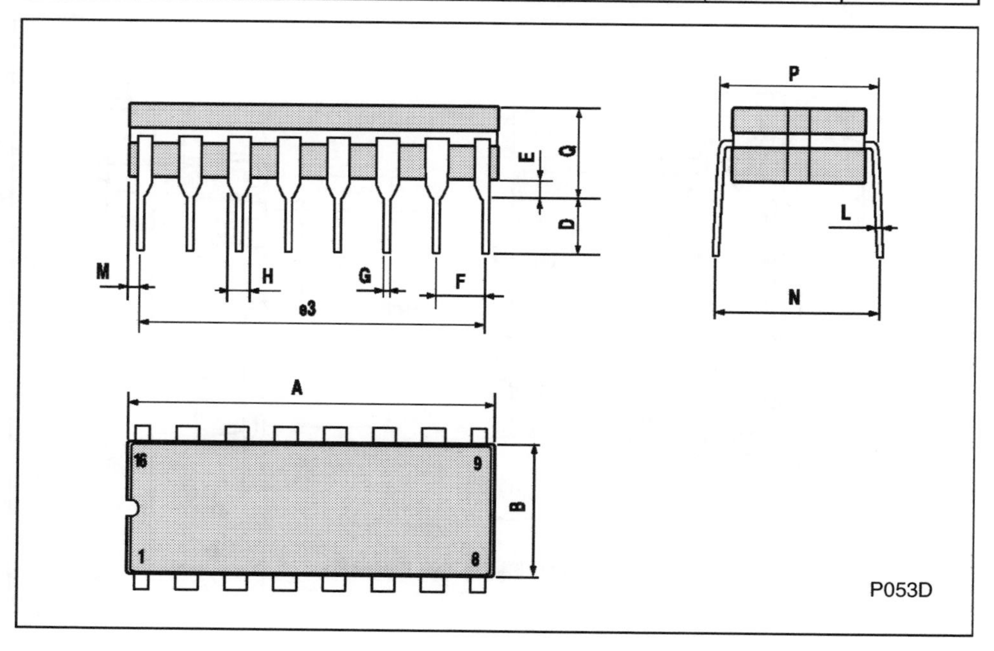

PLCC20 MECHANICAL DATA

DIM.	mm			inch		
	MIN.	TYP.	MAX.	MIN.	TYP.	MAX.
A	9.78		10.03	0.385		0.395
B	8.89		9.04	0.350		0.356
D	4.2		4.57	0.165		0.180
d1		2.54			0.100	
d2		0.56			0.022	
E	7.37		8.38	0.290		0.330
e		1.27			0.050	
e3		5.08			0.200	
F		0.38			0.015	
G			0.101			0.004
M		1.27			0.050	
M1		1.14			0.045	

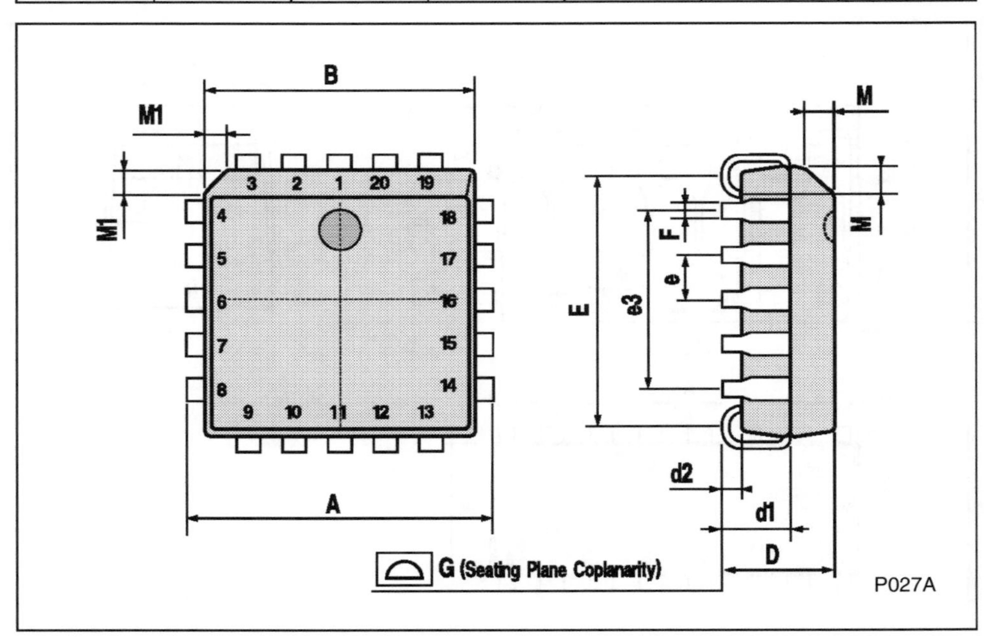

SGS-THOMSON
MICROELECTRONICS

7

Microwave devices and components

7.1 Introduction

A transmission line is a system of conductors which connects two points by means of electromagnetic energy which travels along the system. A system may include wires which carry 240 V/50 Hz mains supplies to consumers, or it may consist of lines which feed thousands of watts to a transmitting antenna. There are various kinds of transmission lines in use – the twisted pair, twin lead and coaxial, to mention a few. Twin-wire lines are often used for carrying radio frequency power to transmitters, while coaxial cables are commonly employed for low-power applications such as coupling receivers to their antennas.

Consider Fig. 7.1 in which a transmission line is connected between a generator and a load. In this circuit a section of transmission line is used and takes on the characteristics of a T-network. The components shown are not lumped components connected into the transmission line as such, but inherent parameters which manifest themselves when energy is propagated along the transmission line.

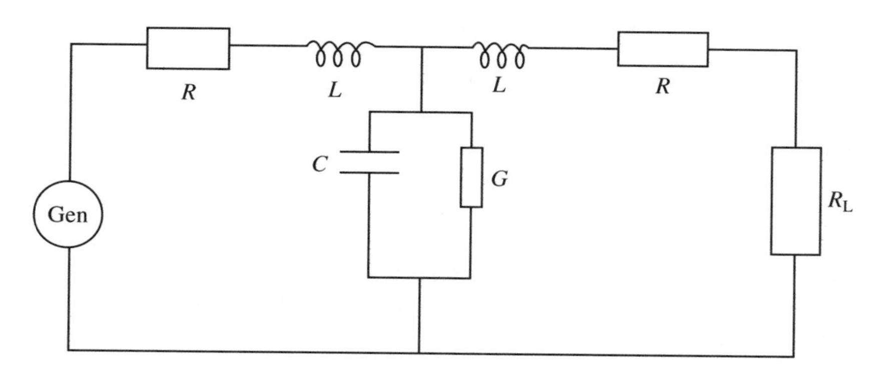

Fig. 7.1

The wave which moves along the transmission line is known as a travelling wave; as it moves from generator to load the capacitance of the line is charged up and the moving charges cause magnetic energy to be stored. Because of this the line does not appear as an open circuit to the generator but presents a finite value of impedance known as the characteristic impedance (Z_o) which is connected across the generator terminals. Each type of transmission line has its own characteristic impedance regardless of the length of the line; a symmetrical T-network like that in Fig. 7.1 is terminated in its characteristic impedance, equal to the network's characteristic impedance. This is irrespective of how many T-networks are connected. Obviously the loss in the line is due to the resistance (R) and the conductance (G).

From Fig. 7.1 there are four parameters connected with such a line: the resistance (R) which is dependent upon the dimensions of the line and its resistivity; the inductance (L) caused by the circular magnetic field around the conductors and dependent on the conductors' separation, radii and permeability; the capacitance due to the electric field between the conductors, also dependent on the dimensions of the conductors and the permittivity; and finally, the conductance due to the insulation of the line which allows some current to leak from one conductor to another. All these constants are referred to as the primary constants.

7.2 Phase delay and propagation velocity

If the losses are assumed to be small, R and G can be removed from Fig. 7.1, so that it simply becomes an LC low-pass filter network. As with any conductor, a current passes through the inductance of this conductor and charges up C. This takes time, and this time will increase as more and more sections of line are added (its length is increased). There will therefore be a phase difference between the generator voltage and the voltage at any point along the line.

The phase delay parameter (β) is given by

$$\beta = \omega\sqrt{LC} \text{ (rad/m)} \tag{7.1}$$

where L and C are the inductance and capacitance per metre of the line. Also as the wavelength (λ) on the line is the distance between two points with the same phase (360° or 2 π radians apart), the phase change per metre is $2\pi/\lambda$. Hence

$$\lambda = \frac{2\pi}{\beta} \tag{7.2}$$

The **phase velocity**, which is always strictly less than the velocity of light, is given by

$$V_p = f\lambda = \frac{2\pi f}{\beta} = \frac{\omega}{\beta} \tag{7.3a}$$

Another velocity, known as the **group velocity** (v_g), deals with the velocity of transferring energy along the line. As the line contains inductance and capacitance, energy is stored along its length and is transferred at velocity

$$V_g = \frac{d\omega}{d\beta} \tag{7.3b}$$

This velocity is normally associated with the group movement of several waveforms which have different phase velocities and is generally applied to pulses which consist of numerous frequencies having different phase relationships. Phase and group velocities are important in microwave devices.

7.3 The propagation constant and secondary constants

As current flows in a transmission line its value decreases because of attenuation in the line. If I_1 is the initial current at the generator and I_2 is the current (say) 1 km along the line towards the load, then the attenuation, which is a logarithmic decay, is given by

$$\frac{I_1}{I_2} = \mathrm{e}^{\gamma}$$

The propagation coefficient γ is a complex quantity given by $\gamma = \alpha + \mathrm{j}\beta$, where α is known as the attenuation constant and β is the phase-shift coefficient. If n sections of transmission line are connected then

$$\frac{I_G}{I_L} = \mathrm{e}^{n\gamma} \tag{7.4}$$

where I_G is the current at the generator end and I_L is the current at the load end. Hence

$$\frac{I_G}{I_L} = \mathrm{e}^{n(\alpha + \mathrm{j}\beta)} = \mathrm{e}^{n\alpha} \angle n\beta$$

from which

$$I_L = I_G \mathrm{e}^{-n\gamma} = I_G \mathrm{e}^{-n\alpha} \angle - n\beta \tag{7.5}$$

The receiving end voltage is similarly given by

$$V_L = V_G \mathrm{e}^{-n\alpha} \angle - n\beta \tag{7.6}$$

Note that Z_o, γ, α and β are known as the secondary constants and impedance and they may be expressed in terms of the primary constants. A full analysis of the following expressions may be found in other texts. The characteristic impedance is given by

$$Z_o = \sqrt{\frac{R + \mathrm{j}\omega L}{G + \mathrm{j}\omega C}} \tag{7.7}$$

Note that this is for a lossy line; for a lossless line the expression becomes

$$Z_o = \sqrt{\frac{L}{C}} \tag{7.8}$$

Also the propagation coefficient is

$$\gamma = \sqrt{(R + \mathrm{j}\omega L)(G + \mathrm{j}\omega C)} \tag{7.9}$$

Once again this is for a lossy line; for a lossless line this equation is

$$\gamma = \mathrm{j}\omega\sqrt{LC} \tag{7.10}$$

The characteristic impedance of the line may also be given by

$$Z_o = \sqrt{Z_{oc}Z_{sc}} \tag{7.11}$$

where Z_{oc} and Z_{sc} are the open-circuit and short-circuit impedance of the line, respectively. This is a practical method of obtaining the characteristic impedance as Z_{oc} and Z_{sc} can both be easily measured.

7.4 Transmission line distortion

The waveform at the receiving end of a transmission line may be distorted for many reasons, but the most common problems are the following:

(a) The terminating impedance is not the same as the characteristic impedance. As the characteristic impedance varies with frequency for a lossy line (equation (7.7)), the terminating impedance may not change with frequency in the same way. If the characteristic impedance is to remain the same for the complete frequency operating range, the condition $LG = CR$ must be achieved.

(b) The attenuation (and thus the propagation coefficient) of a line varies with frequency (equation (7.9)), thus different frequencies are attenuated by different amounts. This is avoided if $LG = CR$ and when this is the case.

$$\alpha = \sqrt{RG} \tag{7.12}$$

and

$$\beta = \omega\sqrt{LC}$$

Hence the attenuation coefficient is independent of frequency.

(c) The velocity of propagation varies with frequency. Once again if $LG = CR$ then the velocity of propagation is given by

$$v = \frac{1}{\sqrt{LC}} \tag{7.14}$$

which is independent of frequency, as will be the time delay.

If the above conditions are applied, i.e. $LG = CR$, then distortion will be kept at a minimum. However, in practice $R/L >> G/C$. Also the inductance is usually low and the capacitance is large. For the condition $LG = CR$ to be achieved either L or G has to be increased since C and R cannot really be altered. Also if G is increased the attenuation and power losses will increase. Thus L is generally altered, and when this occurs **loading** is said to take place. In practice, this is achieved either by inserting inductance at intervals along the line or by wrapping the conductors with high-permeability sheaths.

7.5 Wave reflection and the reflection coefficient

It has already been mentioned that for a lossless line equation (7.8) applies. If the line is increased in length the inductance increases, and the capacitance increases because the capacitance area is greater. So overall the *L/C* ratio remains the same.

When the load and generator impedances are both equal to the characteristic impedance of the line, maximum energy is absorbed by the load. However, when a line is not terminated in its own impedance the load absorbs only a portion of the energy, reflecting the remainder back along the line. Standing waves of voltage and current are set up, their amplitude depending on the extent of the mismatch. The standing waves are smaller in amplitude than the generator signal because the resistance absorbs a portion of the energy, sending the remainder back as reflected waves.

It is instructive at this point to explain what is meant by a standing wave. A wave travelling along a transmission line consists of electric and magnetic components, and energy is stored in the magnetic field of the line inductance ($\frac{1}{2}LI^2$) and the electric field of the line capacitance ($\frac{1}{2}CV^2$). This energy is interchanged between the magnetic and electric fields and causes the transmission of the electromagnetic energy along the line.

When a wave reaches an open-circuited termination the magnetic field collapses since the current is zero. The energy is not lost but is converted into electrical energy adding to that already caused by the existing electric field. Hence the voltage at the termination doubles, and this increased voltage causes the reflected wave which moves back along the line in the reverse direction.

The same reasoning can be applied to a short-circuited termination. In this case the electric field collapses at the termination and its energy changes to magnetic energy. This results in a doubling of the current.

Extreme resistance loading conditions are shown in Fig. 7.2. In Fig. 7.2(a) the line is terminated in an open circuit and, regardless of the length of line, the current and voltage conditions at the termination end are the same. In an open circuit the current is a minimum and so the impedance and voltage are maximum. In Fig. 7.2(b) the line termination is a short circuit resulting in maximum current, while the impedance and voltage are minimum.

Figures 7.3(a) and 7.3(b) show terminations in pure capacitance and pure inductance, respectively. In Fig. 7.3(a) all of the energy is reflected because a capacitor does not absorb energy. The relative phases of the waves shift according to the reactance and the line impedance. It is assumed in this case that the capacitive reactance equals the characteristic impedance simulating a phase condition of 45°.

At the termination end of the line the current and voltage arrive in phase, but due to the current through Z_o and X_c in series, the current leads by 45° while the voltage lags by 45°. So although current and voltage arrive in phase they are reflected 90° out of phase, as shown by the standing waves.

The voltage standing wave is a minimum at a point $\lambda/8$ from the termination end when $X_c = Z_o$. When X_c is greater the arrangement acts more like an open circuit and the minimum voltage point moves further away from the termination end. When X_c is less than Z_o it is more like a short circuit, and the voltage minimum is closer to the end.

In Fig. 7.3(b) the line is terminated in its inductance and the waveforms occur when $X_L = Z_o$. Conditions are similar to the capacitance case, except that the phase shifts occur in the opposite direction. The shift is again 45° but the waveforms at the termination end

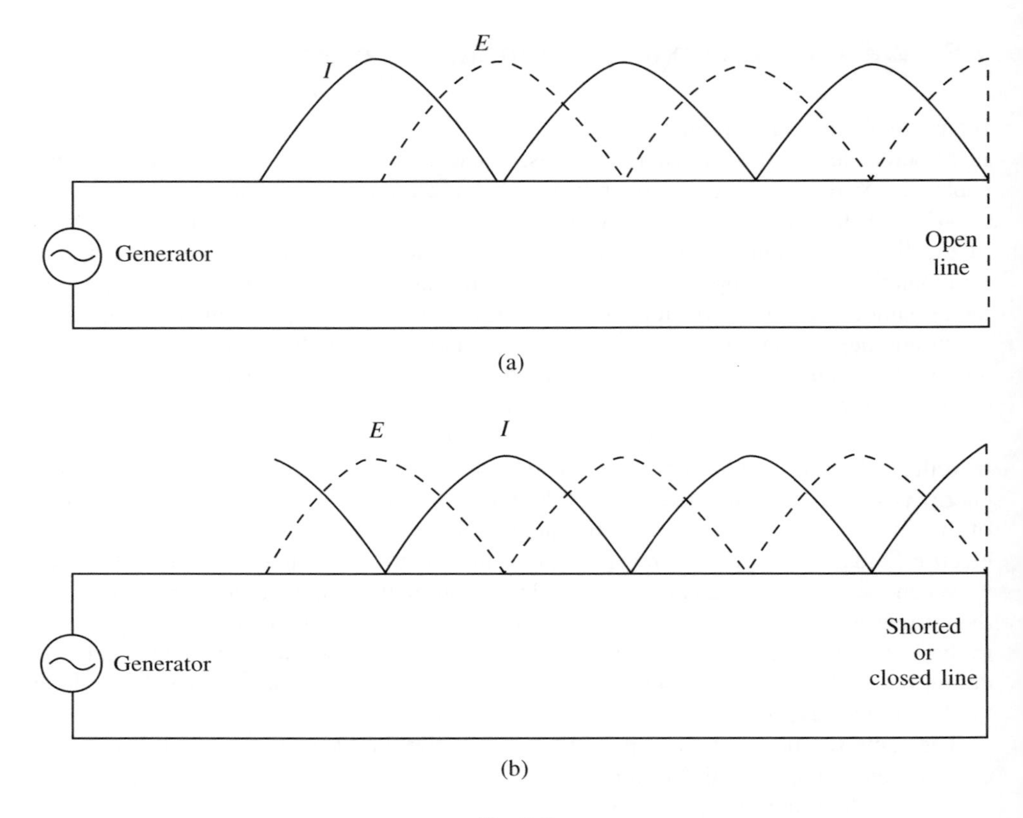

Fig. 7.2

are reversed. The current minimum now occurs at $\lambda/8$ from the end, whereas in Fig. 7.3(a) the voltage minimum occurred at this point.

If X_L is increased the condition is more similar to an open circuit and the current minimum moves toward the termination end, while a decrease in X_L moves the waves in the opposite direction.

A variety of conditions may be obtained using either an open-circuited or short-circuited transmission line cut to the desired length. This can simulate such conditions as series and parallel resonance and pure capacitance and inductance.

The reflection coefficient is an indication of amplitudes of the incident and reflected currents (I_i, I_r) and voltages (V_i, V_r). It is given in several forms, and once again the analysis of these formulae can be found in most common texts dealing with transmission lines. One form is

$$\frac{I_r}{I_i} = \frac{Z_o - Z_L}{Z_o + Z_L} = \rho \tag{7.15}$$

Another form relates the voltages and currents

$$\frac{V_r}{V_i} = -\frac{I_r}{I_i} = \rho \tag{7.16}$$

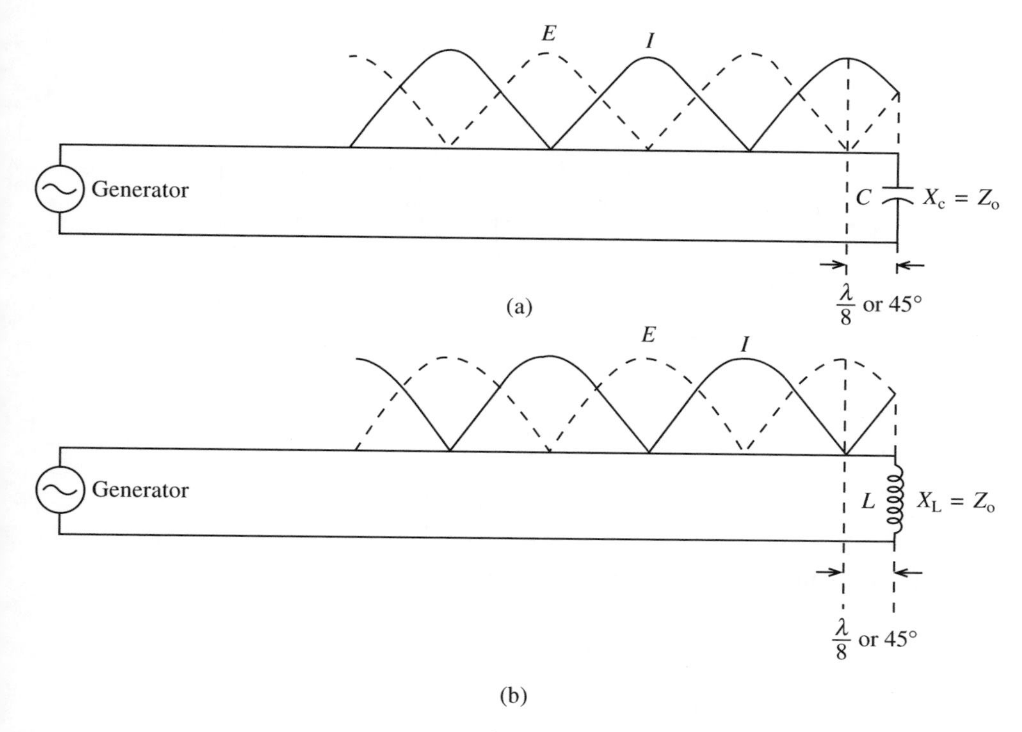

Fig. 7.3

Thus the ratio of the reflected to the incident voltage has the same magnitude as the current ratio, but opposite sign.

7.6 Standing wave ratio

Standing waves have already been mentioned, together with their formation due to load termination. Figure 7.4 shows the result of producing standing waves on a transmission line. Remember that standing waves are produced when two waves having the same frequency and amplitude and travelling in opposite directions are superimposed. The figure shows that standing waves produce points of minimum and maximum disturbance called nodes and antinodes, and this applies to voltage or current waves.

Associated with standing waves is the **standing wave ratio** which indicates the amount of reflected energy relative to the incident energy which is absorbed at the termination. Several forms of this are derived below. At a voltage maximum

$$V_{max} = V_i + V_r$$

while at a minimum

$$V_{min} = V_i - V_r$$

Hence

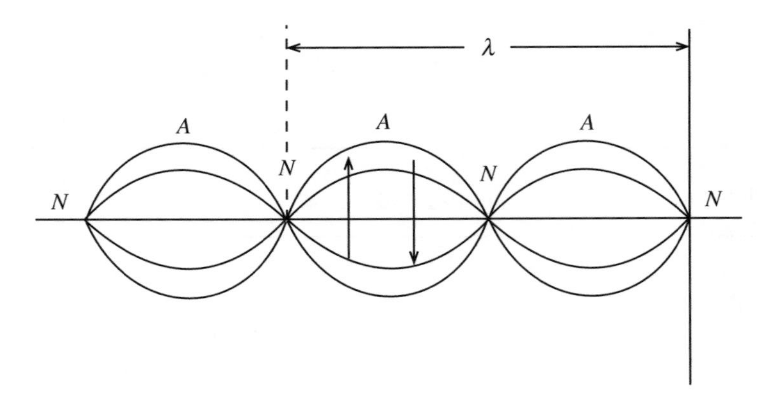

Fig. 7.4

$$S = \frac{V_{max}}{V_{min}} = \frac{V_i + V_r}{V_i - V_r}$$

Dividing numerator and denominator by V_i gives

$$S = \frac{1 + V_r/V_i}{1 - V_r/V_i}$$

It has already been stated that

$$\left|\frac{V_r}{V_i}\right| = |\rho|$$

hence

$$S = \frac{1 + |\rho|}{1 - |\rho|} \tag{7.17}$$

Also the power absorbed in the termination and the power reflected are given respectively by

$$P_i = I_i^2 Z_o \text{ and } P_r = I_r^2 Z_o$$

Hence

$$\frac{P_r}{P_i} = \frac{I_r^2 Z_o}{I_i^2 Z_o} = \frac{I_r^2}{I_i^2}$$

Since

$$\frac{I_r}{I_i} = \frac{S - 1}{S + 1}$$

then

$$\frac{P_r}{P_i} = \left(\frac{S - 1}{S + 1}\right)^2 \tag{7.18}$$

7.7 Fundamental waveguide characteristics

The principles and parameters discussed in previous sections are applicable to waveguides and microwave devices. However, generally waveguides are used over short distances and losses are minimized. If a dielectric waveguide is used then primary and secondary coefficients such as R, G and α have to be considered. In this text air-filled rectangular waveguides will be discussed.

The part of the electromagnetic spectrum referred to as the microwave region loosely includes the range 1–300 GHz for practical purposes. Microwaves can be used for almost any of the applications for which the lower frequencies are used, but some of the advantages of the shorter waves make them more applicable for certain purposes. The two main advantages of microwaves are that the energy can be focused into narrow beams and that large signal bandwidths are possible.

These microwave bands are especially advantageous for relay sevices such as are used in satellite transmission, telephone relay, oil platform transmission and the broadcasting of commercial radio and television. However, special components and devices are required for operation at microwave frequencies, and one of these, a circulator, is shown in Fig. 7.5. The circulator allows the same microwave antenna to be coupled to the receiver and transmitter. This and other components will be discussed in this chapter.

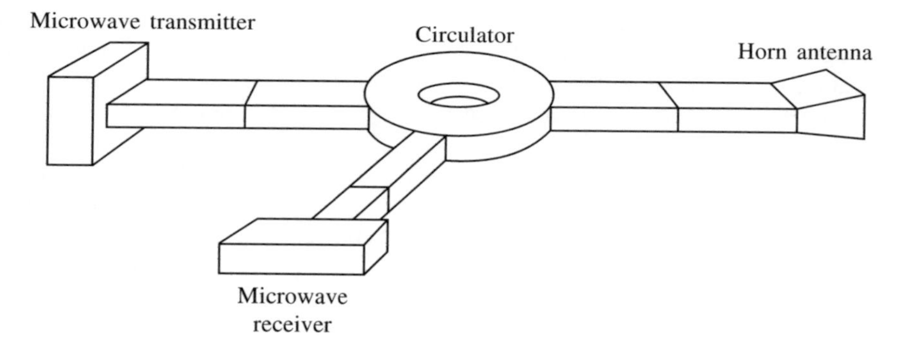

Fig. 7.5

The purpose of this chapter is to give a basic idea of the various microwave generators and devices which are used in the industry, but initially a knowledge of microwave transmission is necessary.

Transmission modes

Electromagnetic fields are formed as voltage and current waves travel down a transmission line. As propagation frequency increases a portion of the associated wavelength becomes comparable to the geometry of the transmission line dimensions. As the frequency increases, different types of propagation modes appear. The principal mode is the one which can carry energy at all frequencies. Higher modes are those that propagate only above a definite frequency range, and the point at which these frequencies start to propagate is called the cut-off frequency for that particular mode. An analogy is shown in Fig. 7.6.

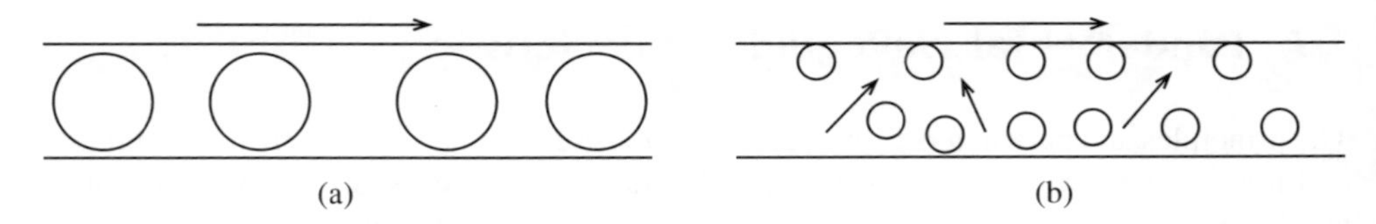

Fig. 7.6

Figure 7.6(a) shows particles moving down a tube. In this case the particles have similar cross-sectional dimensions to the tube, and if the particles are given a constant velocity then a certain propagation will exist. This is analogous to single-mode propagation. If the diameter of the tube is increased or the particle size decreased, then the particles would propagate down the tube in different modes as shown in Fig. 7.6(b). This is analogous to high-order modes. If the particle size is further decreased the number of modes or propagation patterns increases. Hence certain patterns will occur when a definite change in size occurs either in the tube dimensions or in the propagating frequency.

A two-wire transmission line exhibits electric and magnetic fields as current flows along it. Both these fields are perpendicular to one another and also to the direction of propagation. Hence in the principal mode these waves are called transverse electromagnetic (TEM) waves.

If the frequency of transmission is increased so that the wavelength becomes similar to the cross-sectional dimensions of the transmission line higher-order modes can propagate. These higher-order modes will have at least one of their field components showing in the propagation direction, i.e. either the H-wave (magnetic mode) or the E-wave (electric mode). Depending on the mode involved it will be called transverse electric (TE) or transverse magnetic (TM).

The velocity of the TE and TM modes is different from that of the TEM mode. Two types of velocity are involved, namely group velocity (the velocity of the entire moving group) and phase velocity (including all path movements of the particles).

Skin effect

This effect occurs due to conductor losses mainly caused by the series resistance of the conductor. As frequency increases, skin effect becomes more critical and the current flows on the surface of the conductor rather than inside the conductor. Microwave guides are therefore used as hollow pipes with the inner surface coated with a conductive material. Because current penetration due to the skin effect is shallow, the currents have to follow the surface irregularities, thus increasing the length and consequently the losses. This is the reason why microwave components are highly polished in order to erase all surface markings.

The rectangular waveguide

When James Clerk Maxwell developed the theory of electromagnetic waves he imposed certain boundary conditions which are present when electric and magnetic fields travel through one material bounded by another.

Consider Fig. 7.7, in which a wave is guided along a tube. Under the principle of total internal reflection $\theta_i = \theta_r$ and the wave is continually reflected along the metal guide. A phase reversal occurs at each reflection, and for the electric field there is a change of permittivity (ε) at the boundary, while for the magnetic field there is a change in permeability (μ).

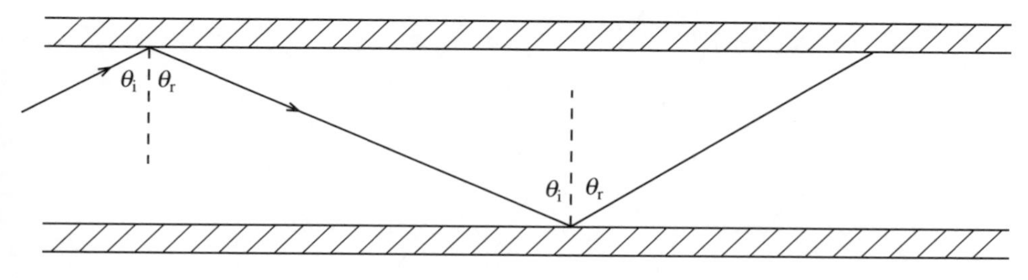

Fig. 7.7

An electric field which is parallel to a conductor will be short-circuited, and for this reason electric lines of force must always be at an angle to the guide walls. The magnetic field creates surface currents in the guide walls. This is the skin effect which determines the waveguide loss, and so it is essential that the propagation occurs through the guide rather than along the walls. This is achieved by using various modes which were explained earlier. For a rectangular guide the mode is described as TE_{mn}, where m and n refer to the dimensions shown in Fig. 7.8. Note that m indicates the number of half-wave variations of the electric field along a, while n indicates the number of half-wave variations of electric and magnetic field along b. If a dimension is less than half a wavelength no propagation occurs.

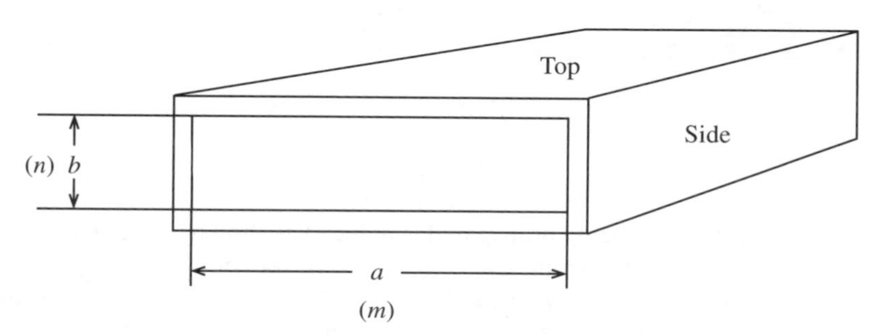

Fig. 7.8

Cut-off conditions

Each mode has a cut-off frequency below which it will not propagate. Single-mode propagation can be achieved by using only the mode with the lowest cut-off frequency. This is the dominant mode, and the waveguide will function at frequencies between its cut-off frequency and that of the mode with the next lowest cut-off frequency.

The concept of the parallel plate waveguide is helpful in understanding how the wave is guided along a waveguide. Figure 7.9 shows a plane wave being reflected from a plane conducting sheet. The angle of incidence $\theta_i = \theta_r$ and the lines denote positions of equal phase. The propagation derection of the wave is perpendicular to the equal-phase lines. If a second sheet is added at *AA* as shown in Fig. 7.9 then a single ray along the guide would be propagated without loss as shown in Fig. 7.10.

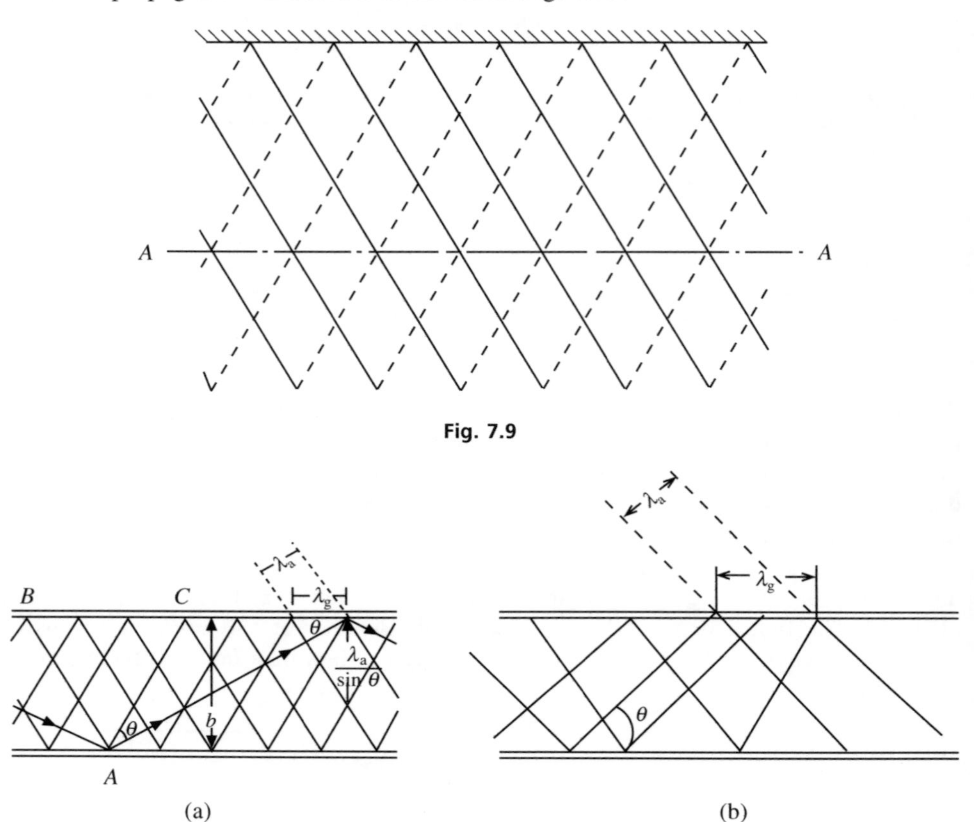

Fig. 7.9

Fig. 7.10

If the frequency which is being propagated is reduced as shown in Fig. 7.10(b) then the wavelength is increased. This would cause the angle of reflection to change in order to accommodate the lower frequency. If the frequency is progressively reduced, the half-wavelength becomes equal to the distance between the reflecting walls of the guide and the wave reflects backwards and forwards without progressing along the guide. This occurs at what is called the **cut-off frequency**. No propagation occurs below this frequency.

Figure 7.10 only shows a single propagation path, and it illustrates the simplest mode which is TE_{10}. The fields for this type of propagation are shown in Fig. 7.11.

For a rectangular waveguide the cut-off frequency is given by

$$f_{cTEmn} = \frac{c}{2}\sqrt{\left(\frac{m}{a}\right)^2 + \left(\frac{n}{b}\right)^2} \tag{7.19}$$

where m and n are subscripts of a particular TE and TM mode. Hence, for the simplest mode, we have

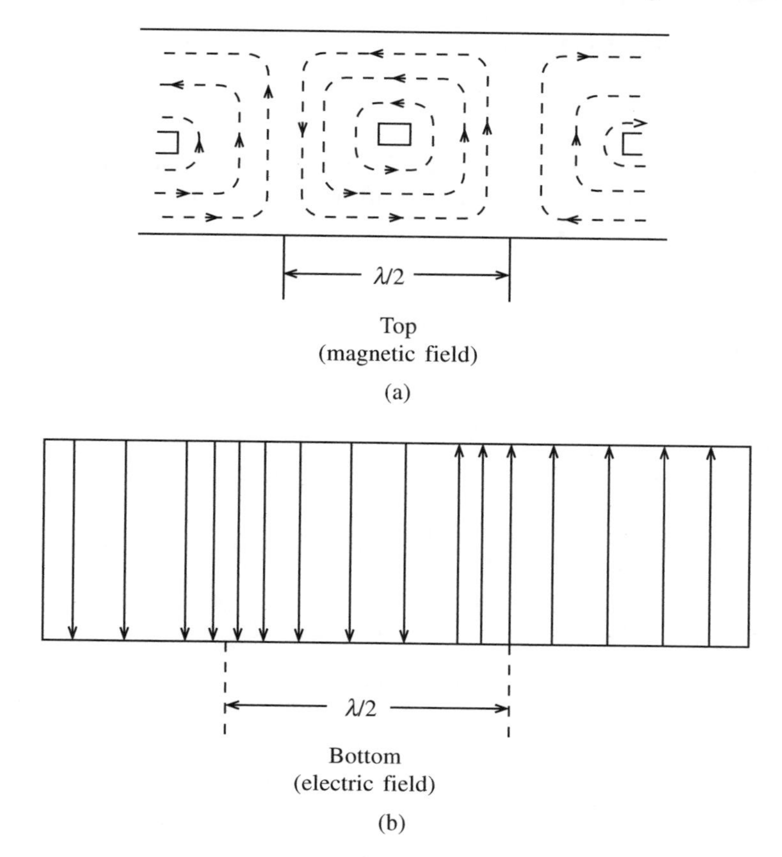

Fig. 7.11

$$\lambda_{cTE10} = \frac{2}{\sqrt{(1/a)^2 + (0/b)^2}} \tag{7.20}$$

This solution depends on the fact that there should be a whole number of half-wavelengths between the plates. If λ_a is the wavelength in air or free space, then the condition would be

$$b = \frac{n\lambda_a}{2 \sin \theta} \tag{7.21}$$

where n is an integer.

Note that the wavelength in the direction of propagation is called the waveguide wavelength (λ_g) and is given by

$$\lambda_g = \frac{\lambda}{\cos \theta}$$

Hence

$$\cos \theta = \frac{\lambda}{\lambda_g} \tag{7.22}$$

If the spacing between the plates is

$$b = \frac{n\lambda_a}{2} \tag{7.23}$$

then $\theta = 90°$ (equation (7.21)), the waveguide wavelength is infinite from (7.22). Hence the wave will not propagate. This is the cut-off wavelength and is given by (7.23).

Since

$$\lambda_a = \frac{c}{f}$$

we can rewrite (7.23) as

$$b = \frac{nc}{2f_c}$$

and

$$f_c = \frac{nc}{2b}$$

hence

$$\lambda_c = 2b \tag{7.24}$$

Combining (7.21) and (7.22), and using the identity

$$\sin^2 \theta + \cos^2 \theta = 1$$

gives

$$\frac{\lambda_a^2}{(2b)^2} + \frac{\lambda_a^2}{\lambda_g^2} = 1$$

($n = 1$ in this mode). Also, from (7.24),

$$\frac{\lambda_a^2}{\lambda_c^2} + \frac{\lambda_a^2}{\lambda_g^2} = 1 \tag{7.25}$$

hence

$$\frac{1}{\lambda_g^2} = \frac{1}{\lambda_a^2} - \frac{1}{\lambda_c^2} \tag{7.26}$$

This equation can also be expressed as

$$\lambda_g = \frac{\lambda_a}{\sqrt{1 - (n\lambda_a/2b)^2}} \tag{7.27}$$

by using (7.24) and assuming $n = 1$.

The following points should be noted regarding waveguide design:

(i) The attenuation loss is greater as b is made smaller.

(ii) The dimension of b determines the voltage breakdown characteristics and hence the maximum power capacity.

(iii) a and b should be made as large as possible for large power applications.

(iv) In practice $a = 2b$.

(v) Dimensions are normally selected so that only one mode of propagation is possible and thus the size of the waveguide is related to the frequency band.

Example 7.1

Determine the cut-off frequency and wavelength of a rectangular waveguide in the TE_{32} mode if $a = 8.1$ cm and $b = 4.3$ cm.

Solution

From (7.19),

$$f_{cTE32} = \frac{c}{2}\sqrt{\left(\frac{3}{a}\right)^2 + \left(\frac{2}{b}\right)^2}$$

$$= \frac{3 \times 10^8}{2}\sqrt{\left(\frac{3}{8.1 \times 10^{-2}}\right)^2 + \left(\frac{2}{4.3 \times 10^{-2}}\right)^2}$$

$$= 8.9 \text{ GHz}$$

and, similarly to (7.20),

$$\lambda_{cTE32} = \frac{2}{\sqrt{\left(\frac{3}{a}\right)^2 + \left(\frac{2}{b}\right)^2}} = 3.4 \text{ cm}$$

Example 7.2

A rectangular waveguide functions in the TE_{10} mode with a free space frequency of 3.2 GHz. If the guide dimensions are given as $a = 6.2$ cm and $b = 3.3$ cm, determine:

(a) the cut-off frequency and wavelength;

(b) the guide wavelength.

Solution

(a)

$$f_{cTE10} = 1.5 \times 10^8 \sqrt{\frac{1}{6.2 \times 10^{-2}}} = 2.4 \text{ GHz}$$

$$\lambda_{cTE10} = \frac{2}{\sqrt{\frac{1}{6.2 \times 10^{-2}}}} = 12.4 \text{ cm}$$

(b)

$$\frac{1}{\lambda_g^2} = \frac{1}{\lambda_a^2} - \frac{1}{\lambda_c^2}$$

Since

$$\lambda_a = \frac{c}{f_a} = \frac{3 \times 10^8}{3.2 \times 10^9} = 0.094 \text{ m}$$

we have

$$\frac{1}{\lambda_g^2} = \frac{1}{(0.094)^2} - \frac{1}{(0.124)^2}$$

$$\lambda_g = 14.32 \text{ cm}$$

Example 7.3

A waveguide is of British Standards size 12 and has the following specifications: a = 17.215 cm, b = 8.315 cm and f_c = 3.15247 GHz. Its recommended operating range is 3.95–5.85 GHz. If it has to function in the TE_{21} mode, calculate the guide wavelength for its operating range.

Solution

We will use equation (7.27):

$$\lambda_g = \frac{\lambda_a}{\sqrt{1 - (n\lambda_a/2b)^2}}$$

For f_a = 3.95 GHz,

$$\lambda_a = \frac{c}{f_a} = \frac{3 \times 10^8}{3.95 \times 10^9} = 0.0759 \text{ m}$$

$$\therefore \quad \lambda_g = \frac{0.0759}{\sqrt{1 - (0.0759/(2 \times 0.08315))^2}} = 8.5 \text{ cm}$$

For f_a = 5.85 GHz,

$$\lambda_a = \frac{3 \times 10^8}{5.85 \times 10^9} = 0.05 \text{ m}$$

$$\therefore \quad \lambda_g = \frac{0.05}{\sqrt{1 - (0.05/(2 \times 0.17215))^2}} = 5.4 \text{ cm}$$

7.8 Microwave passive components

The basic theory of microwaves and their transmission along waveguides has been discussed previously. In order to couple or direct this energy special components are used which operate generally in the TE_{10} dominant mode for rectangular waveguides. This text will only deal with rectangular waveguides, although other geometries exist. The components discussed here are: the directional coupler, waveguide junctions, cavity resonators, probes, and circulators and isolators.

The directional coupler

The directional coupler is a device which, when installed in a waveguide, will respond to a wave travelling in one direction but will be unaffected by a wave travelling in the opposite direction. Directional couplers are four-port devices and the power ratios between the ports are defined by the **coupling factor**, **directivity** and **insertion loss**.

Consider the four-port directional coupler in Fig 7.12. In this case power entering port 1 gives an output at ports 2 and 3 but no output at port 4. This coupler can be inserted in two ways: forward and reverse. Note the following parameters. The coupling factor is given by

$$C_{\mathrm{f}} = \frac{\text{Power in port 3}}{\text{Power in port 1}} \text{ (dB)} \tag{7.28}$$

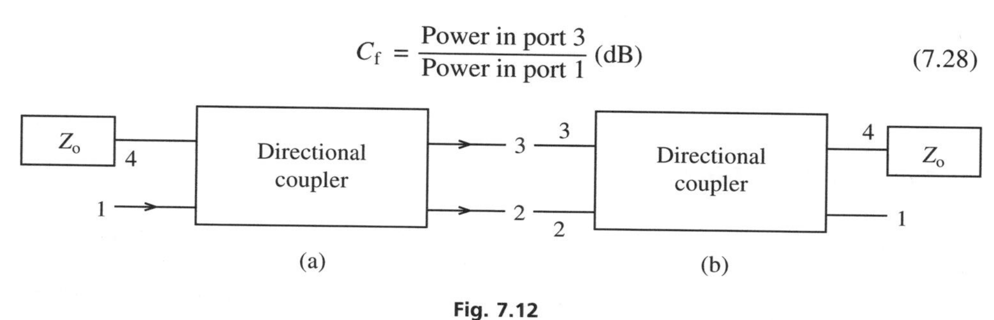

Fig. 7.12

The directivity is given by

$$D = \frac{\text{Power in port 4}}{\text{Power in port 3}} \text{ (dB)} \tag{7.29}$$

The insertion loss is the amount by which the signal in the main guide is attenuated and may be given by

$$I = \frac{\text{Power in port 4}}{\text{Power in port 1}} \text{ (dB)} \tag{7.30}$$

Note that the directivity is defined as the ratio of the power appearing at port 4 when the coupler is in the forward direction to the power appearing at port 3 when the coupler is in the reverse direction and port 4 is terminated in its characteristic impedance (Z_o).

One of the most common directional couplers is the two-hole interference directional coupler shown in Fig. 7.13. This coupler will only work when the holes are a quarter of a wavelength apart. If higher frequencies are required the number of holes is increased. Figure 7.13(a) shows the physical structure, while Fig. 7.13(b) shows how the signal is propagated through the holes. At each hole the signal splits into two components, one travelling in the forward and one in the reverse direction. The waves which travel from input port 1 to output port 3 move the same wavelength distance, and are in phase and add constructively; the two waves travelling from input port 1 to the load move along a path which differs by half a wavelength, and hence destructive interference takes place.

The holes are separated by half a guide wavelength, as shown in Fig. 7.13(a). In this case only two holes are shown, but more may be included to provide greater coupling between guides.

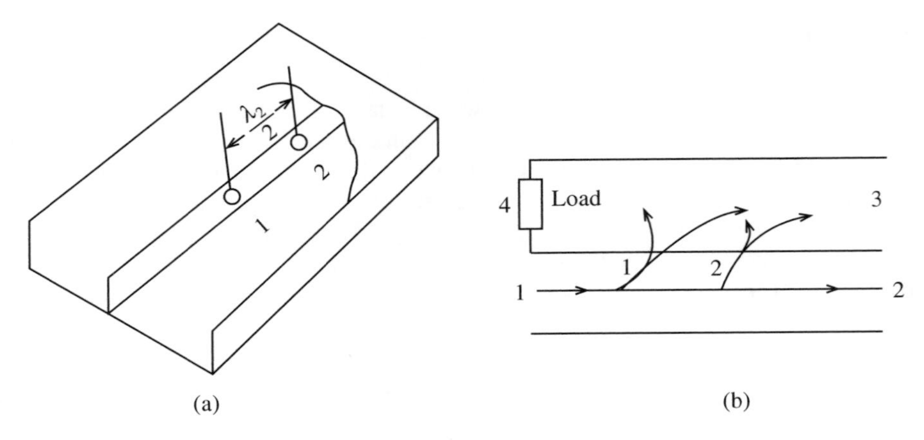

Fig. 7.13

Waveguide junctions

Waveguide junctions are required to reroute power in a similar way to junction devices in optical fibre work. They are also used to mix power from multiple sources. However, anything that alters the geometry of a waveguide will have an effect on the electric and magnetic fields and may change the characteristic impedance of the guide.

If the guide normally works in the TE_{10} mode with E-field and H-field patterns, as shown in Fig. 7.11, then bend couplings, similar to those in Fig. 7.14, should be used.

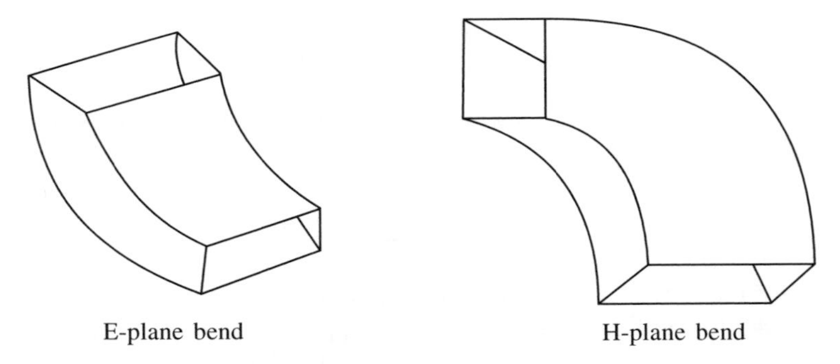

Fig. 7.14

One of the most common junctions is the T-junction shown in Fig. 7.15. This is an E-field T-junction. A signal applied to port 1 appears at ports 2 and 3, both these signals being out of phase. In practice, the T-junction requires some form of matching between the ports; this can be done by using a tuning device such as a brass screw which is inserted into the guide section. It can be adjusted for capacitance, series resonance or inductive loading.

If E-field and H-field T-junctions are combined, the hybrid T-junction is formed (see Fig. 7.16). A wave entering port 1 excites equal waves of opposite phase in 2 and 3. A signal at 4 produces equal signals of opposite phase in 2 and 3. Because of the geometry of the device a signal in 1 will not excite any dominant mode wave in 4. Hence there is no direct coupling between 1 and 4.

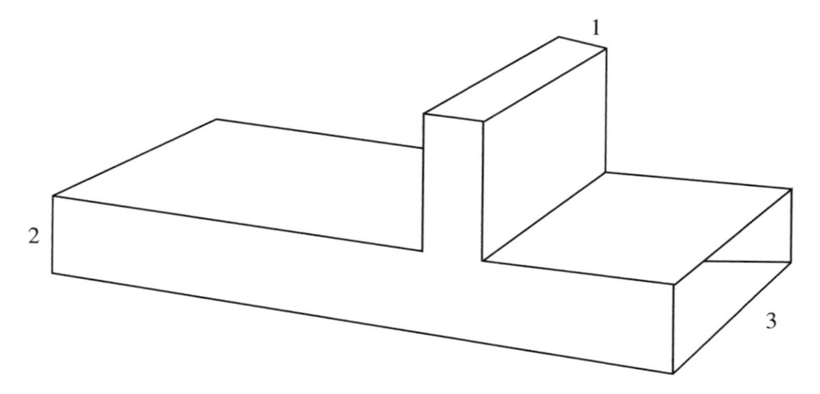

Fig. 7.15

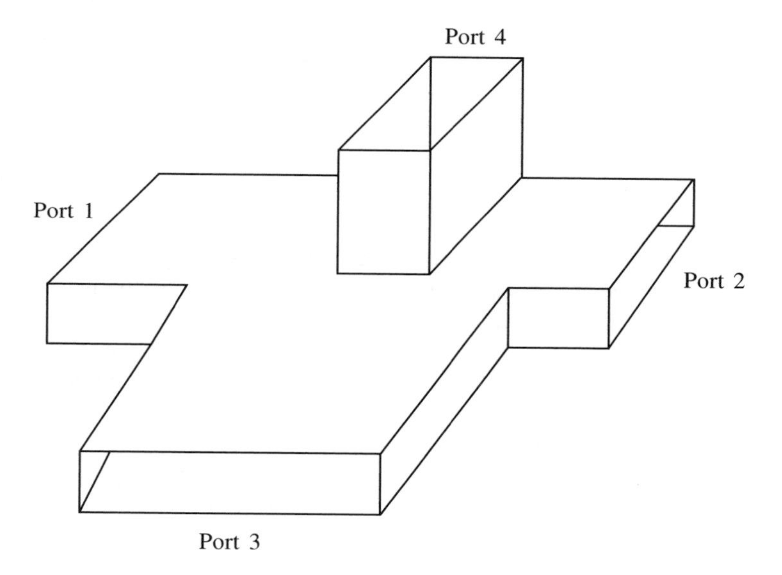

Fig. 7.16

If the E and H branches of the junction are matched, the other two arms are matched and there is no transmission between 2 and 3. Therefore a signal entering 2 will be equally divided between arms 1 and 4, while a signal entering 3 is equally divided between 4 and 1. The converse is also true.

Figure 7.17 shows the coupling between a local oscillator and a mixer. This configuration allows the signal at 4 to be combined with the local oscillator signal at 1. The combined signals then appear at 2 and 3.

Cavity resonators

The progression of the cavity resonator is shown in Fig. 7.18. The common tuned circuit is shown in Fig 7.18(a). Increasing the resonant frequency means that *C* and *L* must be

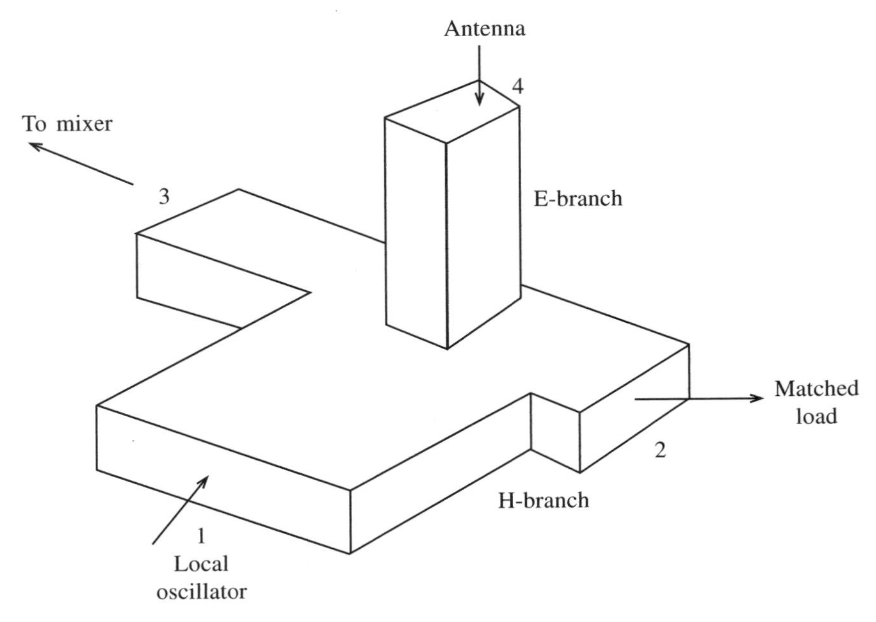

Fig 7.17

reduced until two small plates and a single-turn inductance comprise the *LC* circuit as in Fig. 7.18(b). To reduce the inductance further, several one-turn inductors are placed around the edge of the capacitor plates as shown in Fig 7.18(c). Continuing this process, it can be imagined that the two small capacitor plates are completely enclosed by several inductor bands until essentially a box is formed, known as a **cavity resonator**.

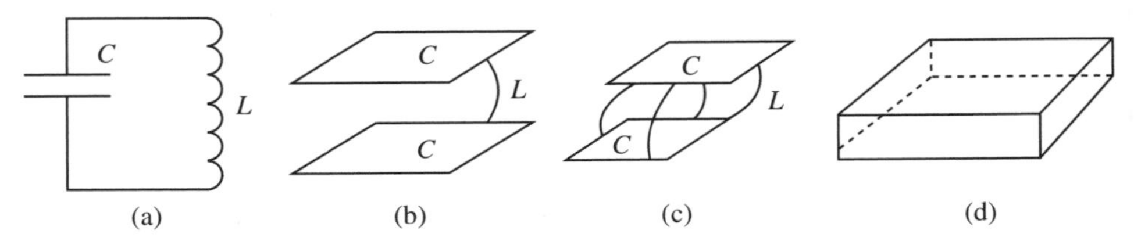

Fig 7.18

Cavity resonators therefore have a natural frequency which depends on the dimensions of the 'box'. At some determined frequency there will be a high-amplitude electric field between the plates of the capacitor and a high current up and down the sides of the box. At some instant the energy is stored in the electric field, while at the next instant it is transferred to the magnetic field.

In order to understand the fields in resonators a three-dimensional effect is necessary, and it is common to use rectangular coordinates as in Fig. 7.19. Here the energy has been introduced at point *A* and standing waves E_y, H_z and H_x have been developed as a result of incident and reflected components of the wave. Here E_y is the electric component and H_z and H_x are magnetic components 90° out of phase with each other and also 90° out of phase with the E_y field.

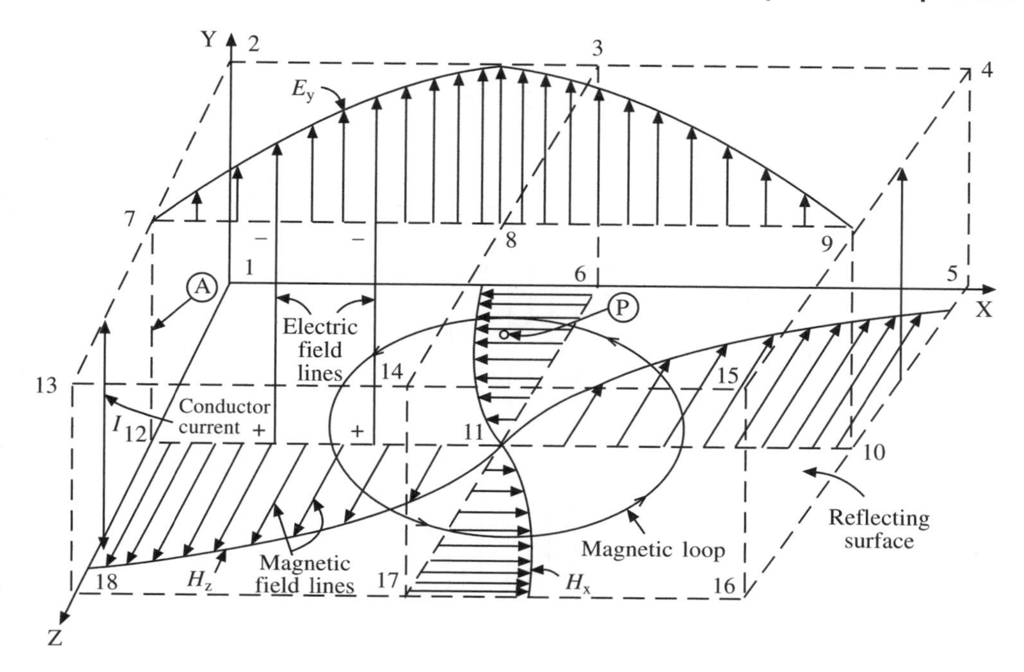

Fig. 7.19

To follow the details of the action more clearly, the corners and centre points of the edges of the three-dimensional enclosure have been numbered in order to identify the location of different two-dimensional areas or planes, e.g. the top plate is indicated by plane 2, 3, 4, 9, 15, 14, 13 and 7.

Figure 7.19 has been redrawn with the electric field distribution as shown with the appropriate polarity. In both Figs. 7.20 and 7.21 only an instantaneous condition of an action in which the magnetic and electric fields are constantly changing is represented, therefore a conduction current does exist mainly on the sides of the box. At the instant indicated in Fig. 7.16 the lower plate is positive and the upper plate is negative, but a half-cycle later their polarities will reverse and these reversals will repeat during each succeeding cycle. As the conduction current tends to equalize the charges on the top and bottom plates it will be alternating as indicated by the double-headed arrows between points 7–12 and 9–10.

The magnetic fields H_x and H_z of Fig. 7.19 can be analysed best by reference to the essential information of plane 1, 5, 16, 18 which is redrawn in Fig. 7.21. Here the non-uniform distribution of the magnetic field is shown by curves H_z and H_x, drawn over the tops of the solid line magnetic fields to represent the variation of the strength of the field from centre point 11 to the corresponding sides of the cavity. If it is assumed further that the extended dashed magnetic field lines are a continuation of the solid lines, a magnetic loop is formed. The intensity of the magnetic loops is greatest near the sides of the cavity and decreases toward the centre until at point 11 the magnetic field is zero.

As explained for Fig. 7.19, the electric and magnetic fields are 90° out of phase, therefore with zero magnetic field at point 11 the electric field must be maximum. This is shown in Fig. 7.22, where the conditions of point 11 extend upward through point P of

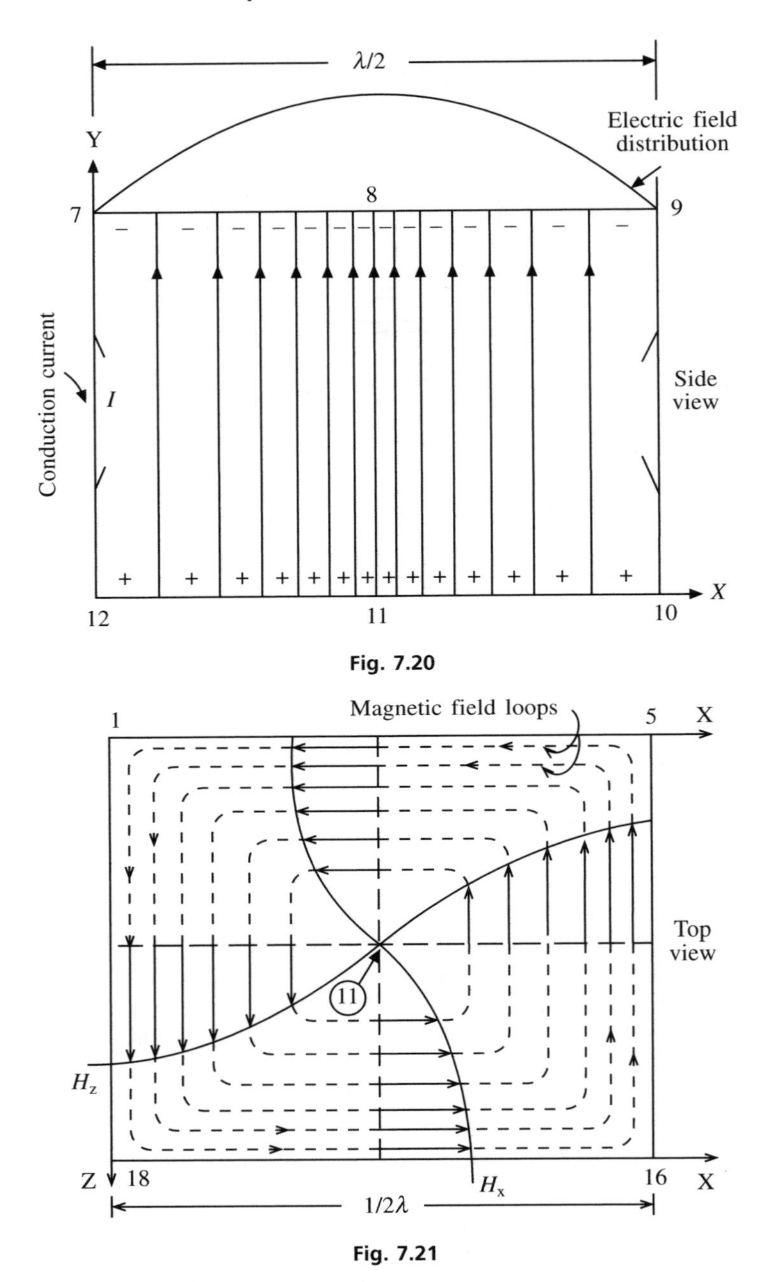

Fig. 7.20

Fig. 7.21

Fig. 7.19 to point 8. Therefore at point P there is zero magnetic field and maximum electric field.

Looking in from the end of Fig. 7.19, the plane bounded by points 14, 8, 3, 6, 11 and 17 appears as shown in Fig. 7.22 with point P at its centre. To show more clearly the direction of magnetic density distribution, solid dots represent vectors out of the plane of the paper and small circles represent vectors into the paper, as indicated by the upper sketch of Fig. 7.23.

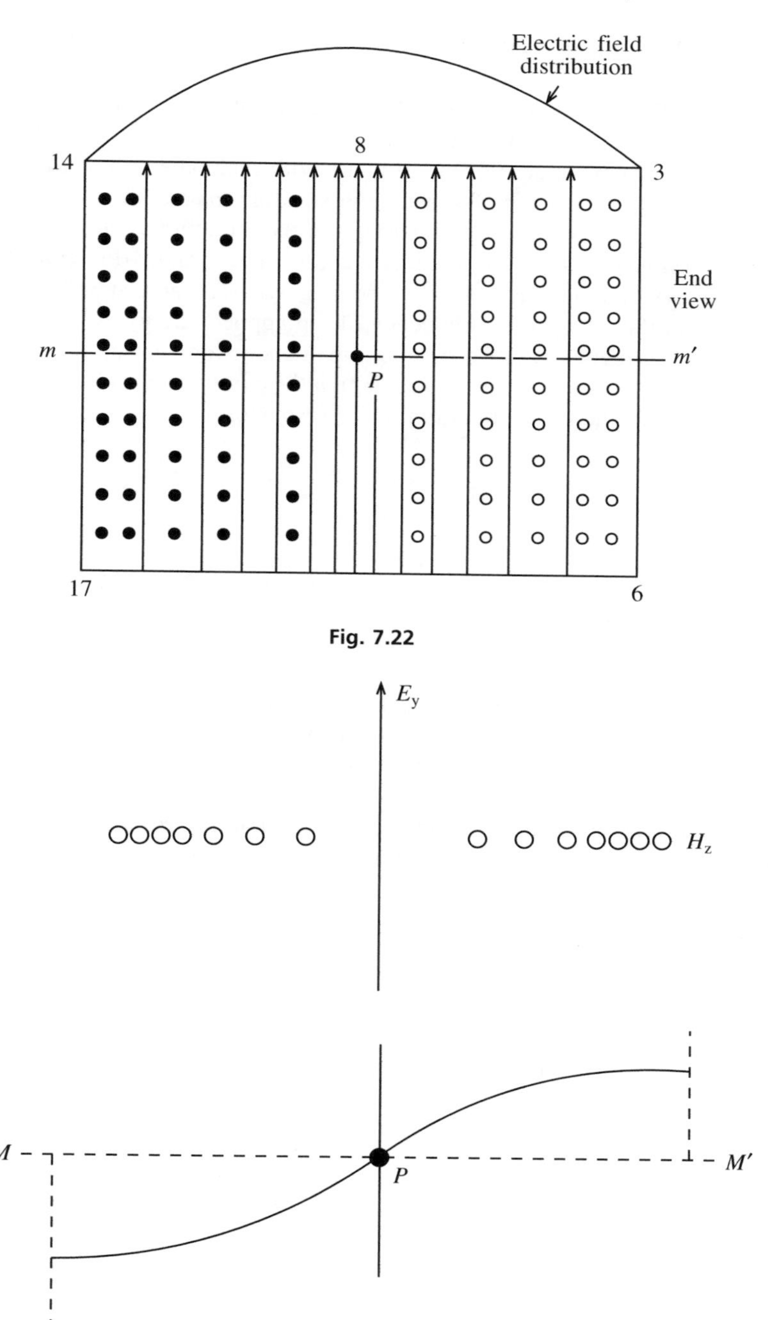

Fig. 7.22

Fig. 7.23

As in Fig. 7.22, the arrowed lines show the electric field distribution with its maximum value in the plane of line 8 and point P, with the electric waveform drawn above line 14, 8, 3 of Fig. 7.22. The magnetic field distribution across section MM' is shown by the curve of Fig. 7.23, but notice that *P* is still the region of zero magnetic field and a point of maximum electric field strength.

Summarizing the important points of these figures, the resonant cavity of Fig. 7.19 shows the instantaneous condition of established electric and magnetic fields with their maximum and minimum values 90° out of phase with each other. The dimensions of the rectangle are such to permit a half-wavelength standing magnetic wave in the *X* and *Z* directions in accordance with the characteristics of the initiating electromagnetic energy, and at the same time developing a standing electric wave in the *Y* direction. The energy is in the electric field at one instant and in the magnetic field an instant later, oscillating from one field to the other at the frequency of the applied energy. Because of the skin effect at very high frequencies of oscillation, the conduction current does not penetrate deeply into the sides of the box but acts as though the current travels back and forth over the surfaces between the top and bottom of the resonator.

Probes

In order to extract energy from or deliver energy to a cavity special methods of coupling are required. Such methods include shooting electrons through the resonator at a point where the electric field is high, or introducing a small conducting loop into the area at a point where the magnetic field is strong. A probe or loop usually is connected to a co-axial cable with the aid of a suitable matching device.

The first method mentioned above is illustrated in Fig. 7.24, which shows a simplified cavity resonator with holes in the near and opposite sides to permit electrons to pass through. As the electrons pass through the holes in the walls of the cavity it is necessary that their transit time be small compared to the duration of a resonator cycle. However, if the electrons pass through in bunches, one to each cycle, the resonator will be excited by the electron beam. When the resonant frequency of the cavity is approximately the same as that of the electron bunches, oscillations of large amplitude will be generated.

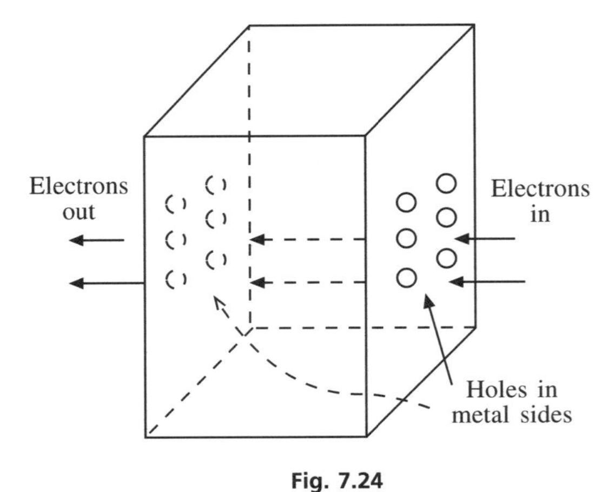

Fig. 7.24

The second method of excitation is illustrated in Fig. 7.25, which shows a small probe placed in the central area of the cavity resonator. The probe is the extended inner conductor of a co-axial cable and should be placed to coincide with the electric field vector. From the explanations given previously, it is known that the greatest strength of the electric field is found near the centre of the resonator, therefore the exciting probe should be placed in the region of point P in Fig. 7.19.

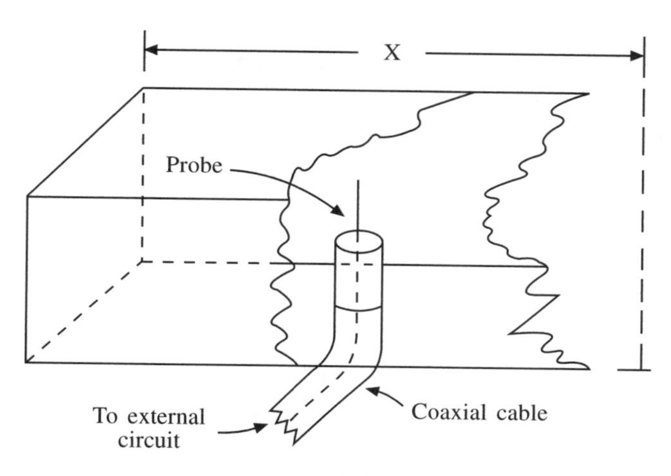

Fig. 7.25

The third method of cavity excitation is that shown in Fig. 7.26, where a loop of arbitrary shape is placed within the cavity in such a position as to envelop the greatest magnitude of magnetic flux lines. The plane of the loop must be at right angles to the magnetic field, as shown at point A in Fig. 7.19.

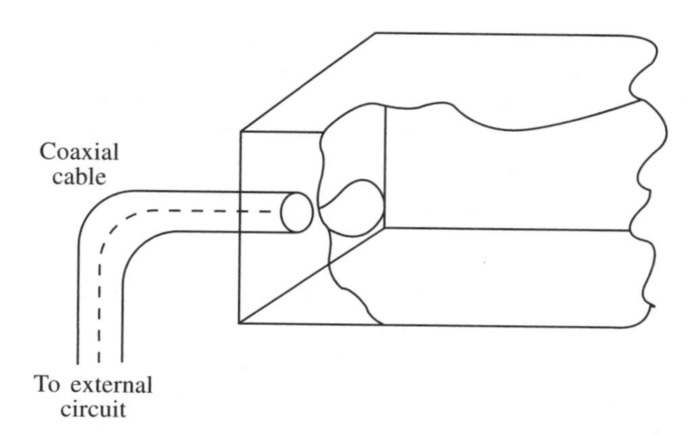

Fig. 7.26

Generally the method of output coupling to the cavity resonator is the same as the methods of injection, and so points of maximum electric and magnetic field strength are chosen for best effects.

Circulators and isolators

An isolator is a device that has the ability to pass a signal in only one direction while the other direction exhibits high attenuation. A typical type of isolator is the ferrite isolator; for the purpose of understanding its operation an explanation of Faraday rotation is helpful.

Any electron has natural spin, and if a d.c. magnetic field is applied to a spinning electron it starts to precess as shown in Fig. 7.27. The direction of precession depends on the polarity of the magnetic field, and its rotational frequency is dependent on the strength of the applied magnetic field.

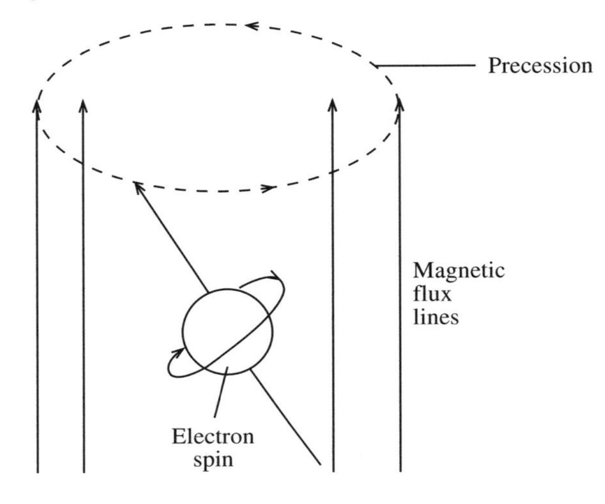

Fig. 7.27

When a circularly polarized wave meets precessing electrons, it is found that interaction only takes place when the wave rotation is in the same direction as the electron spin. Since linear polarization consists of two equal circular polarizations in opposite directions then when the linear polarized wave travels through a ferrite material in the same direction as an applied magnetic field these two circularly polarized waves interact differently with the spinning electron. Hence their phase shifts are not equal and the polarization is rotated.

Figure 7.28 is the functional diagram of a Faraday rotation isolator using these principles. This shows transmission in the forward direction; when the electric field lines of the microwave are at right angles to the attenuation vanes there is little attenuation, but this will increase as the angle decreases from 90° to 0°. The specific action in the isolator is seen in Fig. 7.29. Figure 7.29(a) shows the incident wave with no attenuation, while Fig. 7.29(b) shows the reflected wave which suffers up to 30 dB of attenuation because the vane of waveguide 2 is at 45° to that of waveguide 1.

A circulator allows the separation of signals and once again involves the operation of ferrites. The most frequently used type is the three-port circulator which is shown in Fig. 7.30. Effectively this device consists of three isolators in which power entering 1 is received at 2 only, while power entering 2 is received at 3 only, and power entering at 3 is received at 1 only.

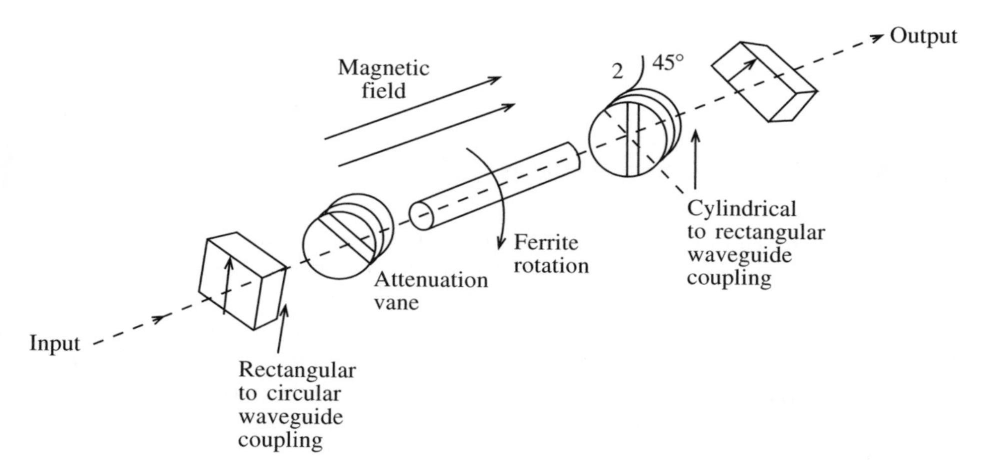

Fig. 7.28

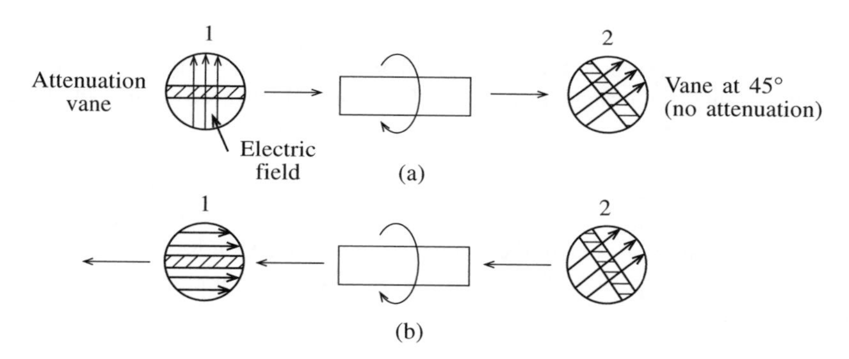

Fig. 7.29

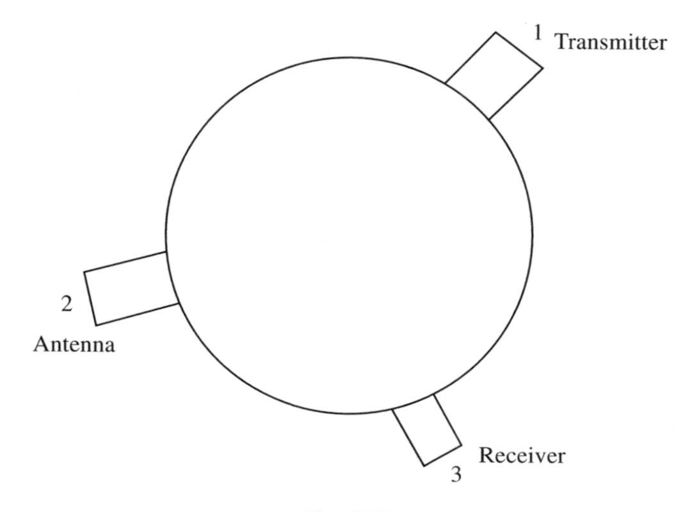

Fig. 7.30

Circulators can have more than three ports and operate without using ferrites, but generally a maximum of four ports are used with a ferrite construction for greater efficiency.

7.9 Microwave active devices

Microwave devices may be categorized into two types: low-power solid-state devices and high-power vacuum-tube devices. Communications involving the gigahertz bandwidths cannot use conventional transistor or integrated circuit design as silicon devices suffer from stray capacitance between leads and the semiconductor elements themselves. Furthermore, the movement of charge carriers from one region to the other takes a finite time called the **transit time**. If the transit time is similar to the period of the frequency being transmitted then phase shifts can occur which affect the signal adversely.

Solid-state devices

Two devices which overcome the major problem of transit time are the Gunn diode and the impact avalanche and transit time (IMPATT) diode. Both these devices use a negative resistance characteristic, i.e. over a portion of the characteristic curve the resistance decreases as the voltage increases.

The Gunn diode is used for oscillator and mixer applications and consists of a thin slice of n-type gallium arsenide between two metal conductors. When a voltage is placed across the slice oscillations occur, the frequency of which depends on the thickness of the slice. Furthermore, the period of the oscillations is equal to the transit time of electrons across the slice. Hence the geometry of the device is fundamental to its operation.

It also operates on several modes, and this allows the diode to be tuned when placed in a cavity. However, like all negative resistance devices, it requires a circulator to separate the input from the output signals.

The IMPATT diode utilizes the delay time required for an avalanche condition and the transit time to produce a negative resistance characteristic. It has a p-n junction and may be a four-layer device. The thickness of the intrinsic region is fabricated so that the transit time of an electron across this region is equal to half the period of the operating frequency.

The diode operates with reverse bias just below the breakdown region, and the fluctuation of the operating frequency, which is superimposed upon this bias, causes breakdown to occur once per cycle. When the junction breaks down, a burst of electrons enters the intrinsic region and appears at the other side half a cycle later. The resultant oscillations are dependent on the dimensions of the diode and the resonant frequency of the cavity in which it is immersed.

IMPATT diodes have higher efficiency and greater power levels than Gunn diodes, but they are noisier because of the avalanche effect.

Microwave tubes

When greater power is required more complex devices must be used. Also the process of miniaturization imposes problems at higher frequencies due to higher power dissipation. Because of these restrictions alternative designs have been developed, among them multicavity magnetrons, klystrons, reflex klystrons and travelling wave tubes.

Multicavity magnetrons

In order to understand the operation of a magnetron, Fig. 7.31 shows the effect of an electric and magnetic field interaction on an electron's motion. The dot in the centre of the figure is the cross-section of a filament in a cylinder. The cylinder is positive with respect to the filament, and the small dots inside the cylinder represent the magnetic flux lines in the same direction as the filament.

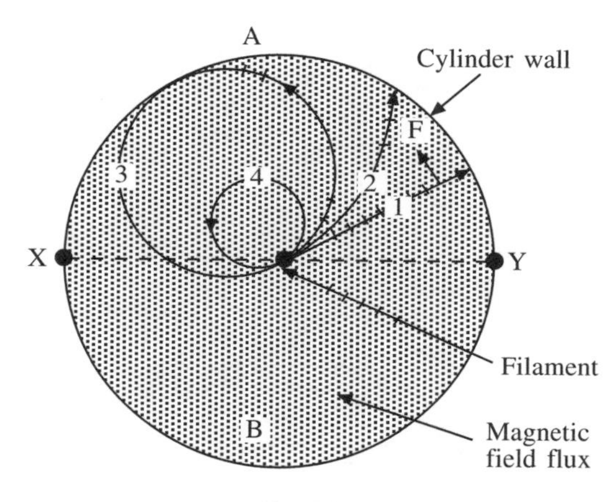

Fig. 7.31

When there are no magnetic lines present the electrons move from the filament to the cylinder wall as shown in path 1. When a weak magnetic flux is applied, the electrons cross the flux lines and, because of the interaction, the electrons travel in a circular path as seen in path 2. If the magnetic flux is increased it is possible to cause the electron stream to follow path 3 and just graze the cylinder wall before returning to the filament. Path 4 shows the movement of electrons under the influence of a stronger field.

Hence if the filament is a hot cathode and the cylinder is the anode, forming a diode, then the electrons can move in curved paths depending on the strength of the magnetic field. For path 2 the average diode current remains constant but for path 3 the diode current suddenly drops to zero or cut-off. Any further increase of magnetic flux merely reduces the excursion of the electrons from the filament source after cut-off has been reached.

A practical magnetron is made with a split anode as shown in the cross-sectional view of Fig. 7.32. If both anodes are positive by the same amount, their force distributes evenly between the anodes and the cathode. An electron in Fig. 7.32(a) will experience the same attractive force at any position on the equipotential lines. The path of an electron is shown for the cut-off condition.

If the voltage on one anode is increased while the voltage on the other is decreased, the equipotentials are distorted as shown in Fig. 7.32(b). At cut-off the electron is attracted,

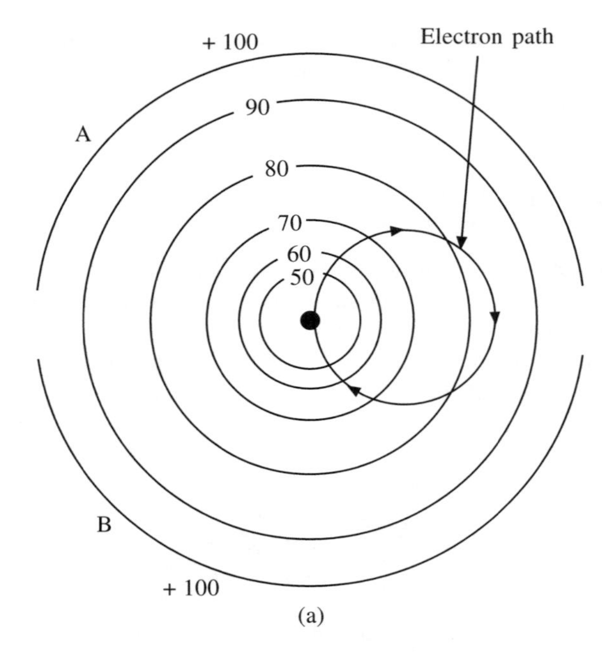

(a)

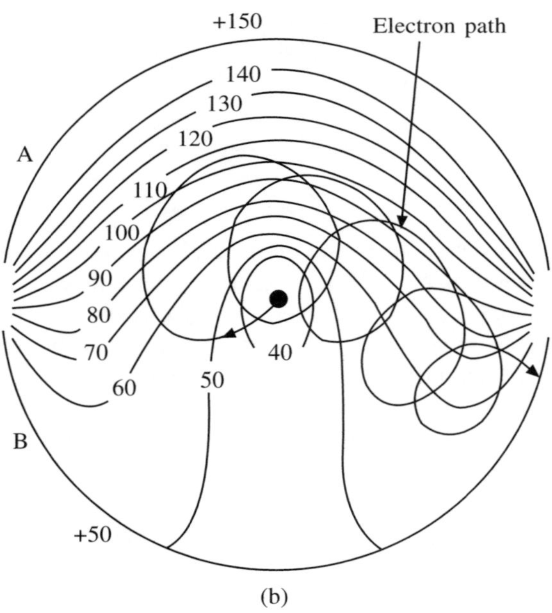

(b)

Fig. 7.32

at first, to the anode with the highest potential. But the magnetic field curves it away from this direction. As its path continues it is once again attracted to the same anode, but the attraction is not as great as before because the anode potential is now less. Hence the electron takes a spiral path, its energy decreasing all the time until it approaches the anode with the lower potential. This potential is enough to attract the electron to the anode with the lower voltage. More electrons arriving at the lower-voltage anode cause the negative resistance effect which was mentioned earlier. This effect is shared by both anodes as an alternating voltage appears across them, and hence this primitive device produces oscillations by altering anode potentials.

The problem of transit time is overcome by adding end plates to the split anode and causing the electrons to spiral around the cathode, as shown in Fig. 7.33. Since the path of the electrons can be adjusted either by varying the anode voltages, the magnetic field or both, the electrons can spiral for the full length of the cathode. There is a mutual action between the anode and an electron. As the electron approaches the anode, the magnetic field curves the electron path away from the anode. The electron stops travelling towards the anode and because of this it loses energy.

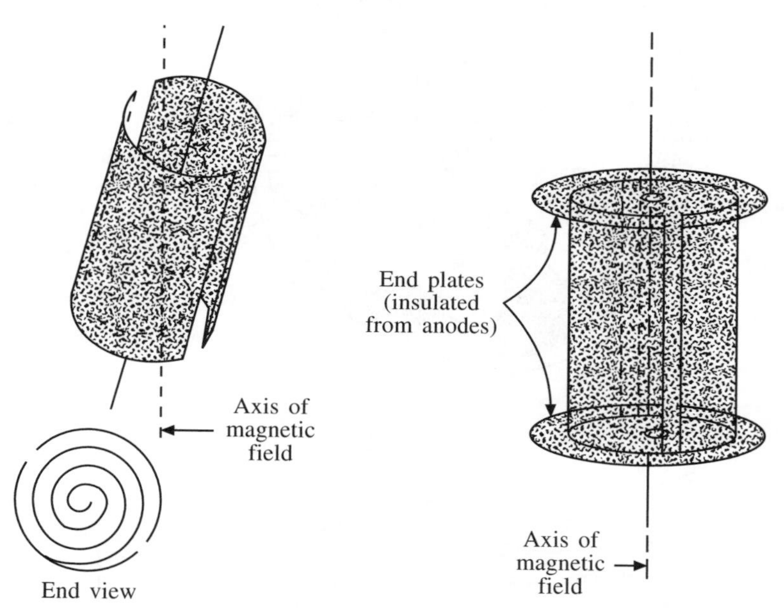

Fig. 7.33

Because the electron travels through the field near the anode, the field for that moment is less by one electron charge. The electron repels one electron charge from the anode to the supply, but as soon as the electron is no longer near the anode, this electron, repelled toward the supply, can return. This back-and-forth motion of the repelled electron on the anode is the a.c. component of the field superimposed on the d.c. field.

The electric and magnetic fields must be so adjusted that the electrons always find a retarding field, one to which they can give up energy. The d.c. component and the magnetic field are so adjusted that the electron can add energy to the a.c. component of the electric field each time around. Thus the transit time of the electron is one complete cycle of the induced field. This is known as the transit time operation of the magnetron.

It should be noted that so long as the fields are present it does not matter if the anode is split or integral. However, to achieve the fields some resonant circuits must be built into the anode and the simplest way to do this is to build cavities into the anode. The transit time between cavities is the time of one cycle.

The structure of a multicavity magnetron, using the principles discussed in the previous section, is shown in Fig. 7.34. In the multicavity magnetron, the cavities are set in oscillation by the rotating electrons in the space between cathode and anode. When the electron path is favourable to adding electron energy to the field, the oscillations are sustained. Other electron paths take energy away from the field. The magnetron operating conditions must be adjusted so that more energy is added to the field than taken away. Maximum output occurs when the added energy reinforces the cavity oscillations.

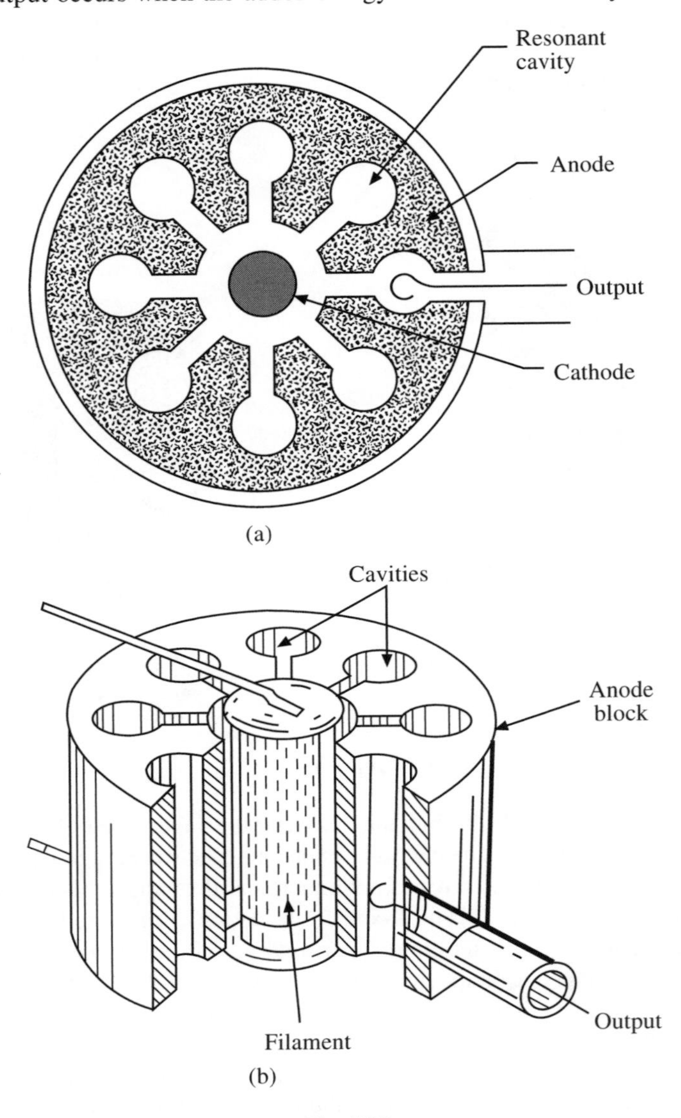

Fig. 7.34

The best time for the electrons to arrive is when there is maximum voltage across the opening in the cavity. The opposite sides of the opening are then at opposite polarities and the electrons approach from the positive side of the opening, as shown in Fig. 7.35. Side

B is negative and contains more electrons than A, so the electrons of the oncoming stream repel those at B, giving them added push to travel to side A around the inner surface of the cavity. In this way the oscillations of the cavity are reinforced.

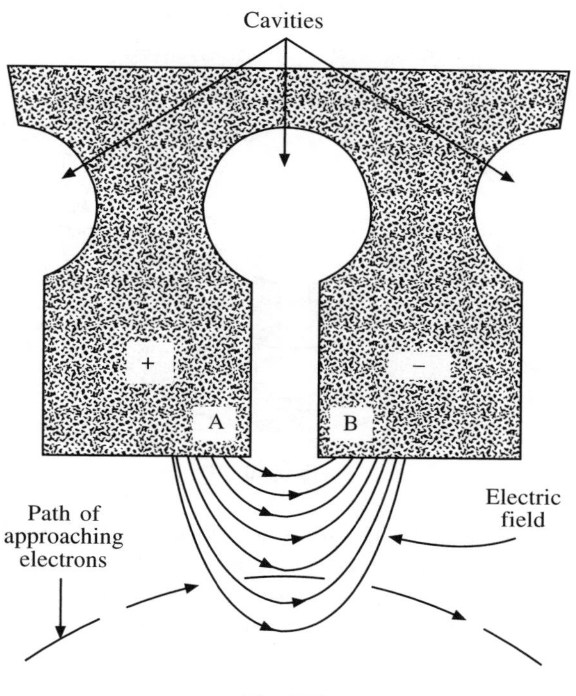

Fig. 7.35

After passing a number of cavities the stream of electrons is slowed down and is attracted by the anode block. The cathode, however, is continually emitting electrons and some of them arrive at the gap at a time when the oscillations are not aided. These electrons, instead of being slowed down, are speeded up by passing the cavity fields. This results in greater action between the electrons and the magnetron magnetic field, and the electrons curve back towards the cathode. Thus, when all the operating constants are correctly adjusted, the electrons which do not aid oscillations are carried off and the magnetron can operate efficiently at high output.

Magnetrons have various modes of operation, each of which gives a different operating frequency. The cavity resonators represent a series of tuned resonant circuits, each coupled to the other. Some phase difference usually exists across the faces of the cavities and, at a given field strength and anode voltage, a phase difference of 360° exists around all of the cavities.

Maximum efficiency may occur for a particular mode. For example, if the phase difference between the oscillations in successive cavities is 180° and there are eight cavities then there are four separate groups of electrons each aiding oscillations. However, when a number of resonant circuits are used, there is a slight tendency for the frequency of each to vary slightly from the other. To prevent this, the coupling between the in-phase cavities is increased by strapping. Figure 7.36 shows the anode straps connected to alternate anode segments. This keeps the anode voltages in phase and stabilizes the frequency.

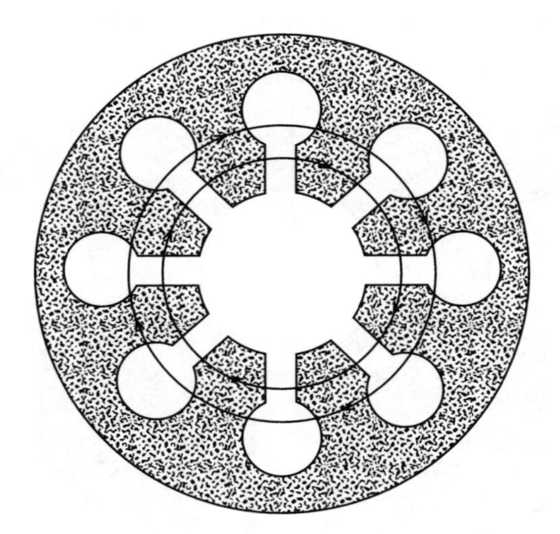

Fig. 7.36

One main disadvantage of the magnetrons described is that it is hard to modulate them by any of the usual modulation methods. Also the transit time sets a limit to the upper frequency which can be used. However, very high-power applications such as modern radar would be impossible without the magnetron, which is capable of producing 10 MW or more of pulsed power.

Klystrons

The problem of transit time can be overcome by use of a klystron. Originally this type of microwave tube was developed as a low-power device, but the power-handling capacity of modern tubes is high. Klystron tubes are superior to magnetrons in that they can function as an oscillator, amplifier or detector.

The klystron uses a process known as **velocity modulation** in which the velocity of the electron stream is varied as it moves through the tube. Figure 7.37 illustrates the basic method of producing this type of modulation. The arrangement consists of an electron beam of constant velocity directed towards two grid meshes. These grids extend out into the electron stream, and a source of RF voltage is connected across the grids.

As the electrons pass through the grids they are affected by the field set up between them by the RF generator. An electron that leaves grid 1 just before that grid goes positive is drawn back to it when the grid becomes positive. At the same time the electron is repelled by grid 2, thus causing it to slow down. An electron that leaves grid 1 when it is just going negative is repelled by grid 1 and attracted by grid 2. This electron is speeded up. Electrons that pass through the grids when the grids are at zero potential are neither speeded up nor slowed down, and they join the slowed-down electrons. The speeded-up ones catch up with those that go through unaffected, and in this way bunches of electrons are formed.

Figure 7.38 shows an industrial velocity-modulated klystron tube. A stream of electrons

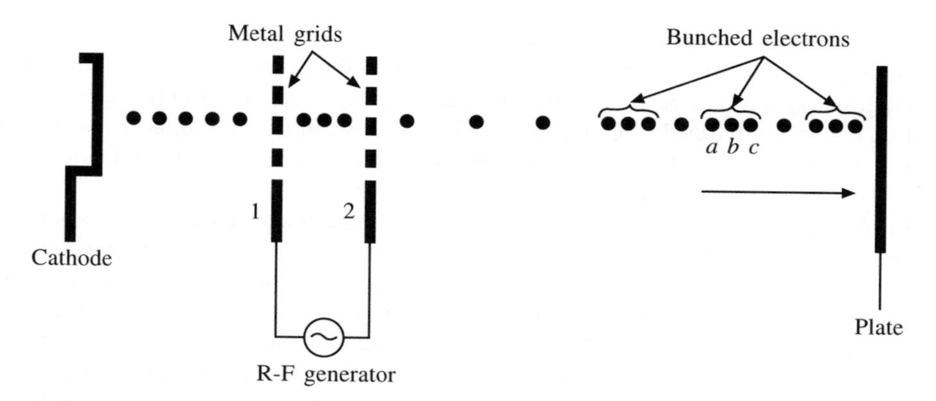

Fig. 7.37

emitted by the cathode is focused into a sharp beam by the action of G_1 which is connected to a suitable voltage. This electron beam is accelerated toward G_2 and G_3 by the high positive voltage of the voltage source V_1 (usually over 1 kV). G_2 and G_3 then act as the grids in Fig. 7.37 to bunch the electrons in their travel towards the collector plate. Since the distance between G_2 and G_3 is short, the change in velocity of the electrons is very slight, but this change is usable if an a.c. voltage V_2 is superimposed on the high d.c. voltage V_1.

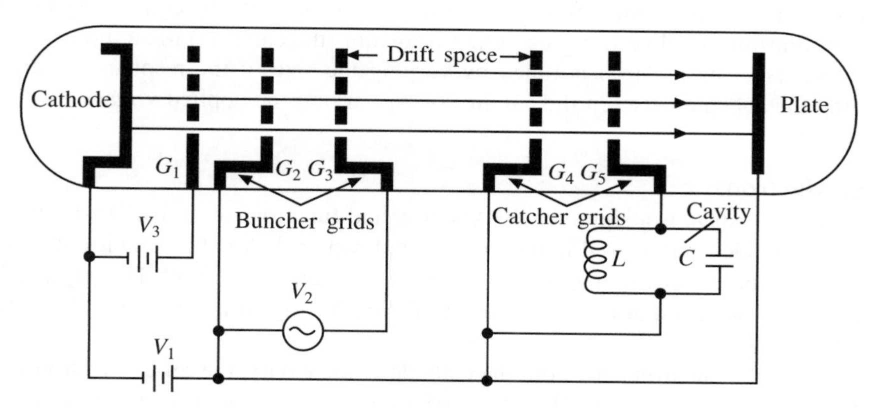

Fig. 7.38

In general, just as many electrons enter the buncher grids as leave, and since the electrons spend only a short time between the grids, there is no resultant current in the grids. Thus, the transit time of the electrons has a negligible effect on the operation.

To be of greater use the electrons are converted into a **density-modulated beam** from which energy can be extracted. There are several conversion methods. In the drift-tube process the electrons pass through a tube that is long compared to the grid spacing. This tube contains no electric or magnetic fields. The electrons accelerated by the buncher grids eventually catch up with those ahead and form bunches as they continue their path through the tube. Referring to the lettered bunch of Fig. 7.37, this means that electron *a* was accelerated, *b* was unaffected and *c* was decelerated to form a density-modulated beam.

The drift tube or space should not be long enough to allow the electrons to be separated by the repelling action between them. In their path to the collector plate these electrons reach grids G_4 and G_5 placed at a suitable distance from G_2 and G_3. In Fig. 7.38 grids G_4 and G_5 connect to a tuned resonant cavity. The electrons form dense bunches, as before, which travel through grids G_4 and G_5, and if this current could be measured it would look as if it consisted of a pulsating d.c.

In order to understand the amplification process, consider a bunch of electrons between the grids, i.e. a strong current. To produce this strong current there would have to be a large potential difference between the grids. As the current flows in the external resonant cavity, this potential decreases. If it reverses while the bunch is between the grids, this will tend to reverse the current. To avoid this, the potential on the grids is reversed at an instant when there are few electrons between them. The bunch must last between the grids G_4 and G_5 just as long as half a cycle. Energy is extracted during this half-cycle and very little is put back during the next half-cycle.

Amplification has taken place because in the buncher grids the electrons are affected very little, but in the drift space this effect is given time to grow and the bunching becomes much more pronounced. Therefore the second set of grids is greatly energized by the bunched electron stream and the RF field is produced on the second set of grids, which are called the **catcher grids**. To operate effectively, the cavity must be tuned correctly and the catcher grids should not be spaced too far from G_2 and G_3.

To change an amplifier circuit into an oscillator, it is necessary to feed back a voltage of proper phase from output to input. This is done in the klystron by connecting a suitable length of transmission line from the output resonator (the catchers) to an input resonator (the bunchers). The klystron is tuned by varying the size of the resonant cavities, by changing the spacing between the cavities or by varying the applied voltage.

Reflex klystrons

This type of klystron is a single-cavity klystron and functions as an oscillator only. It is a low-noise device and is used in low-power applications below frequencies of 30 GHz. Its basic structure is shown in the cross-sectional diagram of Fig. 7.39. This klystron uses a single cavity wrapped around the tube and the output signal is taken from the tube by a coupling loop.

Following the electron path from the cathode to the cavity, the reflex klystron is very much like the two-cavity type; however, beyond the cavity there is a long drift space at the end of which is a repeller plate. When this plate is negative the electrons are repelled, which slows them down and repels them back to the buncher grids, thus causing oscillations.

When an RF voltage exists in the single-cavity resonator, the positive voltage at the cavity attracts the electrons and either accelerates or retards them, depending on the RF cycle. After leaving the buncher field the electrons travel towards the plate, are slowed down, stopped and then reattracted by the buncher. This return action is intensified because the plate is negative, repelling the beam, and the buncher is positive, attracting it. The electrons, which received their initial acceleration from the accelerator grid, eventually return to this positive grid after passing through the buncher grids on their return path.

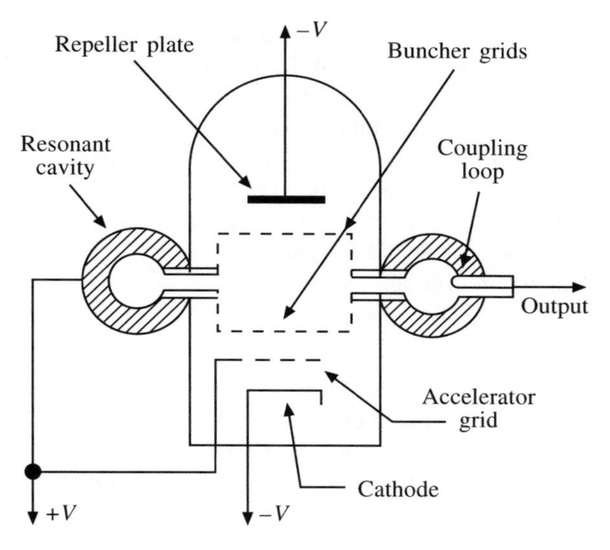

Fig. 7.39

The bunching operation is shown in Fig. 7.40. The lower sine-wave voltage represents the voltage at the buncher grids. And the paths of three electrons that pass through the buncher grids at different instants in the RF cycle are shown at the centre of the diagram.

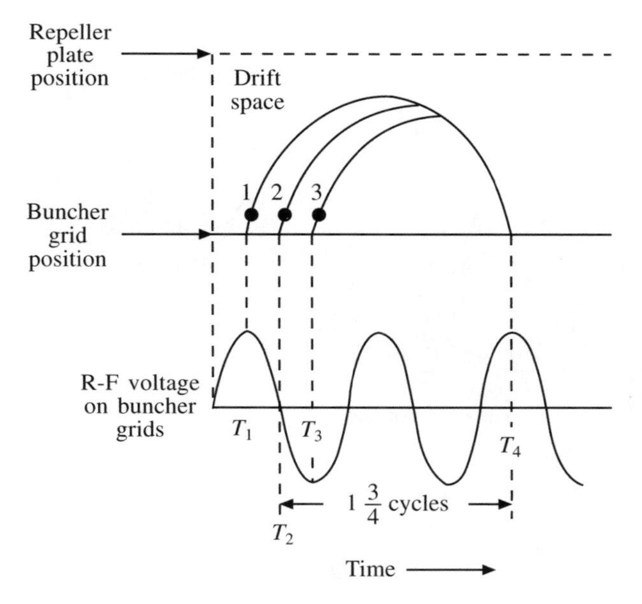

Fig. 7.40

Electron 1 passes through the buncher grids at time T_1 and the positive swing of voltage accelerates it through a large portion of the drift space towards the repeller plate. Electron 2 leaves later, when the RF voltage is zero, so its velocity is not affected. As a result it does not travel as far into the drift space as electron 1.

Electron 3 is decelerated by the buncher because the RF signal is in its maximum negative position, although the voltage is still positive. Electron 3 travels a much shorter distance into the drift space than the other two electrons. When proper adjustments are made, all three electrons arrive at the buncher at time T_4, when the voltage is at a positive peak. This satisfies the two conditions for oscillation where the feedback signal arrives in phase and with the correct amount of energy, i.e. regenerative feedback.

Figure 7.40 shows feedback occurring $1^3/_4$ cycles later if T_2 is taken as the average of T_1, T_2 and T_3. The bunches can spend the period of $^3/_4$, $1^3/_4$ and $2^3/_4$ cycles making the round trip and still maintain oscillations.

Travelling–wave tubes

For very wide-band applications the klystron's operation is a compromise between gain and bandwidth. If the gain is reduced enough to give the required bandwidth, the noise generated within the klystron may decrease the signal-to-noise ratio. Also, in the klystron tube the coupling between the electron beam and the field is limited due to the narrowness of the resonator gaps. Hence the transfer of signal energy is less efficient.

In order to overcome these deficiencies the travelling-wave tube was designed. The construction of this tube is shown in Fig. 7.41. The structure consists of an electron gun which produces a collimated beam of electrons. The electrons are accelerated along the tube towards the collector, which has a high positive potential. The slender section of the tube contains a helix of wire which is rigidly mounted on insulating supports in the long glass tube.

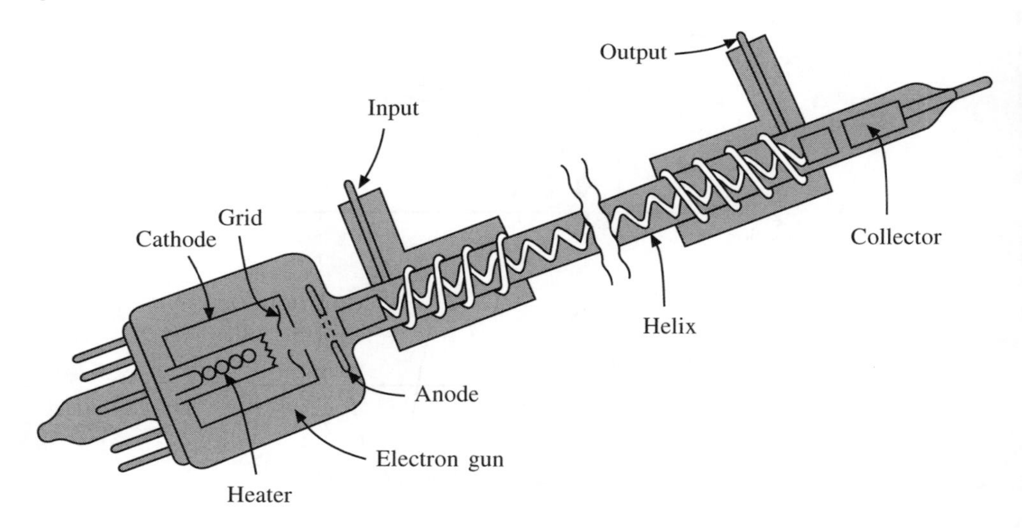

Fig. 7.41

The electron beam travels through the helix to the collector. Outside the helix there are two short spirals wound on the glass at each end of the helix, and these serve as the input and output connections. Outside the tube is a focusing magnet which serves to collimate the electrons.

The helix has inductance and capacitance which are effective in slowing down any wave travelling in it, and hence it functions as a delay line. This avoids one of the problems of the klystron, where the interaction between the field and the electrons is very short. In the travelling-wave tube the interaction can take place through the entire length of the tube.

With a beam of electrons travelling down the helix, a wave is injected on the end of the helix. This wave travels in the same direction as the electrons. By varying the collector voltage, we adjust the speed of the electrons so that the electrons tend to move slightly faster than the wave. Where the wave has a positive maximum it tries to attract the electrons and hence speeds them up. If the wave has a negative maximum it retards the electrons. Hence bunching occurs as in the klystrons. However as the electrons were travelling slightly faster than the wave in the first place, more energy is given up in the retarding than is extracted in the speeding up. This produces considerable gains up to 60 dB for frequencies of 100 GHz or more.

By proper matching and by using directional attenuators which will soak up the backward wave as it occurs, it is possible to avoid backward waves in amplifiers. When operating as an oscillator the backward reflection is permitted. The reflected wave then will travel forward in the right phase, be amplified, and so on, until the circuit reaches its peak amplitude of oscillation. This is termed a **backward-wave oscillator**.

The collector voltage or the voltage applied to the helix controls the speed of the electrons. Thus, when the travelling-wave tube is used as an oscillator, the collector or helix voltage can be used to control the oscillator frequency. Also the tube can be pulse-modulated, which is essential for certain applications.

7.10 Further problems

1. Calculate the cut-off frequency of the following modes in a rectangular waveguide which has the inside dimensions 4 cm by 4 cm for the following modes: TE_{10}, TM_{11}, TE_{01}, TM_{21} and TE_{22}
   Answer: 3.75, 5.3, 3.75, 8.4 and 10.6 GHz

2. A waveguide is used for operation at 10 GHz. If it has dimensions 2.3 cm by 1 cm, calculate the frequency range over which the dominant mode can propogate and explain why the recommended range of operation is in the X-band range of 8.2–12.5 GHz.

3. Calculate the size of a quarter-wavelength section of waveguide which is needed to match a junction between rectangular waveguides having dimensions 2.1 cm by 1.2 cm and 3.8 cm by 1.2 cm. Also determine the frequency at which higher-order modes would cause energy loss.
   Answer: 2.3 cm by 1 cm, 7.8 GHz

4. Determine the cut-off frequency and wavelength of a rectangular waveguide in the T_{21} mode if $a = 3$ cm and $b = 1$ cm.
   Answer: 18 GHz, 1.38 cm

5. A rectangular waveguide functions in the T_{10} mode with a free space frequency of 5.5 GHz. If the guide dimensions are $a = 5$ cm and $b = 2$ cm, determine
   (a) the cut-off frequency and wavelength;

(b) the guide wavelength.
Answer: 3 GHz, 10 cm, 1.02 cm

6. Explain what is meant by E-plane and H-plane bending when dealing with waveguide junctions.
7. Using E-field and H-field diagrams, sketch the amplitudes of the different components of a TE_{11} mode with reference to position inside the waveguide.

Appendix B

Analysis of gain off resonance

Consider the circuit in Fig. B.1, where R_i is the resistance of the source. At resonance, from (4.6),

$$A_{Vo} = \text{Im } R_P\Omega \tag{B.1}$$

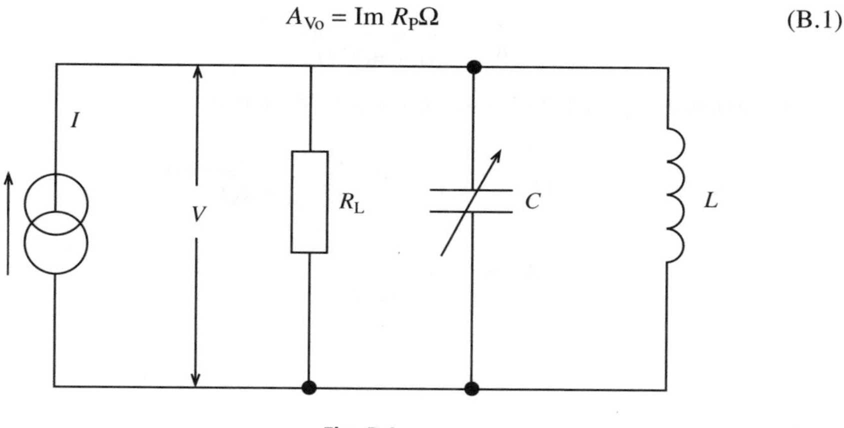

Fig. B.1

Also, from the circuit, the total current is given by

$$I = \frac{V}{R_i} + \frac{V}{i/j\omega C} + \frac{V}{j\omega L}$$

$$= V\left[\frac{1}{R_i} + j\left(\omega C - \frac{i}{\omega L}\right)\right] = \frac{V}{Y} \tag{B.2}$$

From (B.2),

$$Z = \frac{1}{Y} = \frac{1}{1/R_i + j[\omega C - (1/\omega L)]} \tag{B.3}$$

Rearranging gives

$$Z = \frac{R_i}{1 + j\omega CR_i[1 - (1/\omega^2 LC)]}$$

$$= \frac{R_i}{1 + jQ[1 - (\omega_o/\omega)\,2]}$$

Let $D = (\omega - \omega_o)/\omega_o$. Then

$$\left(\frac{\omega_o}{\omega}\right)^2 = \left(\frac{1}{1 + D}\right)^2$$

Expanding this by the binomial theorem gives

$$\left(\frac{1}{1 + D}\right)^2 = 1 - 2D + 3D^2 - 4D^2$$

All higher components can be ignored as D will generally be small. This gives

$$Z = \frac{R_i}{1 + jQ[1 - 1 + 2D]}$$

$$\frac{Z}{R_i} = \frac{1}{1 + j2QD} \qquad \text{(B.4)}$$

$$\therefore \qquad \frac{Z}{R_i} = \frac{1}{\sqrt{1 + 4Q^2 D^2}}$$

At resonance $V_{o1} = IR_i$. Off resonance $V_{o2} = IZ$. Hence

$$\frac{IZ}{IR_C} = \frac{V_{o2}}{V_{o1}} = \frac{A_V}{A_{V0}} = \frac{1}{\sqrt{1 + 4Q^2 D^2}}$$

$$\therefore \qquad A_V = \frac{A_{V0}}{\sqrt{1 + 4Q^2 D^2}} \qquad \text{(B.5)}$$

Appendix C

Circuit analysis for a tuned primary amplifier

This analysis uses the equivalent T-network and a knowledge of two-port networks which may be obtained from any standard text.

Consider the circuit in Fig. 4.11(b), where R_{11} and R_{22} are the primary and secondary load resistances. We have

$$V_{11} - j\omega L_1 I_1 - j\omega M I_2$$

$$O = j\omega M I_1 - (R_{22} + j\omega L_2)\, I_2$$

Hence

$$Z_{11} = \frac{V_{11}}{I_1} = j\omega L_1 + \frac{\omega^2 M^2}{R_{22} + j\omega L_2} \tag{C.1}$$

The secondary is untuned and will have a low reactance so that $\omega L_2 << R_{22}$. Hence

$$Z_{11} = j\omega L_1 + \frac{\omega^2 M^2}{R_{22}} \tag{C.2}$$

This expression represents a series circuit consisting of L_1 and $R_s = (\omega^2 M^2)/R_{22}$. This is shown in Fig. 4.11(c). Using expression (4.3) will give the equivalent parallel resistance at port 1,1:

$$R_{P1} = \frac{\omega^2 L_1^2}{R_s} = \frac{\omega^2 L_1^2}{\omega^2 M^2 / R_{22}} \tag{C.3}$$

The equivalent resistance (R_{P2}) appearing at the input is given by

$$R_{P2} = \frac{R_{11} R_{P1}}{R_{11} + R_{P1}} = \frac{R_{11}}{(1 + M^2/L_1^2)(R_{11}/R_{22})} \tag{C.4}$$

Also $V = IZ = j\omega L_1 I_1$ since $\omega L_1 >> R_s$ (see (C.2)).

From the second hoop in Fig. 4.11(b),

$$V_{be2} = R_{22}I_2 = j\omega MI_1$$

Also since $I = g_m V_{be1}$, the voltage gain at resonance,

$$A_{Vo} = \frac{V_{be2}}{V_{be1}} = \frac{g_m R_{11}(M/L_1)}{(1 + M^2/L_1^2)(R_{11}/R_{22})} \tag{C.5}$$

The effective Q factor (Q_{e1}) is given by

$$Q_{e1} = \frac{R_{P2}}{\omega L_1} \tag{C.6}$$

Hence,

$$Q_{e1} = \frac{R_{11}/\omega L_1}{(1 + M^2/L_1^2)(R_{11}/R_{22})} \tag{C.7}$$

As M actually controls the Q factor, the bandwidth is also affected by M. It is therefore possible to maximize the gain or bandwidth by selecting M. This is achieved by differentiating (C.7) and setting it to zero (maximum turning point in mathematical terms). When this is done the optimum value of M which will give maximum gain is found to be

$$\omega_o M = \omega_o L_1 \sqrt{\frac{R_{22}}{R_{11}}} \tag{C.8}$$

Using this optimum value reduces (C.7) to

$$Q_{e1} = \frac{Q_1}{2} \tag{C.9}$$

where Q_1 is the effective Q factor of the primary circuit ($R_{P2}/\omega L_1$). It can be seen that when the inductivity-coupled circuit is adjusted for maximum gain it will have twice the bandwidth of the tuned primary only ($BW = f_o/Q_{e1}$).

Appendix D

Circuit analysis for a tuned secondary

From Fig. 4.13 the loop equations are formed:

$$V_{11} = j\omega L_1 I_1 - j\omega M I_2$$

$$O = \left[R_2 + j\left(\omega L_2 - \frac{1}{\omega C_2}\right)\right] I_2 - j\omega M I_1$$

At resonance $\omega_o L_2 = 1/\omega_o L_2$ for the secondary tuned circuit,

$$\therefore \quad I_2 = \frac{j\omega M I_1}{R_2} \tag{D.1}$$

Hence

$$Z_{11} = \frac{V_{11}}{I_1} = j\omega_o L_1 + \frac{\omega_o^2 M^2}{R_2}$$

and using current division between r_{ds} and Z_{11} gives

$$I_1 = I \frac{r_{ds}}{r_{ds} + j\omega L_1 + \omega_o^2 M^2 / R_2}$$

As $r_{ds} >> \omega_o L_1$,

$$I_1 = I \frac{1}{1 + \omega_o^2 M^2 / r_{ds} R_2}$$

Using (D.1) gives

$$V_{gs_2} = \frac{-jI_2}{\omega C_1} - g_m V_{gs_1} \frac{\omega_o M}{\omega_o C_2 R_2} \left(\frac{1}{1 + \omega_o^2 M^2 / r_{ds} R_2}\right)$$

Resonance gives

$$\frac{1}{\omega_o C_2} = \omega_o L_2$$

so gain is given by

$$A_{Vo} = \frac{V_{gs_2}}{V_{gs_1}} = g_m \omega_o M_1 \frac{Q_2}{1 + \omega_o^2 M^2 / r_{ds} R_2} \tag{D.2}$$

where $Q_2 = \omega_o L_2 / R_2$. As with primary tuning, the effective Q factor is given by

$$Q_{e_2} = \frac{Q_2}{1 + \omega_o^2 M^2 / r_{ds} R_2} \tag{D.3}$$

If (D.2) is differentiated and set to zero the optimum value for M can be determined from

$$\omega_o M = \sqrt{r_{ds} R_2} \tag{D.4}$$

The bandwidth will therefore be given by

$$BW = \frac{f_o}{Q_{e2}} = \frac{2f_o}{Q_2}$$

This is twice the bandwidth of the resonant circuit.

Appendix E

Circuit analysis for double tuning

In the equivalent circuits shown in Figs. 4.14(a) and 4.14(b), the following points should be noted:

(i) The series and parallel resistances are combined to give:

$$R_{11} = \frac{\omega_o^2 L_1^2}{r_1} + R_1$$

$$R_{22} = \frac{\omega_o^2 L_2^2}{r_2} + R_2$$

(ii) The Norton generator is replaced by a Thevenin generator.

(iii)
$$Q_1 = Q_2 = \frac{\omega_o L_1}{R_{11}} = \frac{\omega_o L_2}{R_{22}}$$

(iv)
$$V_o = -\frac{\mathrm{j}}{\omega_o C_2} I_2$$

Using $I_2 = Y_T V_{11,}$ then

$$I_2 = \frac{\mathrm{j} g_m}{\omega C_1} Y_T V_1$$

and the gain is given by

$$\frac{V_0}{V_1} = -\frac{\mathrm{j}}{\omega_o C_2} \frac{\mathrm{j} g_m}{\omega_o C_1} Y_T = g_m \omega_o^2 L_1 L_2 Y_T \tag{E.1}$$

At resonance the admittance Y_T may be assembled from the z parameters

$$Z_f = j\omega_o M = \mathrm{j}\omega_o k\sqrt{L_1 L_2}$$

$$Z_i = R_{11} + \mathrm{j}\left(\omega L_1 - \frac{1}{\omega C_1}\right) \cong \frac{\omega_o L_1}{Q}(1 + \mathrm{j}2QD)$$

Note that

$$\frac{\omega}{\omega_o} - \frac{\omega_o}{\omega} = 1 + D - (1 - D) = 2D$$

and if D is small then

$$\frac{1}{1 + D} \cong 1 - D$$

When $Z_L \cong -\mathrm{j}/\omega_o C_2$ the following can be deduced:

$$Z_o + Z_L = R_{22} = \mathrm{j}\left(\omega L_2 - \frac{1}{\omega C_2}\right) \cong \frac{\omega_o L_2}{Q}(1 + \mathrm{j}2QD)$$

hence

$$Y_T = \frac{1}{z_f - z_i(z_o + z_L)/z_f} = \frac{kQ^2}{\omega_o\sqrt{L_1 L_2}\,[4QD - \mathrm{j}(1 + k^2Q^2 - 4Q^2D^2]} \tag{E.2}$$

From (E.1) and (E.2) the gain at resonance becomes

$$A_V = \frac{g_m \omega_o k\sqrt{L_1 L_2}\, Q^2}{4QD - \mathrm{j}(1 + k^2Q^2 - 4Q^2D^2)}$$

which in magnitude terms becomes

$$A_V = g_m \omega_o \sqrt{L_1 L_2}\, Q \frac{kQ}{\sqrt{(1 + k^2Q^2 - 4Q^2D^2)^2 + 16Q^2D^2}} \tag{E.3}$$

The frequency deviation D at which the gain is maximum occurs by maximizing (E.3). This gives

$$D(4Q^2D^2 + 1 - k^2Q^2) = 0 \tag{E.4}$$

Two peaks occur from this quadratic:

$$f_1 = f_o\left(1 - \frac{1}{2Q}\sqrt{k^2Q^2 - 1}\right) \tag{E.5}$$

$$f_2 = f_o\left(1 + \frac{1}{2Q}\sqrt{k^2Q^2 - 1}\right) \tag{E.6}$$

Note that when $k^2Q^2 = 1$, critical coupling occurs and

$$k = k_c = \frac{1}{Q} \tag{E.7}$$

The double peaked response is shown in Fig. E.1. Note from this diagram that the two peaks coincide when $D = 0$ and for coupling values greater than $1/Q$ the response shows a double peak and is overcoupled.

The magnitude of the gain at a peak (A_p) is given as

$$A_p = \frac{g_m \omega_o \sqrt{L_1 L_2}\, kQ}{2} \tag{E.8}$$

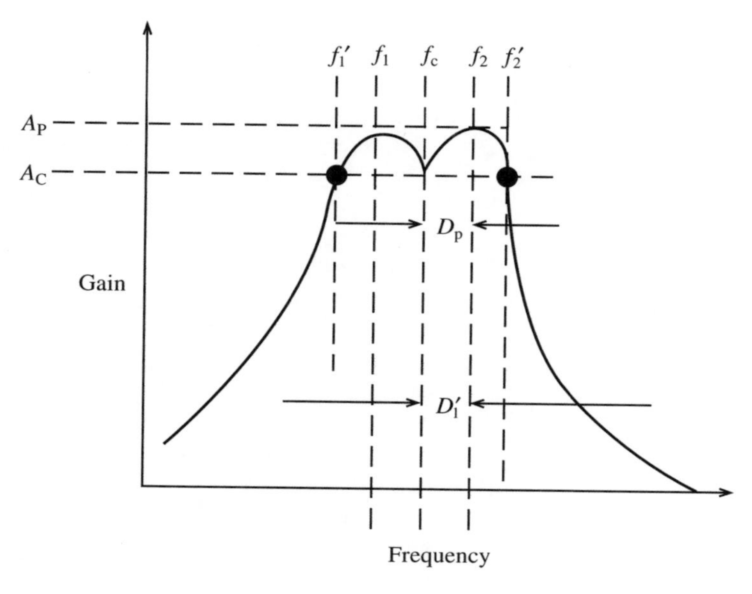

Fig. E.1

When $D = 0$

$$A_c = A_p \frac{2kQ}{1 + k^2Q^2} \tag{E.9}$$

The magnitude of the dip in Fig. E.1 is dependent on transformer design. The ripple ratio G may be given by

$$G = \frac{A_p}{A_c} = \frac{1 + k^2Q^2}{2kQ}$$

hence

$$kQ = G - \sqrt{G^2 - 1} \tag{E.10}$$

Note $kQ > 1$ and the value of (E.10) gives a value of kQ for a given ripple.

The bandwidth between the frequencies at which the gain drops to A_c is the practical bandwidth generally used:

$$2D' = \sqrt{2}\,(f_2 - f_1) \tag{E.11}$$

The 3 dB bandwidth is established between the D' frequencies when $G = \sqrt{2}$ so that $kQ = 2.414$

Also
$$BW = f_2' = f_1' = \frac{3.1 f_o}{Q} \tag{E.12}$$

It can be seen from this that the double tuned circuit can substantially increase the bandwidth while retaining steep roll-off.

Index

Numbers in **bold** face refer to figures; numbers in *italics* refer to tables